Here are your

1994 SCIENCE YEAR Cross-Reference Tabs

For insertion in your WORLD BOOK

Each year, SCIENCE YEAR, THE WORLD BOOK ANNUAL SCIENCE SUPPLEMENT, adds a valuable dimension to your WORLD BOOK set. The Cross-Reference Tab System is designed especially to help you link SCIENCE YEAR'S major articles to the related WORLD BOOK articles that they update.

How to use these Tabs:

First, remove this page from SCIENCE YEAR.

Begin with the first Tab, **Archaeology**. Take the A volume of your WORLD BOOK set and find the **Achaeology** article. Moisten the **Achaeology** Tab and affix it to that page.

Glue all the other Tabs in the appropriate WORLD BOOK volumes. Your set's B volume does not have an article on **Biodiversity**. Put the **Biodiversity** Tab in its correct alphabetical location in that volume—near the **Biochemistry** article.

Special Report
Archaeology
1994 Science Year, p. 116

Special Report
Biodiversity
1994 Science Year, p. 88

Special Report
Bird
1994 Science Year, p. 28

Special Report
Brain
1994 Science Year, p. 146

Science Studies Article
Cosmology
1994 Science Year, p. 174

Science Year Close-Up
Earthquake
1994 Science Year, p. 262

Special Report
Ecology
1994 Science Year, p. 132

Special Report
El Niño
1994 Science Year, p. 72

Special Report
Endangered Species
1994 Science Year, p. 132

Special Report
Indian, American
1994 Science Year, p. 44

Special Report
Materials Science
1994 Science Year, p. 160

Special Report
Motion Pictures
1994 Science Year, p. 58

Special Report
Planets
1994 Science Year, p. 102

New World Book Article
Prehistoric People
1994 Science Year, p. 337

New World Book Article
Satellite, Artificial
1994 Science Year, p. 349

Special Report
Science Fiction
1994 Science Year, p. 58

Special Report
Telescope
1994 Science Year, p. 12

New World Book Article
Telescope
1994 Science Year, p. 330

Science Studies Article
Universe
1994 Science Year, p. 174

Special Report
Weather
1994 Science Year, p. 72

1994 Science Year

The World Book Annual Science Supplement

A Review of Science and Technology
During the 1993 School Year

World Book, Inc.

a Scott Fetzer company

Chicago London Sydney Toronto

The Year's Major Science Stories

The most significant science news events during the 1992-1993 school year ranged from the discovery of the oldest known dinosaur species on Earth to strong evidence of black holes in outer space. On these two pages are the stories that *Science Year* editors picked as the most memorable, exciting, or important of the year, along with details about where you will find information about them in this edition. *The Editors*

Controversial "atom smasher"
Construction of the Superconducting Super Collider began in January 1993, but the project's future was in doubt. In the Science Studies section, see THE UNIVERSE. In the Science News Update section, see PHYSICS. ▼

Devastating hurricanes ▲
Meteorologists in 1992 tracked two of the most destructive hurricanes of the 1900's: Andrew, which battered Florida; and Iniki, which swept across Hawaii. In the Science News Update section, see METEOROLOGY.

World Book, Inc.
525 W. Monroe
Chicago, IL 60661

ISBN: 0-7166-0594-5
ISSN: 0080-7621
Library of Congress Catalog Number: 54-21776
Printed in the United States of America.

Landmarks in genetic science
Geneticists reported many advances in 1992 and 1993. Developments included:

- the completion of detailed maps of two human chromosomes;
- discoveries of genetic defects linked to colon cancer and Huntington's disease;
- genetic testing of human embryos;
- the first attempt at gene therapy for cystic fibrosis.

In the Science News Update section, see GENETICS; MEDICAL RESEARCH.

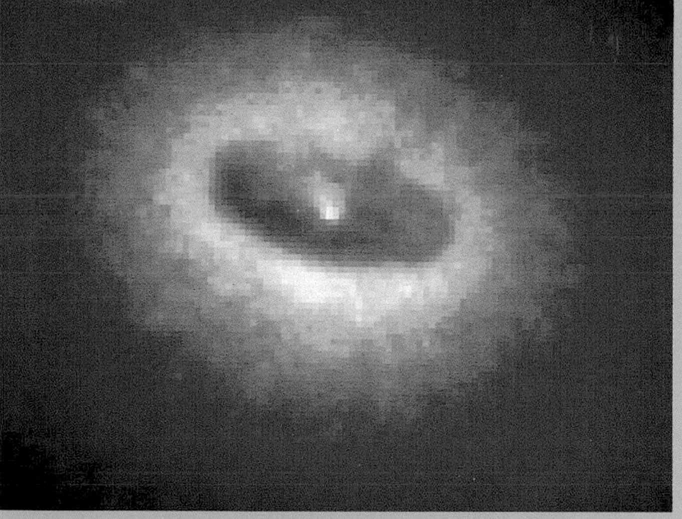

Outskirts of a black hole? ▲

Astronomers using the Hubble Space Telescope in 1992 found strong evidence of matter falling into *black holes,* objects so dense that not even light can escape their gravity. In the Special Reports section, see FIXING HUBBLE'S TROUBLES. In the Science News Update section, see ASTRONOMY, UNIVERSE.

Ancient DNA from amber

Scientists in 1992 and 1993 reported extracting the genetic material of insects that were trapped in *amber* (fossilized tree resin) millions of years ago. The oldest fragments of DNA were relics of the age of the dinosaurs. In the Science News Update section, see FOSSIL STUDIES; GENETICS. ▼

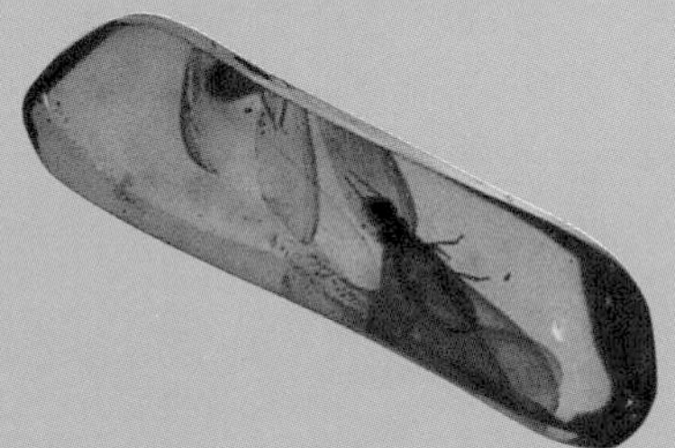

Most primitive dinosaur ▲

Fossils of the oldest known dinosaur species, a predator paleontologists call Eoraptor, were reported in January 1993 and dated to about 225 million years ago. In the Science News Update section, see FOSSIL STUDIES.

Evidence of a new fault?

A large crack in the Mojave desert appeared after a powerful earthquake struck near Landers, Calif., in June 1992. Some geologists claimed the quake indicated that a new fault is forming in the region. In the Science News Update section, see GEOLOGY (Close-Up). ▼

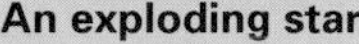

An exploding star

One of the closest and brightest *supernovae* (explosions of collapsed stars) of the 1900's was observed in March 1993. In the Science News Update section, see ASTRONOMY, UNIVERSE.

Contents

Page 102

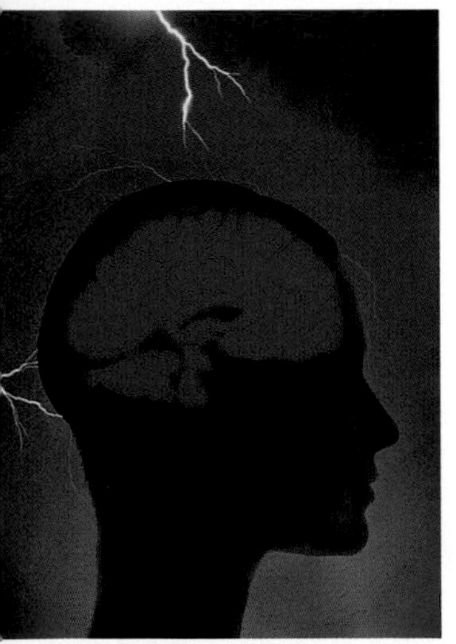

Page 146

Page 211

Page 308

Staff

Editorial Advisory Board

Robert T. Bakker is Adjunct Curator at the University of Colorado Museum in Boulder. He received the B.S. degree from Yale University in 1967 and the Ph.D. degree from Harvard in 1975. He is the author of numerous professional papers and a popular book, *The Dinosaur Heresies*. He has discovered 2 species of Jurassic dinosaurs and 11 species of early mammals. His principal research interests include mass extinctions, the ecology of dinosaurs, and the origin and early history of horses, rhinoceroses, and tapirs in the American West.

William S. Fyfe is Professor of Geology at The University of Western Ontario. He received the B.Sc. degree in 1948, the M.Sc. degree in 1949, and the Ph.D. degree in 1952, all from the University of Otago, New Zealand. He is a companion of the Order of Canada, a fellow of the Royal Society of Canada and the Royal Society of London, an honorary fellow of the Royal Society of New Zealand, and President of the International Union of Geological Sciences. He works on several committees of the International Council of Scientific Unions. He has received the Logan Medal of the Geological Association of Canada, the Arthur Holmes Medal of the European Union of Geosciences, and the Arthur Day Medal of the Geological Society of America.

Susan Armstrong Henry is Professor of Biological Sciences and Dean of the Mellon College of Science at Carnegie-Mellon University. She received the B.S. degree from the University of Maryland in 1968 and the Ph.D. degree from the University of California, Berkeley, in 1971. She is a member of the Genetics Society of America, the American Society for Biochemistry and Molecular Biology, and the American Society of Microbiology. Her primary research interest is genetic regulation of membrane lipid synthesis in yeast.

William A. Lester, Jr., is Associate Dean of the College of Chemistry and Professor of Chemistry at the University of California, Berkeley, and Faculty Senior Scientist in the Chemical Sciences Division of Lawrence Berkeley Laboratory. He received the B.S. degree and the M.S. degree from the University of Chicago in 1958 and 1959. He received the Ph.D. degree in 1964 from the Catholic University of America. He is a fellow of the American Physical Society and the American Association for the Advancement of Science. He has made major contributions to the theory and computation of molecular collision processes and the electronic structure of molecules.

Stephen H. Schneider is Professor in the Biological Sciences Department and the Institute of International Studies at Stanford University and is Senior Scientist in the Interdisciplinary Climate Systems Section of the National Center for Atmospheric Research. He received the B.S. degree in 1966, the M.S. degree in 1967, and the Ph.D. degree in 1971, all from Columbia University. He is a fellow of the American Association for the Advancement of Science and the Scientists' Institute for Public Information and is a recipient of the MacArthur Foundation's MacArthur Prize Fellowship.

Robert G. Wetzel is Bishop Professor of Biology at the University of Alabama in Tuscaloosa and Erlander Professor at the University of Uppsala in Sweden. He received the B.S. degree and the M.S. degree from the University of Michigan in 1958 and 1959, respectively, and the Ph.D. degree from the University of California, Davis, in 1962. He is the recipient of an honorary Ph.D. awarded by the University of Uppsala in 1985 and is a Fellow of the American Academy of Arts and Sciences. He is the author of numerous books and articles on aquatic ecology.

Contributors

Adelman, George, M.A., M.S.
Editor,
Encyclopedia of Neuroscience.
[*Neuroscience*]

Amato, Ivan, M.A.
Staff Writer,
Science Magazine.
[Special Report, *Designing "Smart" Structures*]

Asker, James R., B.A.
Space Technology Editor,
Aviation Week & Space Technology Magazine.
[*Space Technology*]

Baskin, Yvonne, B.A.
Free-Lance Science Writer.
[Special Report, *Mapping the Human Brain*]

Bower, Bruce, M.A.
Behavioral Sciences Editor,
Science News Magazine.
[*Psychology*]

Brett, Carlton E., Ph.D.
Professor,
Department of Geological Sciences,
University of Rochester.
[*Fossil Studies*]

Cain, Steve, B.S.
News Coordinator,
Purdue University School of Agriculture.
[*Agriculture*]

Chiras, Daniel D., B.A., Ph.D.
Adjunct Professor of Environmental Policy and Management,
University of Denver.
[*Environment*]

Dickman, Steven, B.A.
Free-Lance Journalist.
[*Genetics* (Close-Up)]

Donnan, Christopher B., Ph.D.
Professor of Anthropology,
University of California, Los Angeles.
[Special Report, *Solving the Mystery of the Moche Sacrifices*]

Drake, James A., B.S., M.S., Ph.D.
Associate Professor of Zoology,
University of Tennessee.
[Special Report, *Beware the Exotic Invaders*]

Duncan, Jeffrey R., B.A.
Graduate Student,
Department of Zoology,
University of Tennessee.
[Special Report, *Beware the Exotic Invaders*]

Dwyer, Johanna T., D.Sc.; R.D.
Professor of Medicine,
Tufts University School of Medicine.
[*Nutrition*]

Ferrell, Keith
Editor,
OMNI Magazine.
[*Computer Hardware; Computer Software; Electronics*]

Fienberg, Richard Tresch, B.A., M.A., Ph.D.
Technical Editor,
Sky & Telescope Magazine.
[Special Report, *Fixing Hubble's Troubles*]

Goldhaber, Paul, B.S., D.D.S.
Professor of Periodontology,
Harvard School of Dental Medicine.
[*Dentistry*]

Goodman, Richard A., M.D., M.P.H.
Adjunct Professor,
Division of Epidemiology,
Emory University.
[*Public Health*]

Graff, Gordon, Ph.D.
Technical Editor,
McGraw-Hill Incorporated,
and Free-Lance Science Writer.
[*Chemistry;* Science You Can Use: *Repelling Bugs That Bite*]

Hay, William W., Ph.D.
Professor of Geology,
University of Colorado, Boulder.
[*Geology*]

Haymer, David S., Ph.D.
Assistant Professor,
Department of Genetics,
University of Hawaii.
[*Genetics*]

Helms, Ronald N., B. Arch., M.S., Ph.D.
Professor,
University of Kansas.
[Science You Can Use: *Choosing the Right Light*]

Hester, Thomas R., Ph.D.
Professor of Anthropology
and Director, Texas Archeological Research Laboratory,
University of Texas, Austin.
[*Archaeology, New World*]

Holl, Robert C., B.S., M.Ed.
Free-Lance Writer.
[*Zoology* (Close-Up)]

Jones, William Goodrich, A.M.L.S.
Assistant University Librarian,
University of Illinois at Chicago.
[*Books of Science*]

King, Lauriston R., Ph.D.
Deputy Director,
Office of University Research,
Texas A&M University.
[*Oceanography*]

Klein, Richard G., Ph.D.
Professor of Anthropology,
University of Chicago.
[*Anthropology*]

Kowal, Deborah, M.A.
Adjunct Professor,
Division of International Health,
Emory University.
[*Public Health*]

Kristian, Jerome, Ph.D.
Astronomer,
Carnegie Observatories.
[*Astronomy* (Close-Up)]

Lechtenberg, Victor L., Ph.D.
Executive Associate Dean of Agriculture,
Purdue University.
[*Agriculture*]

Limburg, Peter R., B.A., M.A.
Free-Lance Science Writer.
[Science You Can Use: *Chemistry in a Tube of Toothpaste*]

Lunine, Jonathan I., Ph.D.
Associate Professor of Planetary Sciences,
University of Arizona.
[*Astronomy, Solar System*]

Maran, Stephen P., B.S., M.A., Ph.D.
Press Officer,
American Astronomical Society.
[Special Report, *Planets of Other Stars*]

March, Robert H., Ph.D.
Professor of Physics and Integrated Liberal Studies,
University of Wisconsin.
[*Physics*]

Marschall, Laurence A., Ph.D.
Professor,
Department of Physics,
Gettysburg College.
[*Astronomy, Universe*]

McAfee, K. Jill, B.S.
Graduate Student,
Graduate Program of Ecology,
University of Tennessee.
[Special Report, *Beware the Exotic Invaders*]

Merz, Beverly, B.A.
Free-Lance Writer.
[*Nutrition* (Close-Up)]

Meyer, B. Robert, M.D.
Associate Professor of Medicine and Associate Director, Medical Service,
Albert Einstein College of Medicine,
Bronx Municipal Hospital Center.
[*Drugs*]

Moores, Eldridge M., B.S., Ph.D.
Professor of Geology,
University of California at Davis.
[*Geology* (Close-Up)]

Morrill, John H., B.A.
Research Associate,
American Council for an Energy Efficient Economy.
[Science You Can Use: *Clearing the Air About Indoor Air Cleaners*]

Pennisi, Elizabeth J., M.S.
Chemistry/Materials Science Editor,
Science News Magazine.
[*Materials Science; Zoology*]

Pyne, Nanette M., Ph.D.
Director of Development,
Seattle Art Museum.
[*Archaeology, Old World*]

Rankin-Hill, Lesley M., B.A., M.A., Ph.D.
Assistant Professor of Anthropology,
University of Oklahoma.
[Special Report, *Uncovering African Americans' Buried Past*]

Rasmusson, Eugene M., Ph.D.
Senior Research Scientist,
University of Maryland.
[Special Report, *El Niño and Its Effects*]

Rhea, John, B.S.
Free-Lance Writer.
[Science You Can Use: *How Electronic Locks Work*]

Salisbury, Frank B., Ph.D.
Professor of Plant Physiology,
Utah State University.
[*Botany*]

Sforza, Pasquale M., Ph.D.
Professor and Head,
Aerospace Engineering Department,
Polytechnic University.
[*Energy*]

Snow, John T., Ph.D.
Professor of Atmospheric Science,
Purdue University.
[*Meteorology*]

Snow, Theodore P., Ph.D.
Professor of Astrophysics,
University of Colorado, Boulder.
[*Astronomy, Milky Way*]

Stephenson, Joan, B.S., Ph.D.
Chief, Chicago Bureau,
International Medical News Group.
[*Medical Research*]

Tamarin, Robert H., Ph.D.
Professor and Chairman of Biology,
Boston University.
[*Ecology*]

Tobin, Thomas R., Ph.D.
Assistant Professor,
ARL Division of Neurobiology,
University of Arizona.
[*Zoology*]

Turner, Michael S., B.S., M.S., Ph.D.
Professor of Astrophysics,
University of Chicago.
[Science Studies: *The Universe*]

Walter, Eugene J., Jr., B.A.
Free-Lance Writer.
[Special Report, *The Case of the Missing Songbirds*]

Wenke, Robert J., Ph.D.
Professor,
Department of Anthropology,
University of Washington.
[*Archaeology, Old World*]

Weyland, Jack, B.S., Ph.D.
Professor of Physics,
South Dakota School of Mines and Technology.
[Special Report, *Science Goes to the Movies*]

Special Reports

Feature articles give in-depth treatment to significant and timely subjects in science and technology.

Page 22

Page 28

Page 44

Page 137

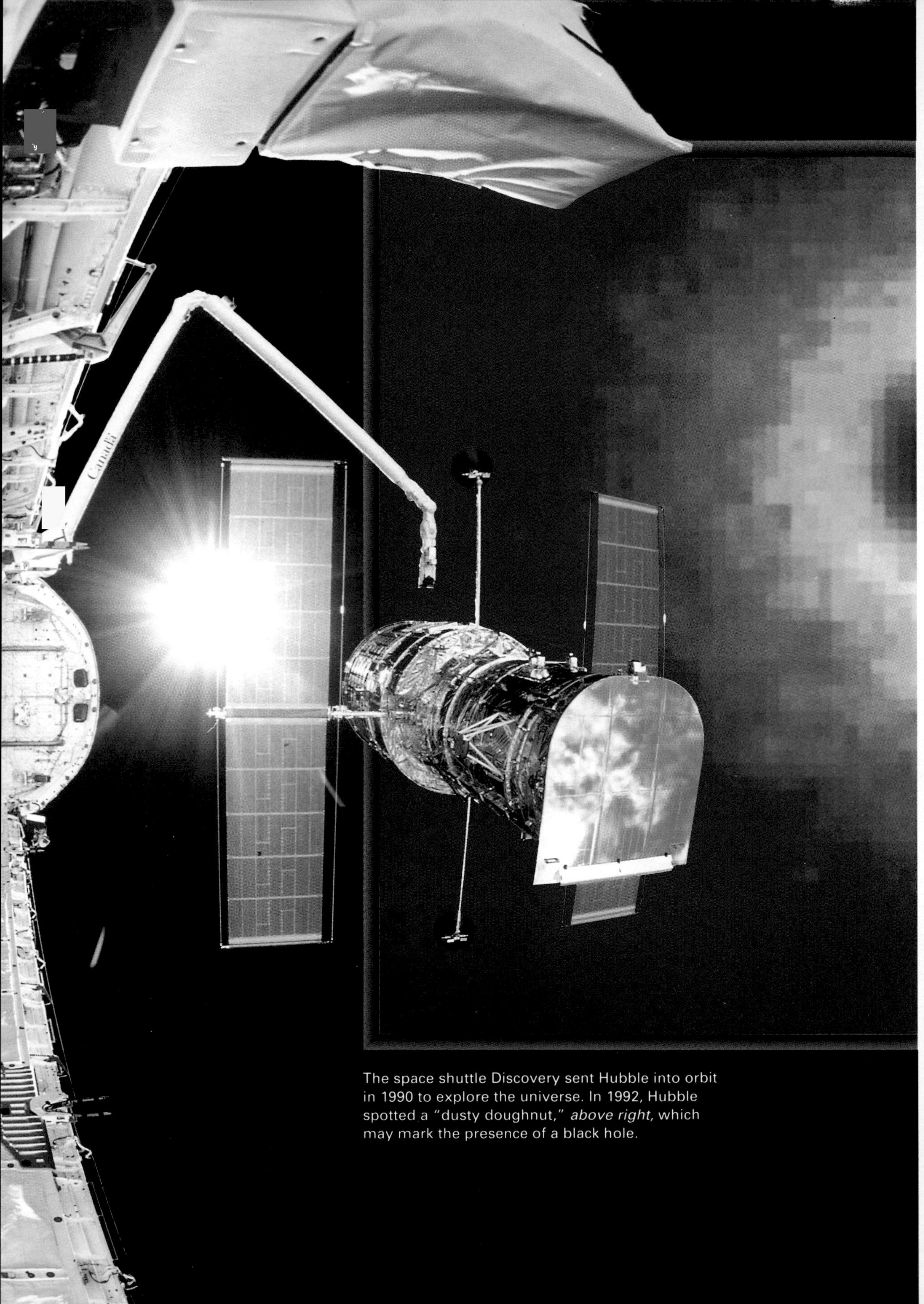

The space shuttle Discovery sent Hubble into orbit in 1990 to explore the universe. In 1992, Hubble spotted a "dusty doughnut," *above right,* which may mark the presence of a black hole.

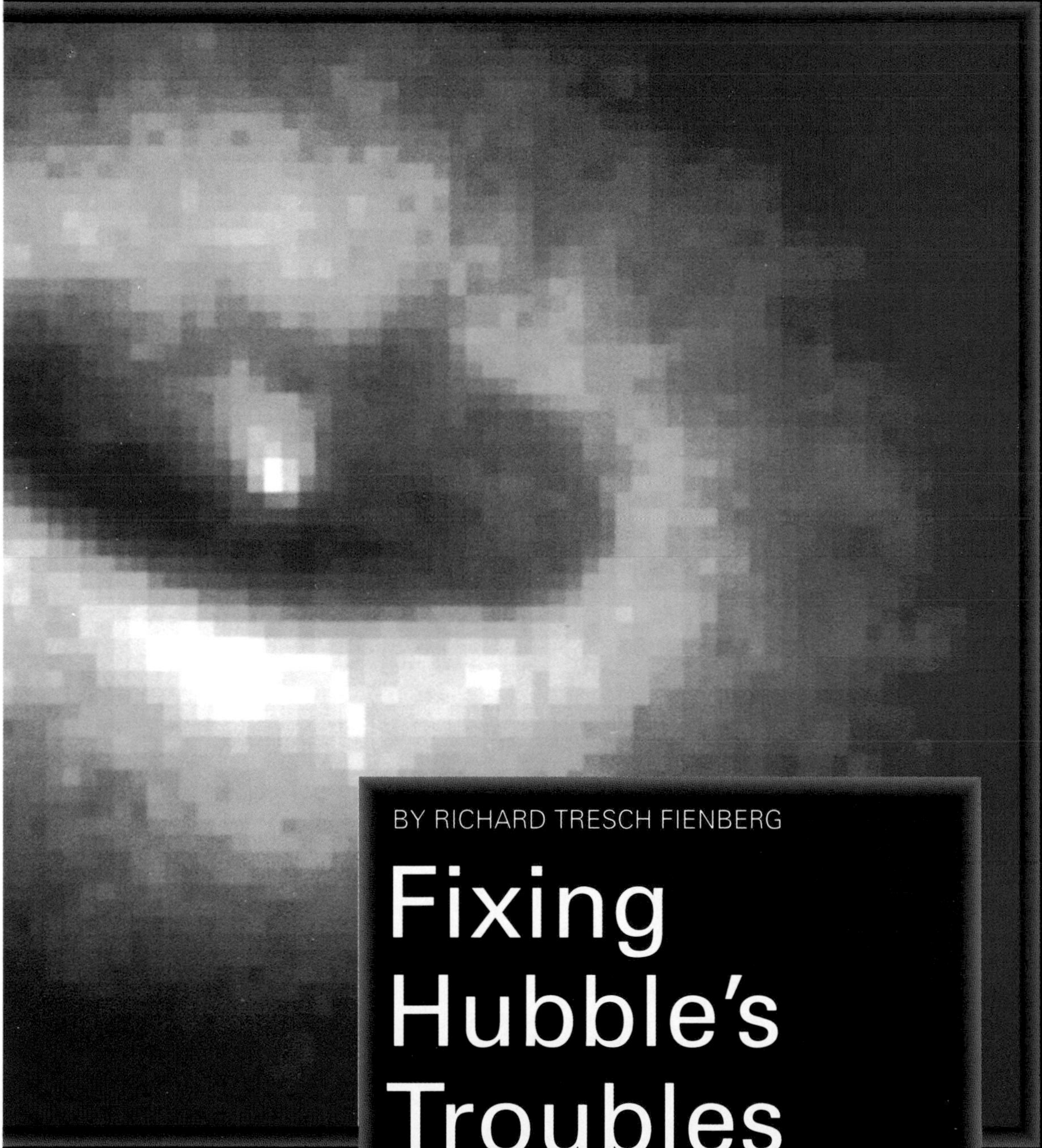

BY RICHARD TRESCH FIENBERG

Fixing Hubble's Troubles

To capture striking new images of the universe, astronomers have compensated for flaws in the Hubble Space Telescope—and planned its daring repair.

Something strange is happening in the galaxy known as NGC 4261. A tiny spot at the galaxy's center outshines tens of billions of stars put together. And two jets of hot gas squirt far out into space from this center, emitting powerful radio waves as they go.

Astronomers wondering what could cause such energetic activity believe the culprit must be a supermassive *black hole* (an object so dense not even light can escape the pull of its gravity). According to scientists' theories about black holes, a glimpse at the center of NGC 4261 should show the fireworks coming not from the black hole itself but from material falling into it. Anything veering too close to the black hole would be compressed and heated to form a thin disk of hot, swirling, shining gas. Some of the gas would end up in the black hole, but some would be flung away from it by the violent acceleration caused by the high pressure of the gas and other forces. A thick "doughnut" of cooler gas and dust encircling the black hole at a safe distance would help channel the castoff matter into two narrow streams.

Although astronomers have believed for some 30 years that galaxies such as NGC 4261 have central black holes, they had not clearly seen the defining traits of these mysterious giants. Even the largest telescopes perched on remote mountaintops could not see centers of distant galaxies clearly enough to find the telltale signs of a black hole. Then, in 1992, astronomer Walter Jaffe of Leiden Observatory in the Netherlands pointed the Hubble Space Telescope at galaxy NGC 4261 and snapped some pictures. On November 19, at a press conference at the United States National Aeronautics and Space Administration (NASA) in Washington, D.C., he unveiled one of his photos. There, at the heart of the galaxy, was a dusty doughnut surrounding a brilliant light, tipped at just the right angle for viewing its characteristics.

Glossary

Black hole: An object in the universe that is so dense nothing can escape its gravitational pull.

Cepheid variables: Stars that brighten at predictable periods.

Electromagnetic spectrum: The range of all the types of radiant energy.

Light-year: The distance light travels in one year, or 9.5 trillion kilometers (5.9 trillion miles).

Nebula: A cloudlike collection of dust and gas in space.

Red shift: A light's apparent shift toward the red end of the electromagnetic spectrum caused by the light's source moving away from the observer.

Reflecting telescope: A telescope that uses mirrors to reflect light to a single focal point.

Supernova: The explosion of a star.

Expectations and frustrations

This finding is a good example of the Hubble Space Telescope's successes—and its failures. Jaffe's image is not the first observation of a dusty doughnut, but it is by far the most definitive. Yet, the Hubble data on the doughnut are not complete enough to prove the existence of black holes, an achievement Hubble probably could accomplish if it were working the way it should.

Astronomers had hoped to prove the black holes' existence by using Hubble to calculate the amount of mass in the galaxy's center. They would "weigh" the center by measuring the speed of the gases and dust swirling around it. A black hole with a mass equal to 10 million suns should accelerate gases a few light-years away from the disk to speeds of thousands of kilometers per second. (A *light-year* is the distance light travels in one year, or 9.5 trillion kilometers [5.9 trillion miles].)

Hubble should be able to detect such high-speed motions, but it can't because of a flawed mirror. Astronomers have found ways to work around the flaw so that the telescope can still discover such things as the dusty doughnut. But the problem limits Hubble's ability to probe the very heart of a galaxy and to complete other planned studies.

The author:
Richard Tresch Fienberg is Technical Editor of *Sky & Telescope* magazine in Cambridge, Mass.

NASA has scheduled a space mission to fix the flaw and make other repairs. Meanwhile, despite its troubles, Hubble is displaying an exceptional ability to reveal important information about the universe.

Hubble's work in space began after it was deployed by the space shuttle Discovery on April 25, 1990. Astronomers had high hopes for the uniquely located telescope. Because it is in space, outside Earth's atmosphere, Hubble's view is not obscured by air turbulence, which makes stars twinkle and limits how clearly Earthbound telescopes see the heavens. At the same time, Hubble can detect radiation that is absorbed by Earth's atmosphere and, therefore, blocked before it reaches Earthly telescopes. Hubble also can see very faint celestial objects because it is above Earth's "airglow," a ghostly light emitted by atoms and molecules high in the atmosphere.

Space Telescope technology

Hubble is a reflecting telescope, a device that focuses light using mirrors. Incoming light from stars, for example, travels down the telescope's large tube to a primary mirror, a *concave* (curved inward) surface 2.4 meters (94.5 inches) in diameter. From there it bounces up to a 0.3-meter (12-inch) *convex* (curved outward) secondary mirror, then back down through a hole in the center of the primary mirror. The light comes to a focus behind the primary mirror and is captured there by light-reading instruments.

The Wide Field and Planetary Camera provides wide-angle views of planets, star clusters, and galaxies. The Faint Object Camera brings the most distant objects into view. Both instruments form images by "seeing" objects in the visible, ultraviolet, and infrared parts of the *electromagnetic spectrum,* the range of radiation from high-energy gamma rays to low-energy radio waves. Visible light is the portion of the electromagnetic spectrum that we can see. The longest wavelengths of visible light appear red, and the shortest, violet. Beyond the visible spectrum is ultraviolet radiation, with shorter wavelengths than visible light, and infrared radiation, with longer wavelengths.

The Space Telescope also carries the Goddard High Resolution Spectrograph and the Faint Object Spectrograph. These devices break the light from planets, stars, galaxies, and *nebulae* (clouds of dust and gas) into its component colors, which astronomers analyze to identify the elements that make up these celestial bodies. Finally, the High Speed Photometer rapidly and precisely measures star brightness, and the fine guidance sensors pick out guide stars at opposite sides of the telescope's field of view. Astronomers use the guides to accurately aim Hubble at specific targets. Several other support systems round out the observatory. Astronomers position the telescope by remote control, and the data Hubble gathers are relayed to Earth as radio waves and processed by computers to create images and provide other information.

The Hubble repair list

Hubble Space Telescope repairs involve replacing broken *gyroscopes* (aiming instruments) and three other types of equipment:

- New solar arrays are needed because the old ones vibrate when the telescope encounters temperature changes as it passes into darkness or light.
- The Wide Field and Planetary Camera is slated to be exchanged for a new model, which has corrective mirrors and increased light-sensitivity.
- The Corrective Optics Space Telescope Axial Replacement (COSTAR) will replace Hubble's *photometer* (light meter). This instrument is designed to correct blurring caused by a flaw in the telescope's primary mirror.

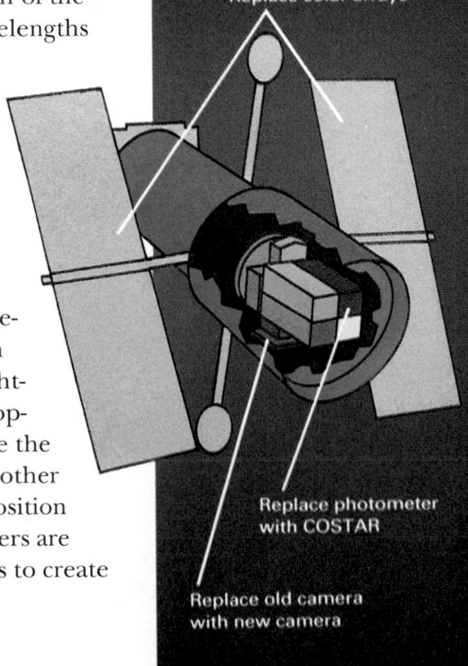

Hubble's vision problem and how to correct it

The outer edge of Hubble's primary mirror was ground down too far by an amount equal to 1/50 the width of a human hair. This flaw causes the telescope's images to be blurred. COSTAR is designed to correct the problem.

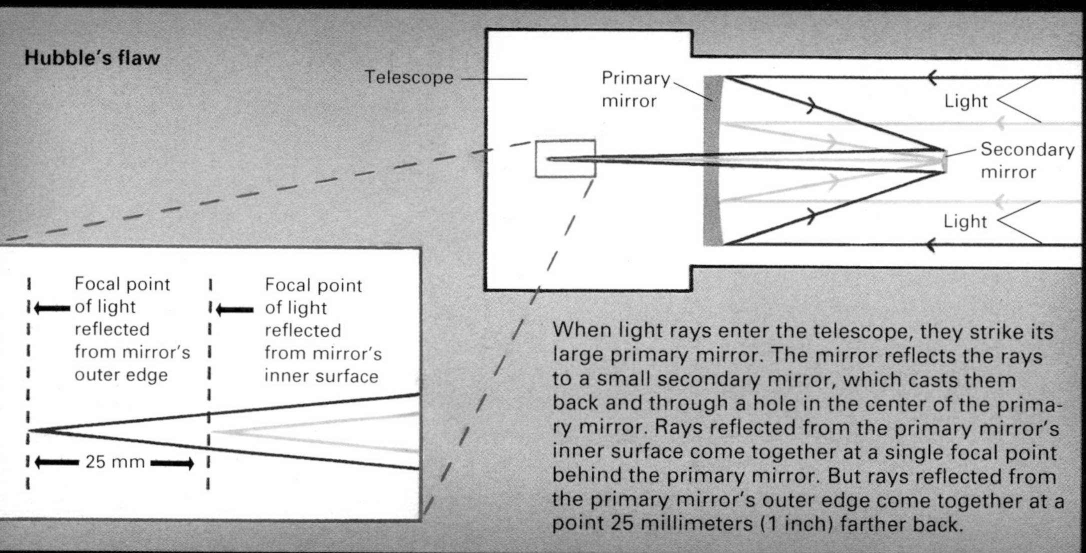

When light rays enter the telescope, they strike its large primary mirror. The mirror reflects the rays to a small secondary mirror, which casts them back and through a hole in the center of the primary mirror. Rays reflected from the primary mirror's inner surface come together at a single focal point behind the primary mirror. But rays reflected from the primary mirror's outer edge come together at a point 25 millimeters (1 inch) farther back.

A simulation of a Hubble image shows how the lack of proper focus smears the light from a cluster of stars, *above left.* Scientists use computers to make the image clearer, *above right,* though it is still not as sharp as it would be if the shape of the telescope's primary mirror were correct.

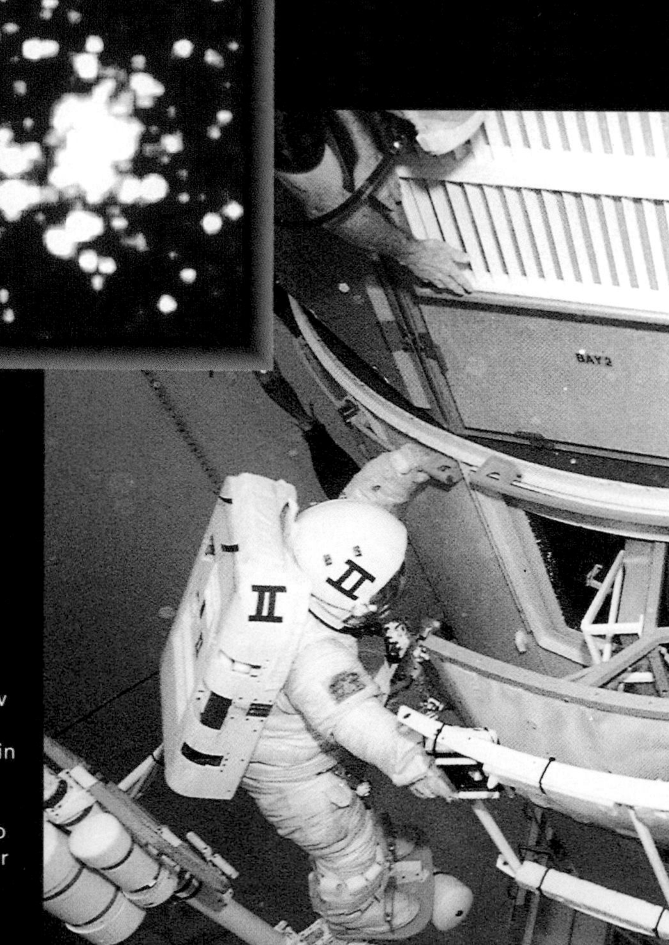

Astronauts practice installing a new model of the Wide Field and Planetary Camera on a Hubble mock-up in a tank of water. Underwater conditions simulate some conditions of space. In space, astronauts will also pull out the High Speed Photometer and replace it with COSTAR.

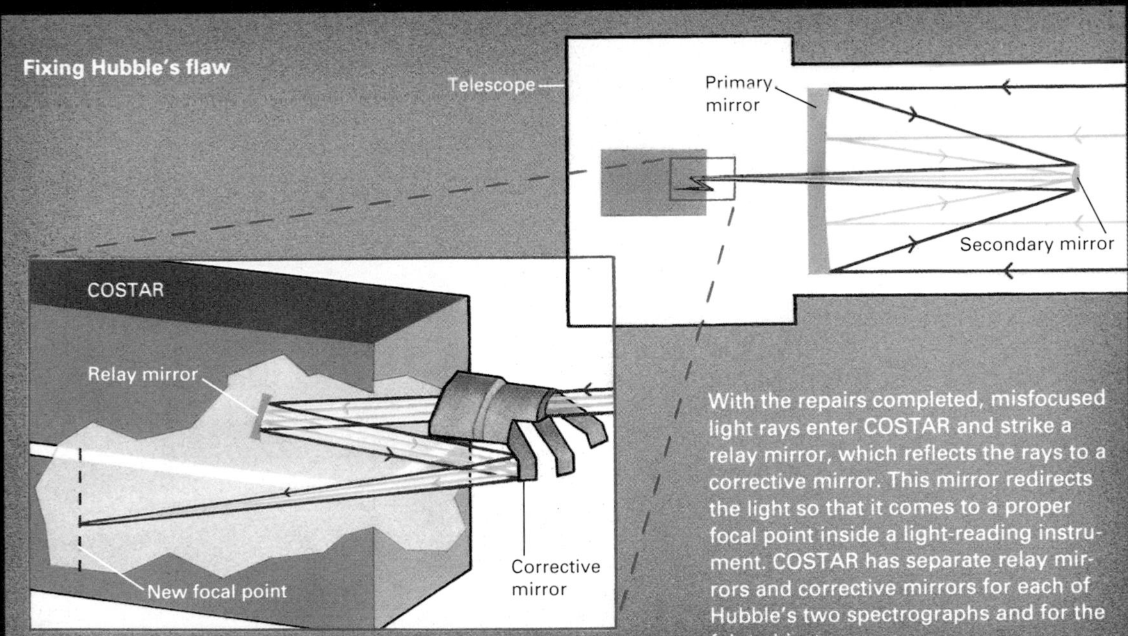

With the repairs completed, misfocused light rays enter COSTAR and strike a relay mirror, which reflects the rays to a corrective mirror. This mirror redirects the light so that it comes to a proper focal point inside a light-reading instrument. COSTAR has separate relay mirrors and corrective mirrors for each of Hubble's two spectrographs and for the faint object camera.

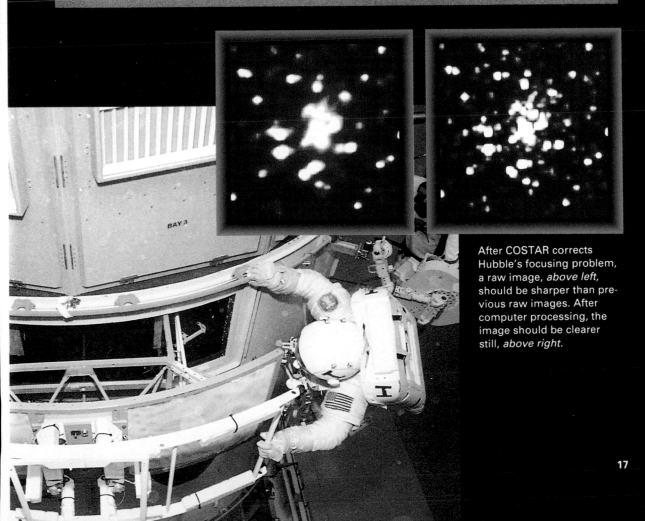

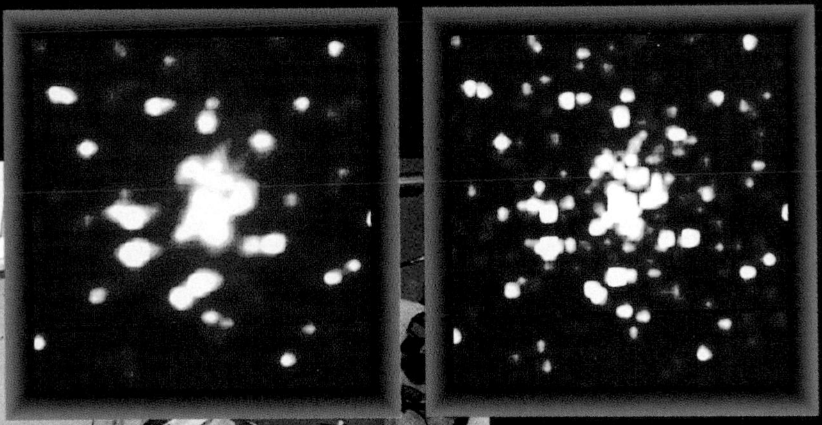

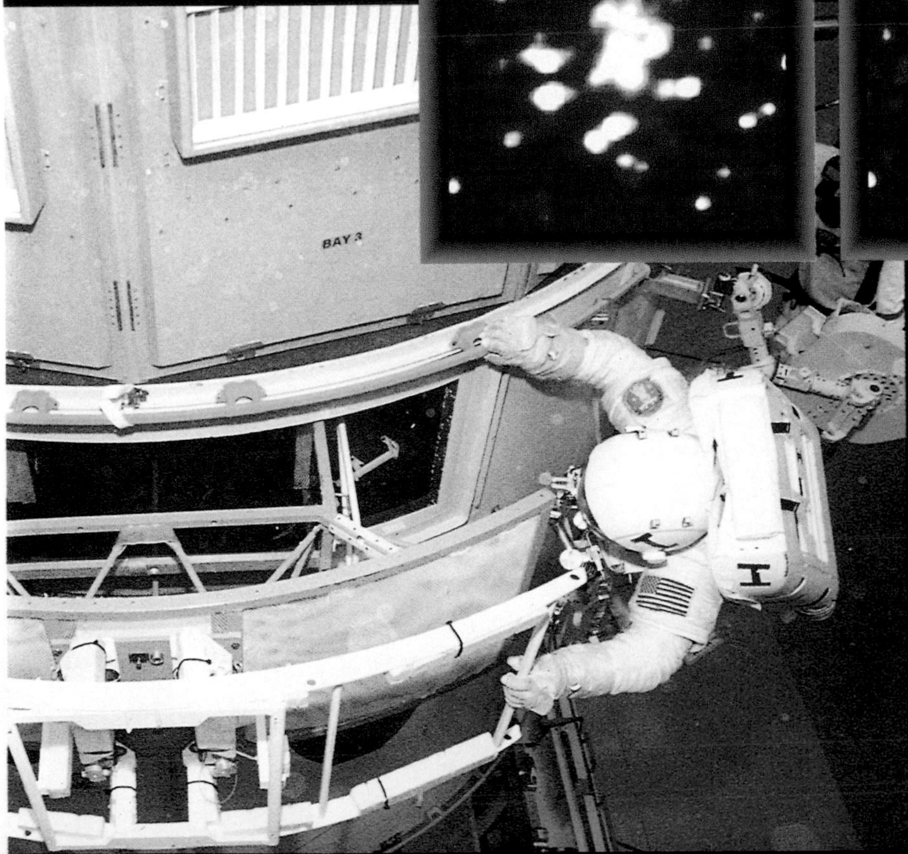

After COSTAR corrects Hubble's focusing problem, a raw image, *above left*, should be sharper than previous raw images. After computer processing, the image should be clearer still, *above right*.

Hubble's view of the solar system

Despite the problems with the Hubble Space Telescope, the instrument began delivering finely detailed images of objects in the solar system soon after its launch.

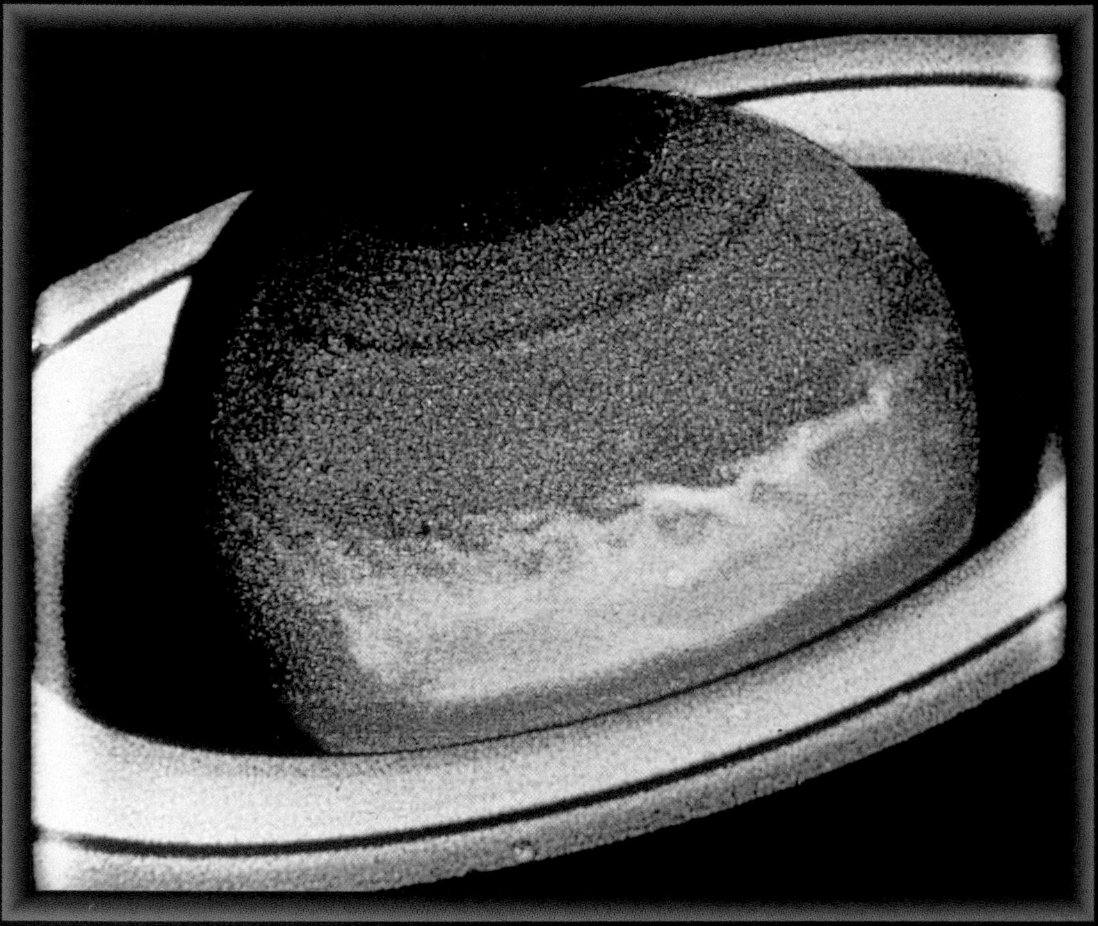

The turbulent white clouds of a Saturnian storm that occurs only once every 30 years appear in a 1990 Hubble image. The Hubble view of the storm helped astronomers study the motions of clouds on Saturn.

An image of Pluto taken with a ground-based telescope, *far left,* fuses the planet with its moon, Charon. But tiny Charon clearly distinguished itself when viewed in 1990 through Hubble, *left*. This distinction allowed astronomers to calculate the motions of both Pluto and Charon.

After Hubble's launch, astronomers spent a month checking all the on-board systems and adjusting the position of the mirrors to set the focus. Then they were ready for "first light." They pointed the telescope at a cluster of stars in the constellation Carina and snapped a picture with the Wide Field and Planetary Camera.

A flaw emerges

At first, astronomers hailed the picture as a major success. Hubble had revealed that what had appeared through ground-based telescopes as an elongated star was really a well-separated pair of stars. However, each star was ringed by a fuzzy halo that made it look like a squashed spider. Despite attempts to sharpen the focus, the halo remained. In late June, after examining more of Hubble's images, NASA admitted the unthinkable: The Space Telescope suffers from an optical error known as *spherical aberration,* in which an improperly curved mirror does not reflect light rays to a single focal point. In addition, the telescope's *solar arrays,* panels that provide the device's power, shake from temperature changes that accompany the transition from day to night and night to day as Hubble orbits the Earth every 95 minutes. This shaking disrupts the telescope's ability to track stars.

An investigation into the manufacturing of Hubble's optics revealed that the telescope's primary mirror had not been ground to the proper thickness at its outer areas, and the error had not been caught during prelaunch tests. The mirror was too shallow by an amount equivalent to 1/50 the thickness of a human hair. Although this error seems exceedingly minor, it causes light rays reflected from the outer surface of the primary mirror to come together at a focal point about 25 millimeters (1 inch) farther back than rays reflected from areas nearer the center of the mirror. This difference in focal points causes about 85 percent of a source's light to be seen as a blurry halo.

Astronomers were able to compensate for Hubble's flaw, however. They developed computer programs to remove the halo and enhance the remaining 15 percent of light that arises from the center of an image. But this method works only on relatively bright objects that give the computer plenty of light with which to work. And it does not work well at all when many images are crowded into one visual field, overlapping in a confusing mess of light. Fortunately, the cosmos has no shortage of bright objects in uncrowded fields.

Spying on Earth's neighbors

Astronomers have found fine targets for Hubble within our own solar system. Hubble cannot look at Mercury and Venus because they never move far enough away from the blindingly bright sun. But all the other planets make good viewing. Although space probes have flown near planets and sent close-up "snapshots" of them back to Earth, most probes are not designed to undertake long-term observations. Hubble, on the other hand, can be used to study planets for extended periods of time to provide detailed information about them.

Astronomers have, for example, used Hubble to study the weather on Jupiter. Jupiter appears to be made mostly of gases and liquids, with little, if any, solid surface. Although Voyagers 1 and 2, space probes launched by the United States in 1977, have obtained sharper views of Jupiter than Hubble ever will, they only swept by the planet once, not staying long enough to study the long-term behavior of Jupiter's turbulent atmosphere. And the probes, equipped with visible-light cameras, could capture only Jupiter's outer clouds. Hubble, with its equipment that detects infrared and ultraviolet radiation, can "see" into Jupiter's atmosphere at depths inaccessible to Voyager's cameras.

Four large moons and many smaller ones accompany Jupiter in its orbit around the sun. One of the large moons, Io, is the most volcanically active body in the solar system. Io is a bit bigger than our own moon, but at a distance of about 1 billion kilometers (600 million miles) it appears as a tiny, featureless disk through ground-based telescopes. Voyager pictures from 1979 showed Io looking like a pizza, with volcanoes spewing lava onto the surface. But, as in the case of Jupiter itself, Voyager's hasty visit did not leave time for careful study.

Astronomers led by Francesco Paresce of the Space Telescope Science Institute in Baltimore used Hubble's Faint Object Camera to examine Io in visible and ultraviolet light. Hubble's visible-light images of Io look a lot like Voyager pictures, suggesting that Io's surface has not changed much since 1979, despite the moon's intense volcanic activity. This finding led Paresce and his co-workers to suspect that some unknown process might remove or cover fresh volcanic debris.

Curiously, the Space Telescope showed that areas on Io that appear bright in visible light are dark in the ultraviolet, leading astronomers to think Io's surface may have vast deposits of sulfur dioxide frost, which absorbs ultraviolet radiation. The Voyager probes gave no hint of such a coating, because they could not record ultraviolet images.

Clearly superior shots

Most of Hubble's observations are planned months in advance by the Space Telescope Science Institute, which approves requests from astronomers who want to use Hubble. But if something new appears in the sky—a comet or an exploding star, perhaps—the schedule goes out the window. Such an event occurred in September 1990 when amateur astronomers using a ground-based telescope noticed a bright white spot on Saturn's normally quiet equator. When astronomer James A. Westphal of the California Institute of Technology in Pasadena, Calif., heard about the finding a couple of weeks later, he convinced the director of the Space Telescope Science Institute to make room in Hubble's schedule for looking at Saturn.

Hubble, pointed at Saturn, treated astronomers to a ringside seat at a Saturnian storm. Scientists think Saturn, like Jupiter, consists mostly of gases, including ammonia, methane, hydrogen, and helium. The white spot forms when blobs of moist gas well up from deep in Saturn's atmosphere, carrying ammonia up to a level at which it crystallizes to form brilliant white clouds.

Astronomers first viewed the Saturnian storm in 1876, and it has reappeared about every 30 years since then. The storm comes and goes in a matter of months, so viewing it can be a once-in-a-lifetime event for some astronomers. Thanks to Westphal's use of Hubble, scientists could examine images of the storm that were clearer than any that had been taken in the past.

Although astronomers already knew a fair amount about Jupiter, Io, and Saturn before they saw Hubble's images, the same cannot be said for their knowledge of Pluto and its moon, Charon. Pluto and Charon usually appear through ground-based telescopes as a single blob because they are so far away, orbiting on the fringes of the solar system at an average distance of 6 billion kilometers (3.7 billion miles) from Earth. But Hubble's cameras easily separated them into distinct points.

Once they could clearly see this planet and its moon, astronomers led by George W. Null of the Jet Propulsion Laboratory in Pasadena, Calif., set out to trace Pluto's and Charon's orbits. The two bodies' movements helped astronomers calculate the best estimate yet of their masses. Pluto's is about one-fifth that of our moon, and Charon's is nearly one-twelfth that of Pluto's.

From this information and previous calculations of Pluto's and Charon's volume, Null's group figured the density of each object. Because different types of matter have different densities—lead, for example, is more dense than water—the astronomers were able to deduce the composition of Pluto and Charon. Pluto's density suggests that it is composed mostly of rock and frozen water, while Charon's density indicates it is made almost entirely of ice. The difference in the bodies' compositions almost certainly rules out a long-standing theory that the two bodies formed together as a "double planet."

All these observations were recorded with Hubble's two cameras. Less picturesque, but just as important scientifically, are data captured by the telescope's spectrographs. These break up light from remote celestial sources into a spectrum and record it with an electronic sensor. Because different elements emit or absorb light at different wavelengths, astronomers can examine the spectrum of light from a star or other object to determine the elements that compose it.

Solar flares seen in new light

Astronomers led by Bruce E. Woodgate of NASA's Goddard Space Flight Center in Greenbelt, Md., used a Hubble spectrograph to confirm a theory about how energy is transported in solar flares. A solar flare is a spectacular eruption in the sun's outer atmosphere, which is called the corona. In a flare, radiation bursts out from the corona and from regions near or on the sun's visible surface. Until Woodgate's study, astronomers could only guess how energy was transported from the corona to the sun's surface.

One proposal, the proton beam theory, held that *protons* (positively

A clear view of the Milky Way

Hubble has penetrated the Milky Way Galaxy to help astronomers better understand its stars and *nebulae* (clouds of gas and dust).

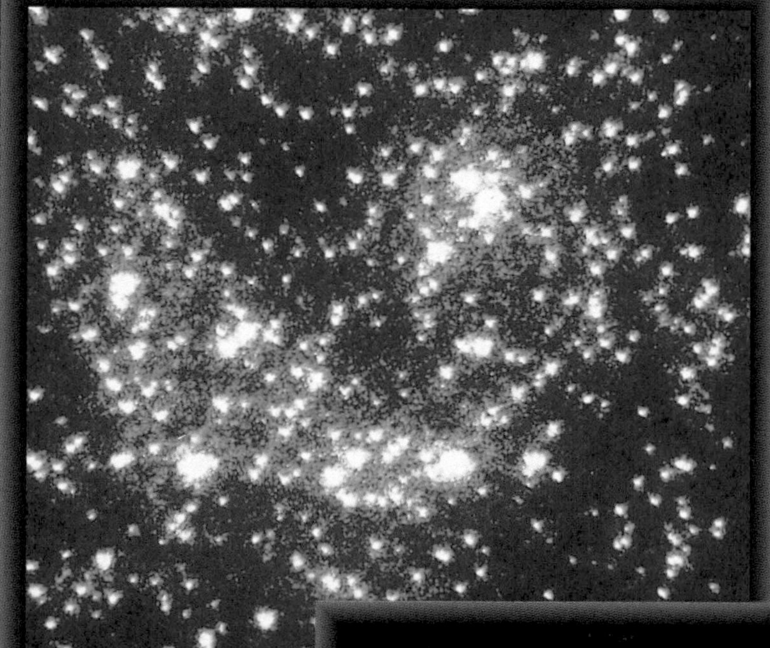

About 600 stars shine at the core of globular cluster 47 Tucanae, a concentration of several hundred thousand stars, in a 1991 ultraviolet Hubble image, *above*. Other telescopes were able to resolve only a few dozen stars. With the Hubble data, astronomers study *blue stragglers,* which appear younger than other stars in the cluster.

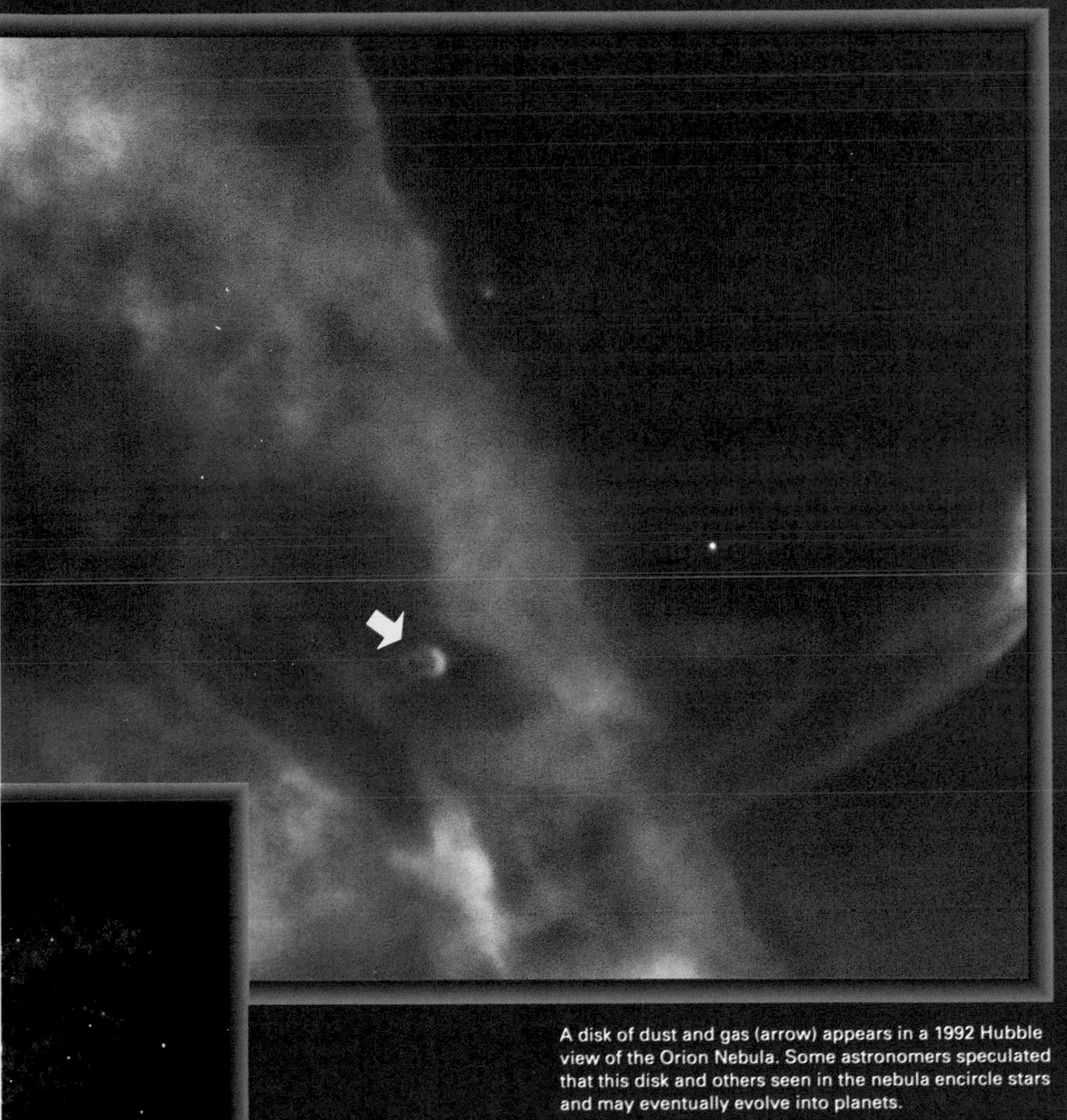

A disk of dust and gas (arrow) appears in a 1992 Hubble view of the Orion Nebula. Some astronomers speculated that this disk and others seen in the nebula encircle stars and may eventually evolve into planets.

At the center of Nebula NGC 2440, one of the Galaxy's hottest stars on record (white dot at center) burns at a scorching 200,000 °C (360,000 °F), according to calculations astronomers made using the Hubble telescope.

charged subatomic particles) moving through the corona pair up with *electrons* (negatively charged subatomic particles) in the flare, to create new, downward-speeding hydrogen atoms. Hydrogen characteristically emits ultraviolet light, but if the hydrogen atoms began moving toward the sun, a spectrum of a solar flare should show the light from them shifting slightly toward the red end of the electromagnetic spectrum. Such a shift, called a *red shift,* occurs when a light source moves away from an observer. The movement causes the light's wavelengths to appear longer. By measuring the amount of red shift, astronomers are able to calculate how fast the light source is moving.

Flares also occur on stars other than the sun, and Woodgate's team chose as their target AU Mic, a star 30 light-years away that is much smaller than our sun but which frequently produces flares that put the sun's to shame. The researchers monitored AU Mic for 3½ hours and, luckily, a flare erupted during the observations. Just as predicted, ultraviolet light intensified for three seconds at the beginning of the flare. Then the astronomers saw the light from hydrogen atoms shift toward the red end of the spectrum with a corresponding speed of 1,800 kilometers (1,100 miles) per second. Thanks to the Space Telescope, the astronomers could conclude that proton beams do indeed carry energy in solar flares.

Solar flares produce bursts of ultraviolet light, X rays, and cosmic rays that can interfere with radio communications on Earth, cause surges in power lines, or harm astronauts in space. By understanding how flares behave, scientists may one day be able to predict when they will occur and thus avoid flares' undesirable effects.

Recalculating the universe's expansion

Although Hubble has offered astronomers a new perspective on our solar system and Galaxy, the telescope's most significant discoveries have come from viewing galaxies outside our own. One of the most important questions such observations can answer is how fast the universe is expanding. Scientists have known since the 1920's that the universe is expanding. At that time, astronomer Edwin P. Hubble—for whom the Space Telescope is named—noted that galaxies are generally rushing away from each other, and that the farther away a galaxy is, the faster it is moving. But how is a galaxy's distance related to its speed? Astronomers have energetically debated theories about this. To determine a final answer, astronomers need to be able to precisely calculate the distance to various galaxies. Then, by examining the degree of red shift, they can determine how galaxy speed is related to distance.

In June 1992, astronomers led by Allan Sandage of the Carnegie Observatories in Pasadena used Hubble to take a step toward calculating the universe's expansion rate. Sandage's team first determined the distance to a galaxy called IC 4182. Then they monitored 27 stars in the galaxy that were of a type called *cepheids,* which periodically burn more brightly. These periods are predictable and they last from 1 to 100 days. Using sensitive light-reading instruments, astronomers have found that the cepheids that have the longest brightening period also have the greatest *absolute brightness* (the true amount of light or energy they give off rather than the amount we can observe from Earth).

By comparing the cepheids' observed brightness with their true brightness, and by applying calculations based on the fact that a light source will appear one-fourth as bright if its distance from the observer doubles, Sandage's team calculated the distance to galaxy IC 4182. It was 16 million light-years away. Sandage's team's next step was to judge the distance to galaxies so far away that any cepheids in them are not visible. To do this, they used information other astronomers had previously gathered on a *supernova* (exploding star) that appeared in 1937 in Galaxy IC 4182. Because a supernova's brightness far exceeds that of a cepheid, it can be seen from great distances. And IC 4182's supernova was of a type that, like cepheids, has the same absolute brightness no matter where it appears. Thus, Sandage's team was able to calculate the absolute brightness of the IC 4182 supernova because they knew how far away it was—16 million light-years.

Then, using the absolute brightness as a reference, Sandage's team analyzed the brightness of the same type of supernova in every other galaxy in which they had been previously detected. This comparison enabled the scientists to calculate the distances to dozens of galaxies. By relating the distance to the speed detectable by examining red shifts, the astronomers learned that for every 3.26 million light-years of distance, a galaxy recedes an additional 45 kilometers (28 miles) per second. This result was especially intriguing because most estimates that had been made using ground-based telescopes were twice as large.

Peering into the past

Hubble's views of extragalactic space have yielded new information not only about the distance and recession speeds of galaxies, but also about how galaxies have evolved over time. Studying distant galaxies is like looking back in time because the light an astronomer sees from the farthest objects in the sky was actually produced several billion years ago. Using Hubble, astronomers led by Alan Dressler of the Carnegie Observatories examined two clusters of galaxies 4 billion light-years away—which thus appear as they were 4 billion years ago.

The clusters contain a total of about 300 galaxies. Earth-based telescopes had imaged these galaxies as fuzzy specks. But Hubble, peering at them for nine hours, produced images that clearly showed their shapes. Dressler found that the clusters contain pinwheel-shaped spiral galaxies, watermelon-shaped ellipticals, and scraggly irregulars. Spirals, which compose 5 percent of the galaxies in similar galaxy clusters closer to us, account for about 30 percent of the galaxies in Dressler's images. Dressler thinks the small number of spirals in today's clusters may be due in part to what he calls the churning "Cuisinart environment" of those early, densely populated clusters. Over time, the spiral galaxies may have been torn apart and dispersed, or they may have merged with other galaxies and formed ellipticals. Using Hubble, Dressler ob-

Looking deep into space

By focusing its gaze on the most distant visible objects, Hubble has helped astronomers gain more understanding about the age and formation of the universe.

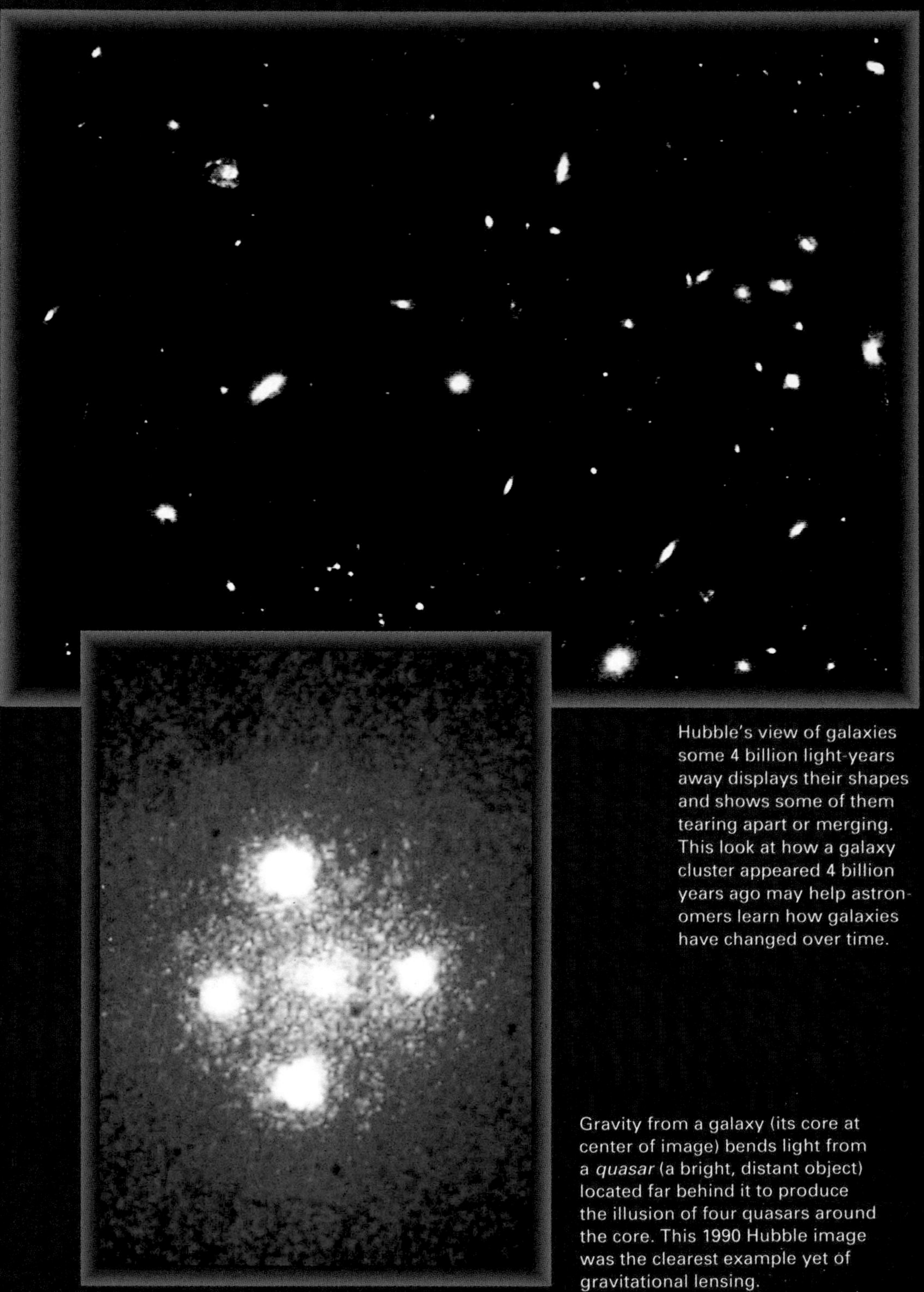

Hubble's view of galaxies some 4 billion light-years away displays their shapes and shows some of them tearing apart or merging. This look at how a galaxy cluster appeared 4 billion years ago may help astronomers learn how galaxies have changed over time.

Gravity from a galaxy (its core at center of image) bends light from a *quasar* (a bright, distant object) located far behind it to produce the illusion of four quasars around the core. This 1990 Hubble image was the clearest example yet of gravitational lensing.

served galaxies undergoing both processes. And although he could draw no definite conclusions about galaxy evolution, the Hubble images show that the cosmos has changed dramatically over time, at least in places dense with galaxies.

Repairs in space

All these findings show that astronomers have made good use of the Hubble Space Telescope despite its flaws. But Hubble may become even more useful. Astronauts aboard a space shuttle are scheduled to rendezvous with Hubble in December 1993 to fix the problems that have short-circuited the telescope's potential.

Astronauts plan to use a robotic arm to snatch Hubble and bring it into the shuttle's cargo bay. Four of the shuttle's seven astronauts will work in pairs to make a record-breaking five spacewalks into the bay to accomplish the repairs. The mission includes installing new solar arrays that will remain stable under changing temperatures; replacing broken *gyroscopes* (instruments that help point the telescope); and making substitutions for two of Hubble's light-reading instruments.

Scientists at the Jet Propulsion Laboratory were already building an improved Wide Field and Planetary Camera when they learned of Hubble's misshapen mirror. It was relatively simple for them to modify the new camera's internal mirrors to compensate for Hubble's flaw. They did this by intentionally giving the mirrors spherical aberration equal to but opposite that of the telescope.

To fix the focus of light entering other instruments, astronauts will remove the High Speed Photometer, which is used less than any other instrument, and insert the Corrective Optics Space Telescope Axial Replacement (COSTAR) in its place. COSTAR is a telephone-booth-sized box containing mirrors that will feed corrected light into three of Hubble's remaining light-reading instruments. Mechanical arms holding corrective mirrors extend from COSTAR in front of the Faint Object Camera and the two spectrographs. Improperly focused light will strike small flat mirrors on COSTAR, bounce to the tiny corrective mirrors, and shoot into the instruments. This mirror trick should bring to a sharp focus 80 percent of a star's light, rather than the 15 percent that Hubble now focuses.

If the mission succeeds, astronomers will embark on the adventure of discovery they had hoped to begin in 1990. But even if the mission is less than successful, scientists can still claim to have used a telescope of unsurpassed power to explore the mysteries of the universe.

For further reading:

Davidson, Greg. "How We'll Fix Hubble." *Astronomy,* February 1993, pp. 42-49.

Fienberg, Richard T. "HST Update: Science Amid Setbacks." *Sky and Telescope,* September 1991, pp. 242-246.

Chestnut-sided warbler

Scarlet tanager

Indigo bunting

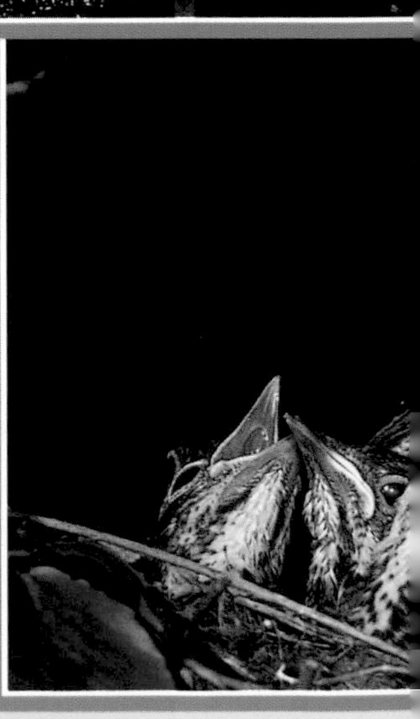

Northern oriole

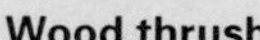

Wood thrush

American songbird populations are mysteriously declining, and scientists want to find out why.

The Case of the Missing Songbirds

BY EUGENE J. WALTER, JR.

In spring, amid carpets of purple violets and yellow lady's-slippers, bird watchers trek into the forests of eastern and midwestern United States and Canada. With binoculars pressed to their eyes, they scan tree branches for a glimpse of some of the millions of songbirds that have migrated northward thousands of miles from their wintering grounds to the south. But each year, bird watchers are sighting fewer such traveling birds.

Scientific surveys confirm what bird watchers have suspected—many songbird populations are declining. Scientists first noticed a downward trend in the 1960's in heavily wooded parks around the Washington, D.C., metropolitan area. Surveys in forest sites in Georgia, West Virginia, Maryland, New Hampshire, Connecticut, New York, Wisconsin, and Illinois all turned up declining numbers of warblers, vireos, thrushes, and other songbird species. In the 1970's, the trend apparently took a steeper slide. At many sites in the United States, some species disappeared entirely. The disappearance of so many birds is a mystery of major proportions. Identifying the causes, or culprits, is a case worthy of the legendary Sherlock Holmes.

Part of the puzzle is that although only 13 species and subspecies are officially endangered, populations of many of the 150 to 175 species of songbirds are declining. Many scientists feel that such a broad-based loss could be a symptom of much larger environmental problems, such as deteriorating soil conditions and perhaps even climate change. And complicating the mystery is the lack of sufficient study data, particularly on how the birds live in their various southern habitats during the winter months and on where they make stopovers during their migratory journeys. Some researchers think that some species declines may be part of natural cycles that are not fully understood.

The threatened birds

Such threatened birds as warblers, tanagers, buntings, orioles, sparrows, grosbeaks, flycatchers, larks, vireos, shrikes, thrushes, thrashers, finches, cuckoos, and hummingbirds share certain characteristics. Many of the birds measure from 10 to 12.7 centimeters (4 to 5 inches) long and weigh less than a 25-cent coin. In spring and summer, they live in the temperate zone of North America, eating caterpillars, flying insects, and spiders. Most of the songbirds breed only in mature forests, though some nest in scrublands and grasslands. In autumn, the birds migrate to their winter habitat, mainly in the tropical areas that extend from about 2,570 kilometers (1,600 miles) north of the equator to about 2,570 kilometers south.

The birds travel varying distances during migratory flights, depending upon the species. Most species that breed in eastern North America spend six to nine months in Mexico, but some spend the winter in the Caribbean islands known as the West Indies, and others in Central America or South America. Western species migrate to northern and western Mexico. Other birds, called temperate-zone migrants, travel only a short distance to the Southern United States. And within some species, part of the population stays in the United States, while the rest goes south of the border. But for most American songbirds, the one-way trip is 3,200 to 4,800 kilometers (2,000 to 3,000 miles).

Biologists estimate that only about one-third of the species winter in forests and woodlands. The rest use nonforested areas, such as the scrub or secondary growth that moves in after forests are cleared.

First evidence of fewer birds

The author:
Eugene J. Walter, Jr., is a free-lance writer.

Counting such mobile creatures as migratory songbirds is not easy. One method is the Breeding Bird Census (BBC), originally a project of the National Audubon Society and now administered by Cornell University's Laboratory of Ornithology in Ithaca, N.Y. At Rock Creek Park in Washington, D.C., the first place where declining songbird populations were noticed, volunteers have conducted the BBC since 1948.

They use a technique called spot mapping or territory mapping to determine populations in the 400-hectare (988-acre) park. Following a grid map of the forest, volunteers fan out to listen for the distinctive songs of male birds. Each bird they hear counts as a breeding pair, according to the survey's guidelines, because females do not sing.

Survey volunteers make counts several times during the breeding season, and ultimately these counts produce an estimated density of the birds of each species. Because scientists studying the data are looking for long-term trends, they analyze the data periodically, rather than every year. Year-to-year analysis would not tell them anything meaningful, as it may, for example, reflect the effect of an unusually abundant or poor food supply.

In 1948, BBC volunteers found 233 breeding pairs of birds per 40 hectares (100 acres), and 85 percent were songbirds that migrated from the tropics. But in 1986, volunteers found only 74 pairs per 40 hectares, and the number of migrants had fallen to 42 percent of the total. The census at other small wooded parks in the U.S. capital yielded comparable declines, sometimes even more severe.

Detecting a 10-year slide

Another important census is the Breeding Bird Survey (BBS), which biologist Chandler Robbins devised for the U.S. Fish and Wildlife Service in 1966. Each year, some 2,000 experienced bird watchers are recruited by state and provincial coordinators in the United States and Canada. Using a method called point count, the volunteers drive along assigned 40-kilometer (25-mile) routes, beginning half an hour before sunrise on a day in May or June, depending on the time migratory songbirds usually arrive in the area. Every 0.8 kilometer (0.5 mile), they stop for three minutes and record every bird species they see or hear.

In 1986, Robbins and his colleagues analyzed BBS data for eastern North America taken during the first 15 years of the survey's existence. They found that from 1966 to 1978 the populations of many species increased each year or remained stable. In 1989, the researchers analyzed data gathered from 1978 to 1987. This time, they found declining populations for 75 percent of the songbird species that migrate to the eastern regions of North America.

Black-throated green warbler

Threatened songbird species

In 1993, the following species of North American songbirds were officially listed as endangered or threatened in the United States:

- Bachman's warbler
- Black-capped vireo
- Golden-cheeked warbler
- Ivory-billed woodpecker
- Kirtland's warbler
- Red-cockaded woodpecker

According to the Breeding Bird Surveys, other songbird species suffering serious population declines include:

- Acadian flycatcher
- Bay-breasted warbler
- Black-billed cuckoo
- Black-throated green warbler
- Canada warbler
- Chestnut-sided warbler
- Common yellowthroat
- Gray catbird
- Indigo bunting
- Northern oriole
- Northern parula warbler
- Olive-sided flycatcher
- Ovenbird
- Rose-breasted grosbeak
- Scarlet tanager
- Tennessee warbler
- Veery
- White-eyed vireo
- Wood thrush
- Yellow-billed cuckoo

For example, the wood thrush population dropped 4 percent each year of the study. The rose-breasted grosbeak was down 4.1 percent each year. The black-throated green warbler dropped 3.1 percent; the chestnut-sided warbler fell 3.8 percent; and the northern oriole was down just under 3 percent each year.

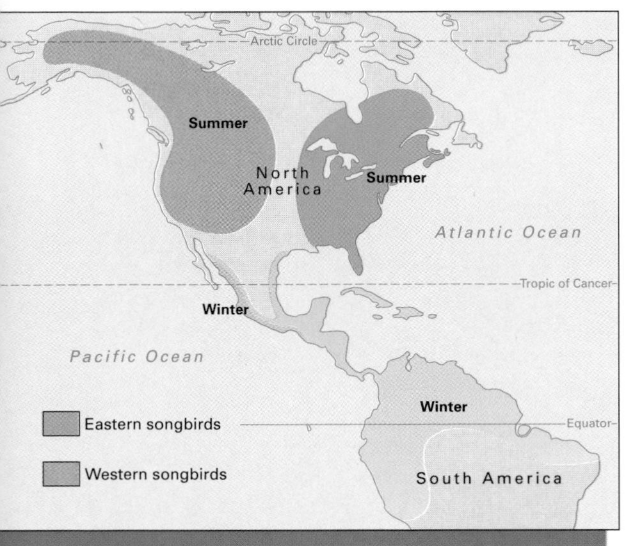

Migration patterns
In the spring, migratory songbirds fly north to the United States and Canada, where they nest and raise young in forests and grasslands. In the fall, songbirds fly south to their wintering grounds. Most migrate to areas relatively close to the United States.

Counting birds with radar

At least one *ornithologist* (person who studies birds) tried to count birds during migration rather than nesting. Sidney A. Gauthreaux, a biologist at Clemson University in Clemson, S.C., used weather radar to detect changes in the number of migrating songbirds. For most songbirds, migration includes a 1,300-kilometer (800-mile) flight across the Gulf of Mexico between the Yucatán Peninsula and the United States. Radar stations have tracked storm systems along the Gulf since 1957, and Gauthreaux reasoned that the thousands of birds crossing the Gulf would appear as blips on radar records. He compared numerous radar images of a three-year period in the 1960's, when he had seen enormous flocks of birds migrating, with images taken in a three-year period in the 1980's, when he had observed much smaller flocks in flight. The radar images indicated that the number of migrant birds dropped by 40 to 50 percent.

Gauthreaux's method has its critics. Some of his colleagues question what the pictures show, because the data are in the form of blips on the radar screen rather than images of single birds. Radar cannot distinguish one species from another. Nevertheless, in terms of sheer volume, far fewer birds seemed to be making the trip each spring.

The smaller the forest, the fewer the bird species

Unfortunately, no matter how information on songbird populations is gathered, it cannot reveal the reasons for the declines. But when some researchers looked at how land use has changed in North America over the years, they thought they had a clue to the mystery. In the 1930's and 1940's, farmland was abandoned, and forests once again took over. But many of these forests were cut down in the 1950's to make way for new suburban communities that sprang up around cities. The forested areas of the Northeastern United States were turned into checkerboards of small wood lots separated by houses, shopping centers, golf courses, roads, and power line rights of way.

Investigators wondered whether this fragmentation of the forest was causing songbird problems. To find out, biologists looked at the relationship between the size of forest tracts and the number of bird species found in them. One of these investigators, ornithologist Robert A. Askins at Connecticut College in New London, took a census of birds in 46 tracts of forest in southeastern Connecticut in the early 1980's. The tracts varied in size from 1.6 to 2,023 hectares (4 to 5,000 acres). Askins then compared the number of species with the size of the tract and its distance from other forested areas. Askins found that smaller, more isolated woodlands contained fewer songbird species.

Askins also analyzed annual bird counts made at the Connecticut College's *arboretum* (garden of trees) since 1953. He found that, after housing and highway construction had eliminated about 15 percent of the forest near the arboretum, songbird populations in the arboretum also began to dip significantly. By 1975, five species had disappeared altogether. But during the 1970's, abandoned farms north of the arboretum reverted to woodland. As of 1985, there was more forest near the arboretum than in the 1950's. Bird populations had increased, and several of the vanished species had returned. It is unclear whether this is a temporary increase or the beginning of an upward trend.

Life on the edge

Researchers also began looking at songbird populations in *edge communities,* the places where two habitats, such as forest and meadow, meet. After a forest is fragmented by, for example, building a road, the same tract of forest has more edges than before, and thus more edge

In tropical America, people burn rain forests to make way for agriculture. Because the tropics are the winter habitat of many songbird species, the destruction of these forests may be one reason for the birds' population decline.

communities. But some investigators found that increased edge habitat of itself did not deter songbird nesting. In 1984, Roger L. Kroodsma of the Oak Ridge National Laboratory in Tennessee examined a forest where rights of way had been cut for power lines. Throughout this area, birds that usually nest in the forest interiors were nesting right up to the edge of the cuts. In the early 1980's, David S. Wilcove, a graduate student at Princeton University in Princeton, N.J., who is now at the Environmental Defense Fund, mapped migratory bird breeding territories in a large Maryland forest surrounding a reservoir. Wilcove found birds nesting right up to the shoreline, another type of edge community.

Development along the Mid-Atlantic Coast of the United States has turned coastal woodlands into resort areas. The development deprives songbirds of natural habitat in which to rest during their long migrations between wintering grounds in the south and nesting sites in the north.

Predators at the edge

But Wilcove found a different situation when he looked at edge communities in the suburbs. Songbirds did not thrive there. Wilcove suspected that two factors could explain the birds' lack of success. First, there are more predators in a suburban edge community than in a natural edge community. For example, raccoons with access to lots of garbage cans grow nearly twice as large as their rural kinfolk, and the raccoon populations become much larger, given such a regular food supply. Other edge-dwelling predators that thrive in the suburban environment include skunks, opossums, blue jays, grackles, and crows. Adding to the problem of wild predators are domestic ones, the dogs and cats kept as pets.

To test his theory that suburban predators were gaining easy access to the nests, Wilcove purchased small wicker baskets to simulate the open-cup nests the songbirds build. He lined the baskets with grass and stocked each with three quail eggs. He placed the fake nests in a range of wooded habitat—from suburban and rural wood lots of only a few hectares to the more than 210,000-hectare (520,000-acre) forest of the Great Smoky Mountains National Park in North Carolina and Tennessee. Half the bogus nests sat on the ground, and the rest were a few meters above the ground in bushes or higher up in trees. Wilcove also placed eggs in wooden nest boxes at the same locations as the wicker baskets. The wooden nest boxes were comparable to tree holes used as nests by year-round resident birds, such as chickadees, nuthatches, and house wrens.

In the small suburban wood lots, predators raided an average of 70 percent of the wicker nests and in some tiny plots nearly 100 percent. But in the Great Smoky forest, predators destroyed only 2 of 100 fake nests. The nests on the ground suffered more than those in trees and shrubs, which indicated that mammals on the ground were the primary villains. Eggs in the wooden nest boxes were virtually untouched, which indicated that the wooden nests provided protection similar to that afforded by the tree-hole nests of resident birds, whose eggs are rarely threatened by predatory mammals.

Wilcove verified his conclusions by spreading modeling clay, soot, and other substances around the fake nests to pick up tracks of predators. Raccoons were the number one culprits. Blue jays came in second. He also found the footprints left by most of the other suspects—skunks, opossums, cats, and dogs.

Why songbirds are vulnerable

Wilcove theorized that songbirds were vulnerable to edge predators for several reasons. First, the birds' own nest design made them easy prey for predators. Most species build open, cup-shaped nests, often right on the ground. Ovenbirds, Louisiana water thrushes, and several warblers—all of which are experiencing population declines in many areas—are among the birds that build this type of nest.

Second, migrant songbirds rarely produce a second batch of eggs, called a clutch. The resident birds, such as robins, mockingbirds, cardinals, and others, often produce a second and even a third clutch if their nests are destroyed. With a late-spring arrival and early-autumn departure, migrant songbirds simply do not have time to lay more eggs. Third, migrant songbirds produce smaller clutches than do the residents—three to four eggs versus five or six.

Finally, warblers, vireos, and most other migrants are tiny birds unable to fight off aggressors, such as blue jays. In contrast, many resident birds, such as robins and mockingbirds, are large enough to drive away jays.

The cowbird threat

Other researchers turned up another natural villain jeopardizing songbirds, the brown-headed cowbird. A small member of the blackbird family, the cowbird is a resident bird that naturally prefers fields and forest edges. The female cowbird lays her eggs in the nests of other birds, frequently nudging other eggs out of the nest, and leaves parenting to the nest builders, which are called "host species." Cowbird chicks hatch earlier than those of the host. Because cowbird chicks are bigger and stronger than most songbird chicks, they grab most of the food and leave the songbird offspring to starve.

The cowbird threat to songbirds is particularly acute in the Midwest. In 1980, Stanley A. Temple and Margaret Clark Brittingham of the University of Wisconsin-Madison surveyed songbird nests along the

Hazards for nesting songbirds

In the Northeastern United States, human activity has fragmented the forest, *right,* where many songbird species build their nests and raise their young. This fragmentation creates more forest edges and exposes songbird eggs and chicks to more dangers than they would encounter deep in the woods.

Cowbird menace

In the forest edges live cowbirds, birds whose nesting behavior sabotages breeding songbirds. Instead of building its own nest, a cowbird lays its large, speckled eggs in the nest of another bird, *above,* and leaves parenting to the nest-builder. A cowbird chick, which is larger and stronger than its songbird nestmates, grabs food from a willow flycatcher parent, *right,* making it likely that the flycatcher chicks will starve.

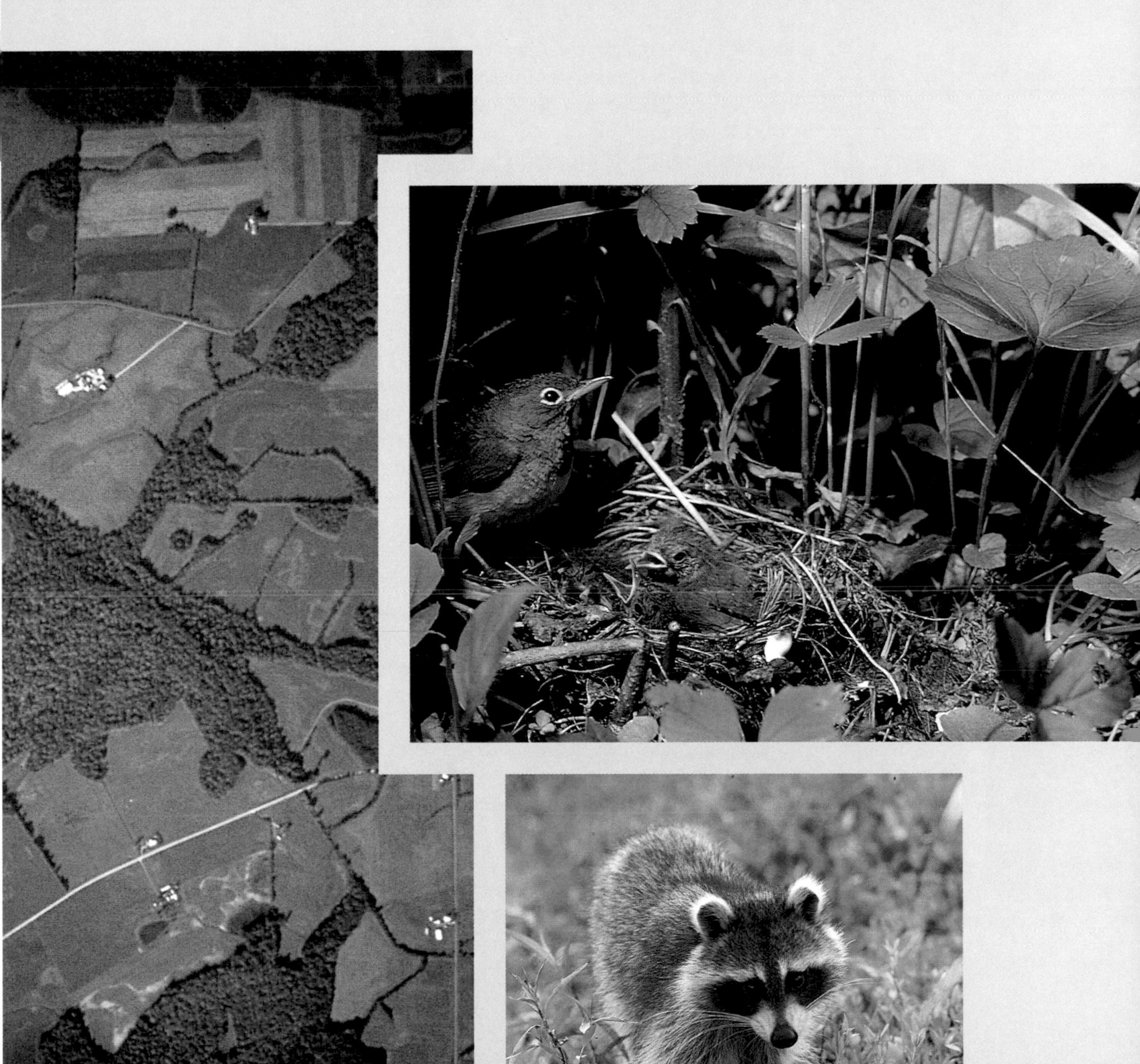

The threat of predators

Many songbird species build open-cup nests directly on the ground, *top*, which makes the nests vulnerable to predators that eat eggs or young chicks. According to a study completed in the 1980's, the most dangerous of these predators in suburban areas is the raccoon, *above*.

edges of a 1,000-hectare (2,470-acre) forest in southern Wisconsin. They found that 65 percent of nests along the edge had cowbird eggs in them. Cowbirds were even taking over nests 300 meters (980 feet) inside the forest.

Follow-up studies in other locations confirmed these figures. The most dramatic data came from a large strip of mature and secondary forest in central Illinois. Biologist Scott K. Robinson of the Illinois Natural History Survey examined songbird nests there and discovered that cowbirds had invaded 80 percent of them. He found that of 30 wood thrush broods, 29 contained cowbirds, which accounted for 75 percent of the eggs.

With the settlement of the United States, brown-headed cowbirds have become an increasing threat to North American songbirds. The cowbirds need open grasslands for their food and their social displays, and in the 1700's, they lived mainly on the Great Plains. But as settlers cut down enormous blocks of Eastern forests to make farmland, the cowbirds began to move east to the newly opened sites. The birds also found an abundant food supply in the fields of grain. Today, the cowbirds have brought several species of songbirds to the edge of extinction. Kirtland's warbler, for example, which nests only in a small area of central Michigan, survives solely because of human intervention. Each spring, wildlife wardens trap and remove cowbirds, so that the warblers are free to raise their own chicks. Even with this help, just 200 pairs of warblers remain.

Dwindling grassland bird species

Although much songbird research centers on the birds of eastern forests, many ornithologists agree that some of the heaviest songbird declines in North America are in grassland species. These species include Baird's, Brewer's, grasshopper, lark, and vesper sparrows, as well as dickcissels, bobolinks, horned larks, longspurs, and prairie warblers. Most of the grassland species migrate only short-distances, many of them spending winters in the United States, or they are partial migrants, which means part of the population stays in the United States while the rest go south of the border.

Scientists believe these species are dwindling because of changing farming practices. The great prairies that once so dominated North America's heartland were taken over by agriculture decades ago, but until relatively recent times, farmers bordered their land with dense plantings of bushes or trees called hedgerows. Farmers also planted stands of trees as shelter belts against wind and snow. Hedgerows and shelter belts were fine bird habitat. But beginning around the 1950's, modern farming machinery and government subsidies led to more intensive farming. Farmers began plowing and planting their fields right to the edge, eliminating the bird habitat.

Farmers also used to allow fields to lie *fallow* (plowed but not planted) for a season or two to recover after the growing season. The practice meant that there was always habitat favorable for birds. With the aid of modern fertilizers and pesticides, however, crops may be grown every season and land no longer lies fallow. In addition, those chemical products could be poisoning birds. And ranching, too, may contribute to habitat loss if cattle are allowed to overgraze and otherwise harm grasslands that used to be good nesting sites.

Questions about western species

Songbirds appear to be vanishing in the western half of North America as well as the east. But less is known about the extent of the problem for western migrants. Many areas in the western United States and Canada are without roads, which has prevented census-takers from making counts in remote regions. There are gaps in census data—both missing years and missing places—and existing data have not been analyzed sufficiently.

Even though some areas are remote, people have changed much of the North American west. Biologists say that logging, livestock grazing, recreation, and water management have eliminated some songbird habitat in western areas altogether. Flood control and irrigation projects alone destroyed more than 90 percent of the original strips of woodland that bordered rivers in the Western United States, according to federal land management agencies. Unfortunately, these woodland strips were where many western songbirds nested.

Loss of winter habitat

Another region undergoing change is the birds' wintering sites. As billions of songbirds fly to the tropics, their numbers thin out the farther south they go. Most stop in lands closest to the United States. The migrants make up 40 to 50 percent of total bird populations in a variety of habitats in Mexico, the Bahamas, and the Greater Antilles. In Central America, they account for 20 to 40 percent of the total bird population; in Colombia, 5 to 15 percent; and in Ecuador, Peru, and Bolivia, they represent only 1 percent. An enormous number of birds are thus funneled into a relatively small area as they migrate south for the winter months. There, songbird densities are five to eight times greater than in the northern summer breeding grounds.

The compression of so many birds in their wintering grounds would seem to make migrants especially vulnerable to changes in their habitat. Unfortunately, most migrant songbirds winter in countries where people are destroying forests at a fast pace—Mexico, the West Indies, and Central America. In Central America overall, an estimated 400,000 hectares (990,000 acres) are deforested each year.

Accurate, on-going surveys of deforestation are unavailable. According to Russell Greenberg, director of the Smithsonian Migratory Bird Center in Washington, D.C., there is no comprehensive project to monitor forest loss in Mexico and Central America. The most conservative estimates of 1993 were that 1 to 4 percent of the forests in the American tropics are lost each year. Without more reliable data, however, researchers cannot measure the impact of tree loss on songbird populations in their winter habitat, though it seems logical to believe that the impact must be devastating.

Fortunately, some migrant species appear capable of adapting to changes in their tropical habitat. Some species spend winter in a variety of habitats. Many seem to prefer scrub or secondary growth. On the other hand, few songbirds adapt to living in areas that are under constant use, such as cattle pasture. And several species appear dependent on native tropical forest. Some birds, such as the wood thrush, are able to survive only in mature forest, both in summer breeding sites as well as wintering grounds.

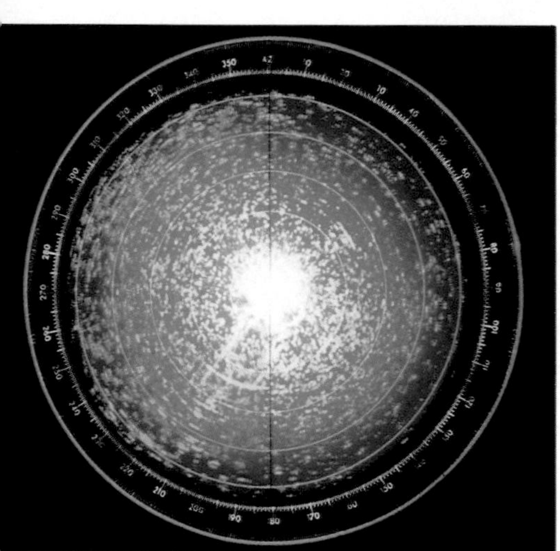

A radar image of a flock of migrating birds over the Gulf of Mexico could be used to help scientists monitor bird population levels. Each dot in the image, which was made in April 1992, represents about 20 birds, according to a South Carolina researcher who compared similar images made in the 1960's with those of the 1980's. His comparison indicated that the number of migrating birds had dropped 40 to 50 percent by the 1980's.

The need for stopover sites

The songbirds may also be vulnerable during their migratory journeys. Most migrants fly at night, at about 64 kilometers (40 miles) per hour. Night flying enables them to avoid predatory hawks. During the journey, however, songbirds need stopover sites to feed and rest. Not enough is known about these sites to assess how changes there affect songbirds. Ideally, the stopovers should be in locations that are secure from predators, though unpredictable weather may force songbirds down anytime, anywhere. And songbirds are at a disadvantage if they must compete with resident birds or other migrants that arrived earlier.

In 1991, scientists with the nonprofit group The Nature Conservancy found that migrants prefer coastal sites for stopovers. The researchers studied songbird routes during autumn migration along North America's Mid-Atlantic Coast. Their study found dense concentrations of songbirds pausing in coastal woodlands. When researchers canvassed areas only 2 kilometers (1.2 miles) inland, they found few birds. But coastal sites are in increasingly short supply because people value them as locations for resorts and vacation homes.

Defining a mysterious problem

With so much of the mystery of the disappearing songbirds unresolved, there is some disagreement in the scientific community about the nature of the problem. Some researchers insist that nobody knows what constitutes normal populations for songbirds. John M. Hagan III, senior research scientist at the Manomet Bird Observatory near Plymouth, Mass., feels that researchers may need as many as 50 years of data to establish, for example, what constitutes a natural fluctuation in a species' population.

Most scientists agree that songbird data are subject to a variety of distortions. The BBS, or any such survey, could be biased by the changing makeup of the pool of volunteer census-takers. Also, current census figures alone do not always tell the full story. Survey data show that some species currently in decline were actually on the rise before 1978.

Another criticism of both the BBS and the BBC is that their counts may even be too high. Both surveys are based mainly on listening for singing males, which are presumed to be mature mated birds who sing to announce their territorial boundaries. But according to some studies, the census figures may be inflated by "floaters," which are young, nonbreeding, nonterritorial males who nevertheless sing. Their presence could skew statistical calculations, giving the appearance of a stable, or even increasing, population when in actuality a decline could be in progress.

Biologist Richard L. Hutto at the University of Montana in Missoula believes that a significant proportion of all American birds are in trouble and that long-distance songbirds are not declining in a disproportionately large number. He thinks resident species are in as much trouble as are the migrant species.

David A. Wiedenfeld, a research associate at the Museum of Natural Science of Louisiana State University in Baton Rouge, also believes that some researchers have overemphasized the decline of long-distance migrants. Wiedenfeld studied BBS data for both forest and grassland

Banding songbirds for study

A researcher, *left,* removes a bird from a *mist net,* a nearly invisible fabric strung up to catch the birds on the fly. The unharmed birds are placed in bags for later banding. A mourning warbler, *below,* sports one of the numbered leg bands put on the birds before they are released. If biologists find the warbler in the future, the number will tell them where and when the bird was banded. Such information helps researchers keep track of various species.

How volunteers can help

The following organizations offer information or programs geared to volunteers interested in protecting songbirds.

- The Cornell Laboratory of Ornithology sponsors several projects in the United States. Project Feederwatch has volunteers set up bird feeders and report on the species and number of birds that visit them. The National Science Experiment has volunteers collect data on species' food preferences. Project Tanager has volunteers record data on the effects of forest fragmentation on tanager breeding. For information, contact Margaret Barker; Cornell Laboratory of Ornithology; 159 Sapsucker Woods Road; Ithaca NY 14850.

- The American Birding Association publishes an annual directory of volunteer bird-related projects. For the 1993 directory, which describes projects sponsored by the U.S. government, send $2.00 to Volunteer Directory; American Birding Association Sales; P.O. Box 6599; Colorado Springs CO 80934.

- The Smithsonian Institutions and the National Audubon Society offer "The Migratory Bird Information Kit," which contains a booklet, *Birds over Troubled Forest;* a species checklist; and ideas for projects to save migratory birds. To get the kit, send $7.50 (check or money order payable to Smithsonian Migratory Bird Center) to Publications; SMBC; National Zoological Park; Washington DC 20008.

songbirds from 1966 through 1990. His interpretation of the data for the period is that populations are actually increasing for many long- and short-distance migrants.

He also found geographic variations. Almost all species are declining in certain areas, such as the Adirondack region of New York, the Blue Ridge Mountains of Virginia, the Mississippi River plain from Missouri to New Orleans, and southern Florida. Wiedenfeld believes such declines are linked to pollution, changing agricultural practices, and real estate development.

John Hagan at Manomet believes short-distance migrants, such as the rufous-sided towhee, are the species with the most serious problems in some parts of the United States. By analyzing BBS data from 1966 to 1992, he found that the towhees have disappeared in southern New England at a rate of 8 to 10 percent a year. The population in 1992 was only 15 percent of what it was in 1966.

Hagan theorized that the decline of this species may be linked to the disappearance of agriculture in southern New England over the same period. By 1992, farming had virtually ceased, and abandoned farmland had passed through a scrubby stage—the habitat towhees prefer—and had reverted to forest. The land probably looks much as it did 250 years ago, when, Hagan speculates, towhees may not have been as abundant in New England. This raises the question of whether such declines in this species warrant conservation concern.

International focus on songbirds

The experts have found that the complexities of studying migratory birds require the cooperation of scientists from many nations. The National Fish and Wildlife Foundation, which the U.S. Congress established in 1984, assists and coordinates much of the international work on songbirds. In 1990, the foundation hosted a workshop that brought together more than 150 scientists and conservation professionals. They concluded there was a need to link the efforts of government, wildlife, and conservation agencies to tackle the task of gathering more complete data on migrant songbirds and their habitats.

The workshop resulted in the creation of an international organization called Partners in Flight—*Aves de Las Americas* (Birds of the Americas). Under the partners program, scientists are working on ways to improve songbird research and monitoring. By developing a standard method of monitoring songbirds, for example, researchers can ensure that songbird data from Canada can be accurately compared with data from Mexico or any other nation where songbirds live. Scientists also are working with industry to find ways of managing forests for profit while still providing suitable habitat for birds.

Such projects underscore the importance scientists place on solving the mystery of declining songbirds. As they see it, the best way to preserve a species is to study it before it becomes endangered—and long before bird watchers in North America are in danger of sighting the last songbird.

For further reading:

Ehrlich, Paul R., Dobkin, David S., and Wheye, Darryl. *Birds in Jeopardy: The Imperiled and Extinct Birds of the United States and Canada.* Stanford University Press, 1992.

Partners in Flight newsletter. Free by sending name and address to Peter Stangel; National Fish & Wildlife Foundation; 1120 Connecticut Ave., NW; Suite 900; Washington, DC 20036.

Terborgh, John. *Where Have All the Birds Gone?* Princeton University Press. 1989.

Questions for thought and discussion

Picture yourself as a member of the city council of a town in New England. Over the past 50 years, most of the farms around your town have been abandoned, and the forest has reclaimed the fields. Nevertheless, the amateur bird watchers in your town say that they have spotted fewer and fewer songbirds in the surrounding forest since the late 1970's. Now a developer wants to purchase the central part of the largest parcel of forest owned by your town. The developer says he will build a golf course ringed with homes on the land. Many citizens appear before the city council to support the sale. Those citizens say that although they enjoy the beauty of the forests, the new development offers an economic boost to the community: The developer will provide well-paying construction jobs for local workers, and, once the homes are built, the town will receive new property tax revenue. After those citizens argue their case, a group of amateur bird watchers appears to say that the sale should not go through. Those citizens claim that keeping the forest intact will help stabilize the populations of songbirds that nest near the town, and that selling off the land could doom the birds.

Questions: What questions would you ask the bird watchers and the the developer before making a decision? Can you think of some way to minimize the environmental damage to the area? If you could select the part of the forest that will be sold, would it be the central portion? Why or why not?

As archaeologists unearthed some of the richest tombs in the New World, they solved a mystery about a remarkable civilization that arose in South America nearly 2,000 years ago.

Solving the Mystery of the Moche Sacrifices

BY CHRISTOPHER B. DONNAN

Painted by the July sun, the rugged foothills of the Andes Mountains towered above Peru's coastal plain, checkered by well-cultivated fields. But we barely noticed the beauty around us. Sunk in a roofless chamber of mud brick, we lay belly-down over the grave of an ancient Andean ruler.

During the past few weeks, working with artists' paintbrushes and dental instruments, we had brushed away the soil and remains of what had been the ruler's wooden coffin, long since decomposed into dust. Slowly, the coffin's contents had revealed themselves: objects of gold, silver, and copper; precious stones; and fragments of textiles. Piece by piece, we had removed the uppermost objects—the rich offerings and garments buried with the tomb's royal occupant.

Previous pages: A Moche warrior brings a bound prisoner before a Warrior Priest, center, and Priestess, right, major Moche religious figures. The Moche killed prisoners and consumed their blood in a ritual known as the Sacrifice Ceremony. Another principal figure in the ceremony, known as the Bird Priest, is not pictured.

Now, the skeleton itself lay before us, surrounded by even more lavish grave goods.

About 1,700 years ago, members of a people called the Moche (pronounced *MO chay*) had buried their dead ruler here, near the modern village of Sipán. The Moche (sometimes called *Mochica*) were one of several important Andean civilizations that preceded the Inca, the best-known Andean culture, which flourished in the 1400's and 1500's. Until the late 1980's, scientists knew little about Moche ceremonial life. The Moche had no writing system, and their archaeology and art—which I have specialized in studying—offered only tantalizing hints of what their culture was like. The excavation we were now conducting would help to broaden our view of this ancient civilization.

Entranced, I stared at the skeleton and its rich possessions, including several sets of gold and silver necklaces, ear ornaments, and nose ornaments, piled one atop another in a dazzling display of wealth. But the object that most captured my attention was what the ruler had held in his right hand. It was a rattle, exquisitely crafted of gold and silver.

The rattle had a boxlike chamber, hammered out of gold. A raised design on the chamber's sides and top showed an elaborately clothed warrior bringing his war club down on the head of a captive. The rattle's silver handle was decorated with images of military equipment: a cone-shaped helmet, a *tunic* (outer garment), a circular shield, armor, and war clubs. The bottom of the handle widened to form a flat blade.

As I stared at the rattle, I thought of the times I had seen just such an instrument depicted in Moche art. Moche artists often decorated their ceramic vessels and other objects with a scene in which prisoners of war were sacrificed and their blood consumed. In this scene, which I call the Sacrifice Ceremony, bound captives are brought before elaborately clothed individuals. The captives' throats are cut and their blood is poured into tall goblets, which the main figures ceremoniously present to one another. One of the main figures in this ceremony is sometimes shown with the rattle.

Questions about the Sacrifice Ceremony

The author:
Christopher B. Donnan is professor of anthropology and director of the Fowler Museum of Cultural History at the University of California, Los Angeles.

We first identified the Sacrifice Ceremony in Moche art in 1974. But for many years afterward, my colleagues and I did not know how to interpret it. Did the Moche actually sacrifice prisoners and consume their blood? Or did the drawings illustrate a scene from Moche mythology, an event performed by supernatural beings in a religious tale? Since all our knowledge of Moche religious beliefs and practices came from the work of Moche artists, we had no way of knowing whether the figures represented real or imaginary individuals.

At first, I considered it unlikely that the Sacrifice Ceremony was really performed. The individuals depicted in the illustrations, if they existed, would have been very important and wealthy officials, who would surely have been buried with lavish grave goods and symbols of their office. No Moche tombs yet excavated suggested that the Moche had rulers of such high status. It was not until 1987, with the excavation of the royal tombs at Sipán, that we began to learn that these individuals were real. By 1992, we had found tombs of three of the major figures at Sipán and the nearby site of San José de Moro. We thus learned that the Sacrifice Ceremony depicted by Moche artists actually took place—a discovery that was to profoundly affect our understanding of the Moche people.

The Moche of Peru

Moche civilization flourished between the A.D. 100's and 700's on the northern coast of what is now Peru. The Peruvian coast is a dry, rugged land that forms a narrow strip between the Pacific Ocean and the Andes Mountains. It consists chiefly of sandy desert, broken only by a series of rivers that flow from the mountains toward the Pacific shore. Human occupation in this arid environment is limited to these river valleys, where irrigation makes agricultural production possible.

Moche occupation extended over about 25,000 square kilometers (10,000 square miles), stretching northward nearly to what is now the Peru-Ecuador border and southward to the modern town of Huarmey. Today, that region is dotted with the ruins of hundreds of monumental structures that the Moche built at their political and religious centers, including mud brick pyramids and massive platforms that probably served as temples. The Moche also dug extensive networks of irrigation canals. Today, Peruvian farmers use what are probably some of these canals. Other canals, now dry, still wind across the landscape.

Scholars are not sure what caused the Moche's decline. One prominent theory is that a series of natural disasters linked to a weather phenomenon known as El Niño forced the Moche to flee their capital and caused social and political upheaval. (See EL NIÑO AND ITS EFFECTS.)

The name *Moche* comes from the site, near present-day Trujillo, where scientific study of the culture began in 1899. Because the Moche had no writing system, we do not know what they called themselves.

Despite the lack of a written Moche language, however, archaeologists working to reconstruct Moche society have found abundant material to help them. The arid climate of Peru's coastal region preserved ancient objects such as plant remains, textiles, basketry, leather, and even human bodies. By sifting through the sand at Moche settlements for plant remains and animal bones, for example, archaeologists have learned that the Moche were primarily a farming and fishing society with a plentiful supply of food. The Moche cultivated a wide variety of crops, including corn, beans, guava, avocados, squash, chili peppers, and peanuts. From the Pacific Ocean, as well as from rivers, marshes, and lagoons, they harvested a rich catch of fish and shellfish. The Moche also raised llamas, guinea pigs, and ducks, which served as additional sources of food.

With an abundant and nutritious diet, the Moche sustained a dense population composed of many social classes. Archaeologists believe the Moche's largest urban centers may together have supported as many as 10,000 people. With such a large population, the civilization had the

Glossary

Bird Priest: A Moche ceremonial figure shown in Moche art as part human and part bird. The Bird Priest is one of the principal participants in the Moche Sacrifice Ceremony.

Moche: A culture that flourished on the northern coast of what is now Peru between the A.D. 100's and 700's. Also called the *Mochica.*

Priestess: A Moche ceremonial figure, one of the principal participants in the Sacrifice Ceremony.

Sacrifice Ceremony: A Moche ritual in which prisoners of war were sacrificed and their blood consumed.

Warrior Priest: A Moche ceremonial figure who is the chief participant in the Sacrifice Ceremony.

Where the Moche lived

Moche civilization flourished from the A.D. 100's to 700's over a 25,000-square kilometer (10,000-square mile) stretch of Peru's northern coast. The Moche farmed a series of river valleys and built pyramids and temples at their political and religious centers.

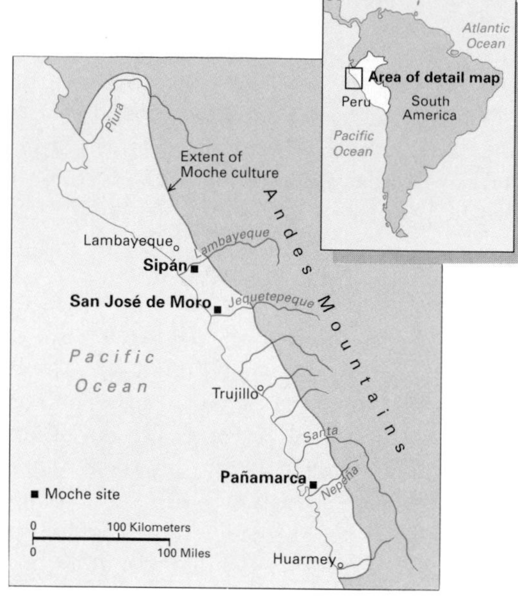

Excavations at Sipán and San José de Moro have yielded clues to Moche political and ceremonial life. At Sipán, *above left,* a badly eroded pyramid towers over the site of several important tombs. At San José de Moro, *left,* archaeologists discovered more tombs.

thousands of laborers needed to build huge pyramids and temples and to maintain the network of irrigation canals on which Moche agriculture depended.

Wealthy Moche nobles may also have supported the skilled craftspeople who produced objects of extraordinary artistic and technological sophistication. With clay, these craftspeople depicted animals, plants, nobles, and supernatural beings and re-created scenes of hunting and fishing, combat, and medical practice. Moche potters painted complex scenes on ceramic vessels. Metalworkers worked with gold, which the Moche probably panned from the rivers, as well as silver and copper, possibly from local deposits, to create jewelry and ornaments for the nobles. Moche weavers created colorful textiles of cotton and wool, often decorated with intricate designs, while other craftspeople carved sculptures out of bone, wood, and stone. The Moche also painted colorful murals on the walls of their religious structures.

In all, thousands of examples of Moche art have survived through the centuries. The Moche thus left a vivid artistic record of their activities, environment, and religious beliefs that provides present-day researchers with glimpses of Moche life.

Since the early 1970's, my colleagues and I have photographed thousands of Moche paintings and objects from museums and private collections throughout the world. The Moche photographic archive, located at the University of California, Los Angeles, now includes more than 125,000 pictures. The archive serves as an important resource for Moche scholars worldwide, enabling them to reconstruct various aspects of Moche culture.

Images of a mysterious ceremony

Among the archive's contents are pictures of some 50 pottery vessels, metal objects, wall paintings, and textiles decorated with illustrations of the Sacrifice Ceremony. Moche artists showed considerable variation in the way they depicted aspects of this ceremony. But the main elements remain essentially the same from one illustration to another.

The Sacrifice Ceremony involves up to four principal ceremonial figures. Each one wears characteristic garments, allowing us to identify the individual figures in different illustrations. Originally, I referred to these individuals as figures A, B, C, and D. However, I now call three of the figures "the Warrior Priest," "the Bird Priest," and "the Priestess."

The most important appears to be the Warrior Priest. He is usually shown larger than the other figures, indicating that he was an individual of higher rank. Often, rays extend from his head and shoulders. The Warrior Priest is sometimes shown with his rattle, and he always wears a crescent-shaped nose ornament, large circular ear ornaments, and a cone-shaped helmet with a crescent-shaped ornament at its peak. He also wears a warrior's backflap, a large piece of metal armor that hangs from the back of the belt.

The second main figure is the Bird Priest. The Bird Priest is always shown as part bird and part human, with a bird's beak, talons in place of feet, and wings projecting from his back and shoulders. Like the Warrior Priest, the Bird Priest may wear a cone-shaped helmet with a crescent-shaped ornament. But he is more often shown wearing a headdress in the shape of a half-circle with an owl at its center. Curved metal branches extend out to both sides. The Bird Priest usually carries a large disk of unknown purpose.

The third important participant is the Priestess. She is always shown wearing a dresslike garment, sometimes together with a cloak. Her headdress has two prominent metal plumes, each ending in three or more jagged points from which small disks may hang. The Priestess also has long, braided hair that often hangs down over her chest.

The fourth figure is as yet unnamed, but he appears to be another

The mysterious Moche sacrifices

Moche artists often depicted a scene in which prisoners' throats are cut and their blood consumed, but scientists had no way of knowing whether such ceremonies actually took place. A pottery vessel, *right,* bears such a drawing, which scholars enlarged for study, *far right.*

priest. Like the Bird Priest, this figure wears a headdress formed of a half-circle with an animal face at its center and curved branches projecting from the top. Long streamers hang from the headdress. Also characteristic of the figure is a sashlike garment with a trailing end.

Clues from the tombs

The story of how we identified the Warrior Priest, the Bird Priest, and the Priestess in Moche tombs began in 1987 in Sipán. Sipán is today a village of some 1,500 people located in the central part of the Lambayeque River valley, in the northern portion of the Moche's former territory. Near the village, three pyramids—the tallest one some 30 meters (100 feet) high—rise dramatically out of the valley floor. The pyramids and neighboring platforms are visible from miles away, their heavily eroded exteriors giving them the appearance of natural hills. But they are made of hundreds of thousands of sun-dried mud bricks, mortared into position centuries ago.

Like other ancient sites, the Sipán pyramid complex has frequently been the target of looters, who dig pits and trenches in ancient ruins in search of gold and other valuable objects. In February 1987, a group of looters working at the smallest pyramid broke into one of the richest funerary chambers ever found. The looters removed sackloads of gold, silver, and copper objects, as well as some ceramic vessels. Many more vessels were broken and left scattered by the robbers.

Peruvian archaeologist Walter Alva soon learned of the looted tomb. Alva, director of the Brüning Archaeological Museum in the city of Lambayeque, near Sipán, quickly organized a team of Peruvian archaeologists to begin excavations at the pyramid complex. I also worked with the group, in my capacity as a specialist in Moche culture.

Alva wanted to prevent further looting at the site and to learn as much as possible about the pillaged tomb. The first stage of the excavation involved surveying the pyramids and what remained of their ramps and neighboring platforms to reconstruct how the site appeared in ancient times. Then, the archaeologists began excavating around the summit of the smallest pyramid. They cleared the looters' tunnel and made a number of modest finds. But for the most part, they unearthed only compacted mud bricks mortared in place when the pyramid was constructed.

In this illustration of the Sacrifice Ceremony, a figure archaeologists call the Warrior Priest takes a goblet from a figure named the Bird Priest. A Priestess holds a goblet with a plate over it. Below the horizontal line, warriors cut the throats of bound captives, whose clothes and weapons lie in nearby bundles. The Warrior Priest's *litter* (carrying chair), with a catlike animal at its base and rays extending from the back, stands to the left.

Then the archaeologists came upon a part of the pyramid where, centuries ago, the bricks had been removed and replaced with dirt. Here, about 4 meters (13 feet) below the surface, they uncovered a human skeleton, accompanied by a shield, helmet, and other warrior's belongings. The warrior's feet were missing, suggesting to Alva that the man might have been buried to guard the tomb of some more important figure. And indeed, digging deeper, the archaeologists found traces of decomposed wooden beams that had roofed a large chamber. Beneath them were eight copper straps that formed the outlines of a rectangular coffin lid, whose wooden planks had long rotted away.

This coffin eventually proved to be the center of a royal burial chamber—the richest burial in Peruvian archaeology known to have escaped destruction by looters. Its principal occupant was a man of between 35 and 45 years of age. Radiocarbon tests later showed that the skeleton had been buried about A.D. 300. (Radiocarbon dating measures the relative amount of a radioactive form of carbon in an ancient material to estimate its age.) In addition, the chamber contained the remains of three adult women; two more adult men, including one warrior; a child; and a dog. All had been ceremonially buried to accompany their dead ruler.

Excavating the main coffin, Alva and his crew uncovered layer upon layer of items, including spear points, the remains of feathered headdresses, and several beaded *pectorals* (garments worn over the chest and shoulders). Many of the items must have been used or worn by the dead man during his lifetime. One of these was a long, shirtlike garment covered with small plates of gilded copper, with a fringe of gilded copper cones along its hem. It must have been dazzling when worn, the plates and cones shimmering and swaying like golden mirrors.

More riches appeared as the fragile skeleton was gradually revealed. Whoever buried the dead man had placed gold ornaments over his nose and eyes and a band of gold over his teeth. Then, they had covered his lower face with a gold mask, hammered into the shape of cheeks, mouth, chin, and neck. Traces of red pigment on the front of

Identifying the Warrior Priest

A royal burial, *below,* found at Sipán in 1987 contained a skeleton covered with gold and other precious objects. Some of the objects—including a nose ornament; a rattle; and a backflap, a piece of armor that hung from the back of the belt—enabled archaeologists to identify the skeleton as a Warrior Priest of the Sacrifice Ceremony.

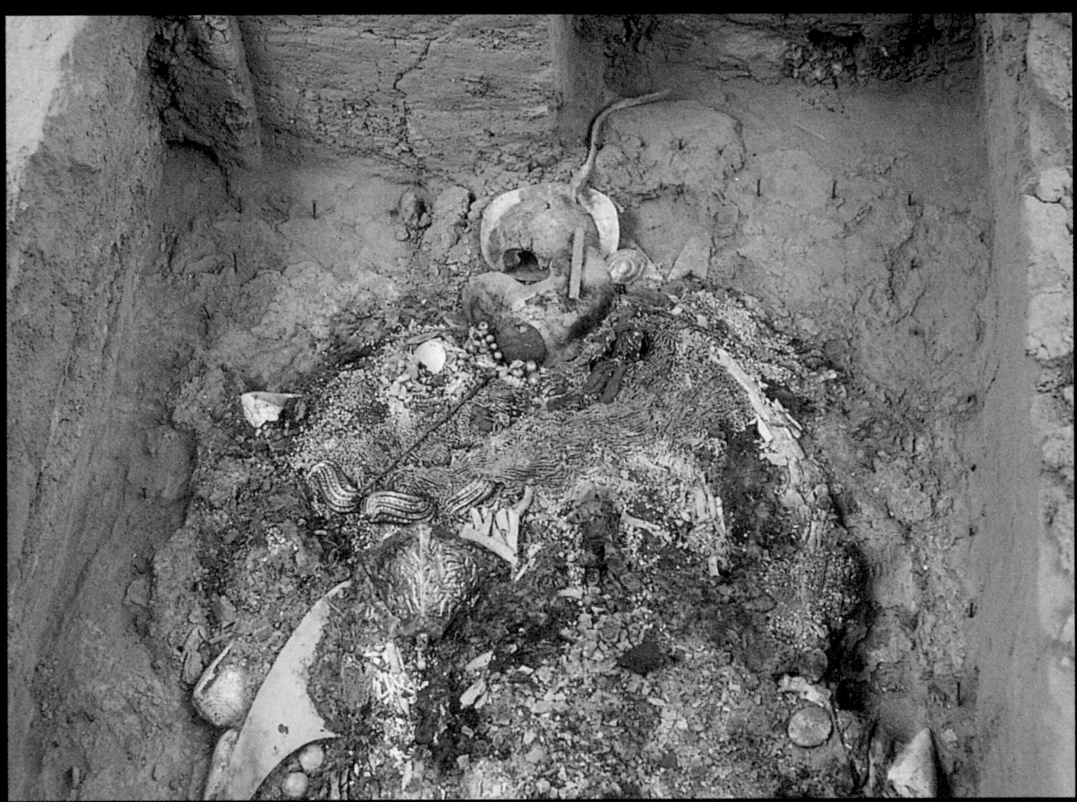

the skull indicated that the dead man wore red face paint at the time of burial. And around the skeleton's neck lay a necklace of gold and silver peanuts, reproduced in minute detail.

Most exciting, however, was the fact that the skeleton was accompanied by objects identical to those worn by the Warrior Priest of the Sacrifice Ceremony. Near the skull lay two crescent-shaped gold nose ornaments, like those typically depicted with the Warrior Priest. At the sides of the head were three pairs of gold ear ornaments, inlaid with blue semiprecious stones—torquoise and sodalite. These ornaments were precisely the size and form of those worn by the figures in the Sacrifice Ceremony. But the most important object was the rattle that lay near the skeleton's hand—the same rattle that we had come

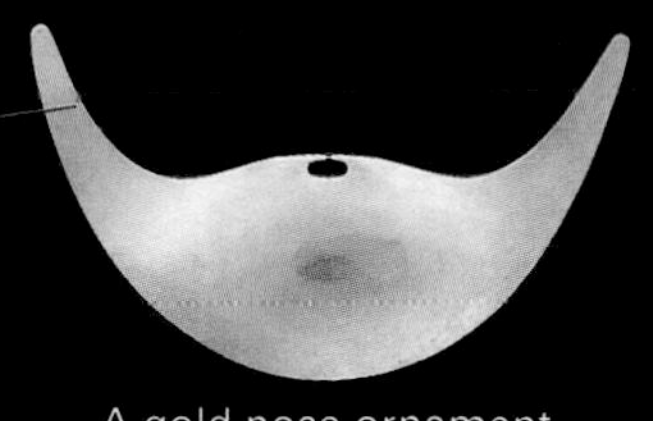

A gold nose ornament like those shown in depictions of the Warrior Priest was one of two found in the tomb. Such ornaments, which dangled over the wearer's mouth and lower cheeks, were apparently worn as a symbol of power.

A solid gold backflap found under the skeleton is the largest Moche gold object known. The backflap is 45 centimeters (18 inches) long and weighs nearly 1 kilogram (2.2 pounds).

A gold rattle, decorated with images of combat and of military equipment, was found in the skeleton's right hand. In Moche art, such a rattle is shown only in association with the Warrior Priest.

to recognize in Moche art as the property of the Warrior Priest.

If the man buried in the tomb was in fact the Warrior Priest, I thought, he should also have a crescent-shaped headdress ornament and a warrior's backflap. And indeed he did. Beneath the body we found a gold headdress ornament and a pair of backflaps, one of gold and one of silver. These backflaps are among the most spectacular objects in the tomb. At 45 centimeters (18 inches) high and weighing nearly a kilogram (2.2 pounds), the gold backflap is the largest and heaviest Moche gold object known today. Both backflaps are decorated with a ferocious figure holding a human head in one hand and a knife in the other. The figure's fanged mouth is set in a snarl, and the eyes stare menacingly beneath heavy lids.

Identifying the Bird Priest

Another rich burial at Sipán, found in 1988, proved to be the tomb of the Bird Priest from the Sacrifice Ceremony, *right.* To the left of the Bird Priest's skeleton is that of a male adult, possibly a warrior, who was buried with him.

Among the objects found in the tomb was a copper headdress with a sculpted owl at its center, *below.* The headdress is identical to that most often pictured in illustrations of the Bird Priest. Other depictions show the priest himself as an owl, *right.*

As the Sipán dig continued through 1988 and 1989, Alva and his team excavated another rich tomb. This burial resembled that of the Warrior Priest, but it was smaller and contained fewer grave objects. This implied that its central occupant, a man 35 to 45 years old, was a noble of lower status than the Warrior Priest.

The contents of this tomb suggest that its occupant was the Bird Priest of the Sacrifice Ceremony. The skeleton was buried with a large owl headdress resembling the headgear frequently depicted in illustrations of the Bird Priest. The tomb also contained a warrior's backflap like the one shown on the Bird Priest in Moche art.

But one more surprise for Moche scholars still lay in store. In 1991, I assembled another archaeological team and began excavations at San José de Moro, a Moche site in the Jequetepeque Valley some 50 kilometers (31 miles) south of Sipán. San José de Moro is a small village today, but in ancient times it was an important ceremonial center, with many pyramids and other structures. Considerable looting has left the ancient structures and cemeteries badly damaged.

During our first field season at San José de Moro, we excavated several large tombs dating to sometime after A.D. 550. Each tomb consisted of a burial chamber that, like those of the Warrior Priest and Bird Priest at Sipán, had once been roofed with wooden beams. Along with its central occupant, each tomb contained several other individuals and a wealth of funerary offerings, including weapons, ceramic vessels, and jewelry.

The principal individual in one of these tombs was a woman, about 35 years of age, who was buried with objects that enabled me to identify her as the Priestess of the Sacrifice Ceremony. Above the skull, for example, were two large plumes made of a mixture of copper and silver. The plumes are identical to those shown on the headdress of the Priestess. The tomb also contained a copper goblet of the type used at the Sacrifice Ceremony. In addition, in a corner of the tomb we found a large black ceramic bowl containing several ceramic cups and another goblet. An identical bowl with cups appears near the Priestess in a wall painting from Pañamarca, a Moche site in the Nepeña Valley.

In 1992, we continued our work at San José de Moro. During this season, we excavated the tomb of a second woman who had evidently served as Priestess of the Sacrifice Ceremony. This individual, who was about 25 years old when she died, was also buried with a ceremonial goblet and the two plumes from the Priestess's headdress. I suspect that this woman inherited the role of Priestess from the woman whose tomb we had excavated the previous year.

A new understanding of the Moche

It will be years before the entire contents of the royal tombs at Sipán and San José de Moro are cleaned, reconstructed, and fully analyzed. Nevertheless, they have already yielded a wealth of information that has dramatically altered our understanding of Moche civilization. For one thing, we now realize that the Moche upper class enjoyed riches

Identifying the Priestess

Archaeologists working at San José de Moro in 1991 uncovered the tomb of a high-ranking female. Objects in the tomb helped identify its occupant as the Priestess of the Sacrifice Ceremony.

Metal plumes found in the grave, *left,* are part of the Priestess's headdress, which appears in a wall painting, *top,* from Pañamarca, a Moche site in the Nepeña Valley. Also found in the tomb were a tall goblet and several cups inside a ceramic basin, *above.* A similar basin and cups appear in the Pañamarca painting.

and status far greater than we had ever imagined. No other known Moche burials have matched the size and complexity of these tombs or yielded such an abundance of gold and silver ornaments.

But the most remarkable aspect of the Sipán and San José de Moro finds is the way they expand our understanding of the Moche culture. The illustrations of the Sacrifice Ceremony have enabled scholars to identify the individuals buried in the tombs and the roles they played in the ceremonial life of their people. At the same time, the excavation of the tombs has helped us understand the part played by the Sacrifice Ceremony in Moche society.

Archaeologists now believe that the Sacrifice Ceremony was performed throughout the Moche kingdom as part of a state religion. As the wealth of the offerings in the Sipán and San José de Moro tombs makes clear, the priests were powerful rulers, and it is likely that the Sacrifice Ceremony served in part to maintain their power. It may be that each river valley had its own ceremonial center, where men and women wearing traditional garments performed the same priestly functions. As members of the priesthood died, other individuals took their place, conducting the rituals as their predecessors had before them.

The age of the tombs also sheds light on Moche culture. Radiocarbon tests show that the Warrior Priest and Bird Priest at Sipán were buried about A.D. 300, while the Priestesses at San José de Moro died sometime after A.D. 550. The long gap between these events—at least 250 years—indicates that the Sacrifice Ceremony remained a part of Moche culture for centuries.

The royal tombs of Sipán and San José de Moro provide a dramatic contrast to the many Moche tombs destroyed by looters. The contents of those tombs are scattered in private collections throughout the world, unavailable for public enjoyment or scholarly research. The Sipán and San José de Moro finds, on the other hand, will be kept together. Many of the most beautiful and important objects will be exhibited in the United States during a two-year tour to begin in September 1993. The exhibition will travel to Los Angeles, Houston, New York, Detroit, and Washington, D.C., before returning to its permanent home at Peru's Brüning Museum.

Museum visitors will see firsthand the priceless treasures found in these tombs. But of even greater value is the information these treasures provide—information that is helping scholars reconstruct one of the grandest civilizations of the ancient world.

For further reading:

Archaeology, November/December 1992. Includes four articles on the Moche, pp. 30-45.

National Geographic, October 1988. Includes two articles on the Moche, pp. 510-555.

National Geographic, June 1990. Includes three articles on the Moche, pp. 2-48.

Science-fiction movies treat us to imaginative visions of different worlds and altered people. But could onscreen events happen in real life?

Science Goes to the Movies

BY JACK WEYLAND

Near the end of the movie *Aliens*, the queen of the alien monsters hitches a ride on a shuttle spacecraft carrying the movie's heroine, Ripley. When the shuttle reaches an orbiting spacecraft, the alien wages all-out war against Ripley. As Ripley battles the alien, she opens a large hatch in her spacecraft to flush her foe into space. But the alien grabs Ripley's leg, dragging her along. The two struggle as a rush of wind threatens to hurl them both into the vast emptiness of space. Finally, Ripley shakes off the alien, and it is sucked into space, where it is doomed to become the newest satellite of a dreary planet.

That was a dramatic ending to an exciting movie. But science-minded viewers may have wondered if it really could have happened that way. Of course, we watch science-fiction movies primarily to be entertained, and we need to suspend our disbelief if we are to truly enjoy films. But after the movies are over, we can appreciate them in a different way—by trying to determine whether the laws of nature would allow certain events onscreen to have happened as they were portrayed.

I became interested in analyzing how science was presented in science-fiction movies after watching *Voyage to the Bottom of the Sea* when I was in college. In that movie, a submarine passes under the polar icecap just as the ice above begins to break up and sink. Large chunks of ice fall through the water and crash into the submerged submarine. This scene is so unlikely that I couldn't stop myself from shouting, "Ice floats!" Since then, I have used science-fiction movies in physics classes to help students understand science. But you don't need to be a physics professor to pick up errors portrayed on the screen. Armed with an understanding of just a few scientific principles, any moviegoer can quickly spot an array of sci-fi bloopers in areas ranging from astronomy to physiology.

Glossary

Atmospheric pressure: The force produced by the weight of the atmosphere pressing upon the Earth.

Density: The amount of matter contained in a given volume.

Gravity: The natural force that causes objects to move toward each other.

Light-year: The distance light travels in one year, or 9.5 trillion kilometers (5.9 trillion miles).

Mass: The amount of matter contained in a particular object.

Vacuum: An area that is completely empty of matter—solid, liquid, or gas.

Volume: The three-dimensional space occupied by an object.

Science and movies in the vacuum of space

Some understanding of how gases behave and of the conditions in space will help us judge the believability of that final battle scene in *Aliens,* in which a human character survives being exposed to space. First, space is, by and large, a *vacuum,* an empty area that has no atmosphere. Earth, on the other hand, is encircled by its atmosphere, the air, which is composed of gases, such as nitrogen and oxygen. In the 1600's, chemists experimenting with gases began to learn how gases behave. They learned, for example, that the more gas in a container, the higher the pressure of the gas will be. These early chemists also found that if a container holding gas under high pressure is opened to a container with gas under lower pressure, the high-pressure gas will rush to the low-pressure area.

This principle can explain why tires go flat. The gases in the upper layers of the atmosphere press down upon the gases below them, creating air pressure. When a tire is inflated with air, the pressure inside is greater than the surrounding atmospheric pressure. If a tire develops a leak, the higher-pressure air inside flows out.

This principle also accounts for why the wind blows. Winds arise when air in high-pressure areas moves toward low-pressure areas. The greater the difference between the high- and low-pressure areas, the more powerful the movement of gas. So let's consider what would happen when air moves into a vacuum, which has no pressure because it contains no gas.

For the astronauts in *Aliens* to survive, their spaceship must have an atmosphere similar to Earth's. We might think of the spaceship as a container of gas under pressure. As Ripley and the alien fight their final battle, air escaping the spaceship into the vacuum of space creates a strong wind that continues for more than a minute. But in reality, the spaceship's pressurized gases would have rushed out into the vacuum of space with nearly explosive force, leaving Ripley with nothing to breathe. (Earth's atmosphere doesn't rush off into space because Earth's gravity holds it down.)

However, the encounter between the alien and Ripley probably never would have taken place. Assuming the alien had physical characteristics similar to most living creatures, it is unlikely that it could have survived the shuttle's trip from the planet to the orbiting spacecraft. For one thing, its lungs would have exploded or collapsed.

The lungs are similar to a balloon filled with air. Under normal at-

The author:
Jack Weyland is professor of physics at the South Dakota School of Mines and Technology in Rapid City.

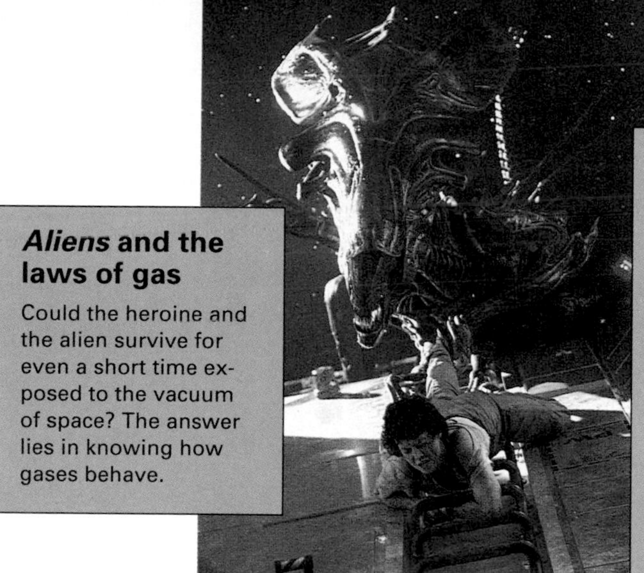

Aliens and the laws of gas

Could the heroine and the alien survive for even a short time exposed to the vacuum of space? The answer lies in knowing how gases behave.

The more gas molecules in a given volume, the greater the pressure of the gas. And gas under high pressure, given the opportunity, will always flow toward an area of lower pressure.

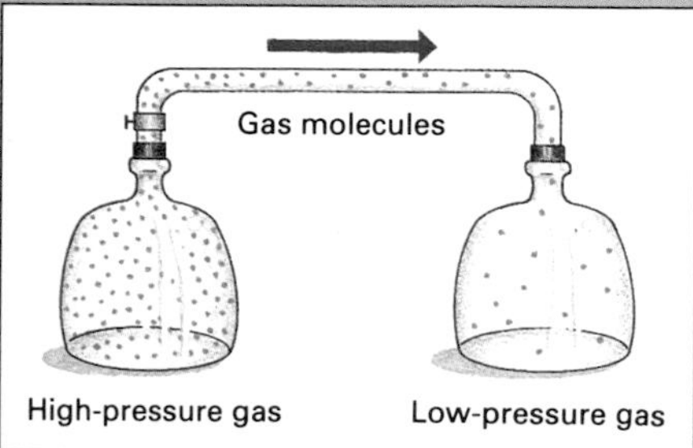

Tires are inflated with molecules of air. The pressure of the air inside a car tire on Earth is greater than the pressure of the air outside the tire. If the tire is punctured, the high-pressure gases inside flow toward the lower atmospheric pressure outside, and the tire goes flat.

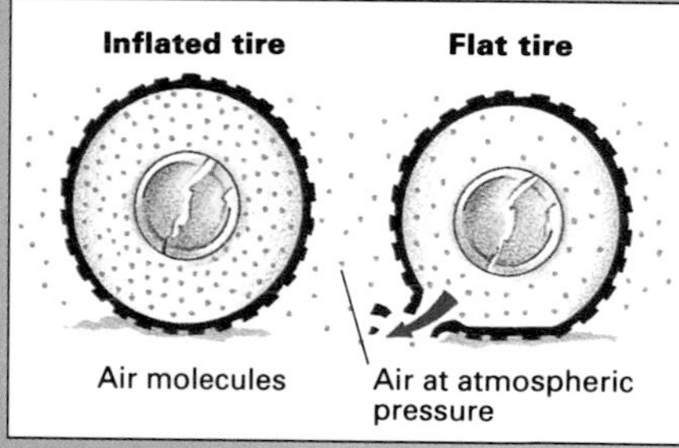

A spacecraft contains pressurized air for people to breathe. But space outside the craft is a near *vacuum,* an area with no gas molecules and, therefore, no pressure. So if a spacecraft's outer door is opened, as it was in *Aliens,* the air from the high-pressure area inside would rush out into the vacuum of space, leaving the people inside with nothing to breathe.

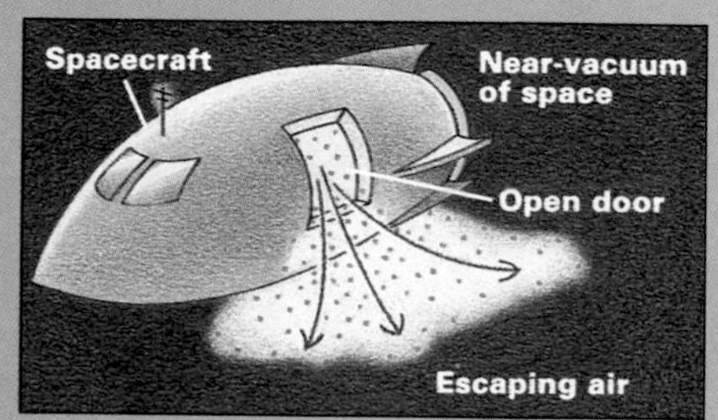

mospheric pressure, the air inside the lungs is balanced by the pressure of the air outside the lungs. The force of each pressure pushes against the other and prevents the lungs from either collapsing or exploding. A creature exposed to the vacuum of space would have two choices: trying to breathe or holding its breath. If it tried to breathe, then all the air in its lungs would rush out, just as the air inside the spacecraft would have rushed into space when Ripley opened the hatch. If it tried to hold its breath, its lungs may have exploded. In the no-pressure vacuum of space, there would be no outside pressure, directed inward, to prevent the lungs from expanding. The lungs would then expand until the pressure inside them was more than their delicate membranes could withstand. Then, they might have exploded like an overinflated balloon. For these and other reasons, astronauts must wear space suits if they leave their spaceship. The space suit maintains a pressure within it that is equivalent to the air pressure inside the spaceship.

The advertising writers for *Alien,* the predecessor to *Aliens,* were absolutely right when they created an advertisement for the movie that said: "In space no one can hear you scream." What some filmmakers forget, however, is that no one could hear the explosions of battling spacecraft, either, because of the properties of a vacuum and the way sound travels.

Sound begins as a vibration that travels in waves through air or another medium. When a guitar string is plucked, for example, the vibration of the string pushes on the air molecules near it. These

Space battles and the physics of light and sound

To determine whether a laser-weapon battle, such as the exciting scene in *The Empire Strikes Back,* could be seen and heard in space, we must first understand how we see light and hear sound.

molecules do not actually move all the way to our ears. Instead, the vibrations are transmitted in waves as the molecules near the guitar string bump into the molecules next to them, which then bump into other molecules, and so on, until this chain reaction reaches air molecules next to the eardrum. These molecules then bump into the eardrum, causing it to vibrate in the same way that the guitar string vibrated. This process is what allows us to hear many different types of sound. In the vacuum of space, though, there are no air molecules to transfer a signal from its source to someone's ear. Thus, people would not be able to hear any explosions, screams, or crashes that might occur outside a spacecraft.

We also would be unable to enjoy such dazzling visual effects as the colorful laser-weapon battles between movie spaceships. Due to the nature of light and how we are able to see, the light from lasers in space would be invisible.

Light is a form of energy that can travel through space. While doing so, it may strike something in its path. The light interacts with the molecules of that object and several things may happen. Some of the light may be absorbed, some may be reflected, and some may pass through. The reflection of light off objects and into our eyes allows us to see things.

For us to clearly see a beam of light—such as a ray of sunshine or a beam from a flashlight—particles such as dust or smoke must be floating in the air to

What it takes to see light

We can see light in two ways. We can see light coming from its source, such as the sun or a light bulb, by looking directly at the source. And we can see light indirectly, when light traveling through the atmosphere strikes molecules. This deflects some of the light in a process called scattering. Scattering allows us to perceive objects and their color.

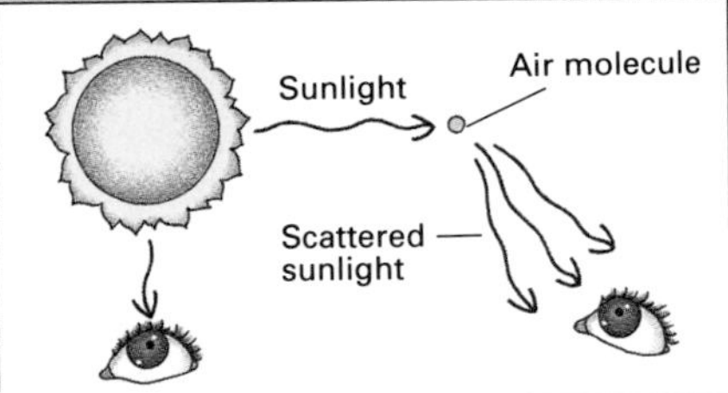

A beam of light, such as that from a flashlight or a laser, is visible if it is scattered by molecules or particles in the atmosphere. At light shows, *laser beams* (intense light of a single color) are typically shone through fog or smoke. Water vapor in fog, and dust particles in smoke, scatter the light, making the beam visible.

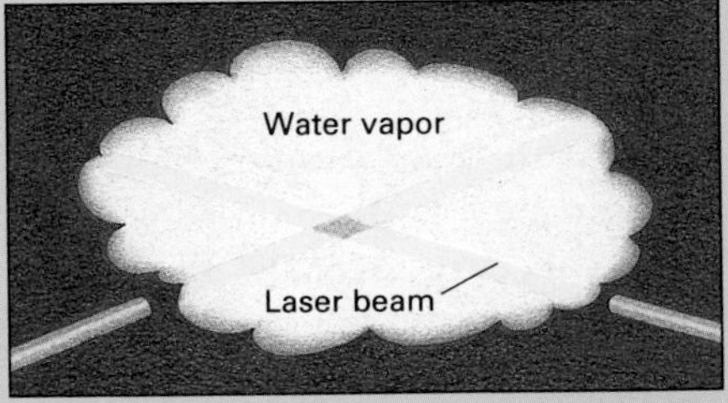

In deep space, there are not enough particles to scatter light. So, in reality, no one could see the spectacular laser battles of science-fiction movies.

What produces sound?

Sound travels in waves, and like most waves, it must have a medium, such as air, through which to travel. Sound waves are set up by vibrations. For example, the explosion of a firecracker causes vibrations in the air molecules around it. These vibrations or sound waves travel outward in all directions. When they reach an ear, they cause vibrations that the brain interprets as sound. But in the vacuum of space, there is no air or other medium through which sound waves can travel. So in space, outside a craft, no one can hear any sound.

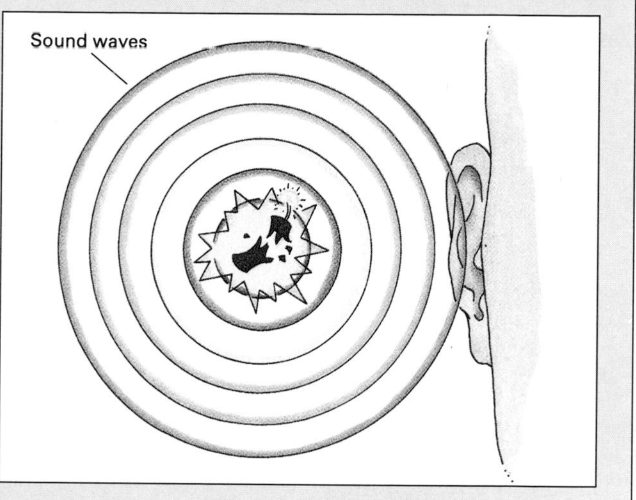

deflect the light into our eyes. If there are no particles in the air, we might see a spot of light where the beam strikes a floor or a wall, but we won't see a ray of light from its source to the object it hits. The intense light produced by a laser can also be reflected by particles in the air. But in space, there are not enough particles to allow us to see a beam from a laser weapon.

Seeing through the notion of invisibility

Light and vision affect another favorite theme in science-fiction movies—invisibility. The first movie to tackle this subject was *The Invisible Man,* made in 1933. In 1992, *Memoirs of an Invisible Man,* starring Chevy Chase, portrayed the problems of not being noticed.

For Chase to become invisible, or transparent, light rays must pass straight through his body as if he weren't there—similar to the way light passes through a sheet of glass. Light passes through glass because the molecules do not absorb the light. But human bodies are made up of many different types of molecules that do absorb light. So one way to make Chase transparent would be to turn him into glass. But if this were the case, he would also have the other characteristics of glass, such as brittleness. Even glass is not invisible, however. Light entering a piece of glass is partially reflected, which allows us to see it. Also, light is bent, or refracted, when it goes through a transparent material. This is why a pencil stuck into a glass of water appears to be bent. So even a glass Chevy Chase would not be invisible.

Even if it were possible for a person to become truly invisible, he or she couldn't see anyone else, either. In order to see, the eye's lens focuses light rays, which then strike the *retina,* a thin tissue at the back of the eye. The retina absorbs the light rays and changes them into elec-

One of the problems of being invisible

In Memoirs of an Invisible Man, an invisible Chevy Chase walks on a beach. But if he could really become invisible, the anatomy of the eye tells us that Chevy would not be able to see anything either.

A person can see because the cornea and lens of the eye focus light rays on the *retina,* a thin tissue at the back of the eye. The retina absorbs the light and converts it to electrical signals, which are sent by the optic nerve to the brain.

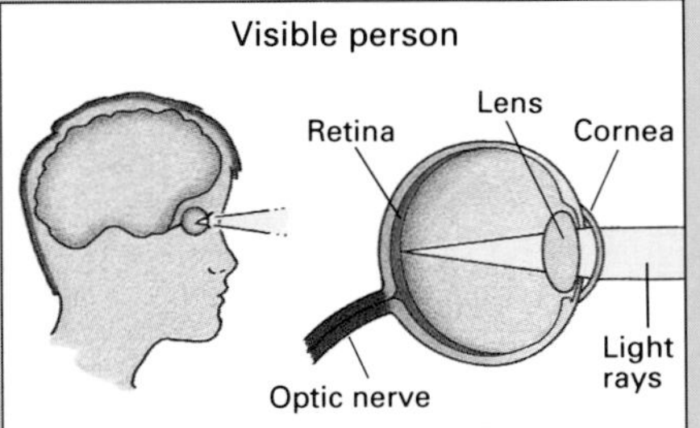

An invisible man would be *transparent,* allowing light rays to pass through his body without being absorbed by any of his molecules. Consequently, light rays would pass right through the eye, the retina would absorb no light, and the invisible man would be unable to see.

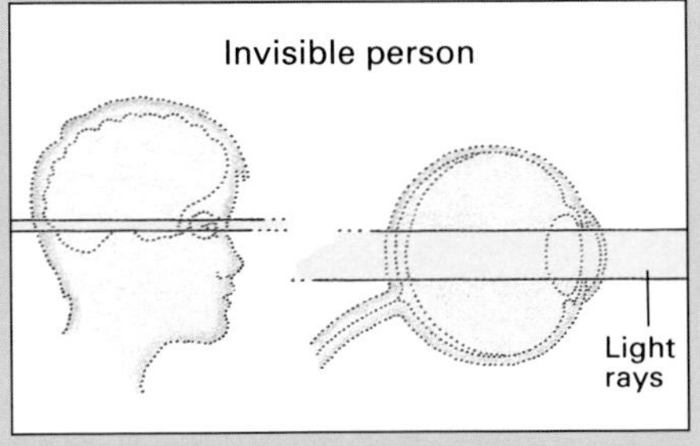

trical signals that travel to the brain. Because an invisible man's retina would be transparent, he would be blind. Light would pass straight through the retina, and none would be absorbed to create the signal to the brain.

Flight and fakery

Some of the most entertaining scenes in movies involve battles between spaceships. The ships are often depicted as high-tech jet aircraft, capable of swiftly turning and banking. But if we consider how airplanes fly, we can see that the vacuum of space would not allow such exciting maneuvers.

Unlike spacecraft, airplanes are designed to move through a gaseous medium and to take advantage of *Bernoulli's principle,* which was named after the Swiss mathematician Daniel Bernoulli. This principle says that the pressure of a moving liquid or gas will decrease as its speed increases.

The design of an airplane wing—which is curved on the top and flat on the bottom—makes air above the wing of a moving plane flow faster than the air below the wing. As a result, the pressure above the wing is less than the pressure below the wing. Two forces act on the wing, but the force of the air pushing upward is stronger than the force of the air pushing downward. This creates *lift,* a lifting force on the wing. By lowering the *aileron* (wing flap) on one wing, the pilot can increase the curve of that wing and, consequently, the amount of pressure or lift pushing up on it. As this wing lifts, the plane banks and turns to the left or right. More lift on the left wing causes the plane to turn right. More lift on the right wing produces a left turn.

Even though some spacecraft, such as space shuttles, are designed to fly like airplanes when they return to Earth's atmosphere, they cannot bank in space because space lacks the air needed to create lift on the wings. A spacecraft could change direction in space by shutting down rocket engines pushing it in one direction and igniting thruster rockets located elsewhere on the spaceship. Unfortunately for filmmakers, such maneuvers would make spaceships easy targets for their enemies.

A scene in *Superman* also poses puzzling questions about flight. In the movie, a plane carrying the President of the United States loses part of its wing and one engine during a violent thunderstorm. Just as the plane seems certain to crash, Superman flies up to it and grabs onto the wing. The movie asks us to believe that Superman is supplying the power of a jet engine to keep the plane aloft and save everyone on board. How could he do that?

Well, the answer is he can't. This is because for each action there is an equal and opposite reaction, according to the laws of motion formulated in the 1600's by the British scientist Sir Isaac Newton. To propel a jet airplane forward, an engine pushes the hot gases of burned fuel backward. But when Superman appears, he's not burning fuel and pushing hot gases backward. Nor is he pushing his feet against something to provide the backward force needed to cause the plane to move forward. He appears to be simply holding onto the wing, exerting no force at all. No matter how strong Superman is, he and the wing should just fall to the ground.

Travel tips for space explorers

The speed at which spacecraft move also provides great adventure for moviegoers. Captain Kirk's speedy Enterprise zips through the galaxy at "warp" speed, hopping from one star system to another in a matter of days or hours. But in reality, the distance between stars is so large that space travel takes an extremely long time. For example, the nearest star to the sun is Proxima Centauri, which is 4.3 light-years away. A *light-year* is the distance light travels in one year at its speed of 299,792 kilometers (186,282 miles) per second. The fastest any astronaut has ever gone is about 40,000 kilometers (25,000 miles) per hour. At this rate, an astronaut would have to travel for more than 100,000 years just to reach Proxima Centauri.

Star Trek's "warp" speed enables space voyagers to travel faster than the speed of light. However, Albert Einstein, the German-born physicist who developed the special theory of relativity in the early 1900's, showed that it is impossible for anything to move faster than the speed of light. And even if a spacecraft could be built that approached the speed of light, distances in the universe are still so vast that it would take years for space travelers to get from one adventure to the next.

In other movies about space voyagers, people act as if they were living under the same conditions as those on Earth. But without gravity, a character in a spacecraft should float all over the screen. Gravity is the force that pulls objects toward one another. Newton proved that the strength of gravity between two objects depends on each object's *mass* (the amount of matter it contains) and on the distance between them. Earth's strong gravitational pull prevents objects from floating off the surface. What we call weight is actually a measurement of the gravitational pull of the entire Earth on the object or person being weighed. One way a space explorer could maintain a gravitational force in space exactly like the one on the Earth would be by carrying around a ball that had the same size and mass as Earth. Of course, this would be a bit awkward.

Some movies have introduced the concept of "artificial gravity" to explain how their characters walk around in a spacecraft as if they were on Earth. Indeed, a force similar to that of gravity could be maintained by rotating a spacecraft. *2001: A Space Odyssey* presented a rotating space station that was based on sound scientific principles.

Rotation produces *centripetal force,* which makes objects follow curved rather than straight paths. You can see centripetal force in action by tying a string to an object and swinging it in a circle over your head. Pulling on the string provides the force needed to keep the object moving in a circle, rather than flying off. Similarly, a rotating space station shaped like a doughnut would exert a centripetal force that would hold objects and people against the inside surface of its outer walls. People in a rotating space station could walk on the curved walls and feel as if they were walking on a floor.

The very small and the very large

Other science-fiction movies that have been favorites with audiences feature either very small human beings or very large monsters. The 1989 movie *Honey, I Shrunk the Kids* turned likable kids into ant-sized creatures. In the 1992 sequel, *Honey, I Blew Up the Kid,* a father accidently turns his son into a Godzilla-sized 2-year-old.

It might seem possible to shrink people to small sizes, since some things in the real world shrink. Wool sweaters shrink if they are improperly washed, and food shrinks when it is *dehydrated* (depleted of water). Let's look at these examples and see if the mechanism involved in each type of shrinking could be used to make ant-sized people.

Fabrics such as wool shrink because washing them causes their fibers to become more tightly packed. The fabric doesn't actually lose any of its mass. It merely becomes more dense, taking up less space. But the human body is not made of fiber that can be compressed in this way. Otherwise, we would shrink up after a hot bath.

But suppose a body could somehow go through a shrinking process. The shrunken person would encounter some very serious problems. For example, an average-sized man who was compressed to the size of a

Fancy maneuvers

In a chase through space in *The Empire Strikes Back*, spacecraft nimbly zoom around obstacles. But are the conditions that allow fighter jets to make fancy maneuvers in Earth's atmosphere also present in space?

Basic principles of aerodynamics govern how airplanes can fly and maneuver through Earth's atmosphere. *Aerodynamic lift*, for example, carries planes aloft. Lift is created as a plane moves forward and air flows over and under the wing. The curved shape of the wing causes air pressure above the wing to decrease. The greater air pressure below the wing tries to move up to the lower-pressure area and, in so doing, pushes up on the wing, creating lift.

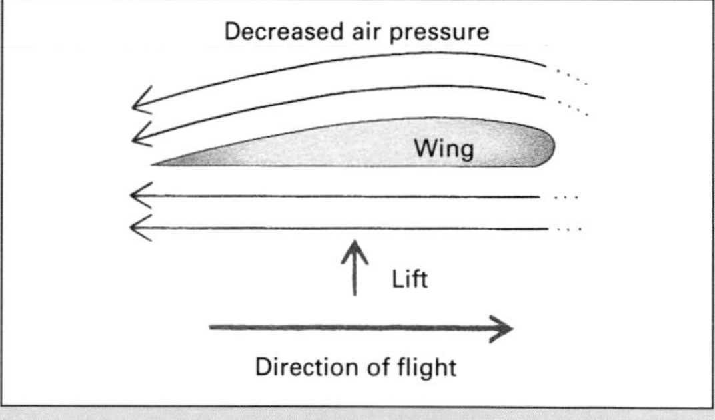

Lift also allows a plane to bank and turn. A pilot banks an airplane to the right by lowering the *aileron* (flap) of the left wing and raising the aileron of the right wing. This makes the right wing dip lower than the left. As that happens, the lift on the left wing increases, pulling the plane around to the right.

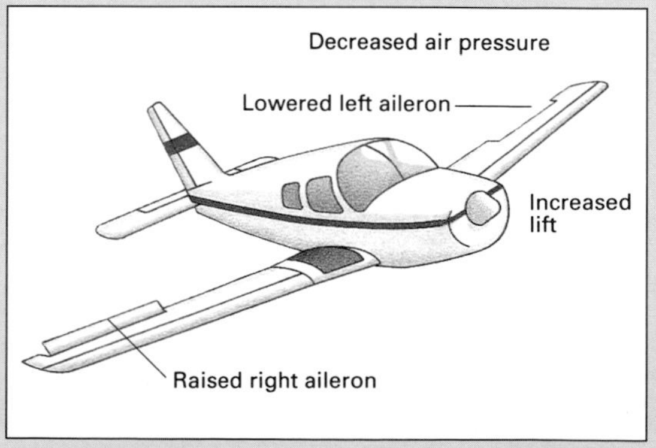

Spacecraft, on the other hand, cannot swiftly bank or turn because space lacks the air needed to create lift. To maneuver, spacecraft generally rely on thrust created by rocket engines. To turn around, for example, a spacecraft would come to a stop, then fire rockets on the side opposite the direction in which it wants to go.

crumb without losing any of his mass would still weigh about 77 kilograms (170 pounds). A man with this weight, no matter how small he was, would be unable to ride on an ant or sit on the leaf of a tree, as the characters did in *Honey, I Shrunk the Kids.*

Another type of shrinking occurs when fruit or other objects are dehydrated. Although a boy or girl would shrivel up by having bodily water removed, he or she would also immediately die. Water, which makes up about 65 percent of a human body, is the major component of blood and is essential for numerous body processes. It dissolves and distributes food and flushes out waste that would otherwise poison us.

To shrink a body and retain its proportions, molecules throughout

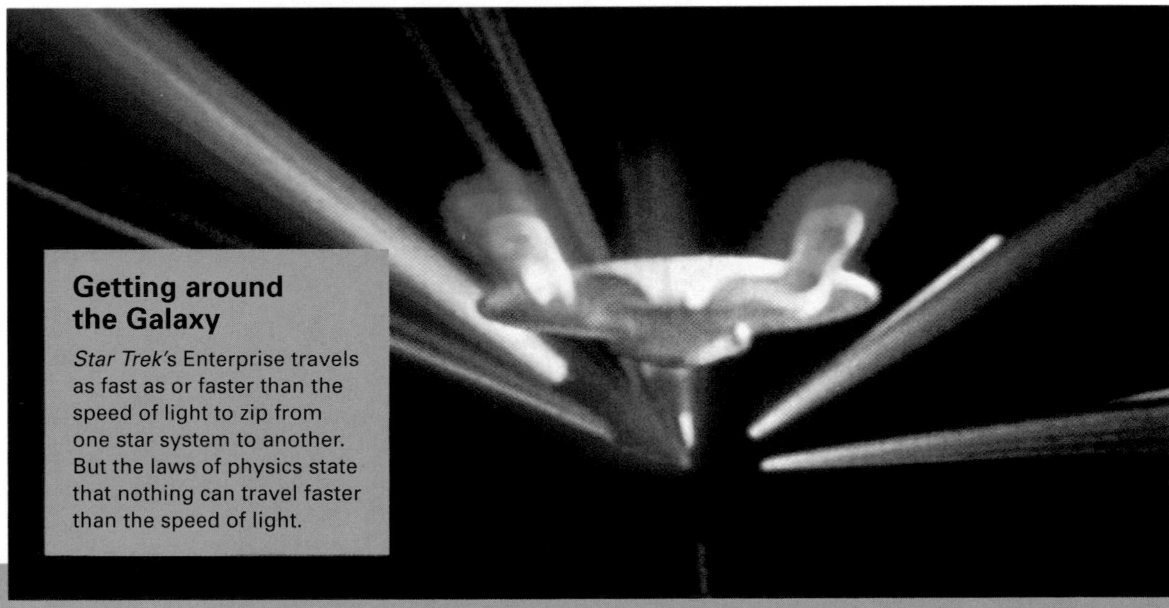

Getting around the Galaxy

Star Trek's Enterprise travels as fast as or faster than the speed of light to zip from one star system to another. But the laws of physics state that nothing can travel faster than the speed of light.

Distances in space are so vast that they are measured in *light-years* (the distance light travels in one year, about 9.5 trillion kilometers [5.9 trillion miles]). Spacecraft travel much slower than the speed of light, so it takes many years to cover just a tiny fraction of the Galaxy. For example, Voyager 1, a space probe launched in 1977, coasts through space at an average speed of about 56,327 kilometers (35,000 miles) per hour. By February 1993, it had traveled about 7.6 billion kilometers (4.7 billion miles), to a point a little beyond the planet Pluto.

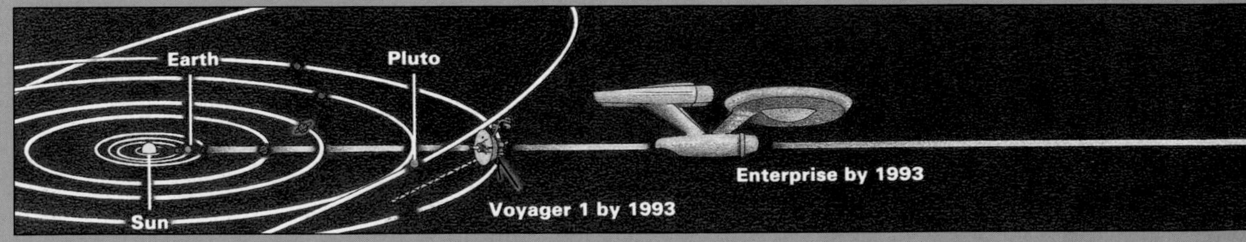

the body would have to be removed. The laws of physics state that energy and mass cannot be created or destroyed by ordinary means but that each can be changed into the other. Perhaps the father's gadget in *Honey, I Shrunk the Kids* eliminated most of the children's mass by precisely converting just the right number of their molecules into energy. But where did this energy go? If it was in the form of heat, it would have gone off into the atmosphere, leaving the children unable to regain their original molecules and return to a normal size.

In the opposite situation, when movie characters are enlarged, the laws of conservation of energy and mass would still apply. If a boy were enlarged by adding extra matter to him, the movie would have to explain where the extra matter came from. In real life, people can become larger by taking in more food than their body needs. But eating more food cannot substantially lengthen an adult's bones. So even if an adult became hundreds of pounds fatter, he or she would not grow taller. Children, on the other hand, do grow in all their dimensions. But their genetic makeup determines when they will stop growing taller. Obviously, the way we grow in reality would not allow us to expand to the size of the 2-year-old in *Honey, I Blew Up the Kid.*

Rather than adding more matter, could a person expand by having the body's molecules pushed apart, increasing the space between them? This process might be similar to what happens when a cake is baked. As a cake bakes, it enlarges, even though matter is not being added to it. Instead, the cake is becoming less dense as air bubbles make the mixture rise and become light and fluffy. Unfortunately, if a person survived being pumped up with a lot of air, he would turn into a huge fluff ball and blow away with the first breeze.

Other problems with reality arise when moviemakers enlarge people

If the Enterprise had been traveling at Voyager's speed since the *Star Trek* television series was launched in 1966, the ship would have traveled about 13.4 billion kilometers (8.3 billion miles)—without stopping for any adventures—by early 1993. But it still would need to travel almost 40 trillion kilometers (25 trillion miles)—a distance that would take 81,000 years to cover—to reach the nearest star to our solar system, Proxima Centauri. Even assuming that Enterprise could travel at the speed of light, it would still take 4.3 years, not just a few minutes, to reach this "nearby" star.

Proxima Centauri

(Not to scale)

A shrunken boy and the matter of density

Could a crumb-size kid bob around in a bowl of cereal and milk as a boy does in *Honey, I Shrunk the Kids*? To figure this out, we must consider how weight, volume, and density are related.

Some materials weigh more than others, even when they occupy the same three-dimensional space, or *volume.* A box of feathers, for example, weighs less than a box of metal balls. This is because metal balls have more *density* (the amount of matter per unit volume) than do feathers.

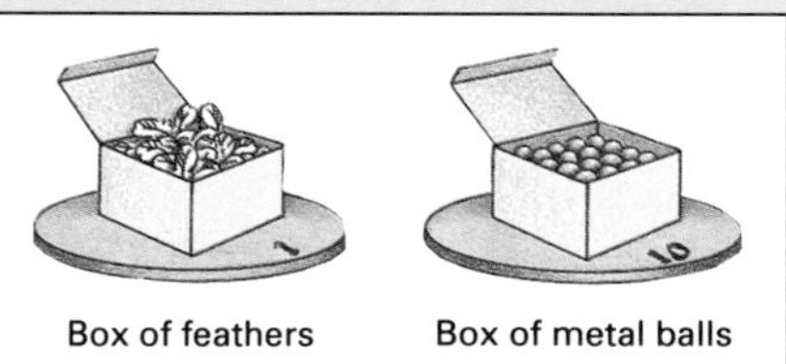

Box of feathers Box of metal balls

It is impossible to shrink a boy to the size of a crumb without losing any of his *mass* (the amount of matter in his body). But if it were possible, his body would be very dense. If he weighed 50 pounds (23 kilograms) before being shrunk, he would continue to weigh that amount at his greatly reduced size. This tremendous increase in his body's density would cause a host of physical problems—and make him sink to the bottom of a bowl of cereal and milk.

or animals simply by scaling them up, making them proportionally bigger. When something is enlarged in all its dimensions, its *volume* (the amount of space it occupies) increases at a much greater rate. Say, for example, that a 2.5-square-centimeter (1-square-inch) box can hold 10 dimes. Does that mean that a 5-square-centimeter (2-square-inch) box would hold 20 dimes? No, it doesn't. A 5-square-centimeter box would hold 80 dimes, or eight times more than the smaller box. Even though each side of the box was only doubled in length, the volume of the box increased eight times because the box was enlarged *in each of its dimensions.* And because the mass of the coins increases in direct proportion to their number, the box's weight would also increase by eight times.

Of course, people are not square boxes, but the same general principle applies. If people could be scaled up to twice their normal size without decreasing the *density* (amount of matter per unit volume) of their body, then they would weigh much more than twice their normal weight. An average 90-centimeter- (3-foot-) tall boy, for example, weighs about 16 kilograms (35 pounds). But an average 180-centimeter- (6-foot-) tall man does not weigh twice as much as the boy. He weighs about 79 kilograms (175 pounds), or five times more. And the greater the increase in size, the greater the rate of weight gain.

This relationship between volume and weight causes serious engineering problems with the concept of huge people and animals. Consider a science-fiction movie ant that is 100 times larger than a regular ant. The ant's proportions would have to change to support all that extra weight. That is why elephant legs do not have the same shape as mosquito legs. Elephants must have thick, sturdy legs to support their heavy bodies. Therefore, an ant the size of an elephant would end up looking a lot like an elephant.

On second thought, Scotty, don't beam me up

Sometimes filmmakers come up with an idea that catches on like wildfire among the public despite its improbability. "Beam me up, Scotty"—Captain Kirk's famous line in the popular *Star Trek* series—became a part of everyday language because being "beamed" to another place is a truly attractive idea. The "transporter" machine that *Star Trek* characters use to move back and forth from the Enterprise supposedly breaks the body into "subatomic particles," which are transported within a beam to their destination, where they somehow reassemble.

A few problems would prevent this scenario in real life. Although human bodies are composed of atoms that theoretically could be broken apart, an extremely strong force holds together the particles that make up atoms. Splitting the matter in a human body into subatomic particles would require an enormous amount of energy—equivalent to that released in the 1980 eruption of Mount St. Helens.

Even if a person were broken into subatomic particles, correctly reassembling those particles would be almost impossible. The body contains a complex assortment of tissues made up of molecules governed by hundreds of thousands of genes. Just getting one or two atoms out of place in reassembling this human puzzle could prove fatal.

Although we cannot beam around, travel faster than light, or become invisible, there is nothing wrong with enjoying the special effects that enable characters to do so. Surely even Newton wouldn't mind seeing a few scientific flaws ignored for the sake of entertainment. So the primary rule for movie watching is this: When you go to a movie for the first time, just enjoy it. Unless you want to end up going to movies by yourself, don't keep telling your friends, "That could never happen." When the movie is over, however, ask yourself "Could it really happen that way?" and see if you can outsmart the movie moguls.

For further reading:

Dubeck, Leroy W.; Moshier, Suzanne E.; Boss, Judith E. *Science in Cinema: Teaching Science Fact Through Science Fiction Films.* Teachers College Press, 1988.

Nicholls, Peter. *The Science in Science Fiction.* Alfred A. Knopf, 1982.

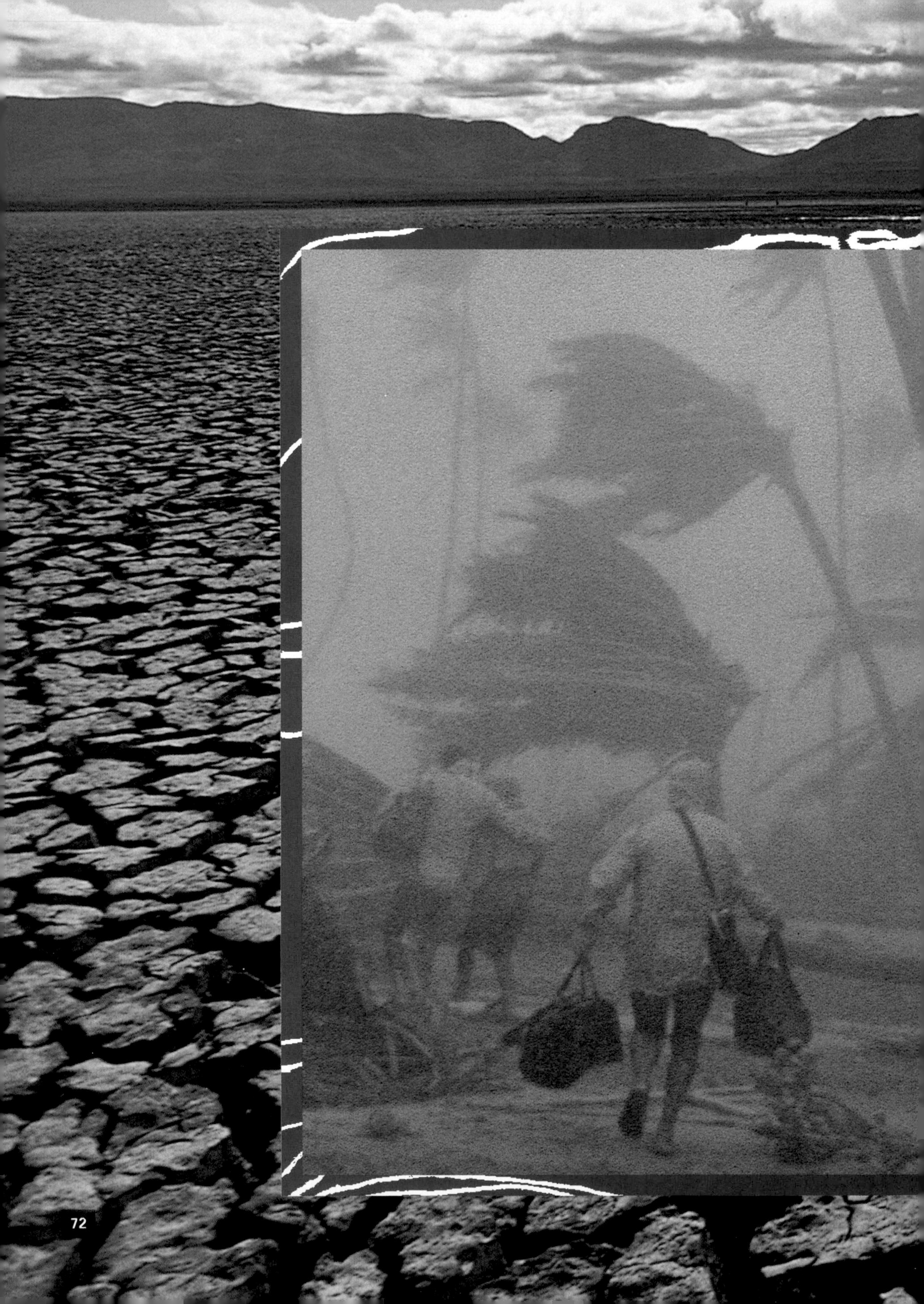

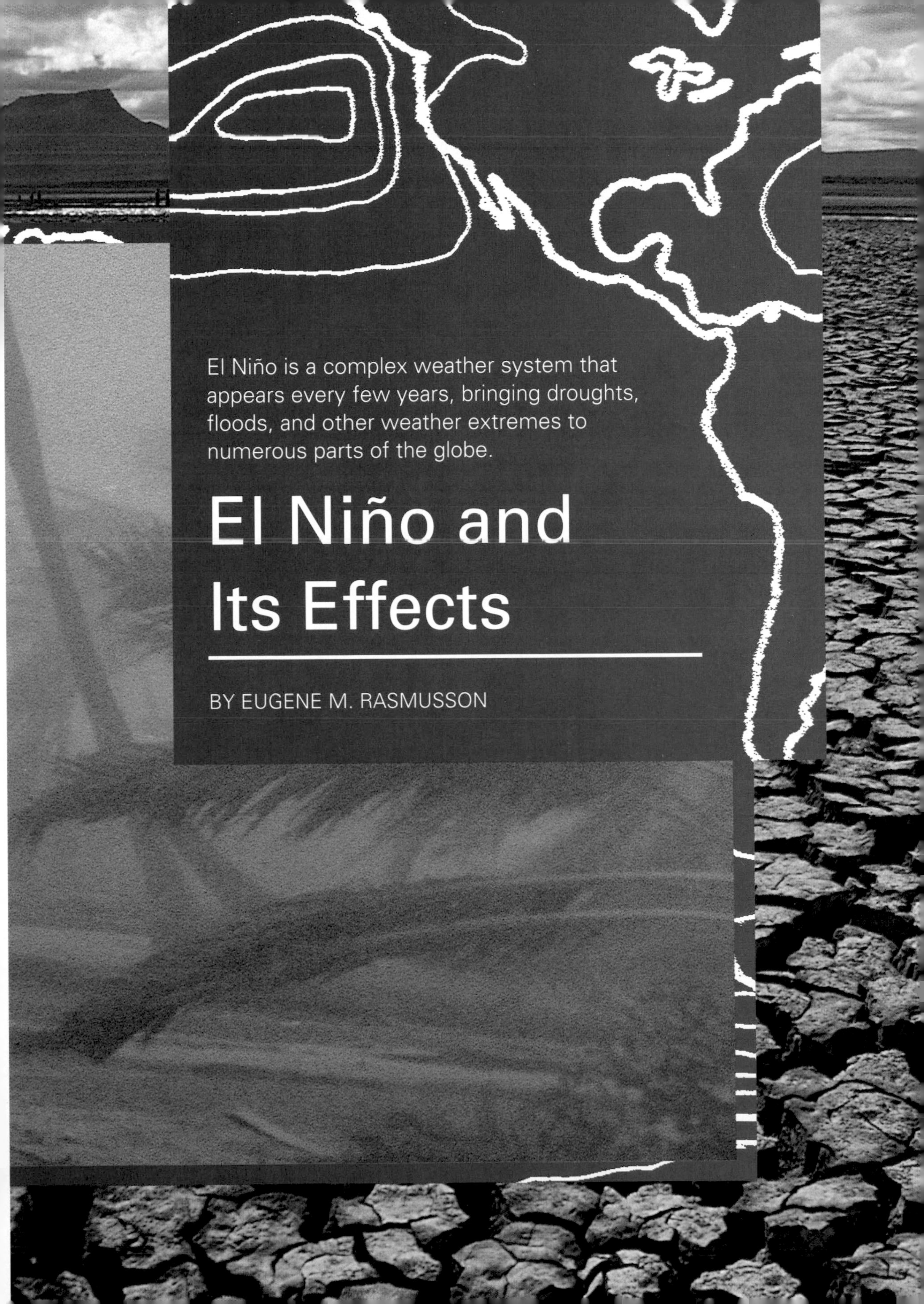

El Niño is a complex weather system that appears every few years, bringing droughts, floods, and other weather extremes to numerous parts of the globe.

El Niño and Its Effects

BY EUGENE M. RASMUSSON

Nearly every river in north Texas swelled beyond its banks with relentless rainfall in December 1991. Storm waters washed toward the Gulf of Mexico and picked up momentum from tributaries in the state's southeastern plains, flooding dozens of communities and causing more than $170 million in damages. Texas was not alone. Heavy rains and storms battered the United States from Los Angeles to Louisiana during much of late 1991 and early 1992. During the same time, just the opposite was occurring in southern and eastern Africa. There, a severe drought further parched a region that had already suffered from several years of sparse rain. Crops withered, and many countries that had been self-sufficient in food production declared food emergencies.

Glossary

Coral: The hard skeletal remains of animals called coral polyps, which live in warm, shallow seawater.

El Niño: The warm phase of an extensive cycle of oceanic and atmospheric conditions called the El Niño/Southern Oscillation cycle. El Niño is marked by a warming of the upper ocean that appears off the coast of Peru every three to seven years.

El Niño/Southern Oscillation (ENSO) cycle: Atmospheric and oceanic conditions that alternate between the Indian Ocean and the eastern Pacific Ocean along the equator. These changing conditions can greatly affect weather around the world.

La Niña: The cold phase of the ENSO cycle. La Niña is marked by cold currents off Peru.

Thermocline: A transitional layer of ocean water between the warm surface water and the colder depths.

Although thousands of miles separate the southern United States and southern Africa, both regions were suffering from the effects of the same massive climate system, popularly known as *El Niño.* This system involves ocean and atmospheric conditions that cover more than a quarter of the globe and affect weather in places as far apart as Australia, the Philippines, Canada, and Peru. Climatologists agree that El Niño, which occurs every three to seven years, is the most important recurring feature of year-to-year variations in global climate apart from the change of seasons.

Until the late 1950's, however, most scientists viewed El Niño simply as a warming of the coastal waters near Peru. Since then, climatologists have learned that the warming is one aspect of a global cycle whose main feature is a slow shift in the pressure of Earth's atmosphere at sea level. As a relatively low average pressure in the Indian Ocean increases, a higher average pressure in the central Pacific decreases. Over a period of years, the cycle reverses itself.

The 1991-1992 El Niño resembled many past occurrences, but it differed in an important way. It was the first that scientists ever predicted. Climate modelers Mark Cane and Stephen Zebiak of the Lamont-Doherty Geological Observatory in Palisades, N.Y., had used computers to analyze information on changing ocean currents, sea surface temperatures, and wind speed and direction in the tropical Pacific to determine in 1990 that an El Niño was likely to occur by late 1991. But after fading in mid-1992, El Niño unexpectedly reappeared in late 1992 and early 1993.

Yet scientists are optimistic about the increased understanding of El Niño they have gained in recent years. Using sophisticated new technology, as well as evidence of past El Niños found in historical accounts of weather and in the growth rings of trees and coral, meteorologists and oceanographers have begun to develop a record of previous El Niños. Scientists hope that their increased understanding can help improve forecasting methods.

The author:
Eugene M. Rasmusson is senior research scientist in the Department of Meteorology at the University of Maryland in College Park.

The El Niño current

Today we know that El Niño is part of a complex system. People first noticed it because of its effect on the currents that flow off the coast of Peru. A cold, broad current called the Peru Current normally sweeps northward along the west coast of South America from southern Chile toward the equator. Off the coast of Peru, just south of the equator, this current is about 8 degrees Celsius (15 degrees Fahrenheit) colder than other ocean waters in the same latitude. The ocean's coolness at this normally warm latitude must have surprised the Spanish conquistadors as they sailed along the Peruvian coast in the early 1500's. They wrote in their diaries of cooling their drinking flasks by lowering them into the water.

In the early 1800's, oceanographers thought that the waters off Peru were so cold because the Peru Current began near Antarctica. In the middle 1850's, the U.S. oceanographer Matthew Maury determined that rising cold waters from the ocean depths near the Peruvian coast also contribute to the low surface temperatures. Trade winds blowing toward the northwest drive warm water away from the coast, allowing the cold water to rise.

At the end of each year, the waters off Peru warm slightly. This warmth comes from a southward coastal current that flows opposite the main Peru Current. As early as the 1700's, the local fishermen named this countercurrent the *Corriente del Niño,* or the *Christ Child Current,* because it usually appeared at Christmastime. Local residents also observed that this annual ocean warming differed from year to year—every few years it was unusually intense. With time, the term *El Niño* came to mean these occasional intense warmings rather than the more typical, annual warmings.

The El Niño warmings change patterns of rainfall in the area. In most years, the low ocean temperatures prevent rain clouds from forming over the eastern Pacific along the equator, leading to desert conditions in the nearby coastal regions of northern Peru. In contrast, during the years of strong warming, the land is swept with torrential rains and severe flooding.

A global phenomenon

Until the late 1950's, climatologists had little reason to believe that the coastal El Niño warming was anything but a local occurrence. A key event drastically changed this picture in 1957, however. During that year, Earth scientists had planned extensive new observations of the ocean and atmosphere as part of an event Earth scientists called the International Geophysical Year. Oceanographers, for example, expanded their studies of the world's oceans, recording and analyzing ocean properties such as surface water temperatures in regions that had not previously been thoroughly examined. As luck would have it, 1957-1958 was also an El Niño year. When scientists monitored the ocean worldwide, they discovered that the ocean warming off Peru was not just a local event. The warm surface waters actually extended westward along the equator more than a quarter of the way around the globe.

This discovery caught the attention of Jacob Bjerknes, a distinguished meteorologist at the University of California at Los Angeles.

How an El Niño develops

El Niño, a warm current that appears off the coast of Peru about every three to seven years, is linked to changes in the atmosphere and ocean over vast areas of the Pacific Ocean along the equator.

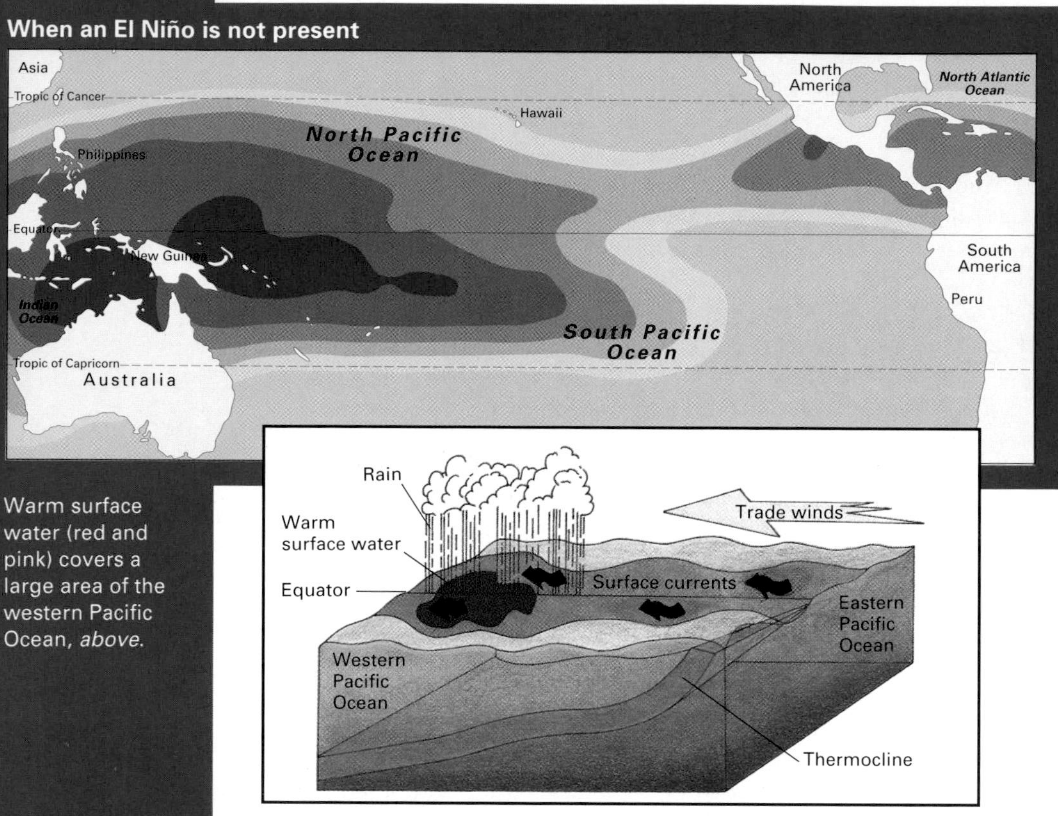

Warm surface water (red and pink) covers a large area of the western Pacific Ocean, *above*.

In the absence of an El Niño, trade winds blow over the Pacific from the east and help create westward currents. The winds and currents help keep the warm water in the western Pacific. The warm water in turn helps create relatively low average atmospheric pressures and frequent heavy rain. A layer of water called a *thermocline,* which separates warm surface waters from the colder depths, also extends deeper in the western Pacific. It inclines toward the surface in the eastern equatorial Pacific, allowing deep, cold water to rise upward.

His studies in the late 1950's and the 1960's revealed that widespread warmings and coolings of the sea surface in the eastern and central Pacific alternated regularly. Bjerknes also discovered that the Pacific ocean warmings usually coincided with the warmings that occurred off the coast of Peru.

In the late 1960's, Bjerknes connected the ocean warming to large shifts in the rainfall patterns across the tropical Pacific and broad changes in the circulation of air currents above the Pacific. Observations from meteorological satellites, the first of which was launched in 1960, greatly aided Bjerknes. These satellites photographed cloud

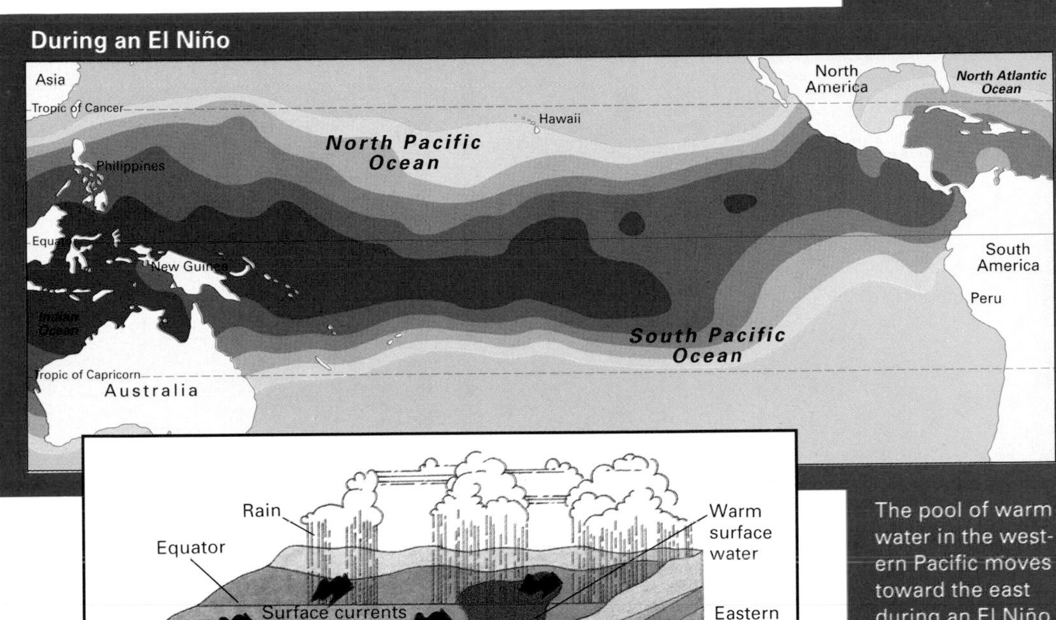

The pool of warm water in the western Pacific moves toward the east during an El Niño, *above.*

Trade winds weaken during an El Niño and the westward ocean currents reverse, allowing the warm water to move eastward. The thermocline in the eastern Pacific deepens as the warm water moves toward the east, thus preventing colder water deep in the ocean from rising to the surface. Weather systems marked by lower atmospheric pressures also move eastward, bringing heavy rain.

cover over vast regions of the world's oceans, and Bjerknes determined that these cloudy areas corresponded to areas of increased surface water temperatures.

From his findings, Bjerknes realized that changes in surface temperatures in the tropical Pacific were linked to a broader global climate pattern. This pattern was discovered by British meteorologist Gilbert Walker almost 50 years earlier. Walker had discovered that year-to-year variations in India's *monsoon* (seasonal wind) rainfall were tied to various climate changes around the world. His main finding was that the average atmospheric surface pressure over the Indian Ocean and the central tropical Pacific changed slowly in opposite directions over periods of a few years. Walker called this seesaw of rising and falling atmospheric pressure the *Southern Oscillation.* Walker found that the Southern Oscillation produced widespread effects. Besides altering atmospheric pressure across thousands of miles, it affected rainfall along the equator in the tropical Pacific and in India, and changed temperatures in places as far apart as southwestern Canada and southeastern Africa.

In 1969, Bjerknes theorized that the recurring large-scale Pacific warmings are part of the Southern Oscillation. He proposed that what scientists had been calling El Niños were simply one feature of the much larger warmings. Since then, studies have proven this, showing that El Niños are part of a huge climate system involving both the atmosphere and ocean. Climatologists now call this global pattern the *El Niño/Southern Oscillation cycle,* or the *ENSO cycle.* Some experts use the term *El Niño* to denote the warm phase of the ENSO cycle, and the term *La Niña* (Spanish for *little girl*) to denote the cold ENSO phase.

How an El Niño develops

As scientists now understand it, the driving force behind the ENSO cycle is the enormous amount of heat that periodically builds in the western Pacific Ocean along the equator. The upper 200 meters (650 feet) of water contain most of the heat in this region. This warm layer becomes thinner in the eastern Pacific, and, in some areas, disappears. This occurs because the base of the warm surface water, a layer of water called a *thermocline,* rises toward the surface in the eastern Pacific. The thermocline is a thin transition zone where warm surface temperatures fall rapidly to the much lower temperatures in the layer of water that extends to the ocean bottom.

As the ENSO cycle progresses, the thermocline sways slowly up and down in opposite directions in the western and eastern equatorial Pacific. This swaying resembles a giant seesaw with its fulcrum in the central Pacific. During El Niño conditions, the thermocline tips downward in the eastern Pacific and rises in the western Pacific.

The ocean and atmosphere regulate the ENSO cycle together. During the cold phase of the ENSO cycle, for example, trade winds blow west from the coast of South America across the Pacific Ocean along the equator. These winds create friction on the sea surface, actually pushing the surface water westward. As the warm surface waters move west, cold waters rise to replace them. The westward trade winds and accompanying surface currents thus keep the warm water in the western Pacific. There, the pool of warm surface water heats the air, which rises and condenses high in the atmosphere into storm clouds that bring torrential rain.

As the ENSO cycle shifts to its warm, El Niño phase, the trade winds weaken. The weaker winds create less friction with the surface, which leads to weaker westward surface currents. As the two forces that keep the warm surface water in the western Pacific falter, the warm water begins to move east, bringing with it clouds and rain. Air currents rising from the warm water alter the circulation patterns of strong winds in the *upper troposphere* (the layer of Earth's atmosphere that extends from about 10 kilometers [6 miles] to about 16 kilometers [10 miles] above the Earth). The winds disturb atmospheric circulation over wide areas of the globe, altering weather in many places.

Scientists still do not fully understand how one phase of the ENSO cycle turns into another. They do not know why the trade winds weaken, for instance, or what triggers particularly severe swings in the cycle. Some experts believe that changes in the ENSO cycle result from the change of seasons or from the complex way in which surface winds affect ocean currents.

El Niño and the world's weather

El Niños are of great interest because of their effect on the world's weather. The 1991-1992 El Niño was relatively mild, but on rare occasions, perhaps every century or two, an unusually severe El Niño occurs. The most recent of these occurred during 1982-1983. Worldwide, it caused an estimated $13 billion in damage. Historical records of flooding, droughts, and crop failures show that the 1982-1983 El Niño was the most prolonged and catastrophic episode recorded since the late 1800's, worse than the two previous most-severe El Niños in 1925 and 1891.

The 1982-1983 El Niño produced devastating weather in many areas of the world. The 1982 monsoon rains, which usually occur from April to October in northern India, faltered, for example. Indonesia, eastern Australia, and the islands of Melanesia suffered severe, in some areas record, drought, which did not break until early 1983. Dust storms and brush fires ravaged southeast Australia, and farmers there had to slaughter thousands of sheep due to lack of feed and water. In Borneo, immense forest fires, among the greatest in recorded history, destroyed vast areas of parched virgin timber. Drought enveloped the southern Philippines and Hawaiian Islands during early 1983, and one of the worst droughts of the century hit southeast Africa.

In contrast to those drought-stricken areas, torrential rain fell along the coast of Ecuador and Peru in October 1982. Between January and May 1983, many parts of Peru received 1 meter (3 feet) of rain or more, resulting in record flooding.

The 1982-1983 El Niño also had a pronounced effect on winter weather over North America. The Pacific subtropical jet stream, a narrow belt of extremely strong winds approximately 13 kilometers (8 miles) above Earth's surface, reached record strength over the eastern North Pacific. These winds extended eastward over Mexico and the southern United States. The result was increased storminess, destructive winds and high tides along the California coast, and unusual wet spells and stormy weather from California to Florida and as far south as Cuba.

El Niño weather

El Niño affects weather and natural environments worldwide. It may lead to floods, droughts, and other disasters that affect wildlife as well as people.

Excessive rains from the El Niño of 1991-1992 flood a highway in Ventura, Calif., *right.*

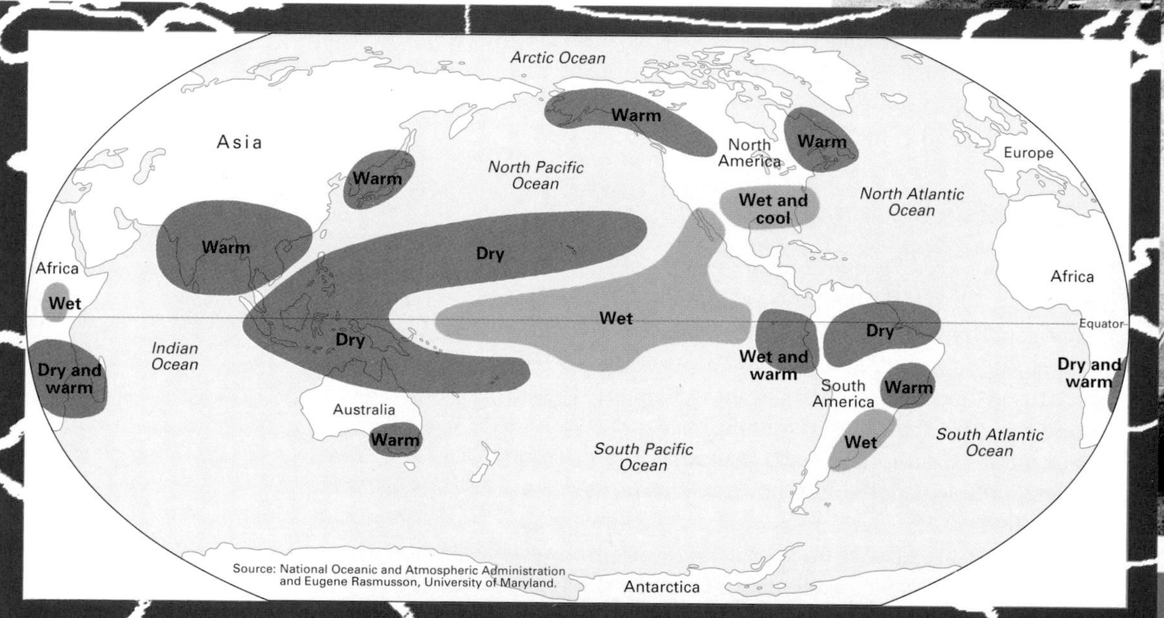

An El Niño causes various regions of the world to be warmer, cooler, wetter, or drier than they would be otherwise, *above.*

White patches of coral in the Pacific Ocean off the coast of Panama, *right,* resulted from dramatically increased ocean temperatures caused by the 1982-1983 El Niño.

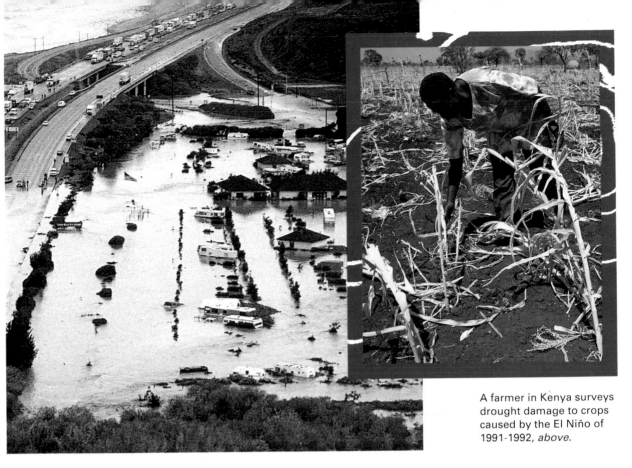

A farmer in Kenya surveys drought damage to crops caused by the El Niño of 1991-1992, *above*.

A huge dust storm descends on the Australian city of Melbourne in February 1983, *below*. The El Niño of 1982-1983 caused severe drought in Australia. As the land became parched, windstorms created immense dust clouds.

Other effects of severe El Niños

Besides affecting weather, severe El Niños such as the 1982-1983 episode can have other effects. These include a small but measurable slowing of Earth's rotation. Although atmospheric scientists do not completely understand how this slowing occurs, they believe that it involves widespread changes in the direction and speed of surface and high-level winds. These changes temporarily disrupt the normal amount of friction that winds create with Earth's surface, altering the pressure of Earth's atmosphere across large mountain ranges and in other areas.

A much more noticeable environmental effect of severe El Niños is the widespread disruption of marine life in the Pacific Ocean. The cold coastal waters that rise from the depths off Peru and Ecuador carry nutrients such as nitrates that support *phytoplankton* (tiny marine plants), which in turn help feed large numbers of animals such as birds and fish. When the thermocline deepens in the eastern Pacific during a major El Niño, however, the colder, nutrient-rich waters cannot rise to the surface. This drastically reduces the food supply in the surface layer. The effect of the nutrient shortage can progress up the food chain through fish, birds, and marine mammals, greatly reducing their populations. For example, a strong El Niño in 1972, followed by the 1982-1983 episode and combined with the effects of overfishing, nearly wiped out the Peruvian fishing industry, once the most productive fishery of the world's oceans.

Severe El Niños can spell disaster for other kinds of plants and animals as well. In November 1982, the El Niño produced heavy rains and high seas that caused serious flooding at Christmas Island in the central Pacific near the equator. The flooding, along with disruptions to marine food supplies, drove away the island's entire sea bird community—about 14 million birds. Only about 1 million of the birds returned; the rest died in heavy storms or from lack of food. Farther east, rising water temperatures killed many communities of *coral polyps*, the animals that make coral.

A look at past El Niños

The continued possibility of such severe consequences from El Niños has encouraged scientists to develop better forecasting techniques. The first step is gaining a better knowledge of how El Niños have occurred in the past. To do this, researchers look for evidence of El Niños in written records, at archaeological sites, and in several types of natural records.

Researchers have found some of the most useful information on El Niño episodes before 1900 in Peruvian historical records. Spanish colonists kept extensive records during their settlement of South America in the 1500's, including many relating to environmental events. Records include descriptions of the destruction of roads and bridges due to flooding, residents' requests for relief from taxes because of crop failures, and special prayers for rain during prolonged

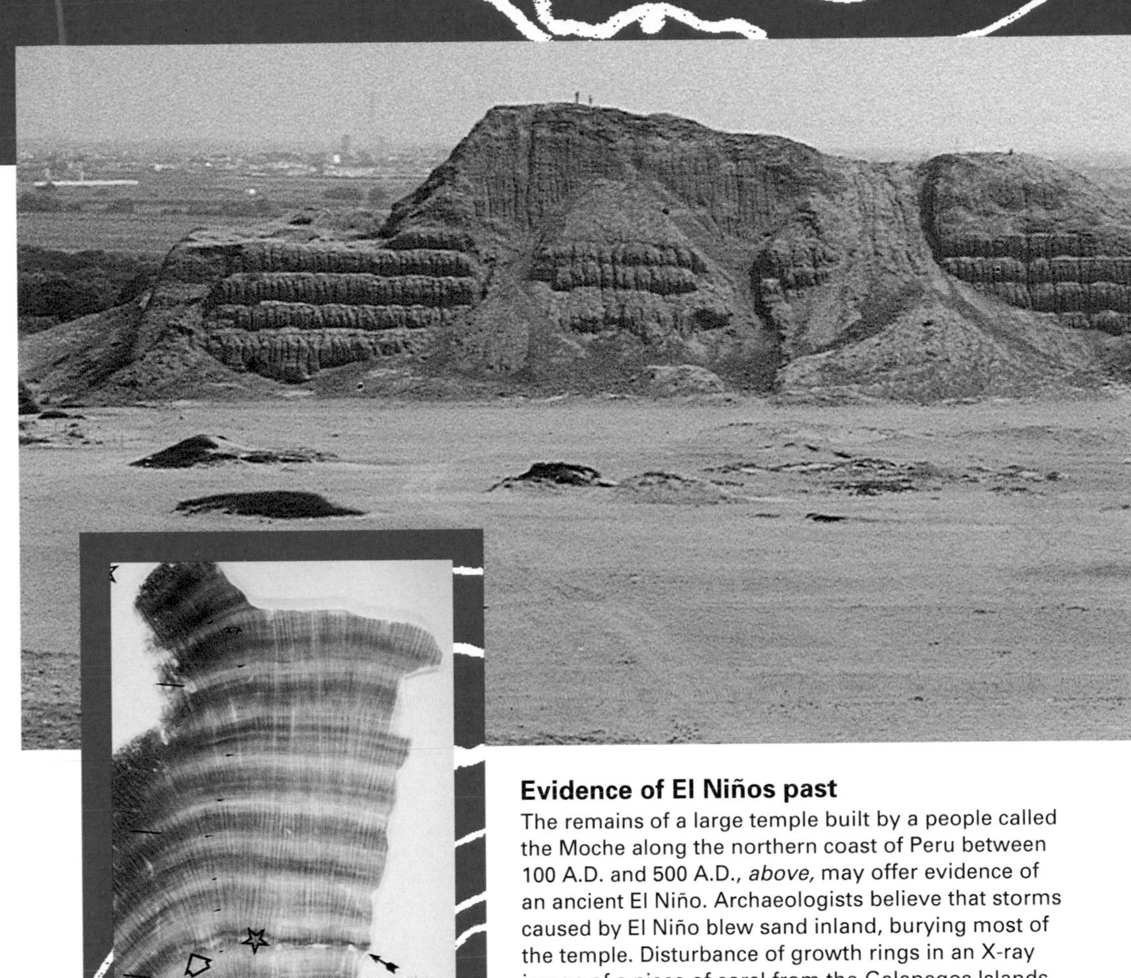

Evidence of El Niños past

The remains of a large temple built by a people called the Moche along the northern coast of Peru between 100 A.D. and 500 A.D., *above,* may offer evidence of an ancient El Niño. Archaeologists believe that storms caused by El Niño blew sand inland, burying most of the temple. Disturbance of growth rings in an X-ray image of a piece of coral from the Galapagos Islands, *left,* is evidence of an El Niño in the year 1720. In proper conditions, coral creates growth rings similar to those of trees. If an El Niño warms ocean water beyond coral's tolerance, growth slows or stops, creating a distinctive band. Scientists determine the year of the event by counting the rings backward from the present.

dry spells. Other records speak of growing crops in areas that were once desert due to abundant rainfall.

In the mid-1980's, oceanographer William H. Quinn of Oregon State University in Corvallis and his colleagues in the United States and Peru assembled hundreds of years of such information from Peruvian historical records dating back to the late 1520's. Quinn's record reveals that the frequency and intensity of El Niños has varied throughout the centuries, but it also clearly shows that El Niño has been a feature of the Peruvian climate at least since the 1500's.

Archaeological studies of the ruins and artifacts in Peru have uncovered evidence of much earlier El Niños. Anthropologist Michael E. Moseley of the University of Florida in Gainesville in 1992 reported on

Studying El Niño

Oceanographers and meteorologists use equipment on land, in the ocean, and in space to monitor El Niño conditions as they develop.

Marine scientists with the United States National Oceanic and Atmospheric Administration adjust a buoy that contains equipment for measuring and recording water temperatures in the central Pacific Ocean along the equator. Such readings help oceanographers track the eastward flow of warm surface water characteristic of El Niño.

evidence that one or more major El Niños contributed to the demise of an ancient culture called the Moche. The Moche lived in Peru between the A.D. 100's and 700's. (See also SOLVING THE MYSTERY OF THE MOCHE SACRIFICES.)

Moseley studied the remains of a large Moche city that archaeologists uncovered in 1972. The city appeared to have been abandoned just before massive sand dunes buried it sometime before A.D. 600. Catastrophic flooding also appeared to have struck the city and stripped as much as 4.6 meters (15 feet) of soil from some areas sometime between 500 and 600.

To find out what caused the flooding and the sand dunes, Moseley

A satellite image of infrared (heat) energy in the Pacific Ocean shows a large pool of warm water in the western Pacific Ocean (red). Scientists use satellite imagery to study the movement of warm water associated with El Niño.

Scientists with the United States National Center for Atmospheric Research set up temperature, humidity, and atmospheric pressure monitors in November 1992 on the island of Kapingamarangi in the western Pacific Ocean to study atmospheric effects that may lead to an El Niño.

considered more recent ecological disasters in Peru—a strong earthquake in 1970 followed by El Niños in 1972-1973 and 1982-1983. Moseley reported that the earthquake dislodged huge amounts of soil and rock in the foothills of the Andes Mountains. El Niños caused floods that carried the rocky debris to the coast. Over the years, this debris was ground into sand, which winds carried inland. Moseley theorized that a similar fate struck the Moche. Although the Moche rebuilt their capital after the devastating flooding, sand dunes gradually buried it.

Scientists can also find records of past El Niños in nature. These records take the form of growth rings in trees and changes in the annual accumulation of snow on glaciers and icecaps in high mountains. Another type of natural record is the growth rate of corals. Scientists have found that coral reefs that surround the widely scattered *atolls* (ring-shaped coral islands) and other islands of the tropical Pacific can provide especially useful information on past ENSO cycles.

Coral reefs contain annual growth rings similar to the growth rings of trees. Coral polyps cannot exist in water colder than about 18 °C (65 °F) or warmer than 30 °C (86 °F). In a year of favorable conditions, the large numbers of coral polyps create a wide band of new coral. Unfavorable water temperatures can slow coral growth or kill the polyps, causing narrow growth rings or none at all. To study El Niño, scientists examine coral from the central and eastern Pacific. They first look for narrow growth rings, which may indicate a year when El Niño warmed the seawater beyond the coral's tolerance. Then, the scientists determine the year of the narrowing by counting the rings.

Toward better predictions

Increased scientific observation of current weather conditions has helped scientists fill in many of the gaps in their knowledge of El Niños. Improved monitoring, data collection, and communications, for example, made the 1982-1983 El Niño the first in which the public learned of developments as they happened, rather than months later.

Meteorological satellites are key to the ability to monitor the atmosphere and ocean over the Pacific. Some satellites peer down continuously over the tropics while others view almost the entire globe twice every 24 hours. Satellites can measure the infrared (heat) radiation emitted by ocean water, as well as detect precipitation and monitor wind circulation over the tropics. Such data enable scientists to track atmospheric and ocean conditions for the first hint that an El Niño may be developing.

Observations on the surface of the ocean are also required to create a detailed picture of a developing El Niño. Until the 1990's, scientists used merchant ships, island stations, and drifting buoys to collect data. During the early 1990's, scientists set up a new network of 70 moored buoys across the tropical Pacific. Instruments on these buoys measure and transmit crucial additional information through communications satellites to climate analysis centers.

The aim of some meteorologists is to learn enough about El Niños to regularly forecast them as much as a year in advance. With such forecasts, governments, businesses, and individuals could develop strategies to cope with crop failures and other economic problems that El Niños may cause.

One method of forecasting is to observe how often certain weather patterns develop in one place after particular ocean and atmospheric conditions occur elsewhere. If certain ocean and atmospheric conditions seem to produce the same weather often enough, scientists assume that those conditions will continue to produce similar weather. This method of prediction does not require a very thorough understanding of why weather in one region of the globe affects weather far removed from that region. It is adequate for predicting typical weather fluctuations but often not very useful for predicting exceptions to normal patterns.

The second and more sophisticated forecast method is based on an understanding of how weather develops. The laws of physics govern changes in Earth's atmosphere and oceans, and they can be expressed in the form of mathematical equations.

These equations are not solvable exactly, however, and even the largest computers cannot process the number and complexity of equations needed to describe a system as vast and complex as the ENSO cycle in detail. Climate modelers build simplified models that capture the most essential aspects of the ENSO cycle. By entering data about actual conditions into these computer models, the experts attempt to predict how climate patterns will develop.

Cane and Zebiak of the Lamont-Doherty Geological Observatory used the most basic aspects of the ENSO cycle in the model they used to predict the 1991-1992 El Niño. These variables included the speed and direction of ocean currents and winds as well as water temperatures at the surface and selected depths.

Climate modelers hope to use the findings from future investigations of the oceans and atmosphere to develop more advanced models capable of more accurate forecasts. During late 1992 and early 1993, hundreds of meteorologists and oceanographers from 19 nations conducted experiments that may provide data for the development of more advanced models. This program, called the Coupled Ocean Atmosphere Response Experiment, was designed to learn more about how the ocean and atmosphere interact over the area of warmest water in the western tropical Pacific. After that information is used to upgrade existing computer models, we may be able to detect El Niños on the horizon long before unusual weather comes ashore.

For further reading:

Canby, Thomas Y. "El Niño's Ill Wind." *National Geographic,* February 1984, pp. 144-183.

Moseley, Michael. *The Incas and Their Ancestors.* Thames and Hudson, 1992.

Philander, S. G. F. *El Niño, La Niña and the Southern Oscillation.* Academic Press, 1990.

Rasmusson, Eugene M. "El Niño and Variations in Climate." *American Scientist,* March 1985, pp. 168-177.

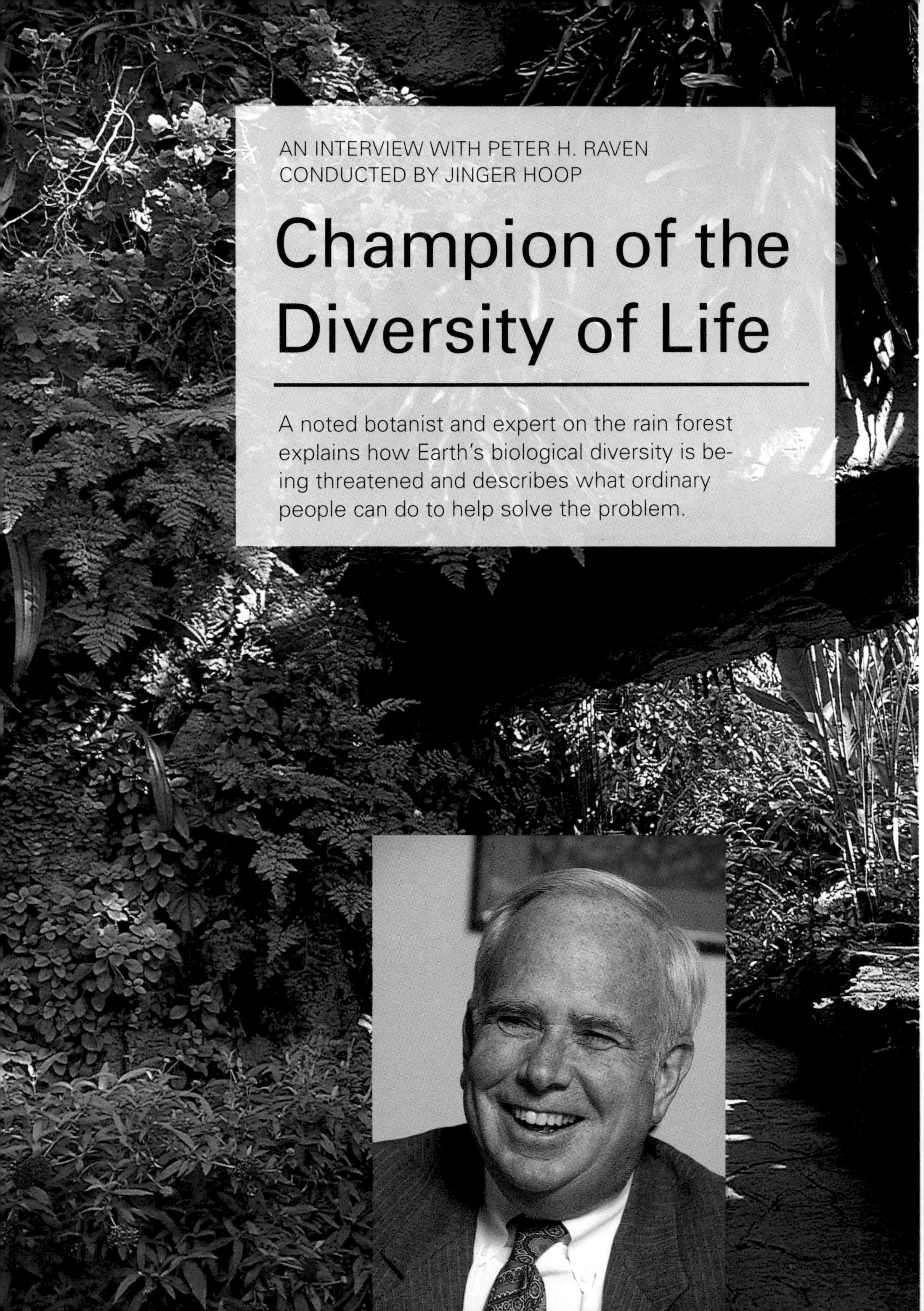

AN INTERVIEW WITH PETER H. RAVEN
CONDUCTED BY JINGER HOOP

Champion of the Diversity of Life

A noted botanist and expert on the rain forest explains how Earth's biological diversity is being threatened and describes what ordinary people can do to help solve the problem.

"Organisms all around you—not just the tigers of the world—are facing extinction."

Science Year: As a distinguished botanist known for your interest in tropical rain forests, you've dedicated much of your life to educating people about the threats to Earth's *biological diversity* or *biodiversity,* the terms scientists use to describe Earth's rich variety of plant and animal species. By now most of us realize that animals such as rhinos, tigers, and spotted owls are in danger of becoming extinct. What else do we need to know?

Peter H. Raven: First, that organisms all around you—not just the tigers of the world—are facing extinction. For instance, botanists predict that nearly 800 plant species in the United States may become extinct during the 1990's unless we take steps to protect them. So you should understand that the problem isn't just that a few large and well-known types of animals are being killed off. Between 20 and 25 percent of *all* the species on Earth may disappear within the next 30 to 40 years.

SY: Why do you think so many organisms are at risk?

Raven: It's simple: Human beings are breaking down the habitats of plants and animals across the planet. Earth no longer has any unoccupied frontiers, and every inch of the planet's surface is affected by human activities. You can see this most clearly in Earth's tropical rain forests—which are found in nations such as Brazil, Zaire, and Indonesia. These forests contain an almost unbelievable diversity of life. About two-thirds of the world's species make their home there. But people are destroying the rain forests so rapidly that only small patches may be left in 60 years.

Previous pages: Plants from the tropics fill the Missouri Botanical Garden's Climatron. As the Garden's Director, Raven has overseen the institution's expanding role in collecting and studying tropical plants.

SY: What's the cause of this destruction?

Raven: The short answer is that the populations in many of the nations that contain rain forests are growing so rapidly that people must exploit the resources of the forest for short-term gain. Actually, the problem is that, on average, 40 percent of the people in these countries are less than 15 years old. Most of the people are choosing to have smaller families. But because the populations are so young, a tremendous number of individuals are reaching childbearing age each year. As those young people start having children, the population rate soars. The result of all this is that many of the people are living in poverty, and up to a third of them are malnourished. These people have an immediate need to use their forests for economic gain. They do this by cutting trees for lumber or clearing land to grow crops on. In addition, the forests are often cut for profit, without considering the region's long-term or general benefits, just as in the United States.

The author:
Jinger Hoop is associate editor of *Science Year.*

SY: When North America was being settled, immigrants cut down forests for lumber and to bring land into cultivation. Is that different from what individuals are doing to the rain forest today?

Raven: The effects are not the same, because clearing land does greater damage to the rain forest. Most of the soil in rain forests lacks the fertility of soil in temperate forests such as those in the United States. Only one or two years after a patch of rain forest is cleared, the land is barren. The rain forest plants cannot become reestablished, and the people must clear yet another patch of forest to get usable cropland.

SY: The rate of rain forest destruction appears to have slowed since the mid-1980's. Do you think the overall loss of biodiversity may soon cease to be a problem?

Raven: On the contrary: It's likely to get worse in the decades to come. The problem, after all, is that human beings are putting too much pressure on their environment. Imagine what will happen as the global human population doubles—as it's expected to—before leveling off at 10 billion or more in about 100 years.

SY: As you know, not everyone agrees that an increase in the human population will by itself lead to ecological disaster. A few economists, for example, propose that a larger worldwide population means more minds at work to solve any environmental problems that come our way. Why do you feel that this view is wrong?

Raven: The loss of one-fifth of the world's topsoil, the pollution of the atmosphere, and the loss of biodiversity—all these things show clearly that even now we are not managing the world in a way that we can sustain. There is no way that adding still more people is going to make these problems easier to solve.

SY: Why do you think it's so important to protect Earth's biodiversity?

Raven: To me, avoiding the destruction of so many of the other organisms in the world is the right thing to do from a moral, ethical, or religious point of view. As far as we know, the living things that share the world with us are the only living things in the whole universe. That alone ought to give us a respect or reverence for life. It ought to make us want to avoid driving the species into extinction permanently.

"As far as we know, the living things that share the world with us are the only living things in the whole universe. That alone ought to give us a respect or reverence for life."

SY: Is the issue purely an ethical one?

Raven: Not at all. There are scientific and practical reasons to protect biodiversity as well. For one thing, having a rich array of organisms on Earth is like having a well-stocked storehouse. Individual organisms provide most of the things we use to support ourselves—our food, the wood we use for shelter, the fibers for clothing, the natural compounds we use as drugs. We can draw on these stores to sustain us in the future. If Earth's average temperature rises due to global warming, for example, we may need to find tropical plants to replace crops now grown farther north.

Peter H. Raven

It must have come as no surprise to Peter Raven's parents that their son became a biologist. Before Raven was old enough to go to school, the boy entertained himself by raising caterpillars, collecting insects, and pressing samples of flowers in the family telephone book. At age 14, he was traveling by Greyhound bus on solo trips to collect plant specimens. He described his findings in his first scientific paper, which was published that year. That was also the year he discovered a new kind of manzanita, an evergreen shrub. The subspecies was named *ravenii* in the boy's honor. "While I did things that typical high school students do," Raven says, "it's obvious that botany was already the organizing principle of my life."

Peter H. Raven was born in 1936 to Americans living in Shanghai, China. His parents moved to northern California soon after his birth, and it was in that state's Sierra Nevada mountains that Raven did most of his early field work. After high school, Raven attended the University of California at Berkeley, at first with the goal of becoming a high school biology teacher. Soon he abandoned that idea in favor of becoming a scientist. It goes without saying that Raven had a head start on the other students. By the time he began graduate school at the University of California, Los Angeles, in 1957, he had already collected more than 20,000 specimens of plants.

After receiving a Ph.D. degree in 1960, Raven worked as a researcher at the Natural History Museum in London and later as a taxonomist and curator at Rancho Santa Ana Botanic Garden in Claremont, Calif. In 1962, Raven took a position as assistant professor of biology at Stanford University. His office at the university was next door to that of biologist Paul Ehrlich, who would later write the influential book *The Population Bomb*, a prediction of dire consequences of continued human population growth. The two men became friends, and Raven credits Ehrlich's views with helping to shape his own ideas about the role of human population growth in affecting world ecology.

Raven made his first visits to the tropics when he was a professor at Stanford. In Central America, he lectured on the consequences of population growth as well as on the effects of chemical pollutants on the jungle. Raven met his wife, biologist Tamra Engelhorn, in Costa Rica during this period. The two were married in 1968. They have two children, and Raven also has two more children from a previous marriage.

In 1971, Raven left Stanford to become Engelmann Professor of Botany at Washington University and the director of the Missouri Botanical Garden, both in St. Louis. The Missouri Botanical Garden, which was established in 1859,

Unfortunately, in many cases we have no idea what wonderful, beautiful, and useful creatures are disappearing forever. More than 85 percent of the organisms we're about to lose are unknown to science. We know next to nothing even about those we have named—a single feature of the species, a place where it's been found, or a very brief description. This helps explain why many scientists find the loss of biodiversity so distressing.

SY: The most common argument for protecting diversity is one you just mentioned—the potential use of organisms as sources of new drugs. Is it the most important reason?

Raven: It is important. Did you know that the 20 best-selling prescription medications in the world are compounds produced by living organisms or modeled on compounds produced by living organisms?

Peter Raven's daughter Katie offers assistance as her father examines a plant's foliage.

is the oldest botanical garden in the United States and one of the most important botanical research institutes in the world. In his years at the garden, Raven has improved the garden's lushly planted public areas, made the garden a major center for computerized databases of information on plants, and overseen an expansion of the garden's role as an international research institute.

Today, the garden's staff of scientists numbers about 50, including several researchers who work in tropical rain forests. The botanical garden's *herbarium*—a collection of dried specimens used as a "library" of plant species—now contains more than 4 million specimens. It includes samples obtained by scientists and explorers of the 1700's as well as the 100,000 or so specimens collected each year in modern times.

Throughout his career, Raven has conducted independent research of his own, focusing on a group of hundreds of plant species—the evening primrose family—that had first attracted his attention in 1950. Over the years, Raven and his colleagues have gathered so much information on the far-flung members of this family that it is used as a model of how flowering plants evolve in response to differing environments.

Now home secretary of the National Academy of Sciences and active in many other scientific organizations, Raven has been the recipient of several honorary degrees and every major award in his field. He is the author of best-selling college textbooks and more than 400 scientific articles.

But Peter Raven is best known for his long-time interest in protecting rain forests. Today he routinely travels throughout the United States—and to Russia, China, and several developing nations—to lecture on the need for international cooperation to protect the planet. In person, he speaks with a rapid-fire delivery, as if he is convinced there is not a moment to lose. [J. H.]

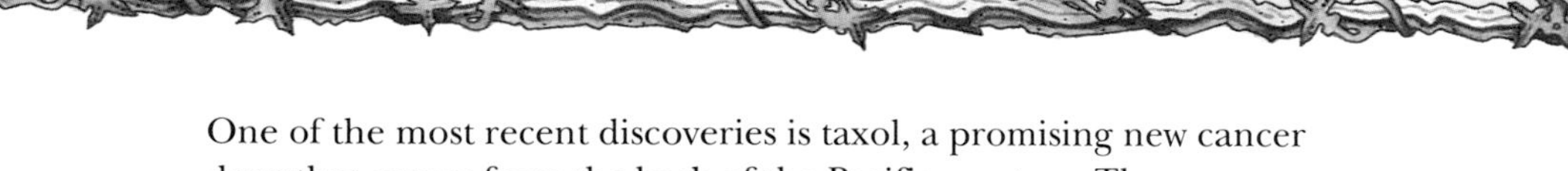

One of the most recent discoveries is taxol, a promising new cancer drug that comes from the bark of the Pacific yew tree. There are many other examples. For instance, drugs used to treat childhood leukemia are extracted from the rosy periwinkle.

Yet that's only one practical reason to protect biodiversity. Another is that collections of living things provide what scientists call "ecosystem services"—protecting soil from erosion, influencing the characteristics of the atmosphere, transforming energy from the sun into food energy, shaping local and regional climates, and other functions. Because prairies, rain forests, and other ecosystems do these things, Earth is a suitable place for us to live, and the planet operates as a self-perpetuating system. Without them, we simply couldn't survive.
SY: You bring up one of the reasons many people are deeply con-

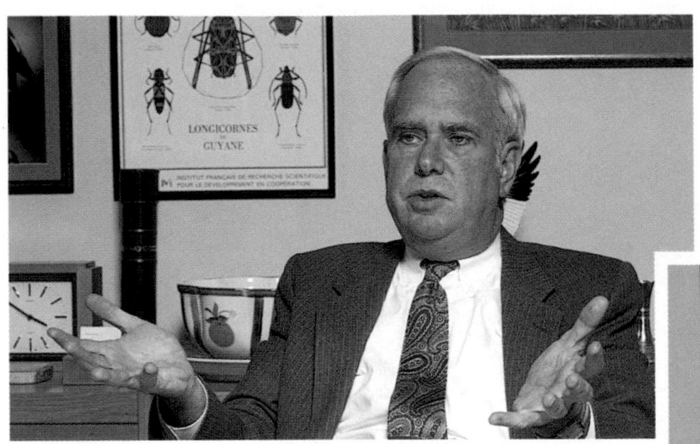

> "The planet will recover from the loss of biodiversity, but whether we will or not is another matter."

cerned about the loss of the rain forests. Not only do the forests contain great biological diversity in the form of millions of species, but they also provide the ecosystem service of removing massive amounts of carbon dioxide from the atmosphere. This gas, a by-product of the combustion of fossil fuels, has been linked with the potential warming of Earth's climate. Does it seem to you that the health of the planet as a whole is now at the mercy of the tropical nations?

Raven: Actually, the planet is at the mercy of the industrialized nations as well. Those of us who live in the United States, Canada, Great Britain, and the other developed countries—which together account for only 23 percent of the world's population—use about 80 to 90 percent of the resources that support life on Earth. We're really having the major effect on the stability of the biological world.

SY: How so?

Raven: We're the ones who have released into the atmosphere tremendous amounts of compounds called chlorofluorocarbons, which are eating away at the ozone layer that blocks harmful ultraviolet rays from the sun. We're the ones who burn most of the fossil fuels and pump most of the carbon dioxide into the air. And we must share the responsibility for destroying communities of plants and animals.

SY: There's nothing inherently "unnatural" about vast numbers of species becoming extinct, however. In fact, there were a half-dozen or more episodes of mass extinction long before humanity's time on Earth. Could it be that today's extinctions are no different from those earlier mass extinctions?

Raven: I see tremendous differences. For one thing, this is the first major extinction in the history of the Earth that is caused by the activities of a single species—ours. It may also kill off more species than did the last mass extinction, which occurred about 65 million years ago. Many scientists think that a huge asteroid hit Earth at that time, casting up a cloud that darkened the sky for many months. This extinction killed off the last of the dinosaurs along with many other organisms. Now, as far as we can tell from studying the fossil record, the number of species in the world has been increasing steadily for the past 65 million years. We may now have an all-time high number. Losing 20 to 25 percent of them within just the next 30 to 40 years will mean that this extinction could easily kill off more species than the last one.

SY: It's true, however, that none of the episodes of mass extinction caused irreparable harm to the Earth. The planet's biodiversity, along with the ecosystem services you talked about earlier, have always been able to recover.

Raven: But not very quickly. It took more than 10 million years for Earth to recover from the extinction that killed the dinosaurs. That's a tremendously long period of time, especially when you consider that our species has been around for only a fraction of it, and that agriculture, for example, is only about 11,000 years old! The planet will recover from the loss of biodiversity, but whether we will or not is another matter.

SY: So people and dinosaurs may end up having a lot in common?

Raven: Indeed—except that dinosaurs could not reason about the world around them, while we have large and complex brains and are supposed to be thinking about what's good for us and taking action based on that. The human brain is the most complex and marvelous product of evolution in many ways. It gives us the ability to understand the nature of our dependence on the natural world. We ought to be using our brains to devise a way to prevent another mass extinction.

SY: You and one of your colleagues, the well-known biologist Edward O. Wilson, have created the framework for such a system.

Raven: Yes. We feel very strongly that there needs to be a world plan for learning about biodiversity and then a mechanism for managing it.

SY: How would that be accomplished?

Raven: First, scientists would learn about the better-known groups of organisms as rapidly as possible. These groups include flowering plants, animals with backbones, and butterflies. We can use our knowledge of those species to calculate the overall biodiversity of the world. Then, during the next 50 years, scientists would try to learn about the other groups of organisms. In the end, we'd have a pretty good outline of the 10 million or so species on Earth. At the same time, we should be conserving Earth's biodiversity. The easiest way is by creating new parks and reserves, but if the areas where certain species live are being destroyed, then we've got to bring the organisms into zoos, botanical gardens, or seed banks.

SY: What do you think about the more fanciful solutions to the problem? For example, do you envision genetic engineers creating new species to replace those we're losing?

Raven: Every species on Earth is a unique product of 4 billion years of

What is biological diversity?

Biological diversity or *biodiversity* are terms scientists use to refer to the richness of the life forms on Earth. The words describe several types of variety.

Earth's biodiversity is most visible in the differences between the communities and ecosystems that collections of plants and animals help make up. Ecosystems include a South American rain forest, *left,* and an African savanna, *above*. Earth contains hundreds of different types of ecosystems, marine as well as land.

evolution. The idea of replacing such a species with one we've concocted is wrong for aesthetic or ethical reasons. Just as an individual human being is not a string of DNA but rather a unique person, so an individual species is more than a collection of genes doing a job.

Besides, we could never replace the losses of whole communities of plants and animals. Scientists first transferred genes from one organism to another fairly recently—in the early 1970's. Imagine the impossibility of trying to put together whole organisms out of strings of genes that we made up, and then somehow grouping those organisms together into communities that would produce something as complex and self-sustaining as a prairie or a rain forest. Remember, we barely understand how existing organisms function in natural communities or ecosystems. So the notion is crazy.

Differences between species account for another part of the diversity of life on Earth. These differences include the unique characteristics of any species of plant, animal, fungus, or microorganism—characteristics such as the smallness of the thumb-sized green tree frog, *top right,* or the vivid coloring of an arrow poison frog, *bottom right.* Earth contains millions of different species.

The differences between individuals of the same species make the life forms on Earth even more diverse. These differences, caused by variations in genetic material, include such things as the variations in height and coloring among human beings. The genetic variations among Earth's life forms are too numerous to count.

What genetic engineering can do is help tremendously in making human-altered communities more productive, taking the stress off natural ecosystems. That is, if we could make plantations more productive by improving the growth rates of individual kinds of trees, then we might not have to cut down the ancient forests in the Pacific Northwest—thus keeping the area beautiful and preserving large amounts of biodiversity. Such techniques applied to farming could also make crops more productive, which would help us avoid destroying more rain forest to get new farmland.

SY: Wouldn't developing nations benefit greatly from this kind of technology?

Raven: Of course. Unfortunately, the industrialized countries have 94 percent of the world's scientists and engineers, and most developing countries have virtually none at all. So fostering scholarly interchanges and building institutions are extremely important for adding to the scientific and technical capabilities in developing nations. With our vast pool of scientific talent, the United States has an enormous contribution to make in that area. And such contributions are entirely in our own interest, because the better equipped these nations are to deal with their problems, the more likely they'll be to act in ways that benefit all of us.

SY: Do you think it's possible to address the loss of biodiversity without dealing with the poverty of many tropical nations?

Raven: No, I don't. Here's one way of looking at the problem: People who live in developing nations—77 percent of the people in the world—have 80 percent of the world's biodiversity but only 15 percent of the money.

> "It's an absolute scandal—or an ultimate act of stupidity—that the nations of the world haven't found any way to get together to protect the threatened organisms and global stability overall."

SY: That fact has prompted so-called *debt-for-nature swaps,* arrangements in which industrialized nations forgive the international debts of developing nations. In return, the debtor nation agrees to protect its tropical forests, for example, by creating national parks. What do you think of these swaps?

Raven: They have limitations. Some nations, for example, are too poor to start creating national parks even if a debt is forgiven. But the industrialized world does hold the developing world in an economic stranglehold. I would be for abolishing developing nations' international debts tomorrow, because anything that would stop or even slow down the flow of money from poor nations to rich ones would help solve many of humanity's problems—such as the fact that 1.5 billion people live in absolute poverty and 700 million people are starving or severely malnourished. Eliminating the poverty would give us a chance to nurture more of the human talent in the world for solving our other problems.

SY: How about encouraging private industry to step in? In 1991, for example, the pharmaceutical firm Merck & Company, Incorporated, paid $1 million to a Costa Rican institute for the right to cooperate with Costa Rican scientists in "chemical prospecting" for new drug compounds in species native to Costa Rica.

Raven: Such arrangements are highly desirable. Unfortunately, extremely poor nations such as Burkina Faso or Burma are not going to be able to make million-dollar deals of their own, because they don't have the means to manage their resources the way Costa Rica has.

Also, it's simply not logical to rely on private companies to make everything all right. It's like the idea that charity or volunteers can solve any problem—something a lot of people in the United States have been saying for the last decade or so. If everyone were charitable and willing to be a volunteer, then that would be marvelous. But, realistically, the most efficient way to provide large amounts of money for any purpose is for people to voluntarily tax themselves. When you're talking about major social problems, we can never deliver more than a small amount of what's needed through volunteerism or private industry. Those things can make important contributions, but only government intervention can deliver the goods.

SY: Many Americans feel that the United States spends too much money on foreign aid already.

Raven: I think they're wrong. The United States is the wealthiest industrialized nation in the world but the stingiest when it comes to the amount of development aid we give per citizen.

SY: What about nations other than the United States? Would you encourage them to shoulder more responsibility as well?

Raven: Yes. It's an absolute scandal—or an ultimate act of stupidity—that the nations of the world haven't found any way to get together to protect the threatened organisms and global stability overall. It's clear that we need to draft international agreements that include legal powers similar to those of the United Nations. World leaders attempted to come to grips with this at the United Nations Conference on Environment and Development in Rio de Janeiro, Brazil, in June 1992. Only a few nations gave any sign they were willing to forge sufficiently strong agreements, but the conference did mark the start of a new world commitment to dealing more effectively with biodiversity. I hope we'll get on with that now.

SY: The United States did not sign the treaty concerning biodiversity when it was presented at the June 1992 Earth Summit, however. It wasn't until mid-1993 that U.S. officials signed the treaty, and in the meantime the United States was sharply criticized for its reluctance to get on board.

Raven: True, but there are two things that need to be said about that. First, the United States provides far more than half the funds spent worldwide for preserving and studying biodiversity. So it's wrong to characterize us as having been unconcerned or not a major participant in this effort.

"Earth has only one atmosphere, one world ocean, one complement of life. Our plans for economic development must keep these facts in mind."

Second, I'm convinced our reluctance stemmed from a misunderstanding concerning the agreement's effect on the U.S. biotechnology industry. Biotechnology involves using the properties of organisms—their genes, the compounds they secrete—to create new foods, medicines, or other useful products. So there is a direct relationship between conserving organisms and helping the biotechnology industry. The treaty called for profits from biotechnology to be shared with the nations in which the useful organisms are found. The effect of this

"You cannot lead an informed life as a citizen of the future, making good decisions for yourself, your company, or your family, if you don't understand global ecology."

would be to encourage nations to preserve their biodiversity. The United States representatives didn't seem to understand such preservation is necessary if our biotechnology industry has any future.

SY: We do tend to see such issues as a confrontation between conserving the environment on one hand and using it to provide economic growth on the other. Is that how you view the matter?

Raven: No, because the two must work together. We can't ignore the fact that the world is limited. No matter what economic model you are using, we cannot keep consuming more and more of our limited natural resources in order to increase our wealth. Earth has only one atmosphere, one world ocean, one complement of life. Our plans for economic development must keep these facts in mind.

SY: The most well-known example of this dilemma is the struggle over the northern spotted owl, which lives in the old-growth forests in the Pacific Northwest. The media have usually described the issue as a battle between loggers, who want to ensure they have jobs for the next few years, and environmentalists, who want to protect the owls at all costs.

Raven: If you look at the issue from either perspective, you've missed the whole point. What we need to consider is how to keep the entire system going. We ought to be thinking about why people choose to live in Oregon and Washington in the first place. For many, it's because the area has beautiful forests and mountains and hills, as well as clear streams and salmon runs and a certain climate and character. We should be trying to find ways for people to live in harmony permanently in this region without destroying its features altogether.

SY: So you don't accept the argument that environmental protection costs jobs?

Raven: No, I don't. And I'm convinced that as the students in school today grow up, they're going to find more and more careers in fields having to do with the environment. Environmental protection won't cost jobs; it will generate jobs.

SY: Not all young people will want to pursue a career in environmental protection, but many of them may be interested in protecting biodiversity. What would you advise them to do?

Raven: The most important thing is to learn all they can about the environment. You cannot lead an informed life as a citizen of the future, making good decisions for yourself, your company, or your family, if you don't understand global ecology. Understanding will make you a better citizen and will motivate you to take action when it's needed.

SY: What's the best way to learn?

Raven: For me, it was getting involved with the California Academy of Sciences when I was about 8 years old. Other young people might choose to join an environmental group, many of which have special sections for students.

And there's one other thing those of us in developed nations need to learn about—the lives of the majority of the people in the world. I encourage young people to try to visit or learn about people from such places as Indonesia, the Philippines, and Brazil. If you don't do that, you may not understand what life is like for most of the people in the world, and there's no way you're going to be motivated to look out for their interests as well as your own.

SY: Are there some concrete actions readers can take?

Raven: Yes. First, we can all look at our level of consumption and think about how we can lead better lives by using less. The U.S. population comprises 4.5 percent of the people in the world, but we use at least 25 percent of everything that the world produces to maintain our standard of living. Per person, we use up more metals, plastics, wool, and food than any other nation. We waste more energy per person, and we produce more garbage per person. I urge families to discuss how they can use their cars less, waste less food, and use less electricity. That doesn't necessarily mean lowering their standard of living. It may mean turning to new, energy-saving inventions or technologies. It definitely means being more thoughtful.

Finally, we have to take seriously our role in the political process in the United States. It's completely foolish to say that a mayor, governor, or President doesn't pay as much attention to the environment as we want him or her to do. After all, you and I are responsible for electing representatives and telling them what issues are important. Too many of us sit around and wait for somebody else to take action. We need to give our elected representatives input so that they'll do what we want.

SY: And what should be the end result of all these efforts?

Raven: Bringing ourselves into balance with what the world has to offer. If we can find a population level that our single, common, planetary home can support, and if we can learn how to manage that planetary home so that it continues to sustain us, then we'll be able to save whatever biodiversity is left at that time. We'd better hope we save a good sample of it, because only biological diversity is going to allow us to continue to exist and improve our lives in the future. That's why blowing away 20 to 25 percent of all the species on Earth ought to be seen as an insane act.

SY: Do you think we'll achieve the goals you've described?

Raven: I have enormous faith. I'm very optimistic about the power of individual human beings. That optimism makes me want to take action and get other people to do so, too. Whatever is left of life on Earth in 100 years will depend directly upon how you and I accepted our responsibility to the planet. Like it or not, we're in the position of Noah just before the flood—looking at an upcoming extinction of enormous proportions and realizing we alone are responsible for saving as many creatures as we can.

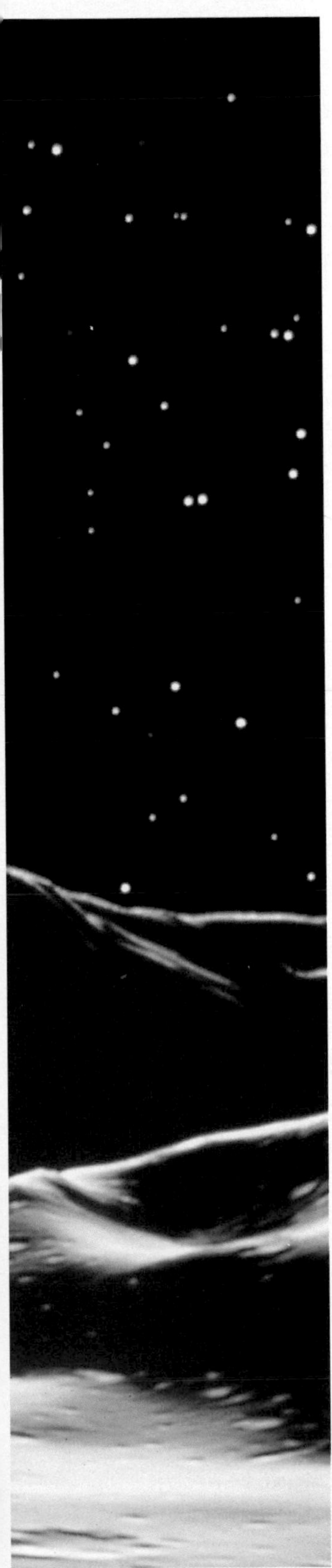

Astronomers believe that hundreds of millions of stars in our Galaxy may be orbited by one or more planets. But finding evidence of them is not easy.

Planets of Other Stars

BY STEPHEN P. MARAN

In the constellation Virgo, 15 thousand trillion kilometers (9.3 thousand trillion miles) from Earth, two dim objects invisible to even the most powerful telescopes hurtle through space. Although astronomers have not directly observed these objects, they believe that they are planets. The two planets, and possibly a third, orbit a tiny star called a *pulsar* (the collapsed remnant of an exploded star). The pulsar, named PSR 1257+12, spins with extraordinary speed, making 161 complete turns each second. It also emits a deadly beam of electromagnetic radiation and high-energy particles that sweep past the planets, boiling away any atmospheres that might exist, and scorching their surfaces.

As of mid-1993, the two planets were the only ones known to exist outside the solar system. Although the planets are almost certainly lifeless, astronomers are excited about the discovery for what it may reveal about the likelihood of life elsewhere in the universe. Astronomers believe that the same processes that formed planets around our sun and around PSR 1257+12 must have created planets orbiting other stars in our Milky Way Galaxy. There are hundreds of billions of stars in the Galaxy, many resembling the sun. There might be billions of planets, and life may exist on some of them.

Previous pages: In a computer generated graphic, the surface of a planet circling a *pulsar* (collapsed remnant of an exploded star) appears as a barren, rocky landscape. A companion planet, seen half in shadow, is the largest object in the sky. The pulsar appears as a large, bright star to the right and above the companion planet.

To test these ideas, astronomers are searching for planets of other stars and even listening for radio transmissions from other civilizations. Because the glare of any stars with planets would hide them from direct view, and because existing telescopes cannot photograph them, no one expects to see a planet outside our solar system soon. Instead, astronomers must use more indirect methods if they hope to detect these planets.

Searching for planet nurseries

One way of looking for planets is to search for evidence of them in the early stages of formation. Astronomers theorize that planet formation begins with star formation. Stars form from enormous clouds composed largely of hydrogen and helium gas and dust. Gravity pulls together a portion of gas that is denser than the rest of the cloud. More and more of the gas and dust accumulate into a dense, spinning clump. Friction from the colliding gas molecules creates heat. As gravity continues to pull the clump of gas together, it collapses further and spins more rapidly. As friction heats the gas even more, the temperature at the center eventually becomes hot enough to start *nuclear fusion reactions,* in which the *nuclei* (cores) of atoms combine. This process creates enormous amounts of energy, and the new star begins to shine.

Astronomers believe that as a star condenses from its birth cloud, a flattened swarm of gas and dust forms around it. Scientists call this a *circumstellar disk.* Young circumstellar disks are the raw material for planet formation. Dust particles in the disk may collide and stick together like falling snowflakes. The particles eventually make snowball-sized objects and then larger ones called *planetesimals* that can be up to a few hundred kilometers in diameter. Planetesimals derive their name from the fact that they are the building blocks of planets. They crash into one another, sometimes breaking apart, but sometimes merging to form planets.

The author:
Stephen P. Maran is press officer of the American Astronomical Society in Washington, D.C., and editor of *The Astronomy and Astrophysics Encyclopedia.*

Of course, no one has witnessed the birth of a planet, but a team of astronomers in the United States reported evidence in January 1993 supporting this scenario of planet formation. Karen and Stephen Strom at the University of Massachusetts in Amherst and K. Michael Merrill of the National Optical Astronomy Observatories in Tucson, Ariz., observed about 3,000 extremely young stars inside a dark cloud called Lynds 1641 in the constellation Orion. The youngest stars among them were in small, tightly packed groups of 10 to 50. The cloud also contained slightly older stars of similar types.

To detect the young stars, the Strom team used a telescope that is able to detect *infrared radiation.* (Infrared radiation is a form of light that has a longer wavelength than visible light and a shorter wavelength than radio waves.) The team's measurements showed more infrared radiation than normal from the youngest stars. This indicated that circumstellar disks surrounded the stars. As a star heats the dust particles in its disk, it makes the disk "glow" with infrared radiation.

Next, the team compared the infrared readings from the young stars with readings from the slightly older nearby stars. The Stroms found that the older stars in Lynds 1641 emitted only slightly more infrared radiation than normal—less than would be expected from stars with circumstellar disks. The team concluded from this that planetesimals must have formed in the older stars' circumstellar disks. Planetesimals orbiting a star produce less infrared radiation than do circumstellar disks without them, because most of the dust particles are collected into the planetesimals as the planetesimals form and grow. This results in less dust to absorb visible light from the star and emit infrared radiation.

Stars most likely to have planets

The Milky Way contains many different types of stars, but all the stars that the Strom team studied were young versions of the most common type of star in the Milky Way—*main sequence stars.* These are stars that are burning hydrogen at their centers in nuclear reactions. The Stroms chose main sequence stars with *masses* (amounts of matter) similar to our sun.

Because the Strom team studied the most common type of star in the Milky Way, their study suggests that circumstellar disks and planetesimals might form around most stars. The Stroms' work also suggests that the disks vanish as planetesimals form. But other infrared studies of young main-sequence stars suggest that new disks may reappear later, once planets have formed.

In 1983, for example, the Dutch Infrared Astronomy Satellite (IRAS) detected more infrared radiation from Beta Pictoris than would be expected for a star of its age. Beta Pictoris is a star similar to our sun about 53 light-years from Earth. (A light-year is the distance light travels in a year, about 9.5 trillion kilometers [5.9 trillion miles].)

The star is about 1 billion years old, and any small dust particles that surrounded it from its birth should have vanished long ago. This effect is due to *photons* (particles of light) from the star that bombard the dust particles, causing many of the smaller ones to lose the momentum that enables them to orbit the star. As they lose momentum, the dust particles spiral inward until they fall into the star. One billion years is ample time for those smaller dust particles to fall into Beta Pictoris. Yet the IRAS readings and subsequent photographs taken from telescopes on Earth show that plenty of dust still exists.

So where does Beta Pictoris' dust come from? Some astronomers, including Stephen Strom, suggest that planets or asteroids orbiting the

Glossary

Astrometry: The measurement of the position and movement of astronomical bodies.

Center of mass: The point around which all objects in a stellar system orbit, including the system's sun.

Circumstellar disk: A flattened cloud of gas and dust that forms around a star as it is born.

Main sequence star: The most common type of star in the Milky Way. A star that generates energy by *fusing* (combining) hydrogen nuclei at its center.

Planetesimal: A small body formed from the dust in a circumstellar disk. Astronomers believe that planets form from planetesimals.

Pulsar: The spinning, magnetized core of a star that collapsed in a violent explosion. Pulsars emit radio waves in beams that sweep around the sky.

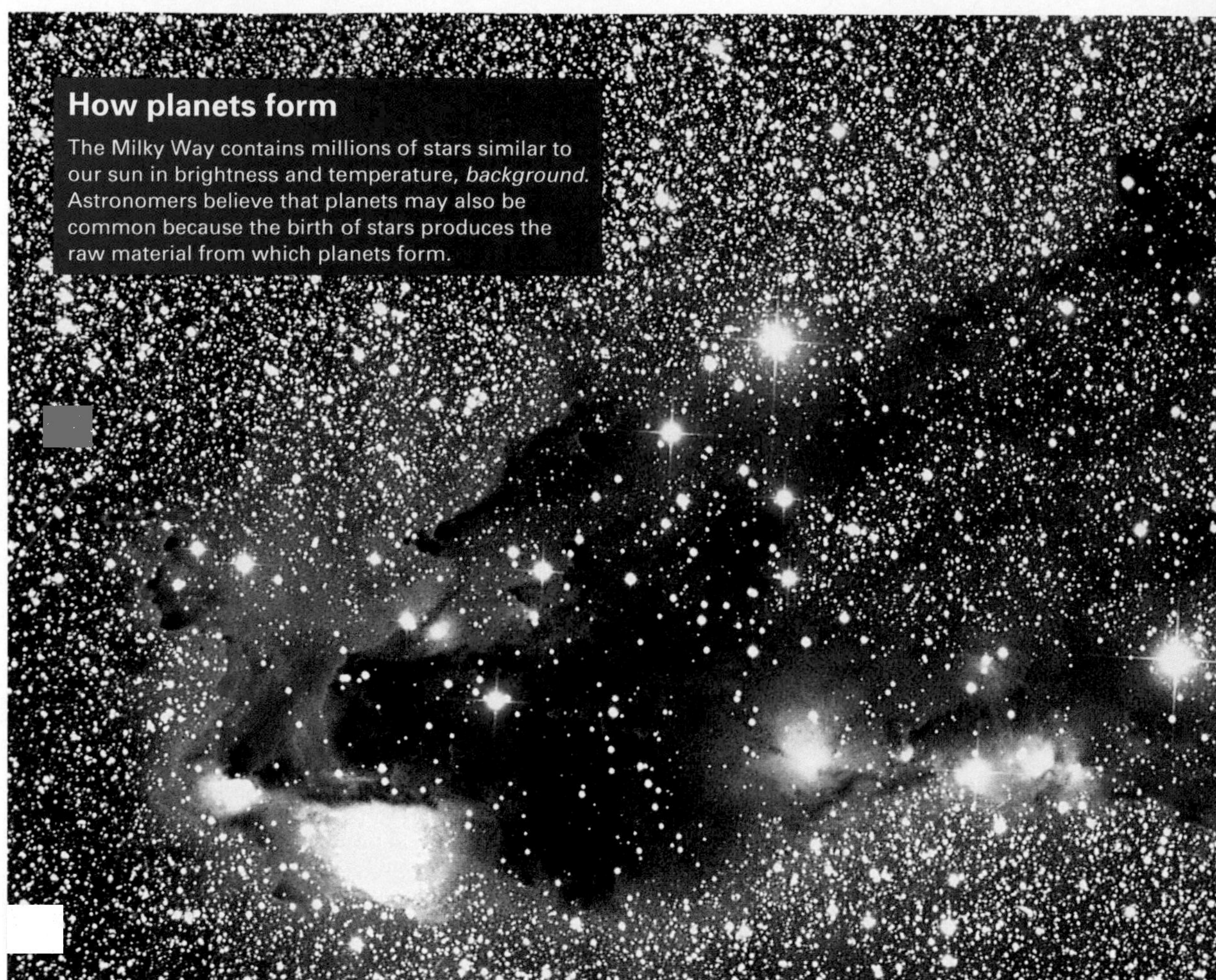

How planets form

The Milky Way contains millions of stars similar to our sun in brightness and temperature, *background.* Astronomers believe that planets may also be common because the birth of stars produces the raw material from which planets form.

star are colliding with one another, grinding themselves into dust that forms a disk. Astronomers know that such grinding produces dust, because in 1983 IRAS discovered bands of dust in our solar system created by recent asteroid collisions. Planets in our solar system have stable orbits, but astronomers point out that some stars may have many planets that travel in erratic, unstable orbits. Such orbits may produce dramatic collisions, resulting in clouds of dust particles.

In June 1992, two astronomers from the Jet Propulsion Laboratory in Pasadena, Calif., reported yet another way in which infrared detection can reveal the possible existence of planets. Kenneth A. Marsh and Michael J. Mahoney examined infrared radiation from the disks around five young stars about 450 light-years from Earth. Lower intensities of some infrared wavelengths from the disks led the astronomers to the suspicion that the disks have gaps where circumstellar dust is absent. These gaps resemble the divisions between the rings of Saturn, though the suspected gaps are much larger.

Marsh and Mahoney suggested that the gaps are evidence of planets.

A star forms when a cloud of gas and dust collapses inward due to gravity.

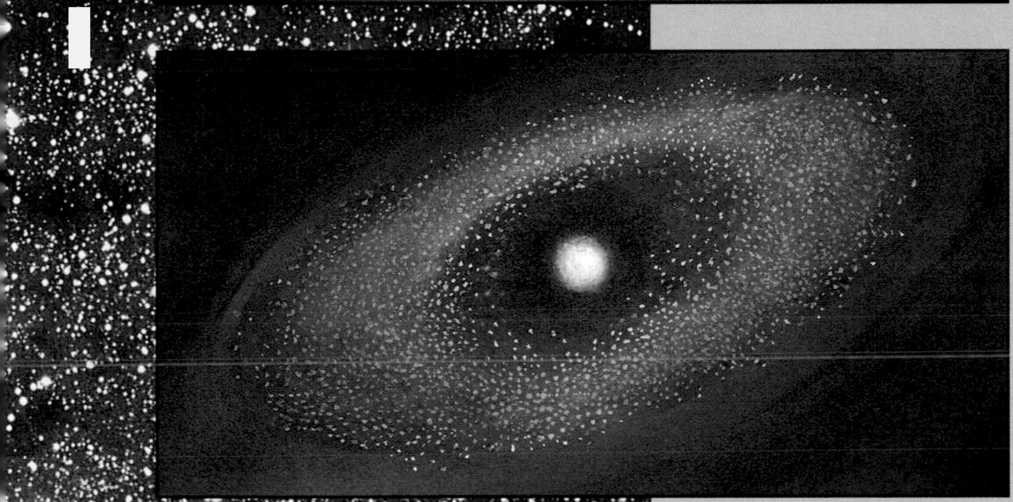

The rest of the cloud becomes a vast swarm of dust particles called a *circumstellar disk*.

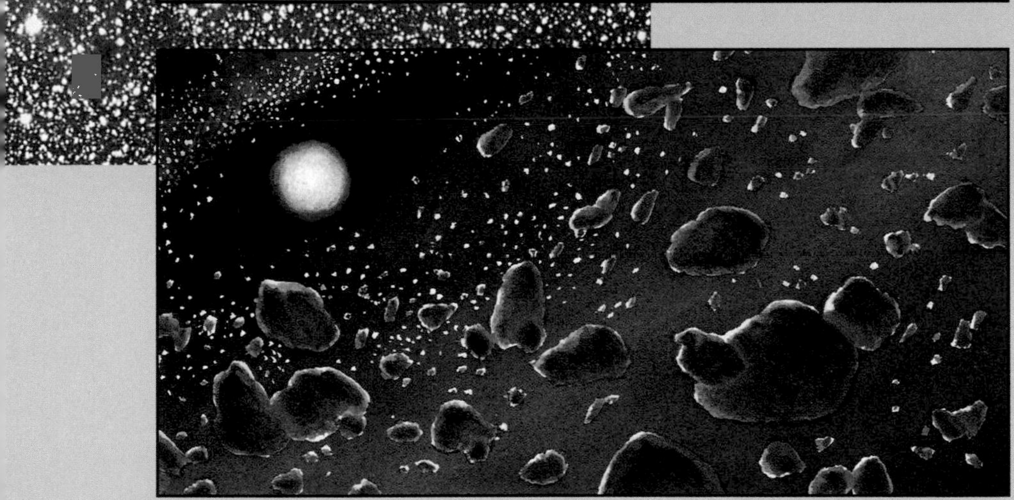

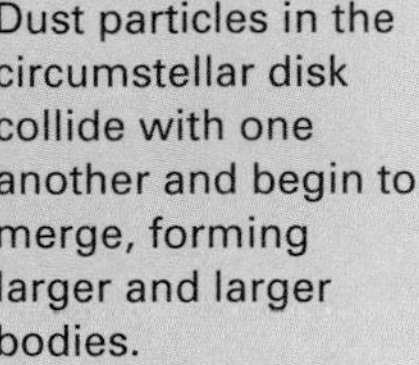

Dust particles in the circumstellar disk collide with one another and begin to merge, forming larger and larger bodies.

After about 10 million years, the dust particles have collected into *planetesimals*—orbiting bodies up to a few hundred kilometers in diameter. Planetesimals may grow into planets as they continue to collide and merge.

Searching for Planets

Astronomers searching for planets outside the solar system use such techniques as analyzing infrared radiation near stars and monitoring star movements.

Looking for wobbling stars

As the stars in the Galaxy orbit the center of the Milky Way, they appear to move across our sky. By measuring the movements of a star, using a technique called *astrometry,* astronomers can look for the influence of any unseen planets.

A star without planets will appear to move across the sky in a straight line.

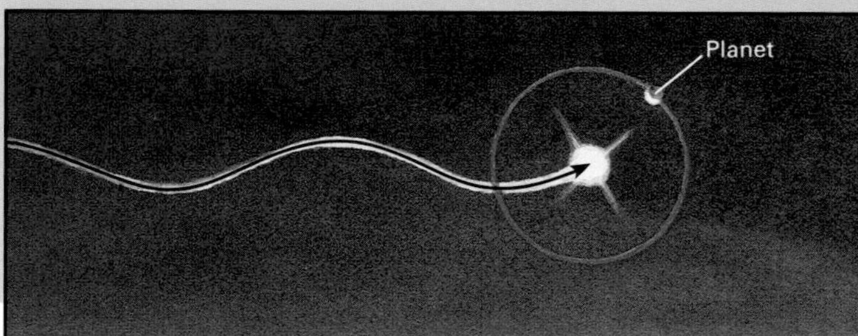

A star that is orbited by a planet may appear to wobble as the star moves across the sky. The wobble is caused by the planet's slight gravitational pull on the star.

They explained that the dust in a circumstellar disk should be spread out smoothly around the star. But when planets form, they sweep out wide, ring-shaped zones in their orbits, much as a vacuum cleaner sweeps a path on a dusty floor.

How planets affect star motions

Another method astronomers use to search for evidence of planets is to examine the star itself for hints that planets lurk nearby. For example, astronomers may examine the speed of a star's rotation

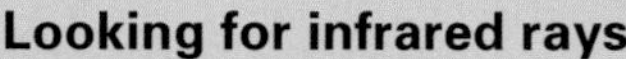

Looking for infrared rays

An infrared image of the star Beta Pictoris, *left,* highlights its surrounding disk of gas and dust, the raw materials of planet formation. Such circumstellar disks give off infrared, or heat, rays because they are warmed by light from the star. Using telescopes that block out radiation from the star itself, astronomers can photograph only the circumstellar disk. Scientists then analyze the disk for unusual temperature patterns, which may indicate planets have formed.

Looking for changes in starlight

Another way of looking for planets is to look for evidence that the star is being tugged toward or away from the Earth. These motions can be visible as changes in the star's light.

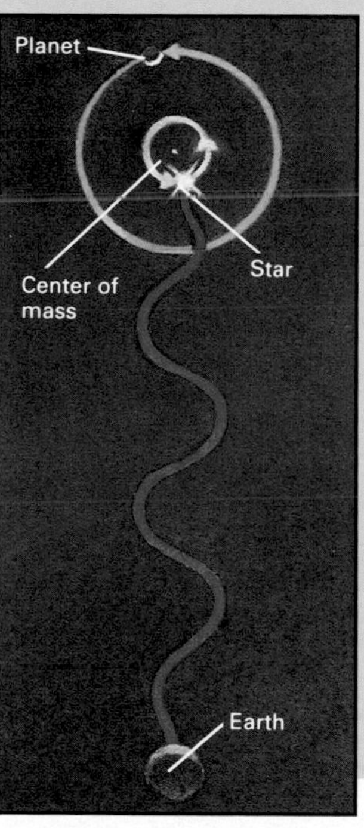

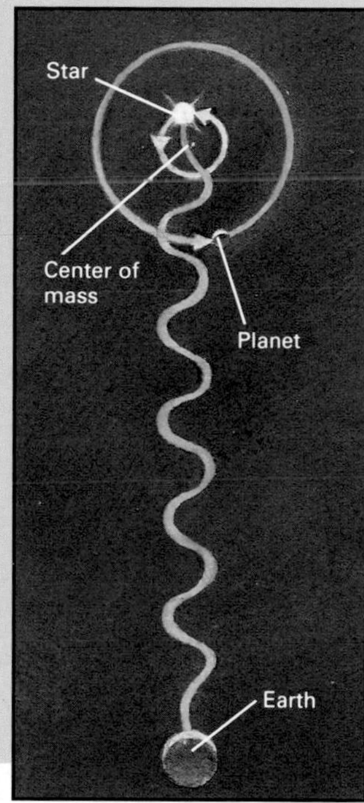

When a planet is at a point in its orbit farthest from Earth, the force of its gravity tugs the star away from Earth. This type of movement causes the star's light to appear shifted toward the longer-wavelength, or red, end of the visible spectrum, *far left.*

When a planet is at a point in its orbit closer to Earth, it pulls the star toward Earth. This movement causes the starlight to appear shifted toward the shorter-wavelength, or blue, end of the spectrum, *left.*

because a slower than expected rate of rotation may signal the existence of planets. Many main sequence stars take no more than a few days to complete one turn on their axis, for instance, but the sun takes 28 days. Physicists believe that the sun's rotation may have begun to slow as the planets formed and absorbed some of the momentum that makes the sun spin. This would have decreased the sun's rotation, much as a rapidly spinning ice skater slows down by extending his or her arms.

Astronomer G. Fritz Benedict of the University of Texas at Austin reported in January 1993 that his ongoing observations with the Hubble

Listening for life beyond Earth

On Oct. 12, 1992, 500 years after Columbus arrived in the New World, radio astronomers from the United States began the most adventurous search ever for life elsewhere in the universe. The new effort uses powerful radio telescopes to listen for radio signals from intelligent life forms that may be trying to communicate with us. The new project is the latest in a continuing worldwide quest for signs of intelligent life beyond Earth.

The search, like previous attempts, is based on the idea that intelligent beings might try to communicate by sending radio signals into space. Because radio waves travel at the speed of light, such transmissions could reach distant planets far more rapidly than any space vehicle could.

To detect extraterrestrial communications, astronomers on Earth are listening for unique radio signals—ones that could not possibly come from distant galaxies or other natural sources in the sky. The researchers must also recognize and discard artificial signals generated by our own civilization, such as radio beacons on space probes. The "alien" signal should come repeatedly from the same direction among the stars, and its wavelength should seem to lengthen and shorten as the planet it comes from orbits its star.

Many practical concerns confront astronomers in this search. Experts speak of trying to cover a four-dimensional "search space," in which the dimensions are time, direction from Earth, radio frequency, and signal intensity. In other words, the astronomers have to decide when to listen, in what direction they should search, what frequencies they should tune to, and how weak a signal they should try to detect. If the listeners make the wrong choice for any of these options, they may listen in vain, even if extraterrestrials are trying to reach us.

Despite the difficulty, astronomers have been listening for radio transmissions that could reveal the existence of other civilizations in space since 1960. That is when Frank Drake at the National Radio Astronomy Observatory (NRAO) in Green Bank, W. Va., launched Project Ozma, a two-month search for radio signals coming from the vicinities of two nearby stars—Epsilon Eridani and Tau Ceti. Drake used NRAO's 26-meter (85-foot) radio telescope, but he detected no unusual signals. There have been occasional false alarms in subsequent searches, but there have been no genuine detections.

The formal name for the new U.S. search for extraterrestrial intelligence is the High Resolution Microwave Survey, but this and similar efforts are collectively known as the Search for Extraterrestrial Intelligence (SETI). The survey began with two radio telescopes—the 300-meter (1,000-foot) antenna at the Arecibo observatory in Arecibo, Puerto Rico, and a 34-meter (112-foot) telescope at Goldstone, Calif., which the Jet Propulsion Laboratory in Pasadena, Calif., operates for the National Aeronautics and Space Administration. Cornell University in Ithaca, N.Y., operates the Arecibo telescope. The telescopes will search the sky at least until the year 2002.

Radio astronomer Jill Tarter of the SETI Institute in Mountain View, Calif., leads the search

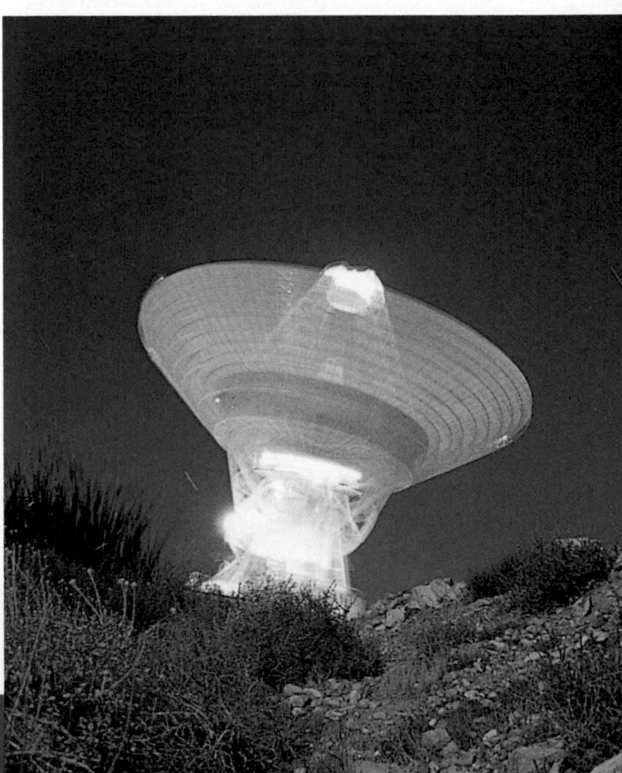

from the Arecibo radio telescope. This search will target about 1,000 stars within 100 light-years from Earth—stars that astronomers believe are most likely to support life. The Goldstone telescope will scan all of the sky visible from that location in search of promising signals. It will not search as long in any direction as the Arecibo telescope, but if detectable intelligent transmissions are coming from any direction that it can survey, it will record them.

Tarter and her co-workers have chosen to listen for short wavelength signals called microwaves. This portion of the electromagnetic spectrum contains the fewest natural signals and thus offers the quietest listening. The astronomers believe that any extraterrestrials intelligent enough to communicate with other planets would also understand the properties of the electromagnetic spectrum and choose to communicate on its quietest section.

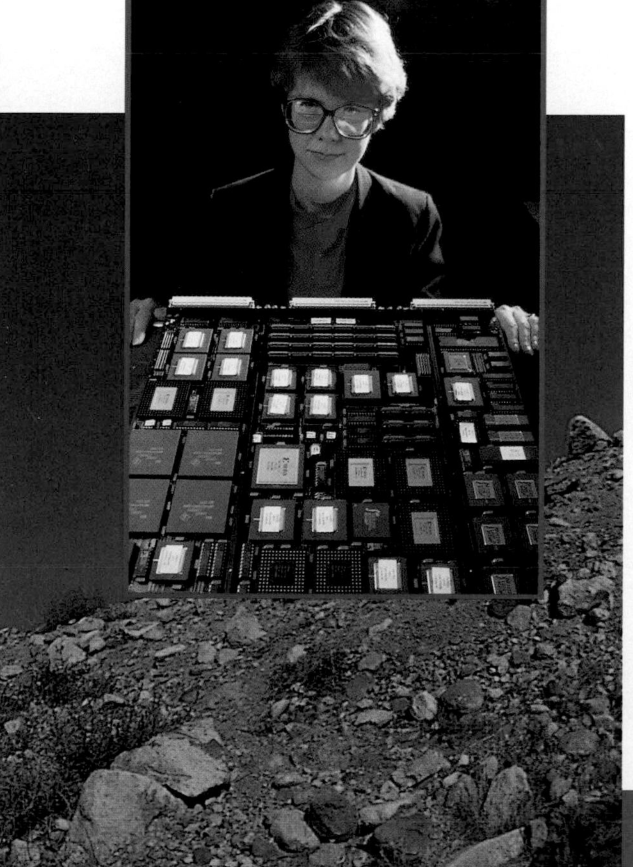

The new survey uses special radio receivers capable of listening to millions of channels at once. The Arecibo telescope will survey 14 million radio channels, and the Goldstone telescope will eventually survey 32 million channels. The two telescopes will search for radio signals of between 1,000 and 10,000 *megahertz.* (One megahertz equals 1 million *hertz,* a unit that describes a wave's *frequency* [number of cycles per second].)

Other searches for extraterrestrial intelligence include the Serendip Project, headed by Stuart Bowyer, a space researcher at the University of California at Berkeley. Bowyer believes that we have no idea where signals from extraterrestrial beings may originate, so he declined to spend precious radio telescope time listening in what may be the wrong direction. Instead, the Berkeley scientists built equipment that attaches to existing radio telescope receivers. As other astronomers observe radio signals from targets in various directions, the Serendip equipment analyzes the signals for evidence of an intelligent source. Such evidence would include a signal strong enough to distinguish it from background noise, regular transmissions from the same direction in space, and cyclic variations in the frequency as the source planet orbits its star.

What will happen if these or other scientists discover a promising signal? After making sure that it comes from an artificial source and is not produced by a natural celestial process or by our own civilization, the astronomers will verify that other observatories on Earth can detect the same broadcast. Only when an incoming signal passes these tests will astronomers dare to thrill and alarm the people of Earth with the news that another world is calling. [S. M.]

The Goldstone radio telescope in Goldstone, Calif., *far left,* is one of several radio telescopes used to "listen" for signals from space. Astronomer Jill Tarter, *inset,* head of one of the search projects, displays part of the computer-processing equipment her group uses to determine whether any of the signals could have an intelligent source.

Space Telescope had revealed that Proxima Centauri, the nearest star to the sun, turns at the very slow rate of once every 42 days. Benedict determined the star's rate of rotation by examining the presence of bright and dark areas on the star. The star brightened and dimmed as these areas turned toward and away from the Space Telescope, offering a signal that Benedict used to time the rotations. Could Proxima Centauri be turning so slowly because it has planets? Benedict's future studies with Hubble may tell.

Much current planet hunting involves another technique, called *astrometry*—the measurement of precise star positions. The stars in the Milky Way orbit the center of the Galaxy. By making highly accurate observations of the position of a star as it moves, astronomers can chart the star's path through the heavens.

Normally, a star with no planets should appear to move in a straight line across the sky. If planets orbit the star, however, their slight gravitational pull should alter the star's movement. Our solar system illustrates this motion. Most people assume that Earth and the other planets orbit the sun, but the sun as well as the planets actually orbit their common *center of mass,* the point around which the mass of the entire solar system is balanced. All objects in the system orbit this point. Because the sun contains most of the mass in our solar system, the center of mass is inside the sun, but not at its very heart. The sun follows a small orbit around that point inside itself. This orbit would make the sun appear to wobble if we viewed it with precise instruments from a great distance.

A rare find: Extrasolar planets

By mid-1993, astronomers had reported convincing evidence of only one stellar system outside our own. It consists of two, perhaps three, planets orbiting the pulsar called PSR 1257+12.

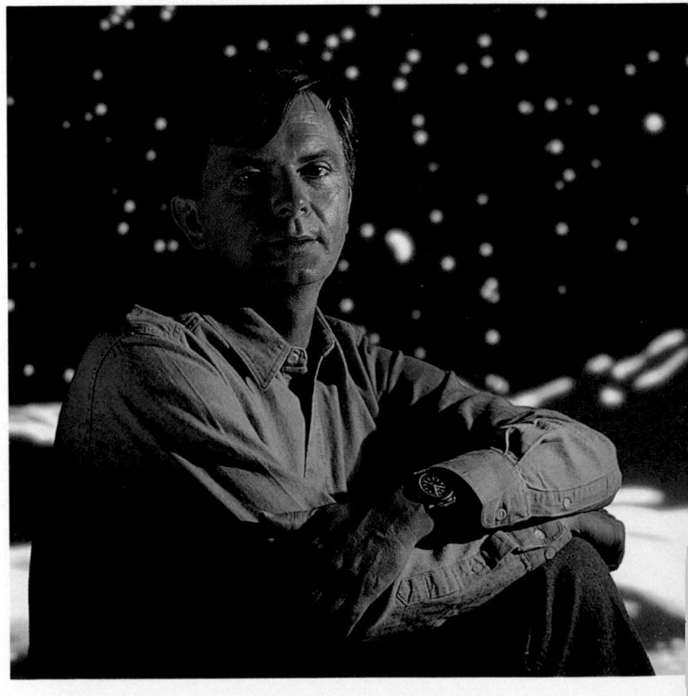

Astronomer Alexander Wolszczan, then of Cornell University in Ithaca, N.Y., was one of the scientists who reported the discovery of the planets in January 1992.

In 1963, astronomer Peter van de Kamp at Swarthmore College in Swarthmore, Pa., reported finding wobbles in the motion of Barnard's Star that suggested the presence of a planet. This star, named for the American astronomer Edward E. Barnard, who discovered it in 1916, is about one-fifth the diameter of the sun and 6 light-years from Earth. After further observations of the wobbling pattern, van de Kamp concluded in 1969 that two planets circle Barnard's star. The wobbling was not great, however, and other astronomers have been unable to confirm the existence of any planets.

Detecting changes in starlight

Another way to tell if a star is wobbling is to monitor changes in the wavelength of its light as it moves toward or away from Earth. Astronomers can view these changes with a *spectrograph,* a device that can be attached to a telescope to break light into the *spectrum,* the colors of the rainbow. When a star's wobble causes it to move closer to Earth, its light appears to shift toward the shorter-wavelength, or blue, end of the spectrum when viewed with a spectrograph. Astronomers call this phenomenon the *blue shift.* When the star moves away from Earth, its

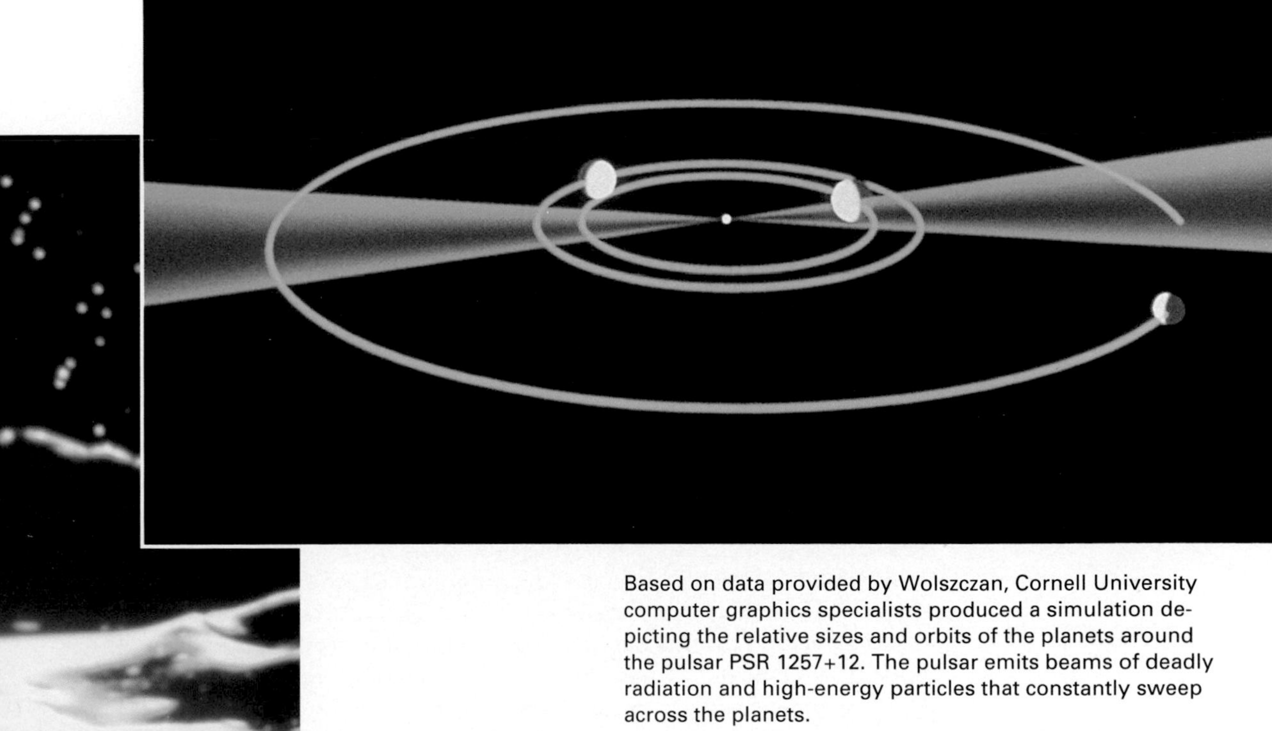

Based on data provided by Wolszczan, Cornell University computer graphics specialists produced a simulation depicting the relative sizes and orbits of the planets around the pulsar PSR 1257+12. The pulsar emits beams of deadly radiation and high-energy particles that constantly sweep across the planets.

light appears to shift toward the longer-wavelength, or red, end of the spectrum. Astronomers call this effect the *red shift.* These two effects can make light from a star orbiting its stellar system's center of mass appear to alternate between longer and shorter wavelengths, shifting very slightly toward the blue and toward the red ends of the spectrum as the star approaches Earth and recedes from it.

In June 1987, a team of astronomers led by Bruce Campbell of the Dominion Astrophysical Observatory in Victoria, Canada, announced results from such an analysis of starlight. The team looked at the changing wavelengths of visible light coming from 17 nearby main sequence stars. Campbell's group said that apparent wobbles in the motions of six of the stars could be due to planets. The best case was Epsilon Eridani, about 11 light-years from Earth. However, in 1992, several Canadian astronomers who had continued to monitor Epsilon Eridani reported that changes in the star's motions did not repeat consistently enough to serve as convincing proof that it was being orbited. Some astronomers believe instead that the star itself may be swelling and contracting, causing blueshifts and redshifts, respectively.

Planets around peculiar stars

Astronomical observations have also revealed the existence of planets where they would once have been unthinkable—around pulsars. A pulsar is a *neutron star,* the tiny, rapidly spinning, highly magnetized remnant of a much more massive star that exploded as a supernova when it ran out of nuclear fuel. The remaining core is made up of densely packed subatomic particles called neutrons. Much in the manner of a lighthouse sending out beams of light, pulsars emit beams of radio waves that sweep across the heavens. On Earth, the passing beam may be detectable as pulses of radiation. To study pulsars, astronomers employ huge radio antennas such as the 300-meter (1,000-foot) antenna in Arecibo, Puerto Rico, to listen for radio waves.

Until 1991, most astronomers would not have looked for planets around pulsars. They thought that the violent supernova explosion would either blast any nearby planets to pieces or hurl them far away. But in July 1991, radio astronomer Andrew Lyne of the University of Manchester in Manchester, England, reported that such a planet seemed to exist. He deduced this by studying the intervals between the radio signals coming from pulsars. He noticed that, unlike the regular pulses of other known pulsars, the signals from a pulsar named PSR 1829-10 appeared to arrive at varying intervals. Lyne timed the arrival of the pulses and determined that the pulsar was moving toward Earth, then away from Earth, and back toward Earth again. Lyne concluded that a planet was tugging on the pulsar, causing it to move farther from Earth and then closer as it orbited its stellar system's center of mass. Unfortunately, Lyne had made an error in analyzing the arrival times of the pulses. In January 1992, he said that corrected calculations showed there was no planet around PSR 1829-10.

Lyne's work, however, helped alert astronomers to the possibility of finding planets around pulsars. Less than a week after he reported his correction, Alex Wolszczan and Dale A. Frail announced their discovery of the planets around the pulsar PSR 1257+12. They used the Arecibo radio telescope to time the pulses from PSR 1257+12 and found that they sometimes arrived more than one *millisecond* (one thousandth of a second) early or late and that this time interval repeated regularly. The astronomers concluded that the pulsar itself was following a small orbit, and at the farthest point of its orbit from Earth, the pulses arrived late. At the point in the orbit closest to Earth, the pulses had a shorter distance to travel, and so they arrived early. Only the gravitational pull of another object or objects could cause the pulsar to follow such an orbit, the astronomers reported.

But when physicists pondered the possibility of planets around PSR 1257+12, they realized that they might be newly formed, not old planets that somehow survived the supernova. Gas and dust left over from the explosion could have remained around the pulsar and could have formed a circumstellar disk. From the disk, the pulsar's planets might have formed.

The existence of at least one star with planets outside our stellar system is encouraging to astronomers involved in this search. Unfortunately, because PSR 1257+12 is the collapsed remnant of an exploded star and therefore not a normal star, it offers little hope of helping us understand how planets form around more common stars. And because the pulsar sweeps its planets with powerful radiation, there is almost surely no living thing on them.

Astronomers nonetheless continue to believe that planets of ordinary stars exist. But their most likely methods of detection—searching for wobbles in star motions or the blue- and red-shift of stars' light as they move toward or away from Earth, had not, as of mid-1993, produced positive results. Hope may lie in larger telescopes. Five such telescopes, all larger than any existing telescope, were under construction in 1993, and astronomers began scientific observations during the year from the 10-meter (400-inch) Keck Telescope at Mauna Kea, Hawaii. After astronauts fix the Hubble Space Telescope's blurred vision, its photographs should help in the search. As the new telescopes increase the planet hunters' ability to scan the heavens, the chances of discovery can only increase.

For further reading:

Bruning, David. "Desperately Seeking Jupiters." *Astronomy*, July 1992, pp. 37-41.

Drake, Frank D. *Is Anyone Out There? The Scientific Search for Extraterrestrial Intelligence.* Delacorte, 1992.

Folger, Tim. "Forbidden Planets." *Discover*, April 1992, pp. 38-43.

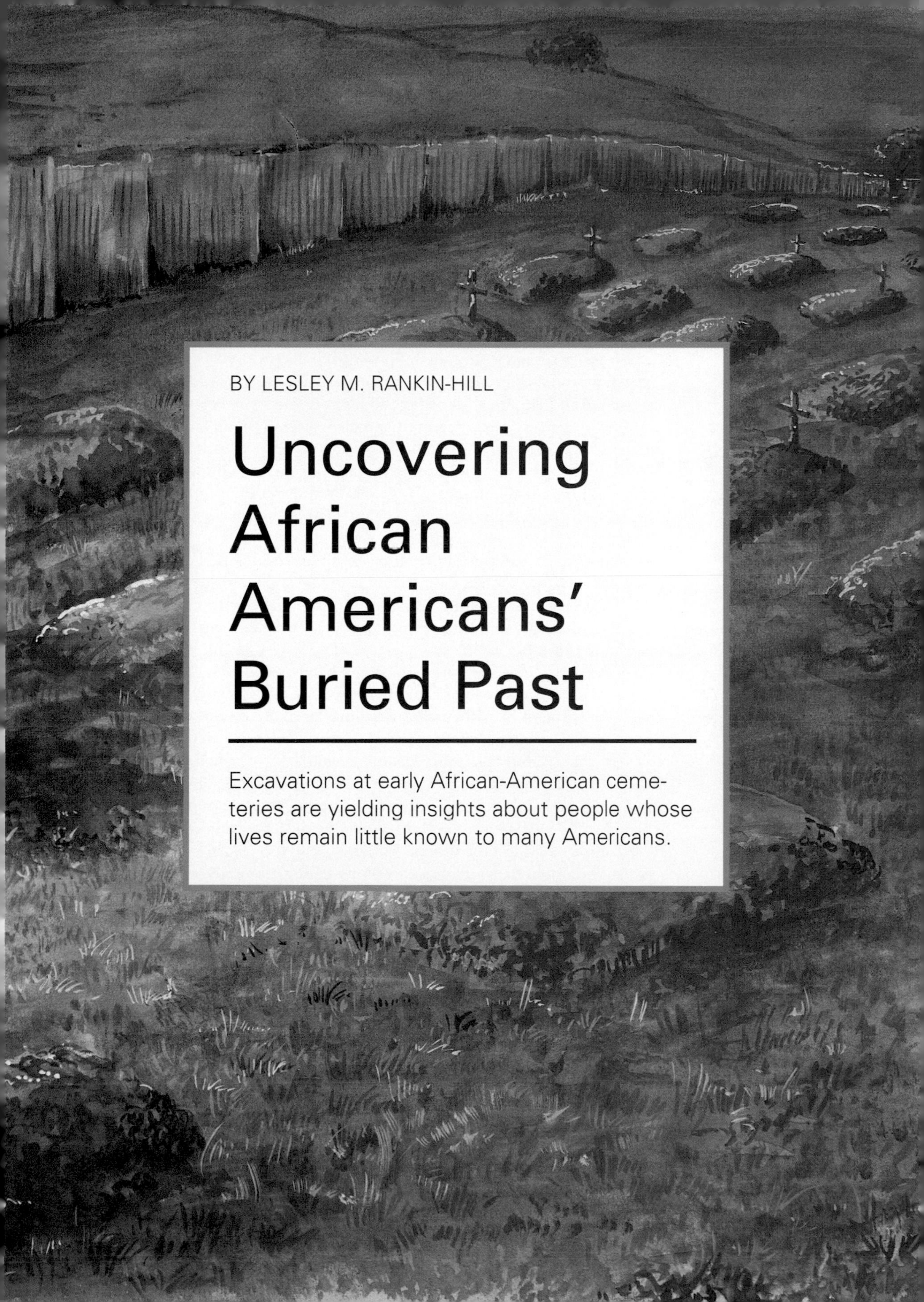

BY LESLEY M. RANKIN-HILL

Uncovering African Americans' Buried Past

Excavations at early African-American cemeteries are yielding insights about people whose lives remain little known to many Americans.

Preceding pages: African slaves gather for a funeral in the 1770's at the Negro Burial Ground, now referred to as the African Burial Ground, in New York City. Archaeologists think that most funerals there took place at night, because slaves were not allowed time off from their work during the day for such occasions. City laws forbade gatherings of more than 12 African slaves, but slaves and freed African Americans apparently violated this ban for funerals. Services were conducted according to Christian rites and the rituals of various African cultures.

The author:
Lesley M. Rankin-Hill is assistant professor of anthropology at the University of Oklahoma in Norman, Okla.

For several days, the two archaeologists had been digging into the grave, carefully lifting out the soil in layers only inches deep until they had dug down more than 1 meter (3.3 feet) below the surface. Nearby, other archaeologists were intent on similar excavations. All were working on one of the most remarkable archaeological discoveries made in the United States—a cemetery containing the oldest and largest collection of African-American skeletal remains ever found. Construction workers found the first of the remains in August 1991 while building a government office complex in a busy area of New York City.

Sheltered by a large plastic tent pitched over the graves, the archaeologists seemed far removed from the skyscrapers towering over the 280-year-old cemetery and the hum of the city beyond. Suddenly, as they removed another layer of soil, pieces of a wooden coffin lid appeared. Although most of the lid had disintegrated, the rectangular outline of the coffin was apparent as a dark stain in the lighter soil. The excavation stopped while the two archaeologists recorded the dimensions of the coffin and drew pictures of its outline and the coffin pieces. Then, the scientists carefully removed the coffin pieces, numbered them, and sealed them in airtight, acid-free plastic bags to prevent further deterioration caused by exposure to oxygen or by contamination with the bacteria in the air or on human hands. A search for other artifacts soon turned up several iron nails that had secured the lid of the coffin. These were also drawn to scale, labeled, and bagged.

Using small bamboo brushes and picks, the archaeologists probed the grave until the outline of a skeleton appeared. Carefully brushing away the soil, they determined where the head and feet lay. Then they exposed, in order, the top of the skull, the legs, ribcage and spine, and then the feet and hands. Finally, they uncovered the fragile face bones and the *pelvis* (the hip bones and end of the backbone), all of which are critical for identifying the sex and age at death of a skeleton.

The skeleton appeared to be that of an African-American woman. But a positive identification and a more detailed analysis of the remains would have to wait until the skeleton could be examined in a laboratory. Although the woman's clothing had long since disintegrated, an ornament had survived—a belt made of glass beads and shells that had circled her waist. This was removed from the skeleton, sealed in a bag, and labeled.

Because the bones were broken and crumbling, the archaeologists carefully removed all the dirt on either side of the skeleton until it lay exposed on a pedestal of soil. Then, pedestal and all, the skeleton was lifted from the grave, wrapped in paper, and placed in a box for shipment to Herbert H. Lehman College, also in New York City, where all the skeletons removed from the cemetery were being stored.

The African Burial Ground

By July 1992, archaeologists had excavated the remains of about 400 people, most of them African slaves, who had been buried between about 1712 and 1790 in what had been known as the Negro Burial

A colonial-era find in modern Manhattan

White plastic tents shelter graves being excavated from an exposed section of the African Burial Ground. The cemetery, discovered in 1991 during construction of a New York City office complex, was used between about 1712 and 1790 chiefly for the city's African slave community—one of the largest in colonial America—and their immediate descendants.

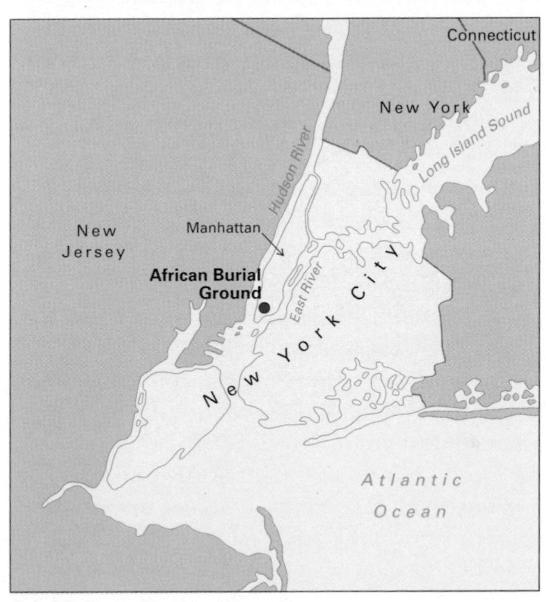

Today, the African Burial Ground lies in the crowded governmental center of New York City. In the 1700's, however, the cemetery, originally known as the Negro Burial Ground, covered 2.4 hectares (6 acres) of land just outside the northern limits of the city. In the early 1800's, the city expanded into the cemetery and the surrounding land, where many of the people buried in the cemetery had lived.

Ground. (About 7 percent of the remains were of poor European-American immigrants who lived with the freed descendants of African slaves in a nearby neighborhood known as Five Points.) During the excavation, the name of the cemetery was changed to the African Burial Ground at the request of New York City's African-American community. The change was intended to clearly identify the site as a burial place for slaves from Africa and their immediate descendants.

Although archaeologists knew from historical records that a cemetery had existed at the site, the discovery of human skeletons there was something of a surprise. In the early 1800's, after the cemetery was closed, the city had expanded northward into the cemetery and the surrounding area. To make construction easier on the cemetery land—2.4 hectares (6 acres) of ground sloping down to a pond—the city had leveled the area with huge amounts of dirt. Eventually, the graves lay from 4.8 to 5.5 meters (16 to 18 feet) below the surface. In addition, at least two buildings with deep foundations had been constructed on the cemetery land. The leveling and the construction had greatly increased the risk of destruction to the graves below. Because of this, archaeologists surveying the site in 1991 before the new construction project began had doubted that any human remains had survived.

Glimpses of early African-American life

The African Burial Ground is only the latest of several African-American cemeteries excavated in the past 30 years. The remains found in these cemeteries have yielded valuable information about individuals as well as communities. For example, by studying a skeleton, scientists can usually determine sex, height, and age at death. But that is only a start. Bones are one of the most active tissues in the body. Throughout a person's lifetime, specialized bone cells called *osteoblasts* form new bone by laying down fibrous protein and depositing hard minerals. Other cells called *osteoclasts* erode the protein and minerals. As living bones are dissolved and rebuilt, they record the injuries and diseases a person suffers, indicate how well nourished that person is, and offer clues to the physical strenuousness of his or her life.

In addition to these findings, however, results from the studies of the remains are giving scientists a broader understanding of living conditions in African-American communities. For example, by studying several individuals, researchers can calculate how widespread disease and malnutrition were in a community. Such findings also enable scientists to compare the lives of urban slaves and freed slaves in the North, about whom little is known, with those of slaves on Southern plantations. Finally, the findings are helping researchers identify elements of African culture that may have survived slavery.

Like the African Burial Ground, most of the other African-American cemetery sites were excavated on an emergency basis because construction projects were underway. For example, excavation of the First African Baptist Church (FABC) Cemetery in Philadelphia began in 1980 after human bones were discovered during the construction of a

Excavating the African Burial Ground

Archaeologists removed the remains of about 400 people—the oldest and largest collection of African-American skeletal remains ever found—from the African Burial Ground before excavations were halted in mid-1992. Portions of the cemetery buried beneath streets and buildings in the area may hold the remains of another 10,000 people.

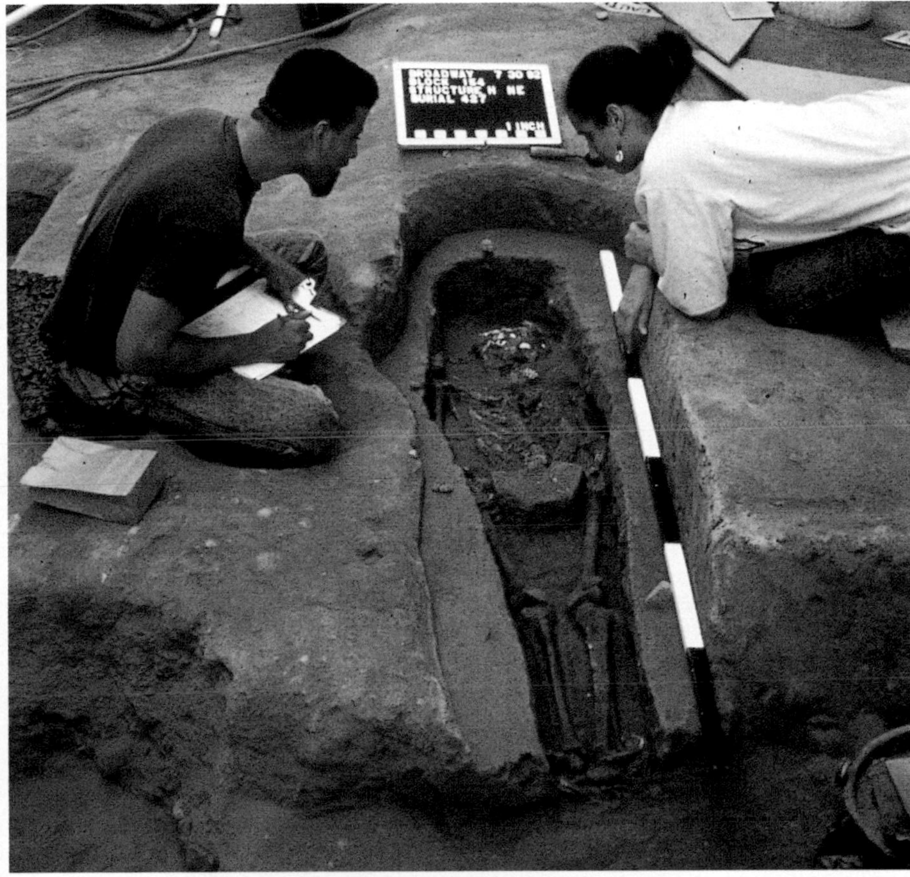

Archaeologists assess the condition of the skeleton of a woman before carefully removing the remains and shipping them to a laboratory for detailed study.

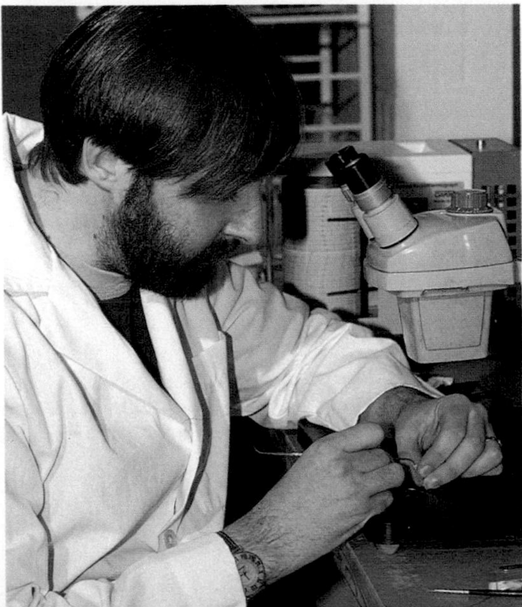

In the laboratory, *left,* a scientist examines a pair of enameled cuff links, *above,* found in a man's grave. Few objects were found in the cemetery because the people buried there were poor. Thus, the cuff links may indicate that their owner had a special status in colonial New York City's African-American community.

commuter rail tunnel. The African-American remains excavated in 1984 at the St. Peter Street Cemetery in New Orleans were found during the construction of a condominium complex. Until the discovery of the African Burial Ground, the New Orleans skeletons, which date from 1720 to 1810, were the oldest known collection of urban African-American remains. In contrast, the exhumation of 101 skeletons from a slave cemetery at Newton Plantation on Barbados in the West Indies in 1972 and 1973 was one of the few excavations that occurred as a planned archaeological project.

The lives of the individuals whose remains lay in these cemeteries for hundreds of years are all but unknown to most modern Americans. Few written accounts describe African-American life in colonial times. In fact, when the African Burial Ground was discovered, many people were surprised to learn that colonial New York City had once had such a large African slave population.

The anonymity of African Americans in the American colonies is partly the result of sketchy record keeping by colonial governments, which generally gathered little information on anyone, whether they were African American, European American, or Native American. Data on African Americans, who were not considered citizens even if they were freeborn, were even less likely to appear in official documents.

Nor are descriptions of African-American life common in the histories or letters, diaries, and other personal documents of the colonial period. Some African slaves, violating bans against reading and writing, kept diaries, as did some freeborn African Americans. But few of these documents have survived. As a result, there are almost no accounts written from the perspective of African Americans or in their words. Moreover, the images of African Americans and their lives in the documents that have survived are usually distorted, if not by racist attitudes then by the tendency of their European-American authors to judge African-American behavior and cultural beliefs by their own standards.

Findings from a Philadelphia dig

Although the voices of African slaves and their descendants may be muted, their bones and teeth and the objects buried with them can speak for them now. The study I and my colleagues made of the skeletons found in Philadelphia, for example, yielded valuable information about both the health and the living conditions of an important part of the African-American community in that city. At the FABC Cemetery, archaeologists recovered 75 adult skeletons—36 males and 39 females. They also found the remains of 60 children. All had been buried between 1822 and 1842, according to church records.

Philadelphia was a major commercial center and port city and offered abundant opportunities for unskilled commercial laborers. By the early 1800's, it had the largest population of free African Americans outside the slaveholding states. According to a census of Philadelphia's African-American community conducted in 1837 by groups opposed to slavery, the FABC congregation, like other African-

American congregations in Philadelphia, included freed slaves and freeborn African Americans as well as some runaway slaves. The survey and documents from the Association of Baptist Churches reported that the female members of the church had worked chiefly as laundresses or as cooks, maids, and nannies in homes. The male members worked as unskilled laborers, waiters, porters, seamen, coachmen, or cart drivers.

One of the most interesting skeletons we examined was that of a man who had probably lived in pain for most of his life. Unable to walk normally, he had, nevertheless, apparently held a physically demanding job. The man, who was approximately 165 centimeters (65 inches) tall, was about 60 years old when he died. Although he had been relatively well nourished as an adult, he had suffered from at least several moderately serious fevers or illnesses during childhood.

Unfortunately, a list of the people buried in the FABC Cemetery had not survived, and so my colleagues and I would probably never know the man's name. We also were unable to determine the cause of his death. But his bones revealed interesting details of his life.

We knew from historical accounts that the skeleton was that of an African American. When such records are not available, assessing the race of a skeleton requires great caution. Even scientists who are highly experienced in distinguishing the subtle differences that sometimes exist between members of various racial groups, such as the size and shape of the nasal cavity and certain dental traits, frequently err.

Determining that the skeleton was that of a man was easier—the pelvis was angular with a backward tilt in the hip sockets, where the top of the thighbone fits into the hipbone. (A female's pelvis is flatter and broader and has a larger central cavity.) The bones in the skeleton's

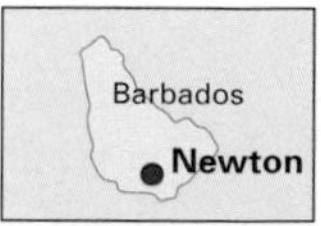

Scientists have excavated several African-American cemeteries dating from the 1700's and 1800's in addition to the African Burial Ground in New York City. These include the First African Baptist Church Cemetery in Philadelphia; the Catoctin Furnace Cemetery in Frederick County, Maryland; the Remley Plantation near Charleston, S.C.; the Cedar Grove Baptist Church Cemetery near Lewisville, Ark.; the St. Peter Street Cemetery in New Orleans; and the Newton Plantation on Barbados in the West Indies.

A coffin handle found at a cemetery for African-American workers on the Remley Plantation was fashioned in a style widely used in the mid-1800's. Discovering the handle helped archaeologists determine that the man whose remains were found in the coffin had been buried during that period.

chin and skull and the teeth were also fairly thick and large, as they generally are in men. To calculate the man's age, we had examined his *pubic symphysis,* the place at the front of the pelvis where the two pubis bones meet. The appearance of these bones changes over a person's lifetime, from an almost frothy texture to rough and grainy. The texture of the man's pubic symphysis indicated that he was about 60.

What set the man apart from the other skeletons we had examined was the terrible condition of his hip joints. The outer surfaces of the man's hipbones and the tops of the thighbones had deteriorated, so that when the man walked, the thighbone had rubbed directly against the hip socket. The man had certainly experienced some pain when he walked. A specialist in bone disorders determined that the man had suffered from a rare condition in which, for unknown reasons, the blood flow to the hips is temporarily cut off, damaging the outer surface of the bones.

Despite his handicap, the man had lived an active life. On his arm bones, we found tremendous muscle attachments, raised areas of thick bony buildup where muscles were once connected. In general, the larger the muscle attachment, the larger the corresponding muscle must have been. We concluded that the man had used his arms to move around—possibly walking with crutches—and probably had held a job that required great upper body strength.

The man's handicap apparently had little effect on his life span. In fact, he had lived much longer than most of the people buried in the FABC Cemetery. We found that the average life span for the men there was about 45 years and for the women about 39 years, fairly typical for African Americans of the time, but about 10 years less than that of European Americans.

Brief and strenuous lives

Interestingly, the discovery that the FABC men had lived longer, on average, than the women paralleled findings at cemeteries used for urban slaves—such as the St. Peter Street Cemetery in New Orleans—and at cemeteries for industrial slaves—such as the Catoctin Furnace Cemetery in Frederick County, Maryland, which was in use between about 1790 and 1840. The women buried in the St. Peter Street Cemetery had worked as cooks, nurses, and housemaids, while the men had worked as butlers and coachmen as well as carpenters, bricklayers, painters, cabinetmakers, and butchers.

The 31 skeletons found at the Catoctin Furnace Cemetery, discovered in 1979 and 1980 as a highway was being widened, belonged to members of a community of slave ironworkers. The women in this community lived an average of 35 years, compared with 42 years for the men.

In contrast, the women buried in the Remley Plantation Cemetery outside Charleston, S.C., had lived longer than the men—an average of 40 years compared with 35 years. The explanation for the difference involves both normal life expectancy and the effects of strenuous work. Historical accounts and studies of African-American skeletons have revealed that although the work of the members of the St. Peter Street and Catoctin Furnace communities was frequently strenuous, it was not as punishing as the agricultural labor of plantation slaves, especially that of males. Thus, among the slaves buried in the FABC, St. Peter Street, and Catoctin Furnace cemeteries, a woman's chances of dying in pregnancy and childbirth—especially given the primitive nature of medical care at that time—were greater than a man's chances of succumbing to the effects of hard labor. On the other hand, in South Carolina, the risk of dying in pregnancy and childbirth was also high, but the risk of being worked to death was higher still.

Signs of disease

The remains of the FABC man with the handicap reflected other medical problems commonly found in African-American communities of the 1700's and 1800's. For example, the presence of scaly patches on the long bones of his legs revealed that he had suffered from at least one generalized infection. Generalized infections are untreated infections that begin in one part of the body—such as a tooth—and then spread throughout the body. They may remain low-grade for many years or they may quickly prove fatal. Bones, especially the long bones of the arms and legs, can provide a record of such infections. Infections can affect bones in several ways. One of the most common is scaliness. Blood circulates through the veins in the thin outermost layer of tissue on the shafts of the long bones. In this outer layer, new layers of bone are constantly being laid down. The presence of bacteria or viruses in the blood causes this outer layer to create bone at an accelerated rate, and this extra bone appears as scaliness.

We found that about one-fourth of the FABC adults had experienced a generalized infection. This rate was relatively high compared with that of European Americans of colonial times but relatively low for colonial African Americans. Scientists studying the South Carolina skeletons, for example, found that 64 percent of the adults had suffered from a generalized infection. The bones of nearly half the adult skeletons excavated from the Cedar Grove Baptist Church Cemetery in Cedar Grove, Ark., in 1982, also revealed a history of generalized infections. The people buried in the Cedar Grove cemetery, which was in use between 1890 and 1927, were the free descendants of local plantation slaves.

The high rate of infection among African Americans was probably related to undernourishment, malnutrition, and poor living conditions in those communities. The immune systems of people eating an inadequate diet, especially one poor in animal protein, have a diminished ability to mount a defense against infection. Most of the FABC skeletons with evidence of generalized infection also showed signs of iron deficiency, a condition that may result from a lack of animal protein in the diet. But the somewhat lower rate of infection among the FABC adults supports the idea, based on historical documents, that African Americans in urban areas had a somewhat better diet than did African Americans in rural areas.

Examining teeth for evidence of malnutrition

We found more evidence of malnutrition, undernourishment, and illness in the teeth of the disabled FABC man. This evidence appeared in the form of *hypoplasias.* Hypoplasias, which appear on the surface of teeth, range from off-white to yellow-brown horizontal lines to grooves and pits, depending on their severity. They are formed during periods before the age of 6 when the adult teeth still forming in the gums stop growing. The interruption in growth may last from one day to several months. The longer the period, the wider or deeper the hypoplasia.

One of the most common reasons for this halt in growth is fever, particularly in people who are poorly nourished and, therefore, more likely to contract infections. When a person contracts an infectious disease, the teeth and bones temporarily stop growing as the body conserves its energy resources to fight the infection. The fevers that left the hypoplasias may have resulted from common "childhood" diseases, such as measles and mumps, or from more serious infectious diseases, such as cholera and dysentery, common in the 1700's and 1800's.

We found hypoplasias on the teeth of all the FABC adults. Such lines also appeared on more than 70 percent of the teeth of skeletons found at the Catoctin Furnace cemetery and 77 percent of the skeletons found in South Carolina. The teeth of most of the FABC adults had several hypoplasias, indicating that they had experienced more than one bout of fever. During the 1700's and 1800's, European Americans also experienced fevers, of course. But the few scientific studies of their remains have revealed many fewer hypoplasias, except among the very poor. As a result, scientists believe hypoplasias are much more likely to form in people with an inadequate diet. The high number of multiple hypoplasias in the teeth of slaves on the Newton Plantation in Barbados indicated that the people there had experienced periodic episodes of near-starvation.

Stories the bones tell

By studying skeletons from African-American cemeteries, scientists have learned about the health of individuals and gained insights into the living conditions in their communities.

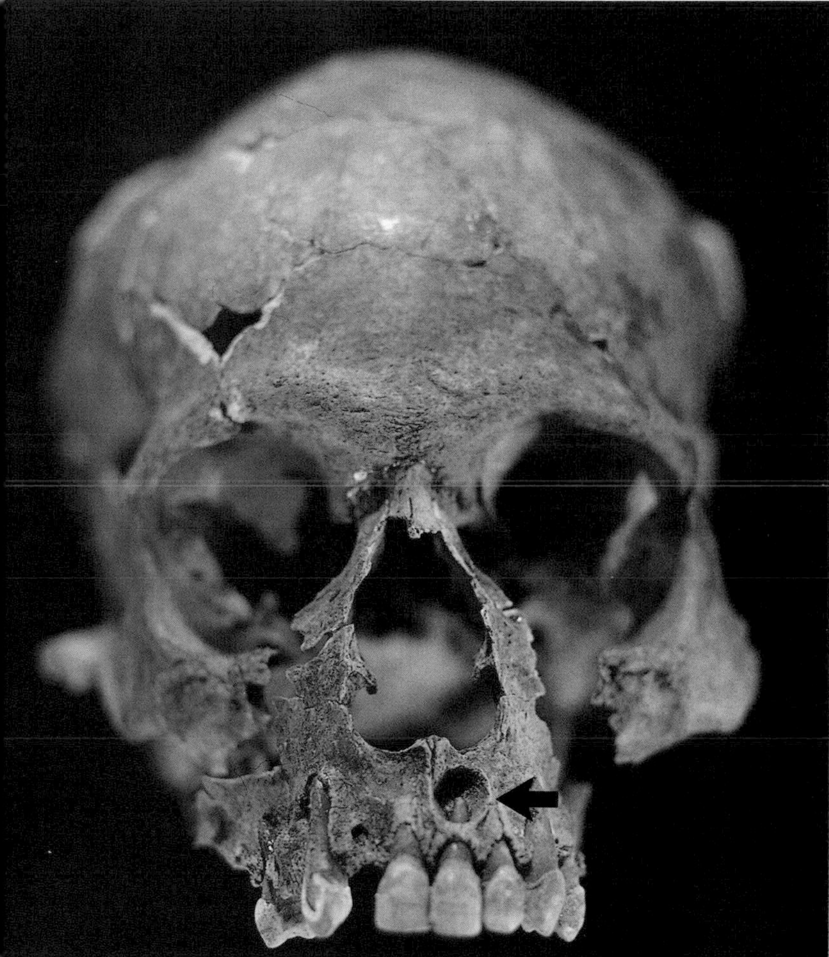

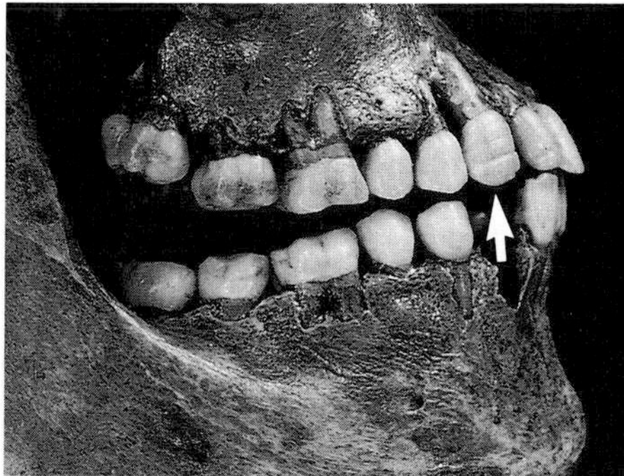

A skull excavated from the Remley Plantation Cemetery, *left,* reveals a hole (arrow) in a man's upper jawbone left by an infection in the root of a tooth. Grooves, called *hypoplasias* (arrow), are visible in the tooth of a man buried in the First African Baptist Cemetery in Philadelphia, *above*. Such grooves form before age 6 if the growth of teeth temporarily slows. Fever is one of the most common causes of such slowing, particularly in malnourished people, as the man apparently was.

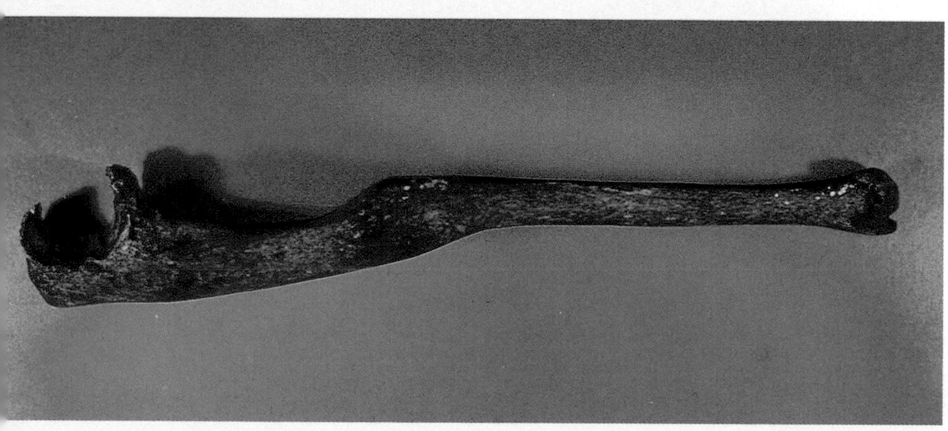

A bulge in the forearm of a skeleton found in the Cedar Grove Cemetery resulted from a fracture that was improperly set. Forearm fractures were found in a number of skeletons of African-American slaves and plantation workers.

Clues to African-American culture

Copper bracelets of African design and a clay pipe (arrow) made in Africa were among the artifacts found with the skeleton of an older man in a slave cemetery at the Newton Plantation. Such artifacts suggest some personal attributes of the buried person. The man, who was buried in the late 1600's or early 1700's, was probably a medicine man, an important person in the slave community. Medicine men diagnosed and treated illnesses and prepared charms used to protect people from harm.

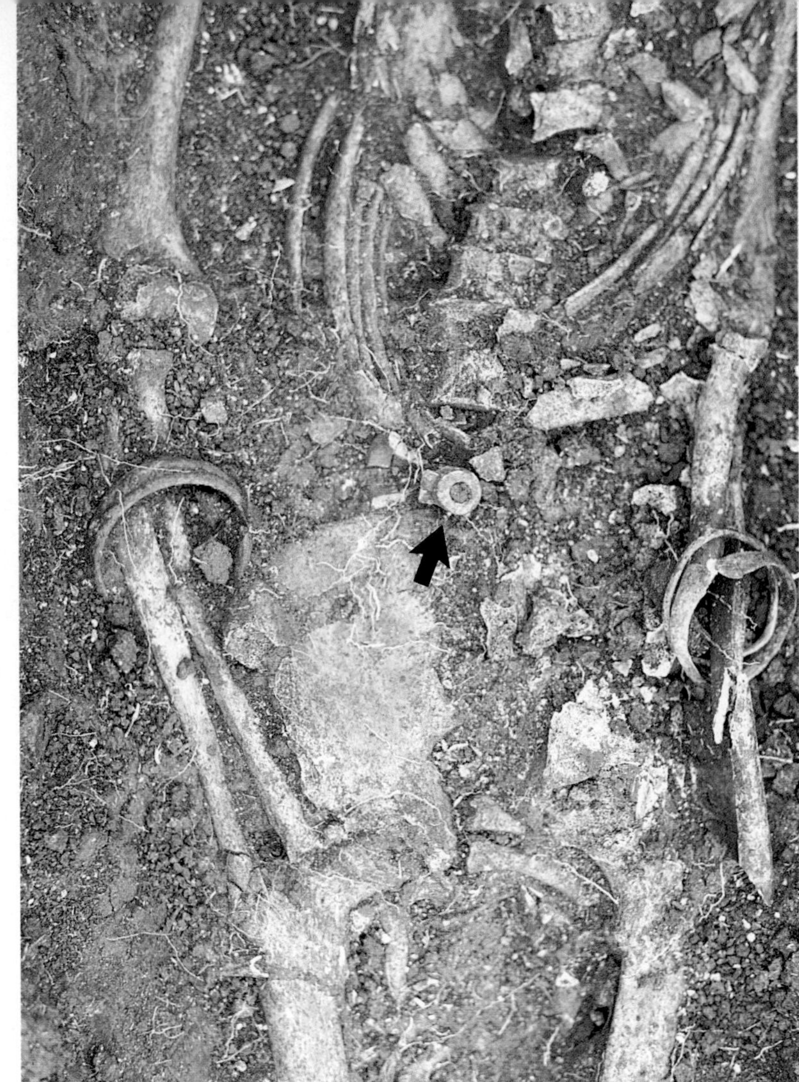

Lives of the children

The nearly universal presence of hypoplasias was sad evidence of the fragile lives of African-American children in the 1700's and 1800's. Our finding that most of the hypoplasias evident in the adults had developed between the ages of 2 and 4 led us to conclude that the children's diet had deteriorated once they were weaned, which commonly occurred at about age 2. Starker evidence of the toll of poverty on the FABC children, however, lay in the cemetery. About 45 percent of the skeletons were those of children. More than half had died before their first birthday. Another 25 percent had died before their fifth birthday. Excavations at other African-American cemeteries have uniformly revealed high death rates for infants and children.

We believe the children died chiefly of infectious diseases. In many cases, the children's ability to fight infection may have been weakened from birth. We found, for example, that most of the hypoplasias in the teeth of the FABC children had developed on that part of the enamel that forms while the fetus is still in the womb. (Teeth start to form in

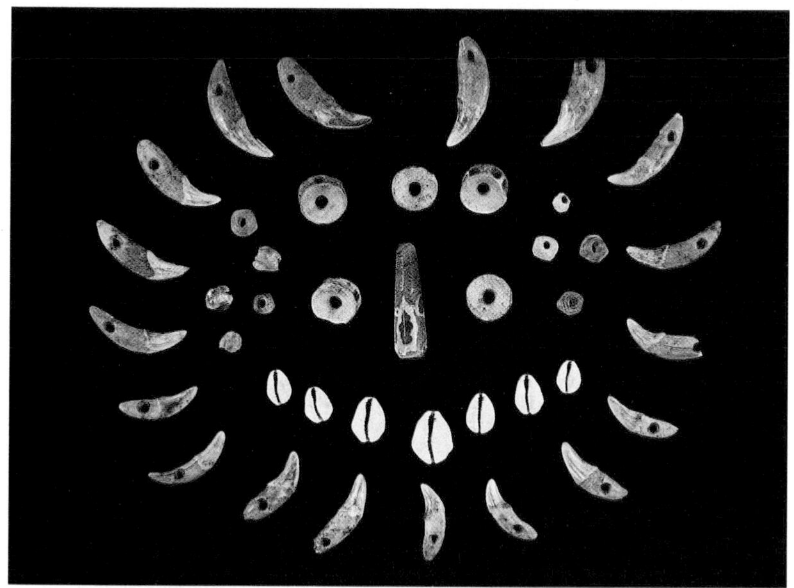

Circling the man's neck was a necklace of African design (shown disassembled), the only one of its kind found at an African slave cemetery. It consisted of drilled dogs' teeth, cowry shells, spinal bones from a large fish, and a reddish-orange bead, probably from India.

The clay pipe, *above left,* which was made in West Africa in the late 1600's, has a short stem into which a reed or hollow piece of wood was inserted for smoking. The coiled bracelet, *above,* found on the man's left wrist, was made of copper alloy.

the gums about 7½ months before birth.) This indicated that their mothers were undernourished when they became pregnant and had not received adequate nourishment during their pregnancies.

Evidence of hard labor

The hard labor of the FABC man with the handicap was evident in his skeleton's joints and muscle attachments. The presence of knobs of bone in the man's joints indicated that he, like most of the other FABC adults, had suffered from osteoarthritis and other forms of degenerative joint disease, conditions in which the tough, elastic cartilage in the joints disintegrates. While these conditions often develop in old age, their appearance in younger people may be the result of intense labor.

Interestingly, degenerative joint disease was more common among the FABC women, especially women under 30, than it was among the men. This suggests that the manual labor of the women was actually more strenuous than the domestic or craft work of the men. According to historical accounts, female domestic workers, were required, for example, to boil heavy cotton sheets and hang them out to dry and to press clothes with heavy iron-metal irons. Because domestic workers were in such demand in colonial Philadelphia, African-American women often began this work at a young age.

Researchers studying remains from other African-American cemeteries have reported similar levels of joint degeneration. The male skeletons found at the Catoctin Furnace Cemetery, for example, revealed severe degeneration of the *vertebrae* (the bones in the spine), evidence of the intense physical labor of these ironworkers.

The male skeletons found at the Catoctin Furnace Cemetery as well as those from South Carolina and the Newton Plantation showed additional evidence of strenuous labor—large muscle attachments. Large muscle attachments were common among skeletons of both sexes at all locations, however. Their presence has, at times, complicated attempts to determine the sex of skeletons.

Artifacts found in the graves

Not unexpectedly, archaeologists have found few objects in the graves of African slaves and their close descendants. These people were poor and any possessions were usually passed along to their survivors. The most common artifacts found in the graves include pins for *shrouds* (cloths in which bodies were wrapped for burial) and buttons. One grave in the African Burial Ground contained cuff links. In a few graves, excavators found simple metal finger rings. The presence of more valuable items in a grave may indicate that the person buried there, such as the woman with the bead-and-shell belt, had special status in the community.

The artifacts found in some cemeteries, however, have provided tantalizing clues to the religious and cultural practices of early communities. Some graves in the FABC Cemetery, for example, contained dinner plates, which may have represented something believed to be needed in the next life or may have been the last object used by the deceased. Most intriguing, however, was the presence of a single shoe on the lid of some of the coffins. In folklore, shoes symbolize power, and so the shoe may have been buried to help keep the Devil away from the deceased. Or the shoe may have been a good luck symbol.

The controversy over examining remains

Although the findings from African-American remains and graves have begun to reveal lives previously unknown to us, the excavations have stirred up considerable controversy. Some African Americans have argued that exhuming the remains is disrespectful and even sacrilegious. Feelings ran particularly high in New York City, where many African Americans viewed the cemetery as a powerful symbol of their heritage. Protests from the community led the U.S. House of Representatives' Subcommittee on Public Buildings and Grounds, which oversees federal construction projects, to halt the excavations in July 1992, after about 400 skeletons had been excavated. Archaeologists estimated that the remains of another 10,000 people were left in the cemetery.

On Oct. 6, 1992, President George Bush signed a law that ended all construction in the cemetery. At that time, the General Services Administration, the government agency overseeing the construction of the New York City office complex, agreed to create a museum at the site and to establish a museum of African-American history in New York City. Meanwhile, the 400 skeletons were being stored at Herbert H. Lehman College while scientists devised a plan for studying them.

Scientists involved in excavating cemeteries and studying the remains agree that human skeletons should be treated with the greatest respect throughout the process. They point out, however, that only two collections of African-American remains—those from the Catoctin Furnace Cemetery and the Newton Plantation—have not been reburied. In both cases, no descendants or members of the African-American community requested reburial. Scientists affiliated with the African Burial Ground project say they will rebury the remains after their studies are completed.

Scientists also argue that the insights their work provides are helping African Americans reclaim their history, a history that has frequently been ignored. Reconstructing the lives of these early Americans, they say, will give all modern Americans increased pride in the contributions African Americans have made to the history of the United States.

Questions for Thought and Discussion

Suppose that you belong to a culture that has few written records of its early history and that has been largely ignored by historians. Archaeologists have found a very old cemetery containing the remains of some members of your culture, which may include some of your distant ancestors. The archaeologists want to excavate the remains so that physical anthropologists can study the skeletons to learn more about the health and living conditions of these people. They agree to rebury the remains after they finish their research. Some descendants of those buried in the cemetery strongly oppose the plan. Other descendants, however, support the anthropologists' argument that studying the remains would be worthwhile.

Questions: Do you think the skeletons should be removed for study, or should they remain undisturbed? What factors would you take into consideration while making your decision? What arguments would you use to support your position?

BY JAMES A. DRAKE, JEFFREY R. DUNCAN, AND K. JILL MCAFEE

Beware the Exotic Invaders

Introduced species—plants and animals purposefully or accidentally transported to new habitats—can cause ecological disasters.

Preceding pages: Kudzu, a species of vine brought to the United States from Asia, blankets a North Carolina landscape. In some parts of the Southern United States, the vines grow wildly, overrunning even trees and buildings.

In June, a pine and beech forest in the White Mountains of the Northeastern United States should be green and lush. But in June 1993, the leaves of many trees in the region's forests were already gone. The view from an airplane was of vast tracts of skeleton trees. This altered landscape was the work of gypsy moth caterpillars. Tremendous numbers of them had devoured the vegetation, killing many of the trees.

Gypsy moth caterpillars have been eating their way through U.S. forests for decades. Until the 1800's, the gypsy moth was found only in Europe and Asia, where natural enemies such as certain wasps and beetles kept it in check. But in 1869, a scientist in Massachusetts who was studying the moths as a possible source of silk accidentally released some of them from his laboratory. Far from their natural enemies, the moths reproduced and quickly spread throughout New England. By the 1980's, the moths had been found in the Southeastern United States and in California, Oregon, and Washington. In 1992, officials in the Great Smoky Mountains National Park in North Carolina and Tennessee were alarmed to find substantial moth populations there.

Biologists call the gypsy moth an *exotic invader*, though it is hardly "exotic" in the popular use of that term. The scientific term indicates that the insect is a species that has been transported from its native habitat to a new environment, where it reproduces and spreads. The moths are by no means the only such intruders. "Killer" bees are exotic invaders, as are zebra mussels, the kudzu vine, and many other pest species. The number of exotic invasions is on the rise, and scientists are concerned about the increase for several reasons. Exotics can cause many ecological problems, from altering the natural balance among species living in a region to driving one or more of those species extinct. Unfortunately, controlling and reducing the impact of exotic invasions are complex problems without easy solutions.

Natural exotic invasions

Some movement of species into new areas has occurred ever since life first appeared on Earth more than 3.5 billion years ago. In some cases, these invasions happen by chance. For instance, a floating log containing seeds may wash ashore on an island, where some of the seeds may sprout, or a natural disaster such as a prairie fire may destroy a habitat and force animal species there to invade new territory.

The authors:
James A. Drake is associate professor of zoology at the University of Tennessee in Knoxville. Jeffrey R. Duncan is a graduate student in the department of zoology, and K. Jill McAfee is a student in the graduate program in ecology, also at the University of Tennessee.

Other invasions occur by nature's design. For example, some plant species evolved special seed structures that promote *dispersal* (scattering). The winglike structures on maple tree seeds and the fluffy fibers on dandelion seeds are two structures that enable the wind to carry seeds to distant locations. Sticky seeds, such as burs, cling to animal fur or bird feathers and hitchhike to new territory. Migratory birds transport exotic plant species great distances by eating berries in one location and dropping seed-containing waste elsewhere.

Natural invasions are relatively rare, however, because barriers exist in nature that thwart the movement of species. Such huge barriers as oceans and mountains place limits on the natural range of many animal species. For example, it is impossible for freshwater fish found in Lake Victoria in east-central Africa to swim across the ocean and reach lakes in North America on their own. Climate differences also prevent many species from successfully invading. If a migratory bird dropped the seed of a tropical plant in a desert, the seed could not take root because the soil would lack sufficient moisture.

Glossary

Biological control: The use of natural enemies to control another species.

Ecological community: The plants and animals living in the same geographic area.

Exotic species: An animal, plant, or microbe that is not native to a habitat in which it is living. Also called *alien species* or *introduced species.*

When people transport species

On the other hand, human activities have helped thousands of species invade regions where they could not otherwise have entered. Early explorers sailing the world's oceans in the 1500's carried with them on their ships rats and mice, which had been attracted on board by the grain in the ships' holds. When a ship reached an island, the rodents scurried ashore. There, without natural enemies, the rats and mice became successful exotic invaders and quickly reproduced.

While explorers introduced species accidentally, the settlers who followed intentionally brought plants and animals to the new regions. Settlers of the Mascarene Islands in the South Pacific, for example, eventually brought cats to control the mice and rats. And the settlers of countless lands brought plants from home to grow for food in the new territory. Many fruits and vegetables currently grown in the United States are in effect exotics, introduced long ago as crop plants. Because crops are controlled and nurtured, they rarely have been destructive.

In some cases, however, species intentionally introduced for food ended up seriously damaging their new environment. One example is the common carp, brought from Germany to the rivers of the United States in the late 1870's. The United States Fish Commission (now the U.S. Fish and Wildlife Service) sponsored the move, believing the carp would be a good game fish, as it was a popular food in Europe and Asia. By 1900, officials had established the carp in most waterways east of the Rocky Mountains. But Americans did not like the carps' taste and declined to fish for them. The carp populations grew, crowding out native species of many rivers and lakes. Controlling the carp populations was still a problem in 1993.

Another reason people have introduced plant species is simply for their beauty. But beauty can become a beast. For example, two vines introduced from Asia as ornamental plants quickly became destructive invaders in the United States. One, vine honeysuckle, which was introduced because of its pink flowers, now overruns gardens and infiltrates forests. It grows over trees and shrubs, in many cases killing them. People brought the other vine, the kudzu, to the Southern United States in the late 1800's because of its lovely purple flowers. The species made an excellent porch vine, providing beauty and shade. It was also used to feed livestock. Then, in the 1930's, farmers in the South began to plant kudzu to stop soil erosion because its long roots helped to hold soil in place. But the vine spread rapidly into fields and forests, smothering native plants. A kudzu vine may grow 18 meters (60 feet) high. The vines overrun everything in their path, including entire buildings.

What is an exotic invader?

Exotic species are plants, animals, or microbes that are transported to new territories. Many pest species are exotics that managed to reproduce, establish colonies, and spread in the new habitat.

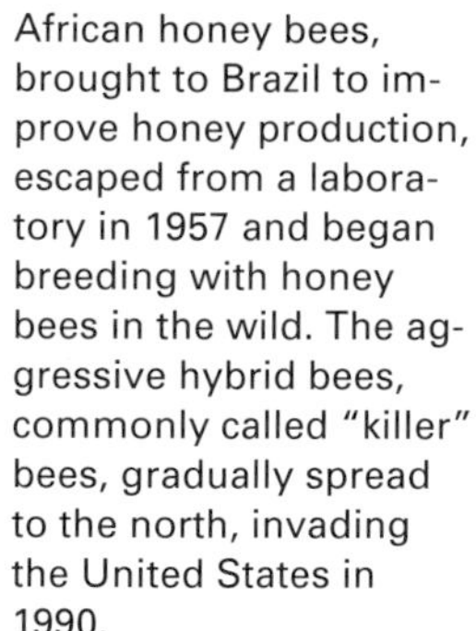

African honey bees, brought to Brazil to improve honey production, escaped from a laboratory in 1957 and began breeding with honey bees in the wild. The aggressive hybrid bees, commonly called "killer" bees, gradually spread to the north, invading the United States in 1990.

The brushtail possum was brought to New Zealand from Australia in the mid-1800's. By 1993, the New Zealand population had grown out of control, with 70 million of the small mammals devouring the leaves of the island's unique trees, thus killing them off.

The stages of a successful invasion

Not every plant or animal introduced to a new habitat, whether accidentally or intentionally, becomes a successful invader. Certain crop plants, for example, have never been successfully cultivated beyond their native lands. Scientists also know that many unintentional introductions take place, but conditions at the new site apparently prevent many species from reproducing and establishing new colonies. One factor that seems to set the stage for a successful invasion is ecological disturbance. Biologists say that human activities such as logging or dam building may alter an area so greatly that an invader can readily take hold. Regions that have few native species are also vulnerable.

Biologists say that successful invasions share certain characteristics. First, the new environment must have a climate and plant and animal life that allow the invading species to survive. Tropical plant species that are accidentally introduced into northern, temperate climates or deserts have little chance of survival. Animal invaders encountering a vigorous predator that it never encountered at home would quickly be wiped out. (Generally speaking, a predator is a species that eats another. The species that is eaten is called *prey*.)

Blue water hyacinths are lovely to look at, which is why people brought them to the Southern United States from South America. But the plants reproduce rapidly, doubling in number in as few as 10 days, and have clogged waterways throughout the South.

Successful exotics must be able not only to survive in their new environment, but also to reproduce. Usually, only a few individuals of a species colonize a new area. If, by chance, all invaders are males, the new species cannot persist even though it may adapt easily to the new environment. And even if both males and females of a species invade, the new territory may not encourage them to reproduce. Another possibility is that so few individuals invade that males and females cannot find one another to mate.

Finally, members of the species must spread out over a wide area. A single, small, isolated population of invaders can be wiped out by drought, fire, disease, or predation. When the population increases, however, members begin competing for food, space, and other resources. This competition spurs individuals to move away from the rest of the group. After the species is widely distributed, it is unlikely that drought, disease, or predation can kill them all at the same time.

How exotics cause ecological damage

After successfully invading and spreading out, exotic species can cause many problems for the *ecological community* (the plants and animals in the same geographic area). Because native species have lived in close association with one another for thousands of years, they have *co-evolved* (changed together) to thrive in those circumstances. Predators evolve to become more successful at catching prey, while prey evolve characteristics that help them avoid predators. But if an exotic invader begins to hunt the same species as a native predator, the prey of both species can quickly be driven to extinction. The native predator species in turn may perish, because its food resource is now gone.

Some intentionally introduced species

Species	Origin	Introduced	Date	Reason	Damage
Dandelion	Europe	East Coast of the United States	1600's	As a food resource	Spreads into lawns, competes with grass for nutrients
Common carp	Germany	United States	1870's	As a game fish	Kills off native species by competing for food and space
Kudzu (climbing vine)	Asia	Southern United States	Late 1800's	As an ornamental vine and as a cattle feed	Spreads rapidly, smothers other plants
Rainbow trout	Western United States	Eastern United States	Late 1800's	As a game fish	Competes with brook trout for food and space
Asiatic fresh-water clam	China	Inland waters of the West Coast of the United States	1930's	As a food resource	Clogs water intake pipes to power-plant cooling systems
Cannibal snail	Florida	Hawaii	1950's	To control the giant African snail, another exotic invader	Has killed off native snail species instead of the giant African snail
Banana poka (vine)	South America	Hawaii	Unknown	As an ornamental plant	Strangles native plants

Critically reviewed by James A. Drake, University of Tennessee in Knoxville.

An invading predator can cause extinctions even more quickly if it begins to feed on animals that do not have a native predator. A good example of this is the brown snake's invasion of Guam, an island previously without snakes. The 1.5-meter- (5-foot-) long snake, native to the Solomon Islands, New Guinea, and northern Australia, was discovered in Guam in the early 1960's. Scientists think it was carried there by accident on U.S. military airplanes or ships. Because the birds and small mammals on the island had never developed defensive strategies to avoid being eaten by snakes, the intruders literally found easy prey. By 1993, some areas of Guam had as many as 11,600 snakes per square kilometer (30,000 snakes per square mile), and they had killed off 9 of the island's 11 species of native birds. Furthermore, by mid-1992, six brown snakes had been found in airports in Hawaii, a chain of islands without native snakes. Airport workers discovered the brown snakes in the wheel wells and cargo bays of jet planes arriving from Guam. Scientists said the discoveries pointed out an ominous threat to Hawaii's rare native bird species.

A different type of chain reaction is set off among native species by invaders that have physical characteristics that allow them to avoid being eaten. Such a species is a water flea called *Daphnia lumholtzi*, a native of New Zealand and Australia now found in many lakes of North America. *Daphnia* are *zooplankton,* tiny aquatic organisms that feed on *phytoplankton,* small, floating plants that form the basis of the aquatic food web. In turn, *Daphnia* and other zooplankton are the food of many fish. But *Daphnia lumholtzi* has large spines that prevent its being

Some accidentally introduced species

Species	Origin	Area invaded	Detected	How introduced	Damage
Sea lamprey	Atlantic Ocean	Great Lakes	1830's	Through canals connecting the Great Lakes	Has killed most lake trout
Gypsy moth	Europe, Asia	Massachusetts	1869	Escaped from a laboratory	Destroys trees
Chestnut blight	China or Japan	Northeastern United States	1904	In trees imported by the Bronx Zoo in New York City	Has killed most American chestnut trees
Mediterranean fruit fly	West Africa	Florida California	1929 1975	Probably in imported fruit or vegetables	Attacks 200 species of nuts, fruits, and vegetables
Dutch elm fungus	Europe	New York	1930	In lumber from Europe	Causes Dutch elm disease, which has killed millions of American elm trees
Asian tiger mosquito	Japan	Texas	1985	Probably in old tires from Japan	May carry dengue, an acute viral infection
Sweetpotato whitefly	Europe, Asia	Florida	1986	Unknown	Attacks 600 species of plants, including fruits, vegetables, and cotton
Zebra mussel	Eastern Europe	Lake Ontario	1988	In ship ballast water	Filters nutrients from lake water, clogs water intake systems of power and water filtration plants, and kills native mollusks

Critically reviewed by James A. Drake, University of Tennessee in Knoxville.

eaten by small fish, which simply cannot get the spiny creatures in their mouths. Thus protected from predation, this species of *Daphnia* has produced large populations that eat the lion's share of phytoplankton. Unable to compete with *Daphnia* for food, other zooplankton have died off, leaving species of small fish without food.

Zebra mussels and other costly invaders

Exotic invaders can cause damage that directly impacts human life. Scientists estimate that exotic insects, weeds, and organisms that cause plant diseases together account for a decrease of greater than 50 percent in crop productivity worldwide. Rice, for example, which accounts for nearly a third of the world's food, is continually affected by exotic weeds that reduce its yield by 30 to 35 percent annually. The quickly growing weeds compete with the rice plants for space in the field as well as for nutrients in the soil. Similarly, an exotic grass species native to Europe and Asia is responsible for decreasing wheat and barley production throughout the world by as much as 11.8 million metric tons (13 million short tons) per year—enough to feed 50 million people.

Invasions by some aquatic species have harmed the means of supplying water to human populations. The Asian freshwater clam is such a

The damage that invaders cause

Once an exotic species takes hold in a new territory, it may cause ecological and economic damage.

A *feral* (wild) pig, *left,* digs up roots in Hawaii. The animals, descended from pigs Polynesians brought to Hawaii before the 1500's, have eaten many of the island's native plants nearly to extinction.

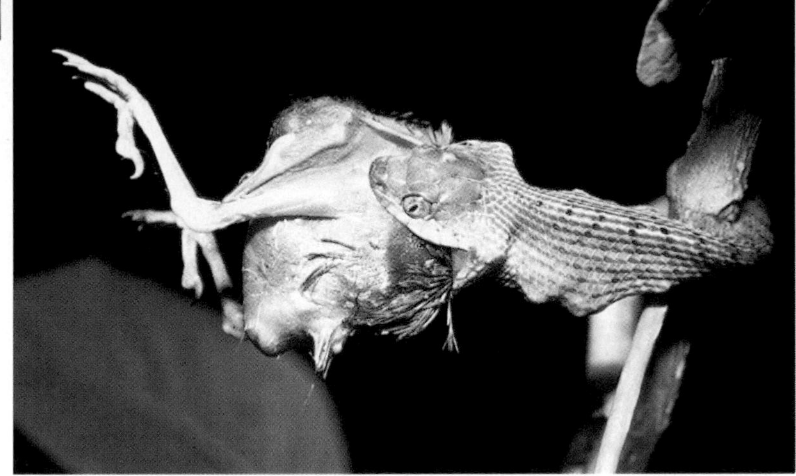

A brown snake swallows one of Guam's native birds, *right.* Since being introduced in Guam in the early 1960's, brown snakes have killed off 9 of the island's 11 native bird species. By 1993, some areas of Guam had as many as 11,600 snakes per square kilometer (30,000 snakes per square mile).

species. In the 1930's, Chinese immigrants brought the clam to the West Coast of the United States to establish it as a food resource by putting a relatively small number of clams in irrigation ditches. The Asian clam reproduces only once a year, so its numbers grew slowly as the clams spread in the mud or sand at the bottom of the irrigation ditches. But 20 years later, the clams had grown to sufficient numbers to cause serious damage to North American waters. By the 1950's, clam invasions had clogged the flow of water in the cooling systems of power plants. In the early 1990's, U.S. electric utility companies spent more than $1 billion a year attempting to control clam populations.

The zebra mussel is another invader that is wreaking havoc with water systems, but it is spreading at a much quicker pace than the Asian clam. Scientists believe that the initial invasion of mussels must have occurred in 1985 or 1986 in Lake Ontario, one of the Great Lakes. The

Workers remove zebra mussels from a water screening facility in Lake Michigan, *above.* The tiny mussels, discovered in Lake Ontario in 1988, have spread throughout the Great Lakes and down the Mississippi River as far as Louisiana. They attach themselves to any underwater object, from native mussel species to ship bottoms. They are also invading water treatment plants and causing extensive damage.

Skeletons of trees on a Virginia hillside, *above,* mark the feeding grounds of gypsy moth larvae. The caterpillars eat the leaves of trees, in some cases stripping the trees so completely that they die.

invasion went unnoticed until 1988, when the tiny mussels had developed to a size large enough to be detected. But how did this animal get from its native habitat in Europe's Caspian Sea to the Great Lakes of North America? Biologists think that an oceangoing ship unloaded its cargo at an inland freshwater port in Europe, then pumped in ballast water containing zebra mussel larvae. Because the ship was fast and could cross the Atlantic Ocean in a few days, the larvae were able to survive in the ballast water until the ship reached Lake Ontario. There, the ship pumped out the infested water.

Zebra mussels differ from most other species of freshwater clam in that they produce larvae capable of swimming much like a small fish. This mobility allowed them to spread rapidly. Within two or three years of the invasion, adult zebra mussels had spread along the shores of Lake Erie and Lake St. Clair near Detroit. From there, they quickly established colonies throughout the Great Lakes and other waterways of the Midwest. In spring 1993, zebra mussels were even reported in the

Weapons to combat invasions
Government officials and scientists agree that preventing invaders from entering new habitats is the best way to control the problems they cause. Many localities examine goods and baggage brought into the area to make sure that a potential pest is not a stowaway. An inspector, *right,* examines imported logs at a seaport, looking for signs of insects.

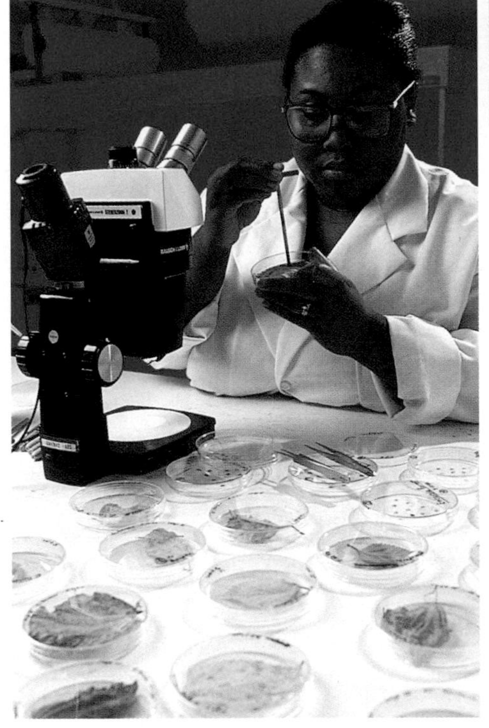

If a destructive invasion does occur, scientists look for ways to control the alien species, often by introducing another species that feeds on or kills the first. A technician experiments with ways of controlling the sweetpotato whitefly, *above,* an insect that ravaged California crops in the early 1990's.

Mississippi River as far south as Louisiana and in the Arkansas River in Oklahoma.

The zebra mussel causes damage by secreting stringy threads, called byssal fibers, that serve to tightly fasten the animal to any underwater object. The mussel threatens to drive many native mussel species extinct because it attaches to and grows over the native mussels, eventually killing them. The zebra mussel also attaches itself to the inner surface of water pipes for public water treatment stations, electric power plants, and other facilities that pump water from lakes and rivers. The huge numbers of mussels reduce water flow through the pipes, at times by as much as 20 percent. In 1989, the invasion seriously threatened drinking water supplies to more than 45,000 people in Monroe, Mich. Scientists estimate that removing the mussels from intake pipes in the Great Lakes alone will cost industry, and ultimately consumers, more than $5 billion by the year 2000.

Exotic marine animals are not the only species that can damage aquatic environments. Plant invasions can be just as menacing. The water hyacinth is an excellent example of an aquatic plant that overran its new habitat. The plant was introduced to the Southern United States from South America as an ornamental because of its lovely blue flowers. But the water hyacinth can double in number every 10 days, and it grows to a height of about 60 centimeters (2 feet) above the water, blocking sunlight from plant and animal life below. The plant clogs waterways, restricting boat traffic and shutting down irrigation pumps.

Elsewhere, the water hyacinth causes serious problems, too. For example, it has blocked canals in India. And it has choked lowland waterways in Bangladesh, causing them to overflow and flood rice fields, resulting in severe crop failure.

Other exotic invasions have resulted in profound ecological change and economic loss on a huge scale. The epidemic of Dutch elm disease, caused by an invading fungus, that began in the 1950's has killed off millions of elm trees shading U.S. streets nationwide. And chestnut blight, another fungus that struck between 1905 and 1940, all but destroyed the American chestnut tree, once the most important forest tree in the Eastern United States. The tree grew from Maine to the Mississippi River. Because the wood resisted decay, it was used for everything from railroad ties and telephone poles to furniture and woodwork. Then, in the 1890's, chestnut blight entered the United States from China or Japan. It destroyed almost all American chestnut trees, and caused a ripple effect. Birds lost habitat, shade-loving plants died out, and shrubby vegetation moved in and took over.

Controlling invasions

Because the cost of exotic invasions is clearly high—in both an ecological and economic sense—scientists today attempt to control an invasion as soon as possible. Often, the first step is to import natural enemies from the species' home territory, a technique first used in the late 1800's. The cottony cushion scale, an insect native to Australia, had invaded California's citrus orchards and nearly destroyed the state's citrus industry. But in 1880, an insect taxonomist brought to California 12,000 parasitic insects called *Cryptochaetum* and 500 ladybird beetles from Australia. The beetle population exploded when it was introduced to such an abundant food supply, and initially, it controlled the scale. Although it had little effect at first, *Cryptochaetum* eventually became the major control agent.

Fortunately, neither *Cryptochaetum* nor the ladybird beetle has caused unexpected problems of its own. But biologists realize that the introduced predators are themselves invaders with the potential for causing environmental damage. To avoid this, scientists first study such predators in the laboratory before releasing them into a new habitat.

Many federal, state, and university researchers in the United States are currently undertaking such studies to discover which of a number of insect parasites might attack the sweetpotato whitefly, another exotic that invaded California fields and orchards in 1991. Biologists hoping to thwart the bug are in the process of determining if this whitefly is a new species or a strain of whitefly that invaded the United States from Europe and Asia in the 1890's. Certain parasitic wasps will attack many species of whitefly, but others are very selective, attacking only a few specific species.

Other types of control measures have been used to try to combat the Mediterranean fruit fly, popularly called the Medfly. The insect has attacked more than 200 species of fruits and vegetables, destroying crops worth millions of dollars. This invader from tropical west Africa first threatened crops in Florida in 1929. State authorities quickly ordered a quarantine—an attempt to isolate an area to make certain no insect pests could spread to other regions. Although quarantines are a relatively safe and economical means of controlling exotic pests, they are not always completely effective. Thus, several infestations occurred after 1929, both in Florida and Texas.

Then in 1975, the Medfly arrived in California. Since 1975, Medflies have invaded California 12 times and cost more than $170 million to destroy, according to state authorities. California's ongoing program to detect and track the Medfly includes placing thousands of traps across the state. The traps are baited with a sex hormone to lure the insects inside. If biologists determine that the Medfly has invaded, the California Department of Food and Agriculture uses three means of control. First, infested fruit is stripped from the trees. Second, the insecticide malathion is sprayed on the trees. Third, sterilized Medflies are dropped on the infested site, where they mate with wild populations. The eggs this mating produces are infertile, incapable of producing larvae, so the populations have difficulty reproducing.

Biological controls and quarantines are not always successful, however. Neither practice has prevented the Africanized honey bee from invading the United States. The aggressive Africanized honey bee, more commonly known as the "killer" bee, invaded southern Texas in October 1990. It poses economic problems for beekeepers of the Southern United States because it is an unreliable pollinator. It poses a health problem because each bee may sting people or livestock many times.

Devising a way to control the Africanized honey bee is complicated by several factors. First, its habit of burrowing into the ground makes locating it very difficult. And because it closely resembles the domestic honey bee, identifying the invader is not easy. Finally, the Africanized bee is able to interbreed with domestic honey bees, thus passing on its aggressive behavior to domestic species.

Stopping invaders at the border

Experts say that perhaps the most economical and environmentally safe means of controlling exotic invasions is simply to prevent organisms from reaching new areas in the first place. Because of its extensive agriculture industry, California has an ongoing inspection program to do just that. Anyone driving into California must stop at a checkpoint near the state border for a vehicle inspection. Fruits and vegetables with high potential for being infested with exotic pest species cannot be transported into the state.

The U.S. government has also passed laws to prevent exotic invasions from outside the United States. The U.S. Congress passed the Nonindigenous Aquatic Nuisance Prevention and Control Act, for example, in 1990. One of the first laws under this act went into effect in November 1992. As of that date, ships were required to discharge their ballast water in the open ocean before entering the freshwater rivers and lakes of the United States. The law's intent is to prevent the transportation of another species similar to the zebra mussel. The Brown

Tree Snake Control Act, signed into law in 1992, calls for eradicating the snake in Hawaii and for more research on its life cycle and habits.

Another government strategy involves creating public awareness of the dangers of bringing nonnative species into a new region. Hawaii, for example, perhaps the most remote islands in the world, has a particularly pressing need to educate the public. Scientists say the islands now have fewer native species than species introduced by human activity—from early exploration to modern tourism. The state has several programs aimed at controlling and preventing invasions, including displaying posters asking the public to report sightings of plants on a list of alien, noxious weeds. Officials have authority to eradicate any plant on the list. The state also produced a video to show on incoming airplane flights that depicts the harm exotic invaders have already caused on the islands. And in 1993, Hawaii was training dogs to sniff out animals and plants, including the brown tree snake, at airports.

Clearly, controlling exotic invasions requires public awareness as well as scientific research. People need to learn how damaging alien species can be and that individual travelers have a vitally important role to play in preventing new invasions. And scientists say they need funding for studies to better understand the ecological effects of an alien invasion in progress and to find ways to discover such invasions in their earliest stages. Then, biologists hope, scientists will be able to develop controls that target an invader before it does its greatest harm.

For further reading:

Ezzell, Carol. "Strangers in Paradise." *Science News*, Nov. 7, 1992, pp. 314-319.

Mooney, Harold A., and Drake, James A. "The Ecology of Biological Invasions." *Environment,* June 1987, pp. 10-15 and 34-37.

Questions for thought and discussion

Imagine that you own and operate a citrus grove in southern California that you inherited from your parents in 1992. Your parents had bought the grove in 1974, the year before the first outbreak of the Mediterranean fruit fly infestation. You remember how they struggled through 11 more infestations in the succeeding years, each threatening to wipe out the citrus crop and put them out of business. They saved their crops by spraying chemical pesticides. Now, your grove shows signs of sweetpotato whiteflies. You are planning to spray your trees with a chemical pesticide when an agricultural researcher asks you to take part in a field test of a beetle imported from Asia. The researcher explains that the beetle is a natural predator of a species similar to the sweetpotato whitefly. He cannot guarantee that the beetle will control your grove's whiteflies, but he says he has evidence that it probably will. If you agree to take part in the test, you must not use chemical pesticides this season.

Questions: What are the benefits and pitfalls of testing the beetle? What are the benefits and pitfalls of using chemical pesticides? What evidence would you ask the researcher to provide you with before you made your decision?

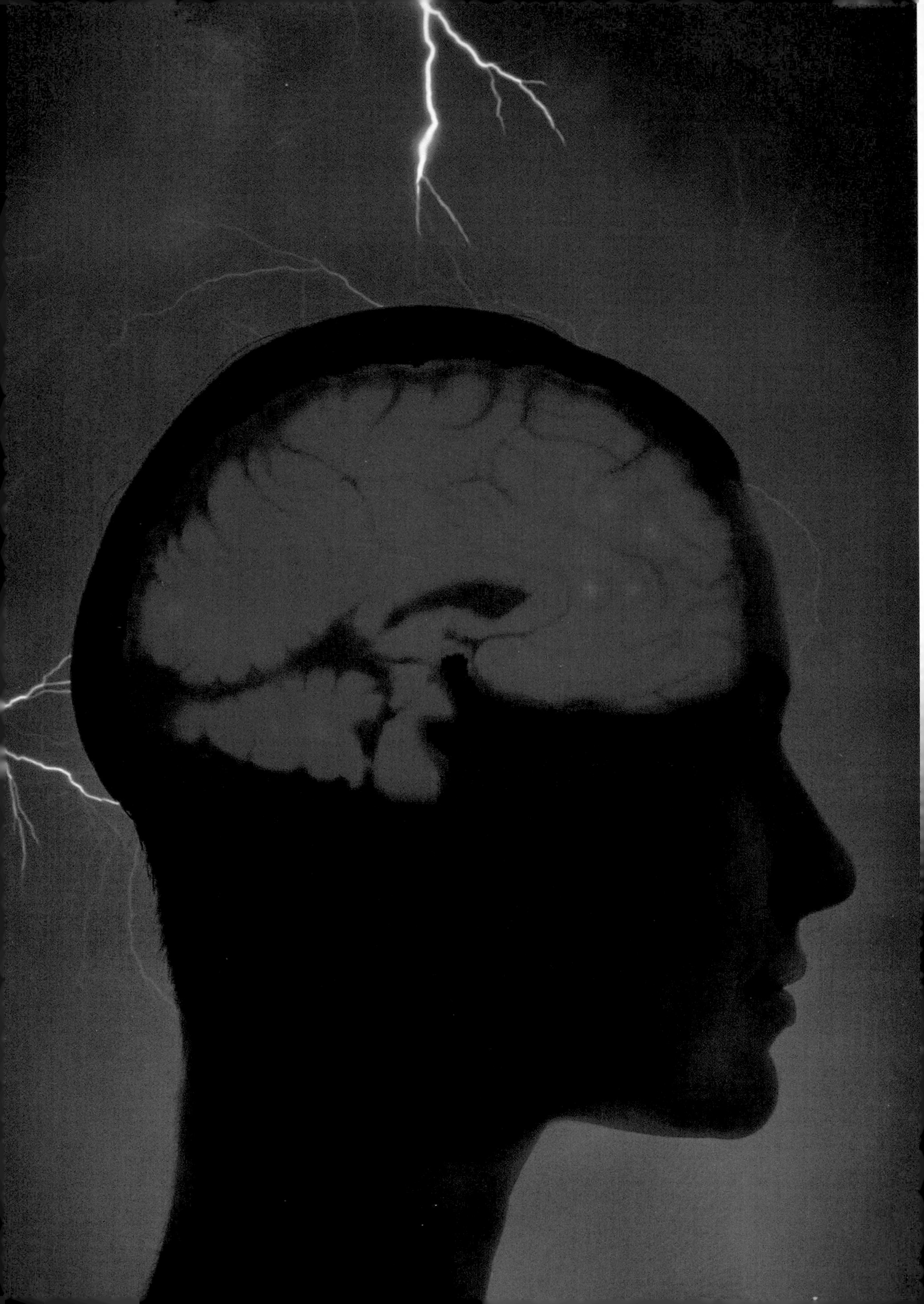

Techniques that can track brain activity are helping to expand our understanding of the body's most mysterious organ.

Mapping the Human Brain

BY YVONNE BASKIN

A woman lies on a hospital table, her head poking eyebrows-deep into the hollow cylindrical center of a huge machine that fills up most of the little room. She seems unaware of the equipment surrounding her as she stares at a computer screen directly above her. The screen flashes the names of familiar nouns, one per second. The woman's job is to say aloud the first related verb that comes to mind for each noun. For example, the word *car* prompts the woman to say *drive*, and *telephone* leads to *call*.

After the woman becomes at ease with the procedure, a technician injects a fluid into her bloodstream. The fluid is water bearing a special "tracer" form of oxygen. As the woman continues the word-association game, the active parts of her brain increase their blood flow and take up more of the "tagged" water. Meanwhile, detectors in the huge machine, called a positron emission tomography (PET) scanner, locate the parts of the brain where the tracer chemical ends up. The PET scanner sends this information to a computer in a nearby room, where researchers watch a colorful image of the woman's brain on a screen. In the image of her brain, various areas "light up" in flares of yellow and orange. These scientists in Michel Ter-Pogossian's PET laboratory at Washington University in St. Louis, Mo., are seeing the woman's mind at work.

Glossary

Cerebellum: The part of the brain, located below the back part of the cerebrum, that handles balance and physical coordination.

Cerebrum: The part of the brain that handles the most advanced mental tasks. It consists of two large halves called *hemispheres*.

Cortex: The layer of gray matter that covers the surface of the cerebrum.

Electroencephalography (EEG): A technique that measures the electrical activity of the brain.

Magnetic resonance imaging (MRI): A medical imaging technique that uses magnetic fields and radio waves to produce images of soft tissue.

Magnetoencephalography (MEG): An imaging technique that detects the magnetic fields produced by the brain.

Medulla: The part of the brain, at the top end of the spinal cord, that handles unconscious body functions.

Neuron: A nerve cell.

Positron emission tomography (PET): A technique that uses radioactive molecules to track the body's consumption of a particular substance.

The rise of brain mapping

Since the 1970's, PET and a handful of other multimillion-dollar technologies have given researchers their first chance to examine the workings of the human brain and not just its form and shape. The human brain—that 1.4-kilogram (3-pound) mass of soft tissue inside our skulls—contains 100 billion *neurons* (nerve cells) wired together in a spider web of connections that carry the pulses of our mental lives. Until recently, scientists had few practical ways to determine which clusters of neurons are responsible for recognizing a face, appreciating music, or understanding the meanings of words. With their new high-tech tools, however, researchers can record flashes and blips from the mental landscape, pinpointing the changes that accompany specific brain functions. The goal of all their research is to create a precise *functional map* of the human brain—a blueprint not of the different areas of the brain but of the different functions of those areas.

The prospects for discoveries are so exciting that the Congress of the United States declared the 1990's the "Decade of the Brain." And a committee appointed by the National Academy of Sciences (NAS) concluded in June 1991 that "neuroscience stands at the threshold of a tremendous opportunity to unlock the mysteries of the brain and its functions." However, said the NAS group, one giant obstacle stands in the way: No one was assembling a single map of the brain from the dizzying array of research.

So the group called for an ambitious 20-year Brain Mapping Initiative. The centerpiece of the project is the development of the National Neural Circuitry Database, a vast set of computer files about brain-function findings along with a three-dimensional computer model fitting together all the pieces of the puzzle. When complete, this atlas of the brain will help scientists understand how our brains gather information from the outside world through our senses; how our brains process, ponder, and react to the information; and how our brains respond to the information by triggering movement, speech, and emotion. Mapping efforts may also yield clues to the origin and treatment of medical conditions such as Alzheimer's disease, blindness, drug and alcohol addiction, schizophrenia, and other illnesses. Almost three dozen neuroscience teams worldwide are now working to achieve the NAS's objective.

The author:
Yvonne Baskin is a free-lance science writer.

The structure of the brain

Scientists have long known the physical layout of the brain. By the 1800's, anatomists had probed, dissected, and described most of the structures of the human brain portrayed in modern anatomy textbooks. Human beings, like other *vertebrates* (animals with backbones), have three major brain regions. One is the *brain stem*, the extension of the spinal cord that runs up into the skull. The brain stem includes the *medulla*, the "housekeeper" that oversees our heartbeat, breathing, digestion, and other body functions that do not require conscious control. Other parts of the brain stem are responsible for controlling eye movements and regulating body temperature. Lying behind the brain stem near the rear of the skull is the *cerebellum*. The cerebellum coordinates physical movement and deals with matters such as balance and posture.

Engulfing and dwarfing these two regions is the *cerebrum*, the organ-

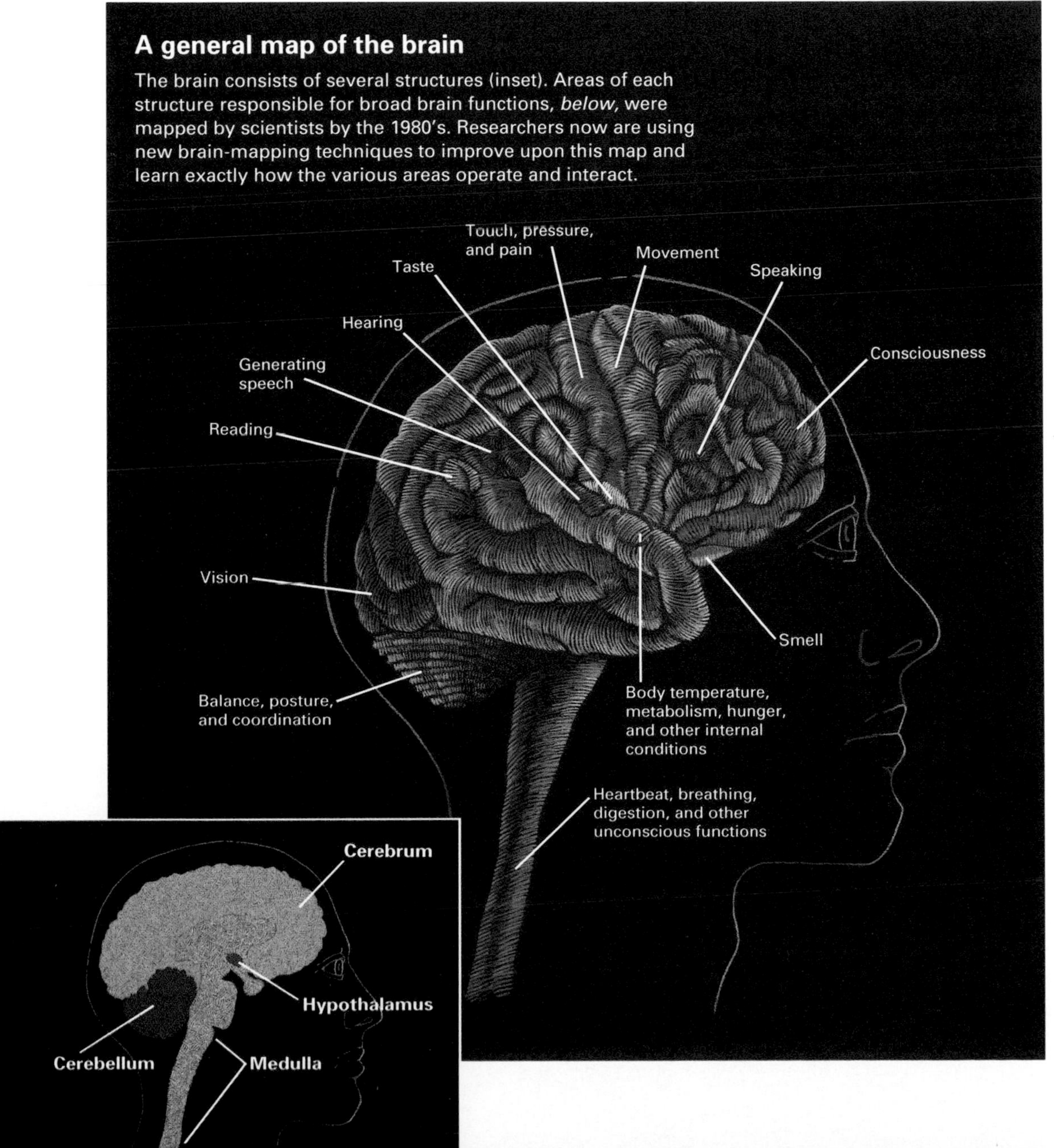

A general map of the brain

The brain consists of several structures (inset). Areas of each structure responsible for broad brain functions, *below,* were mapped by scientists by the 1980's. Researchers now are using new brain-mapping techniques to improve upon this map and learn exactly how the various areas operate and interact.

izer of the most complex mental tasks, such as interpreting vision and creating consciousness. The cerebrum constitutes about 85 percent of the human brain. It is divided into right and left *hemispheres* (halves), which are linked largely by a bundle of nerve cables called the *corpus callosum.* The crowning glory of the cerebrum is the *cerebral cortex,* a 0.6-centimeter (0.25-inch) layer of highly folded gray matter that covers both hemispheres like the rind of an orange. Most of the cerebrum underneath the cerebral cortex consists of nerve fibers connecting parts of the cortex with each other as well as with the cerebellum, brain stem, and spinal cord.

Early theories about brain function

People have pondered the question of which brain areas control which mental abilities for thousands of years. But only in the 1800's did physicians at last begin to pin down the functions of some brain areas. They reached their conclusions by studying patients who had suffered strokes or head injuries. In the 1860's and 1870's, for example, two neuroscientists, Paul Broca in France and Karl Wernicke in Germany, reported that patients with damage to their left cerebral hemisphere lost some of their language skills, whereas the language skills of patients with damage to their right hemisphere were unaffected. This research led to the idea, which persisted for almost a century, that the left hemisphere was the dominant half of the brain.

Then in the 1960's, neuroscientist Roger Sperry at the California Institute of Technology refuted that notion. Sperry worked with so-called "split-brain" patients—people who had undergone surgery to sever their corpus callosum in an effort to prevent seizures caused by *epilepsy,* a disorder involving electrical activity in the brain. By separately testing each half of a patient's brain, Sperry showed that the right hemisphere was just as complex as the left hemisphere and in fact was superior at mentally manipulating shapes and images. Unfortunately, Sperry's research gave birth to a mistaken notion of its own that people can be categorized as either "right-brained" or "left-brained," the former more creative and the latter better at using logic and language. Scientists using new brain-imaging techniques have since come to realize that all mental endeavors, including creativity and logic, require the efforts of multiple brain regions in both hemispheres.

Experiments on the brains of patients who have suffered strokes or head injuries still provide valuable information about brain functioning. The most direct insights come during brain surgery to relieve epilepsy in patients whose seizures cannot be controlled with drugs. This type of surgery mushroomed in the 1980's.

The brain itself cannot feel pain, so neurosurgeons can operate on patients who remain awake and alert and who can communicate their experiences to the doctors. Some such patients volunteer to take part in brain-mapping research. In a typical experiment, the patient says aloud the names of objects pictured on a screen while the surgeon stimulates various parts of the patient's exposed brain using an elec-

Tracking the brain's electrical activity

Brain cells communicate by sending and receiving electrical impulses. Using electroencephalography (EEG), researchers can monitor the brain's electrical activity and determine which areas control certain tasks.

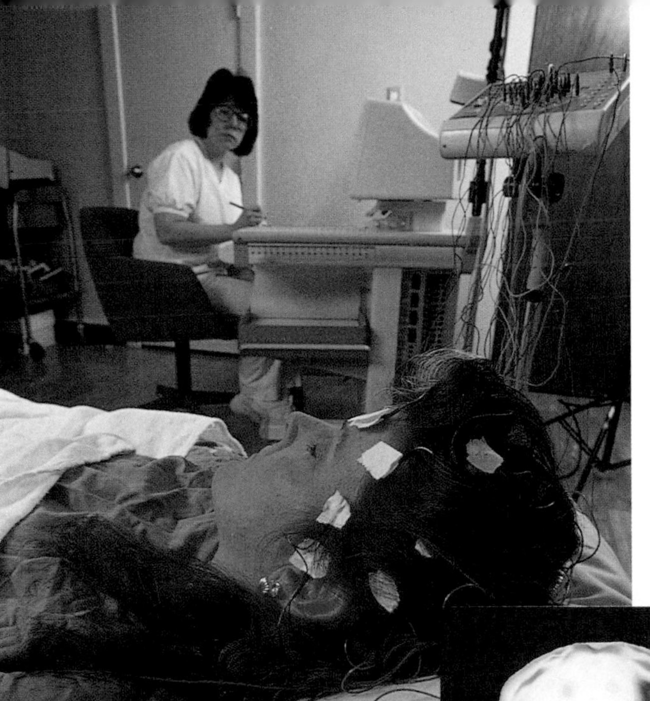

During electroencephalography, electrodes are attached to a patient's scalp, *left.*

Moving right index finger

Moving right middle finger

Moving left middle finger

EEG data superimposed on a computer image of a patient's brain show areas that are active when the person moves three fingers (red patches). The areas lie in the *motor cortex,* a strip of tissue on the top of the brain that controls physical movement.

trode. Neurons communicate with each other via electrical impulses, so small bursts of electricity from the electrode disrupt their normal coordination. For example, if the surgeon applies a weak electric current to an area of the brain that is crucial for language, the patient may temporarily become unable to give the name of an object they can otherwise identify. Surgeons can also stimulate neurons with a weak electric current that mimics a real nerve impulse. The patient may then "experience" a sensation or memory. The patient's responses not only help the surgeons locate the part of the brain responsible for the seizures but also add to the general body of knowledge about the workings of the human brain.

Images of working cells

Positron emission tomography (PET) detects blood being drawn into each brain area. Active areas draw more blood than do inactive ones, so a PET scan can help map brain activity.

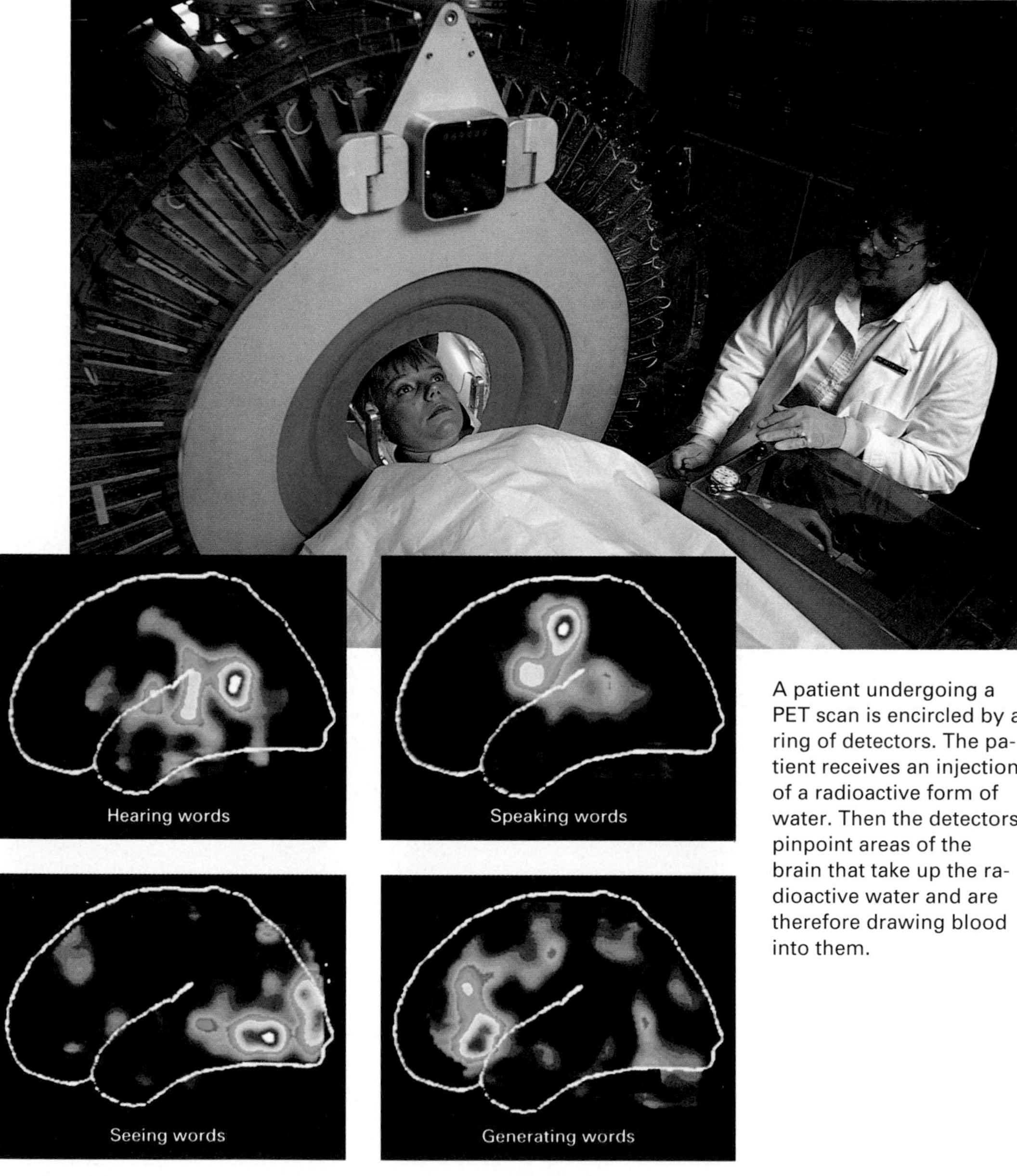

A patient undergoing a PET scan is encircled by a ring of detectors. The patient receives an injection of a radioactive form of water. Then the detectors pinpoint areas of the brain that take up the radioactive water and are therefore drawing blood into them.

Regions of brain activity light up in reds and yellows on PET scans during tasks involving language. Two different areas in the middle of the cerebrum are active when a person hears words and speaks them. Seeing words uses mostly brain cells in the visual cortex, at the back of the cerebrum. Generating words uses the front of the cerebrum, an area responsible for conscious thought.

Examining the brain without surgery

Nonsurgical brain-mapping techniques are becoming more and more helpful for brain mapping. These methods have been used in hospitals since the mid-1950's to help diagnose brain tumors and mental problems. By the 1980's, researchers had started to apply the technologies to study brain functioning in healthy people.

One such technique, the PET scan, was developed in 1972. The PET scanner identifies active areas of the brain by measuring the amount of blood being drawn into each brain area. To perform a PET scan, scientists inject a radioactive form of water into a patient's blood and then ask the patient to perform some mental task. Areas that are active during the task will soon take up more of the radioactive water in the blood than will inactive areas of the brain. Inside the neurons, the radioactive water emits subatomic particles called *positrons* that collide with nearby atoms and destroy themselves in tiny bursts of energy. The PET scanner detects these bursts from neurons throughout the brain and assembles them into a complete picture of the brain's use of fuel and thus the activity of the neurons.

Despite its usefulness, PET has drawbacks. PET scans are too blurry to make out small structures. Also, the scans cannot reveal the order of events among various parts of the brain, because changes in neural activity begin before the neurons take up more blood. Finally, the radioactivity poses a slight health risk to the patient, so to keep the exposure to a minimum, scientists cannot take more than a few pictures of a single patient's brain.

Another brain-mapping method is electroencephalography (EEG), which was developed in 1929. EEG uses electrodes pasted onto a person's scalp to monitor *brain waves.* Brain waves are slight electric fields produced in the space around the brain as a result of the electrical messages traveling between neurons. An EEG scanner monitors the changes in these electric fields in an effort to reveal the inner electrical workings of the brain as it performs some task.

EEG is still used today to diagnose epilepsy and strokes. However, its role in determining brain functionality is limited. The skull is a good insulator, so it dampens the brain's electric field. Consequently, EEG readouts tend to be very crude. An EEG scan can detect large brain abnormalities and can distinguish mental alertness from sleep, but it cannot detect neural activity with the precision needed to map the workings of the brain.

However, scientists in the 1980's used sophisticated computer equipment to develop a more sensitive form of EEG called event-related potentials (ERP). From the cloud of electrical activity sensed by EEG, ERP can isolate slight electric currents produced by neurons in response to specific mental tasks. ERP has proven far more valuable in mapping the brain than has EEG.

In the late 1980's, scientists also began constructing devices to measure the brain's magnetic fields. These devices take advantage of the fact that, like all electric currents, the signals produced by neurons generate both electric and magnetic fields. Although the skull dampens electric fields, magnetic fields pass through it unhampered. Techniques for tracking magnetic fields are called magnetoencephalography (MEG) or magnetic source imaging (MSI). These methods use sensors that are sensitive to extremely faint magnetic fields.

The main advantage of MEG or MSI over PET scans is speed. The magnetic field created by a firing neuron travels outward at the speed of light, so the magnetic detectors can follow changes in brain activity extremely rapidly—essentially as they occur. This feature allows researchers to track the order in which brain areas become involved in a mental task.

New techniques for the 1990's

Finally, since 1990, scientists have added two more mapping techniques to their list. Neurosurgeons George Ojemann and Michael Haglund at the University of Washington in Seattle have made actual photographs of the cerebral cortex at work using a new technique called *optical imaging.* Optical imaging was first applied to brain mapping in primates by neuroscientists studying the visual system. Doctors now perform this procedure, which requires exposing the brain, during epilepsy surgery. They shine near-infrared light, a form of electromagnetic radiation, on the exposed brains and ask patients to perform various mental tasks. Near-infrared light cannot be seen by human beings, so a special camera records the light being reflected off the brain. A computer then enhances the results so that scientists can view them.

The computer images show that certain regions of the cortex reflect more of the near-infrared light when the brain works on specific tasks. The increased reflection indicates a change in the chemical makeup of the brain tissue. Scientists are not yet sure whether the change is related to blood flow, oxygen level, or chemical activity in the neurons. Nevertheless, researchers think optical imaging holds great promise for brain mappers. It generates pictures of brain changes almost instantly, does not require applying electric currents to the brain, and offers high resolution. So neurosurgeons continue to explore and refine the technique.

The other technique developed in the 1990's is based on magnetic resonance imaging (MRI). Researchers created MRI in 1972 to provide images of "soft" tissue, such as organs, which cannot be seen with X rays. The technique takes advantage of the fact that the soft tissues of the body are rich in water.

To be scanned, a person lies inside the cylindrical heart of an enormous circular magnet. The magnet generates a magnetic field that causes the hydrogen atoms of water molecules inside the body to line up like little bar magnets. Generators in the tube then emit pulses of radio waves that rattle the hydrogen atoms like water waves rocking buoys on an ocean. When the radio waves subside, the hydrogen atoms realign themselves under the magnetic field. As they do so, they give off radio signals of their own. The MRI equipment can detect these signals and construct images of the soft tissues in the body.

The brain consists entirely of soft tissue, so MRI can be an effective mapping technique. In fact, of all the available brain-mapping methods, MRI provides the crispest and most detailed images. But ordinary MRI produces pictures only of brain structure. It cannot track changes in brain activity.

In 1991, however, a team of neurobiologists including Jack Belliveau and Kenneth Kwong at the Massachusetts General Hospital in Boston applied a modified form of MRI called "functional" or "fast" MRI to brain mapping. The scientists discovered that changes in blood flow and in the level of oxygen in the blood alter the radio signals of the hydrogen atoms in brain tissue. Because increased blood flow and reduced blood-oxygen levels indicate active neurons, scientists had found a way to use MRI to monitor changes in the brain. The Massachusetts researchers used functional MRI to map the areas of the brain related to vision, limb movement, and language.

Detecting the brain's magnetic fields

When brain cells transmit electrical impulses to one another, they create tiny magnetic fields. Magnetoencephalography (MEG) maps brain activity by tracking these magnetic fields.

An MEG scan reveals areas of the brain that become active when a person listens to a musical tone. (The circles show the active areas, and the arrows show the direction of electrical impulses when the image was made.) The active areas lie toward the back of the brain, above and behind the ears.

Researchers, *below,* position MEG scanners against the head of a patient. The scanners contain sensors capable of detecting extremely weak magnetic fields.

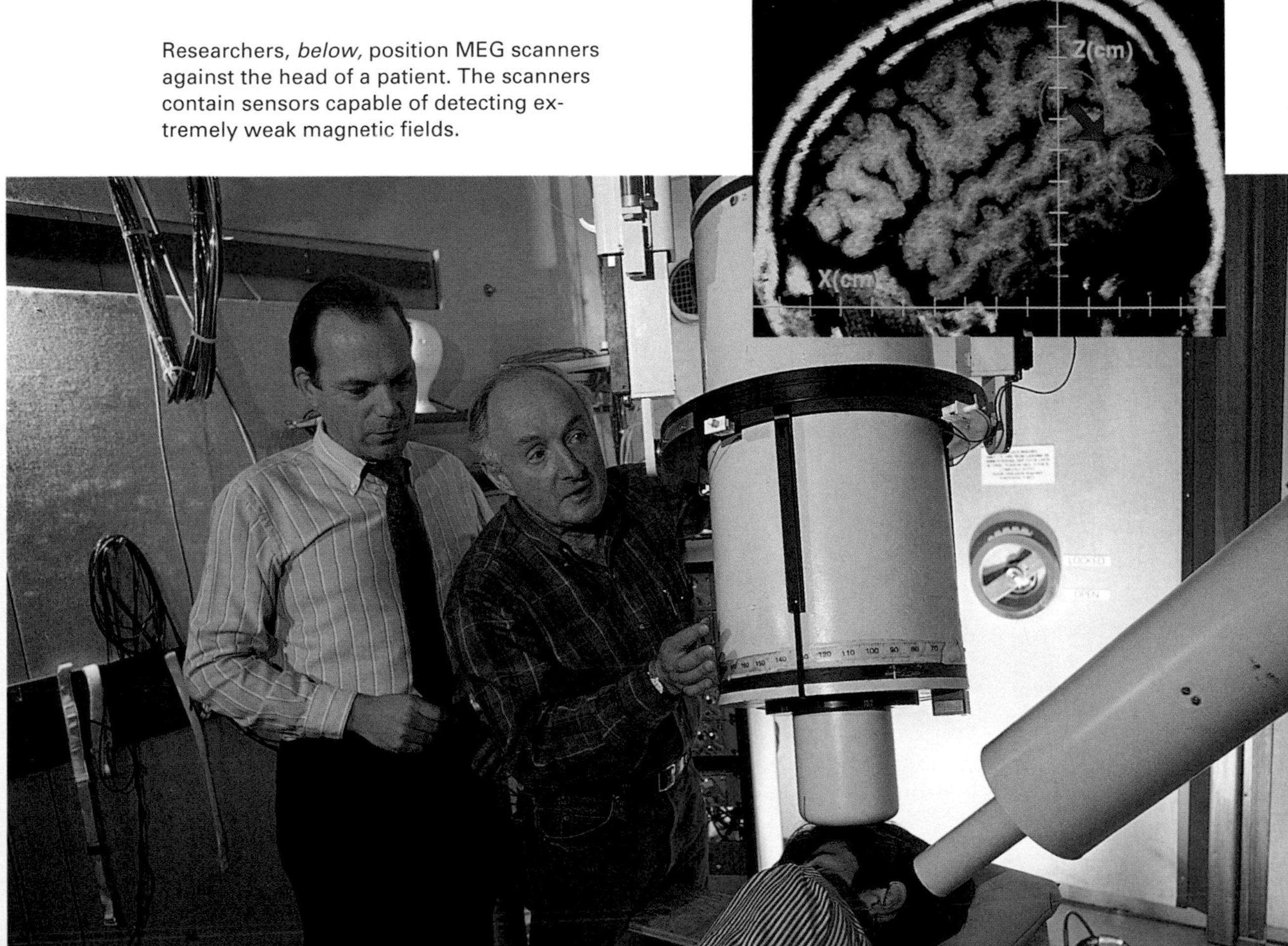

Optical imaging, a technique first used in human beings in the early 1990's, reveals a part of a patient's cortex near the ear that is active during speech (red and yellow). The image was made while the patient was undergoing surgery to treat epilepsy. After the patient's skull was cut open, surgeons shined light on the exposed brain. Changes in the amount of reflected light indicate areas of brain activity.

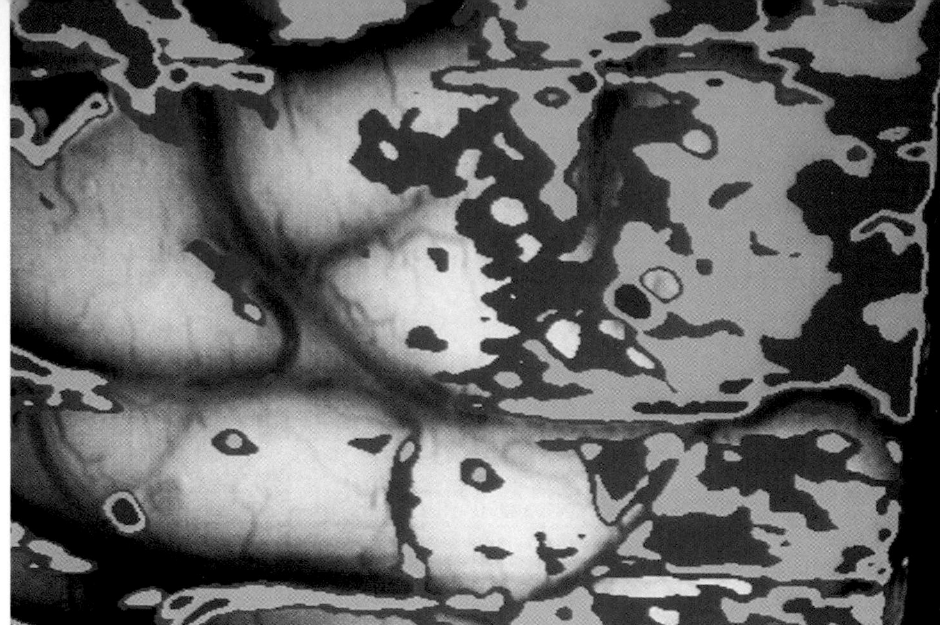

Mapping the senses

Other scientists have been using brain-mapping techniques to study the sense of hearing. Researchers at New York University in New York City in 1987 pointed an MEG scanner at the heads of volunteers listening to musical tones. The scientists discovered a region of the cerebrum just above the ear where bands of neurons that perceive the notes in the musical scale are arranged in a row, similar to the layout of the keys on a piano. Another region nearby perceives the loudness of a sound, with different sets of neurons responding to loud and soft sounds.

Scientists have also been reexamining our understanding of touch and movement. Previously, they had thought that the brain areas responsible for physical movement and for touch sensation consist of two 1.3-centimeter (0.5-inch) wide strips of tissue in the cerebral cortex running from one side of the head to the other. The strips are separated from each other by a groove that splits each hemisphere into the *frontal* (front) and *parietal* (rear) lobes. The strip running across the frontal lobes controls movement and is called the *motor cortex.* The strip on the parietal lobes controls sensation and is called the *sensory cortex.*

But in 1992, a team of neurosurgeons at Johns Hopkins Medical Institutions in Baltimore mapped the brains of 80 people with epilepsy and found a more complicated pattern. The neurosurgeons had opened each patient's skull and placed a plastic membrane studded with electrodes directly onto the brain in the hope of locating the area that was causing the patient's seizures. The scientists then sent an electric current through one electrode at a time and noted each patient's response, whether a sensation, an involuntary body movement, or something else.

The team learned that the regions responsible for sensation and movement could be as wide as 5 centimeters (2 inches) in some people, not just half an inch. And the two functions were not always strictly separated between the frontal and parietal strips. Furthermore, some patients reported sensations or experienced movement when the scientists stimulated regions of their brains outside their motor and sensory cortexes. The Johns Hopkins team's research made clear that scientists still have much to learn about how touch and movement are processed by the brain.

Clues to language processing

Language is the most sophisticated task of the human brain, yet scientists using brain-mapping techniques have learned that, like touch, it is not directed by a single processing center in the brain. After probing the brains of more than 100 epilepsy patients with electrodes, for example, Ojemann at the University of Washington found numerous grape-sized clusters of neurons necessary for language near but outside the primary language area at the back of the brain. And each person has a unique pattern of clusters. Other researchers have spotted such clusters lighting up on PET scans as healthy volunteers tried to understand various parts of speech.

Researchers believe the clusters are responsible for different aspects of language. Some clusters light up for adjectives, others recognize proper nouns such as *Fido,* and still others spot common nouns such as *dog.* Separate clusters handle rules of sentence structure and verb conjugation. For example, one cluster lights up when volunteers are asked to give the past tense of regular verbs such as *wait,* whereas another cluster handles irregular verbs such as *go.* In addition, Ojemann found that the clusters responsible for a person's first language tend to be more tightly grouped than are those for a second language. This difference in organization may explain why people can recall words more quickly for their first language than for their second language.

Tracking down consciousness

Studies showing widely scattered brain functioning have led neuroscientists to ponder the larger issue of consciousness—how our minds assemble the activity of various brain areas into the seamless whole that we perceive as our thoughts. Researchers have found no "television screen" in the brain where neurons direct all the fractured bits of a memory or perception. But scientists have developed many theories for how the firings of billions of brain cells can generate comprehensible thought.

One idea belongs to neuroscientist Antonio Damasio of the University of Iowa College of Medicine in Iowa City. Damasio theorizes that some of the areas that show up on brain maps as active during a task do not handle a specific category of information but are instead what he calls convergence zones. According to Damasio, neurons in these zones are responsible for taking note of activity in distant regions of the brain and integrating the scattered information to produce conscious thoughts.

Monitoring the blood in the brain

A technique called "functional" magnetic resonance imaging (MRI) can pinpoint areas in the brain where blood flow is increased and blood-oxygen levels are reduced. Both are indicators of active cells.

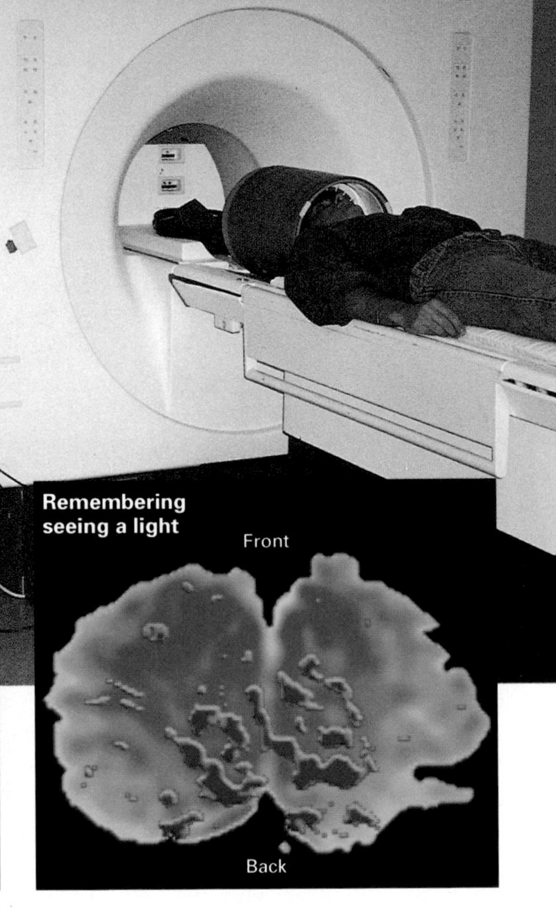

A person waits to be inserted into a device for functional MRI. The machine detects radio signals emitted by hydrogen atoms in water molecules in the brain. Changes in blood flow and blood-oxygen levels alter the radio signals.

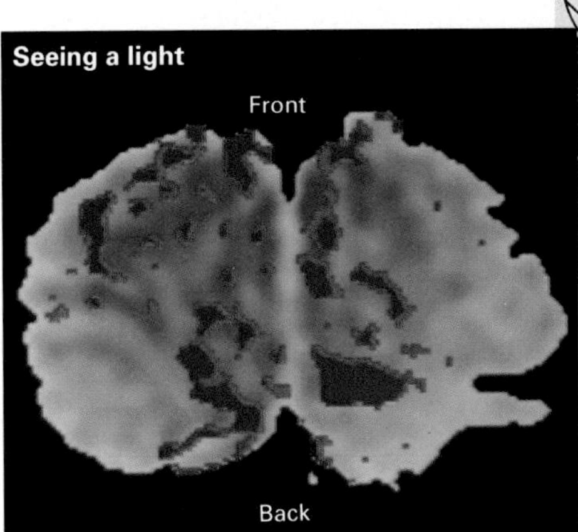

Areas of the brain that are active when a person sees an image and then remembers seeing it appear as red splotches in functional MRI scans. The two tasks use many of the same areas, especially in the region that processes vision at the back of the brain.

Damasio bases his idea on experiments performed by him and his neuroscientist wife, Hanna Damasio, in the 1980's and 1990's. They studied the brain damage of patients by putting the patients into a PET scanner and flashing pictures of famous people on a computer screen. Groups of neurons involved in recalling the names of things showed up as active on the PET scans of all the patients. However, some patients said they did not recognize the people. Because the PET scans revealed that the pictures triggered memories, Damasio blames not damaged memories for the patients' failure to identify people but faulty convergence zones. He believes that, in these people, the neurons that hold the names of famous people are intact, but that damaged neurons in a convergence zone keep the names from being connected with the faces. The Damasios' experiments suggest that our brains possess separate convergence zones for many different categories of things, such as the names of people, animals, and artificial objects.

Another conscious brain activity under study is the act of paying attention. In one experiment conducted in 1987, the researchers at New York University played two different tones at the same time but asked volunteers to pay attention to only one of them. An MEG scanner detected two signals from the part of the brain—above the ear—that is responsible for processing sounds. However, the act of paying attention appeared to have "turned up the volume" of one of the tones so that the MEG scanner picked up a stronger signal from the neurons responsible for recognizing that tone. The MEG scan also showed that the neurons boosted their signals mere fractions of a second after the volunteers heard the tone, before they had time to consciously recognize it. This finding contradicted a belief held by some neuroscientists that people have to recognize what something is before they can pay attention to it.

A different mechanism for mental attention showed up in 1991 in studies by neuroscientists Peter T. Fox at the University of Texas Health Science Center in San Antonio and Marcus Raichle at Washington University. Fox and Raichle asked volunteers to stare straight at a dim dot on a computer screen and alert the researchers if the dot dimmed. The dot never dimmed, but the mental work of directing attention to the task lighted up areas in the right frontal and parietal lobes. PET scan images of the volunteers' brains showed activity among some neurons that could not be attributed to the processing of sight. Instead, the scientists say, these areas allow a person to become more sensitive to an anticipated signal amid a din of other sights and sounds—giving us the ability to listen for a microwave oven timer to go off while watching television in another room, for example, or to "keep an ear open" for a parent's call to dinner while playing outside.

Researchers agree that they have a long way to go before they can present a final view of how the brain works. However, new brain-mapping techniques such as optical imaging and MRI will help clear the path, because researchers can now test their ideas using the images produced by several different types of detectors. And neuroscientists expect researchers to devise even more mapping techniques before the puzzle is fully solved. All in all, say neuroscientists, the most exciting frontier for scientific exploration in the coming years may turn out to be the one inside our heads.

For further reading:

August, Paul Nordstrom. *Brain Function.* Chelsea House, 1987.
Hooper, Judith, and Teresi, Dick. *The Three-Pound Universe.* Dell, 1987.
Metos, Thomas H. *The Human Mind: How We Think and Learn.* Watts LB, 1990.

Engineers are beginning to make buildings and other structures that monitor themselves and their surroundings and alter their shape to deal with changes—all without human intervention.

Designing "Smart" Structures

BY IVAN AMATO

The pilot steers the jet onto the runway and revs the engines. As the craft picks up speed, its wings slowly change shape in order to generate the lift needed to get the plane airborne. But there are no flaps or fins on these wings. Instead, the wings themselves distort, curves appearing in their flexible plastic surfaces to create the proper aerodynamics. There is a sudden wind gust, but it does not concern the pilot. The gust, along with changes in air pressure and temperature, merely cause the wings to adjust their shape. The jet is poised to leave the ground when a warning message flashes on a computer screen in the cockpit. The message reports a slight fracture is forming in a section of the left wing near the engine. The pilot aborts the take-off and returns the jet to a hangar for repair.

Airplane wings like this—capable of monitoring their own structural condition and adopting different shapes as situations demand—were under development in 1993 at research centers such as the Massachusetts Institute of Technology (M.I.T.) in Cambridge. They are but one example of a growing wave of *smart structures*—manufactured creations that can, with a stretch of the imagination, be called "intelligent." A global community of scientists and engineers say they are taking the first steps toward stocking the world with structures as helpful, responsive, and efficient as these self-monitoring airplane wings.

Glossary

Actuator: A component that creates movement in a structure.

Controller: A component that interprets information from sensors and sends appropriate commands to actuators.

Electrorheological (ER) fluid: A liquid containing many tiny particles that stiffens into a gel in the presence of an electric field.

Optical fiber: A thin tube of glass that conducts light.

Piezoelectric material: A substance made of layers of electrically charged molecules that converts physical pressure into electrical signals and vice versa.

Polymeric gel: A substance composed of long, thin molecules that turns from runny to stiff in the presence of an electric field.

Sensor: A component that monitors the conditions within a structure or its surroundings.

Shape-memory metal: A metal that takes one of two forms, depending on its temperature.

The capabilities of smart structures

Of course, smart airplane wings and other smart structures are not intelligent in the way human beings are. People can react to an enormous number of situations, even ones they have never experienced before. By contrast, smart structures can be designed and programmed to perceive only certain conditions and to make only a certain number of responses.

Craig A. Rogers, director of the oldest United States academic center devoted to smart structure research—the Center for Intelligent Material Systems and Structures (CIMSS), founded at the Virginia Polytechnic Institute and State University in Blacksburg in 1987—says the goal of smart structures research is to "mimic the materials and structures in biological organisms." Researchers hope that ultimately smart structures will share three general components with people: "nerves," "brains," and "muscles."

For "nerves," smart structures will have *sensors*—detectors capable of providing information about the structure and its environment. Sensors can be either attached to the outside of the structure or incorporated directly into the structure during its manufacture. The sensors will send their findings to *controllers*—tiny electronic "brains" like the microchips that provide the thinking power for computers. These controllers will interpret the data from the sensors and select a course of action. They will then transmit appropriate instructions to *actuators*—mechanisms that, like the muscles in the human body, have the ability to alter the shape or operation of the structure. Carefully designed systems of these sensors, controllers, and actuators—known collectively as *smart materials*—will make the new structures seem "smart."

The rise of a new field

Most of the materials being used as sensors and actuators have been around for many years. However, their full potential began to emerge only in the late 1980's, when researchers in disciplines such as chemistry, physics, biology, and engineering started putting their heads together. The first scientific conference on smart structures and materials was held in 1988, and the first scientific journals devoted to the subject sprang up in 1989—unmistakable signs that a new branch of scientific research had been born.

The United States was the first country to establish research centers for smart structures. The most well-known site is the CIMSS. Centers at the University of Michigan in Ann Arbor and the University of Vermont in Burlington have since entered the arena, along with private

The author:
Ivan Amato is a staff writer for *Science* magazine.

firms in the aerospace, defense, and construction industries. After the United States, Japan has been the most aggressive country to pursue research in the field, though scientists and engineers in European nations, Canada, and Australia are also working on smart structures.

Why build smart structures?

The researchers hope their engineering feats will make the world safer. Today, a single flaw in a skyscraper, superhighway, or jumbo jet could result in injury or death to hundreds or thousands of people if the structure collapsed or the plane crashed. Building a kind of self-awareness into these structures would reduce the chance of disaster and cut down on the costly need for human inspection and maintenance. Some smart structures may even be able to monitor themselves in ways that people cannot. For example, construction workers cannot easily determine whether there are cracks or corrosion inside a concrete wall, but a slab equipped with self-monitoring sensors could.

Smart structures could also soften the blows of natural disasters. For example, researchers in 1989 unveiled an 11-story building in Tokyo that can brace itself during an earthquake to prevent its becoming damaged. Other researchers are considering the design of a smart building that could sense a high wind and then tilt itself slightly into the wind to better withstand the force of the gale.

Finally, smart structures could help engineers use less raw material in their designs. Today, for example, engineers add tons of "dumb" steel to reinforce concrete structures such as bridges in order to ensure that the structures can withstand almost every conceivable environmental change or stress. Then, human workers must monitor concrete for cracks, which are often spotted only after they have become serious problems. Lighter designs featuring smart materials that activate strengthening components only where and when they are needed might be able to do the same job for less money. Engineer Carolyn Dry of the Architecture Research Center at the University of Illinois in Urbana-Champaign and her colleagues are developing a concrete with components that sense when cracks form and automatically repair the defects. The researchers created a concrete laced with adhesive-filled fibers. When a crack forms, the movement of the concrete splits open some of the fibers, spilling adhesive into the crack and preventing it from widening further.

Nerves of glass and light

A favorite material among researchers for the nerves of smart structures are *optical fibers*—hair-thin glass tubes that carry light. The fibers are clad in a material that keeps the light from leaking out. Optical fibers can be used to detect structural changes or problems in a host of smart structures. In the simplest application, a network of fibers runs along the outside of a structure. A stress on the structure that produces a crack in a vital support would also cause an optical fiber to snap, and

A smart plane's "nerves" of light

To create sensitive "nerves" for structures such as airplane wings, many researchers are experimenting with *optical fibers,* thin glass tubes that transmit light. The fibers can be used to sense changes such as dangerously increasing pressure.

light shining from the broken fiber would serve as a beacon to the site of the crack for human technicians.

But optical fibers can perform a more sophisticated role when they are embedded directly into the skeleton or skin of the structure. Scientists know that changes in temperature, pressure, and vibration in the surrounding material may alter the characteristics of the light traveling inside the fibers—for example, the light might become less intense. Electronic controllers monitoring the light as it exits the fibers would detect the difference and thus be tipped off to internal changes. In this way, optical fibers can produce an intimate picture of the internal health of a structure.

Engineers also can modify the sensitivity of optical fibers by physically weakening them before embedding them in a structure. More critical areas of a structure can thus be planted with fibers that are inherently more susceptible to damage or stress.

Pressure detectors

Another popular class of sensors are *piezoelectric* materials. *Piezoelectric* is a term derived by combining the prefix *piezo,* meaning *to press,* with the word *electric.* Piezoelectric materials, which typically take the form of thin films or plates, can act like the nerves in a person's fingers, responding to pressure with small electrical impulses. Inside the material, the molecules line up so that the positive electric charges face one direction and the negative electric charges face the other. When pressure is applied and the two sides of the material are pushed closer together, the approach of the positive and negative charges generates an

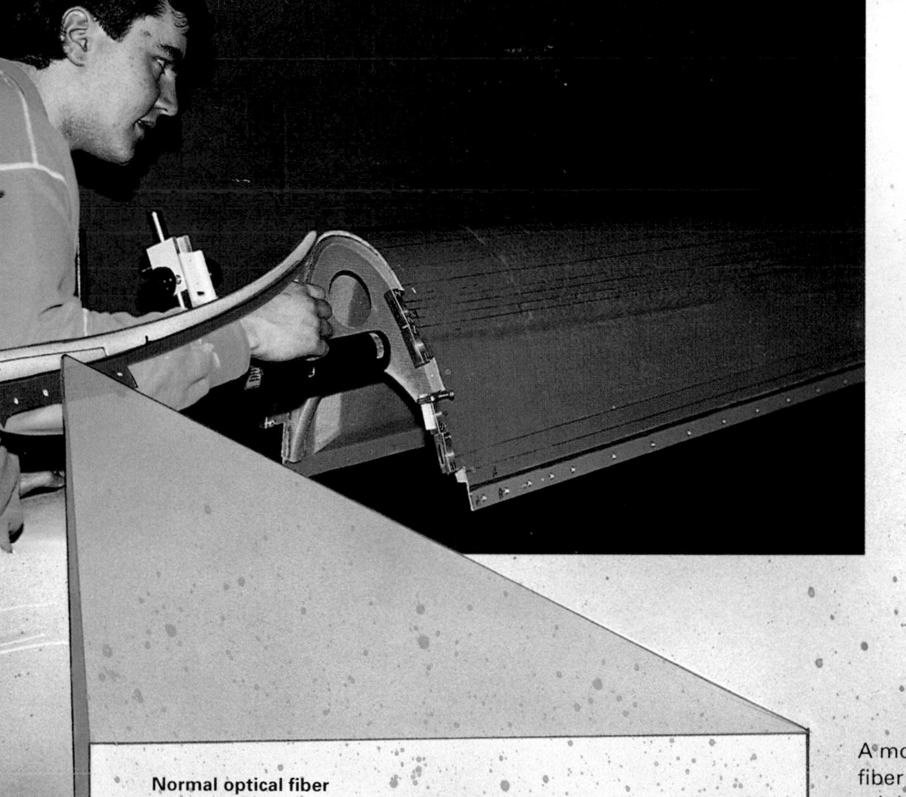

In a model airplane wing containing a simple optical fiber sensing system, light traveling through rows of embedded optical fibers is visible as red lines. Stress on the wing has caused some fibers to break open, allowing light to bleed to the surface as red spots. In a real wing, such spots could help workers find the damaged areas.

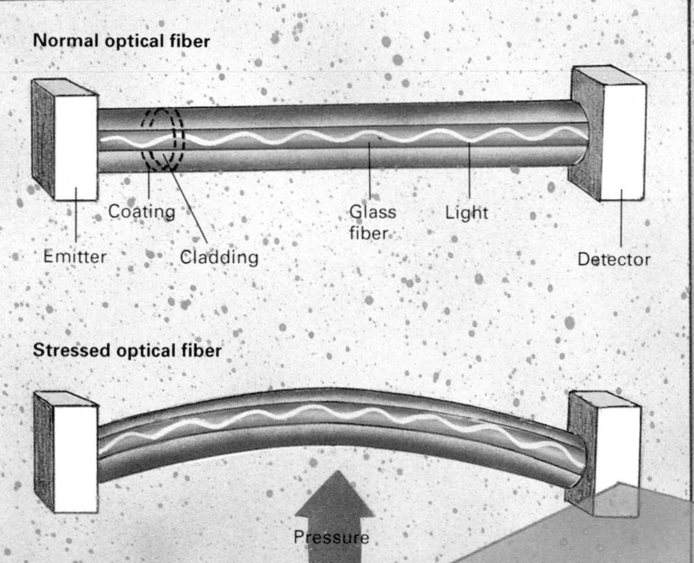

A more sophisticated optical fiber sensing system, *left,* might consist of a light-emitting device, such as a tiny laser; a light detector; and one or more optical fibers (top). The glass fiber is surrounded by opaque cladding and coating to keep the light from "leaking out" the sides. Pressure on the wing would bend the optical fiber (bottom) and thus create subtle changes in the light emerging at the detector.

Light travels through the fiber as a wave, *right,* with crests and troughs (top). If pressure on the wing causes the fiber to stretch slightly, the light will have a greater distance to travel. It therefore emerges from the fiber at a later-than-normal point along its wave (bottom). The detector monitoring the light exiting the fiber sends this finding to a computer, which analyzes the data to determine how much stress the fiber is undergoing.

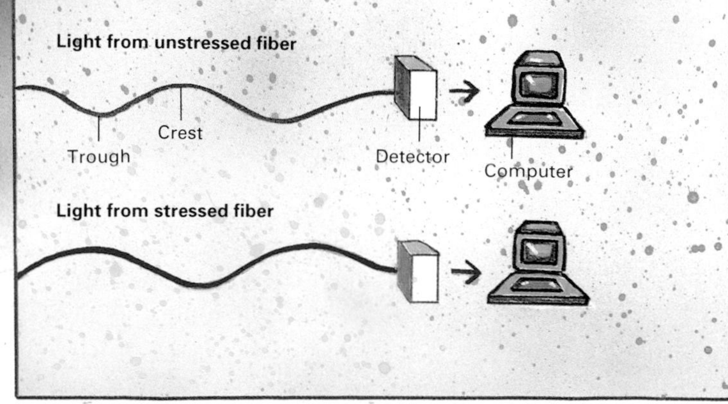

electric signal. The signal can be relayed to a controller along wires attached to the piezoelectric material.

Piezoelectric materials can also perform the job of actuators by turning electrical signals into physical movement. When an electric current runs through a metal wire or plate, an electric field is generated. A piezoelectric material inside this field will expand or contract as the field tugs on the positive and negative electric charges on the material's two faces. Scientists have created piezoelectric actuators that respond to electrical fields with movement in various directions.

One use of piezoelectric materials being investigated by military researchers is a smart hull to enable submarines to evade enemy sonar. Sonar is a method of detecting objects underwater by sending out sound waves and then analyzing the waves reflected off any objects in the area, such as a submarine. Military engineers hope to defeat sonar by placing a thin coating of piezoelectric materials on a submarine's hull. Sound waves traveling through the water and reaching the submarine will register on the piezoelectric coating as slight changes in water pressure. These changes will cause the piezoelectric materials to generate electrical signals, which will be transmitted to microprocessors located in the hull. The microprocessors will instantaneously analyze the signals and then relay electrical orders back to the piezoelectric materials. These signals will cause the material to vibrate in such a way as to create new sound waves that are the opposite of the waves being reflected off the hull. The effect is that one set of waves cancels out the other and no echo reaches the enemy ship from the submarine. By effectively blocking sonar, the sub would be far stealthier than any operating today.

Shape-shifting metals

Piezoelectric actuators respond to electrical fields quickly, but they cannot exert much force. When engineers desire strength rather than speed, they often turn to *shape-memory metals.* Shape-memory metals take two different forms, depending on the material's temperature. For example, below a certain temperature, a wire made of shape-memory metal may be shaped like a coil. But when the wire is heated above that temperature, the coil slowly straightens out or takes some other shape. If the temperature drops back below the critical temperature, the wire "remembers" its first shape and coils up again. These movements occur because the heat changes the arrangement of atoms or molecules in the metal. Scientists refer to this kind of transformation as a *phase change.*

Many phase changes are familiar, such as those that take place between ice and water at 0 °C (32 °F) and between water and steam at 100 °C (212 °F). A shape-memory metal's change from one shape to another is simply a more unusual example of a phase change. By adjusting the composition of the elements in the metal, scientists can cre-

A submarine's sonar watchdog

To detect or dampen vibrations in a smart submarine or other structure, researchers are exploring the use of *piezoelectric materials.* These materials convert physical movement into electric signals and vice versa.

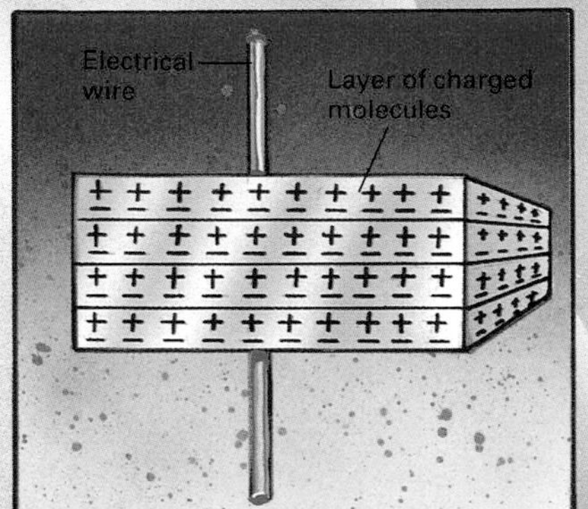

A piezoelectric material is made up of layers of electrically charged molecules. In each layer, the positive and negative electric charges line up on opposite sides. Physical pressure pushes the positive and negative charges closer together, and the movement of the charges creates an electric current in an attached wire. A current sent along the wire has the reverse effect. The current creates electric fields throughout the piezoelectric material that push the positive and negative charges together and apart.

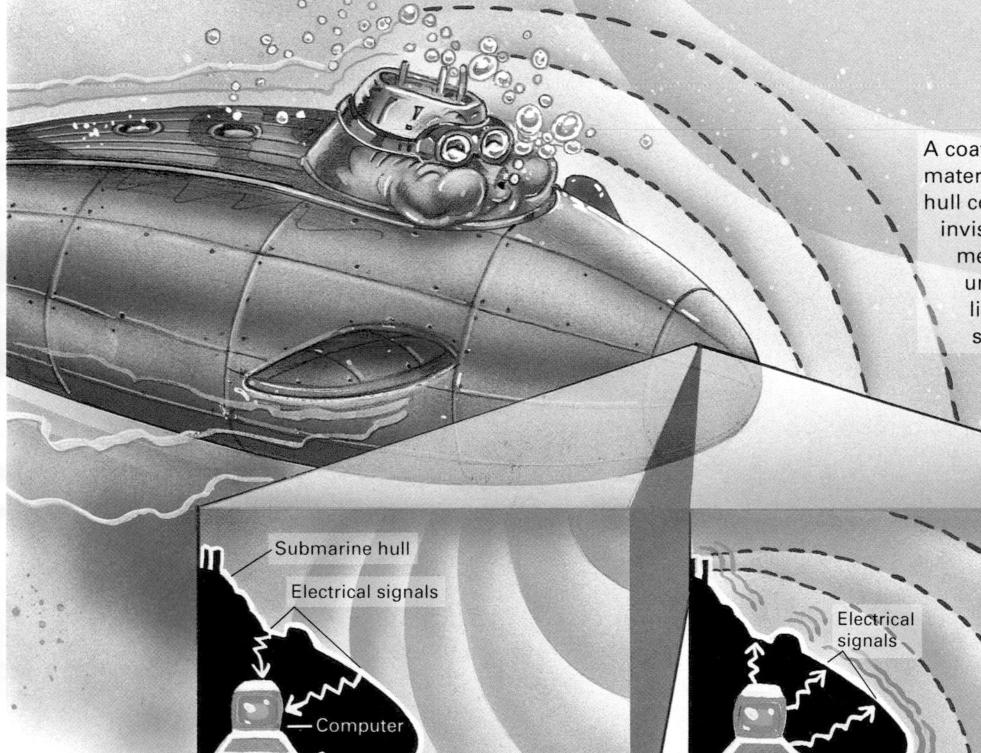

A coating of piezoelectric materials on a submarine's hull could make the sub invisible to *sonar* (the method of detecting underwater objects by listening for reflected sound waves).

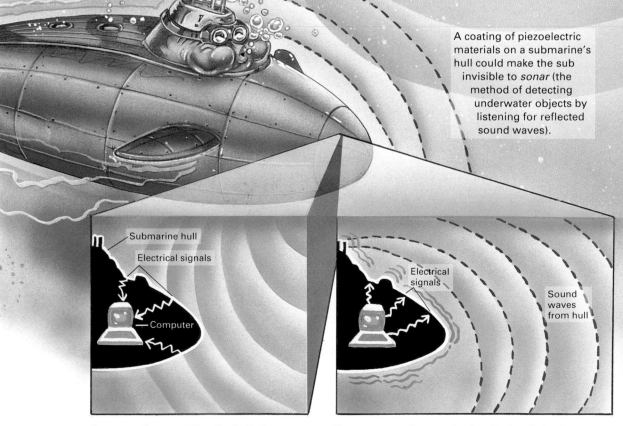

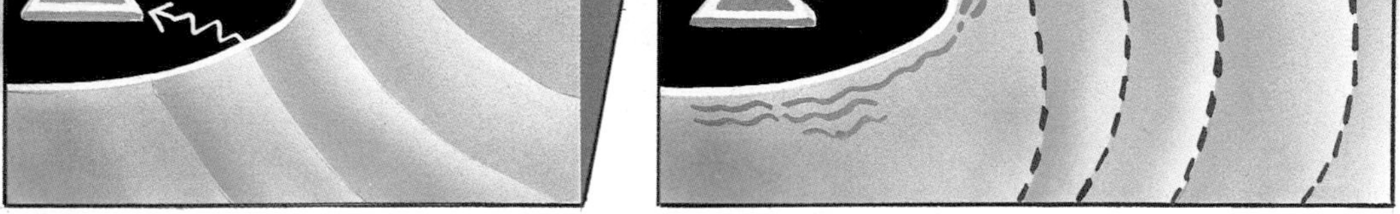

As a sound wave strikes the hull, the coating transforms the pressure of the sound waves into electric signals. These signals travel to a computer, which analyzes them.

The computer then sends electric signals back to the piezoelectric material, causing the hull to vibrate in such a way that it creates sound waves that are exactly opposite the waves being reflected off the hull. The two sets of waves cancel each other out, enabling the submarine to move through the water without producing a sonar echo.

ate shape-memory metals that change shape at different temperatures.

One of the most popular shape-memory metals, Nitinol, is a compound of the elements nickel and titanium. Researchers like Nitinol because it resists corrosion, because it remains sturdy after repeated movement, and because it can be heated simply by passing electricity through it.

A shape-memory metal embedded in a material can function as an actuator. When it is heated or cooled past its critical temperature, it attempts to assume its other shape, exerting a strong force on the surrounding material. This force causes the structure to move in a particular direction. Researchers program controllers to selectively heat and cool the shape-memory metals throughout the structure in order to bring about a range of motions.

Researchers use shape-memory metals to move large, heavy structures. For example, engineers are considering embedding rods of shape-memory metal into the supports of bridges to help prevent them from swaying during storms. A computer that monitors meteorological activity could heat the rods when it detects a storm moving through the area. The rods would become more rigid, bracing the structure. When the storm passed, the computer would turn off the heating elements, and the rods would return to their previous shape.

Fluids that turn into gels

Certain special fluids and gels can also serve as actuators. An *electrorheological (ER) fluid* changes from a runny fluid into a stiff, almost solid substance in the presence of an electrical field. An ER fluid consists of fine particles suspended in a liquid such as silicone oil. When subjected to a strong electric field, the particles come together to form a network of filaments, and the material stiffens into a gel. When the electric field is removed, the filaments break down, and the material becomes runny again. Similar in behavior to ER fluids are *polymeric gels,* in which the electric field turns runny clumps of *polymers* (long chainlike molecules) into stiff filaments.

ER fluids and polymeric gels do not change shape as forcefully as shape-memory metals do. But researchers can make different mixtures designed to expand or contract in response to different stimuli. These stimuli include not only electric fields but also such factors as temperature, acidity, and light levels. These materials can thus play roles in many different smart structures.

Researchers at the CIMSS and elsewhere are developing smart buildings that use ER fluids to help them survive earthquakes. A thin layer of ER fluid would lie between the base of the building and the ground. Normally, the fluid would be in the presence of a strong electric field and so would take on a stiff form, essentially holding the building to the ground. But when an earthquake began, the building would decrease the electric field surrounding the fluid so it would change to a looser, more slippery state. When the ground slid back and forth during the quake, this slippery layer would prevent the building from mov-

A bridge's metal "muscles"

The "muscles" of bridges and many other large smart structures under development consist of *shape-memory metals.* A rod or other part made of these metals will have two different shapes, depending on its temperature. When the part changes from one shape to another, it exerts a strong force on the structure in which it is embedded.

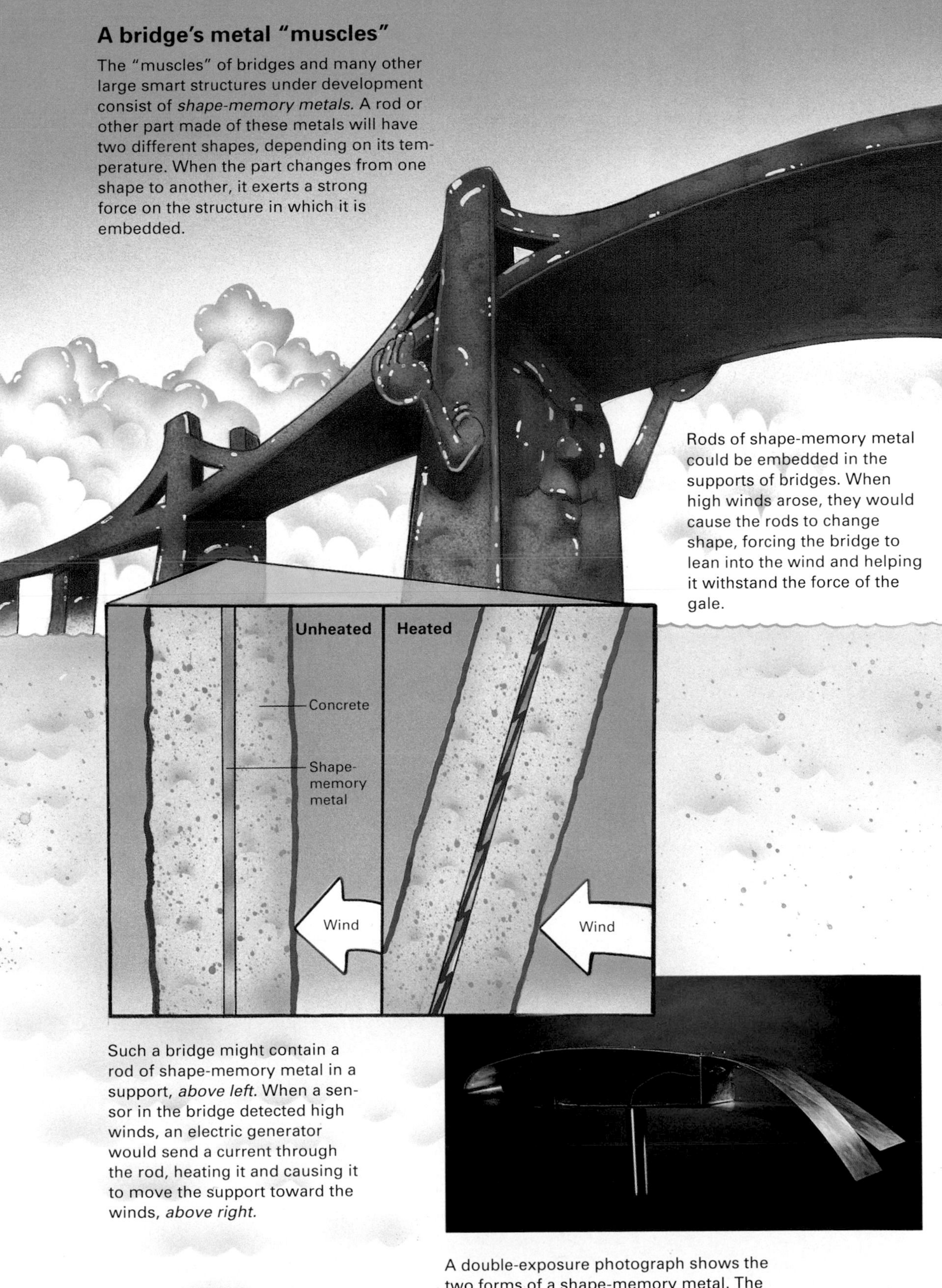

Rods of shape-memory metal could be embedded in the supports of bridges. When high winds arose, they would cause the rods to change shape, forcing the bridge to lean into the wind and helping it withstand the force of the gale.

Such a bridge might contain a rod of shape-memory metal in a support, *above left.* When a sensor in the bridge detected high winds, an electric generator would send a current through the rod, heating it and causing it to move the support toward the winds, *above right.*

A double-exposure photograph shows the two forms of a shape-memory metal. The metal is slightly curved at room temperature but takes a more bent shape when heated.

A building resting on fluid "feet"

Electrorheological (ER) fluids could be used in smart structures such as earthquake-resistant buildings. ER fluids normally take a runny form. But in the presence of an electric field, the fluid stiffens into a gel.

An ER fluid consists of fine particles suspended in a liquid. When the fluid is surrounded by an electric field, the particles in the fluid link into strands, stiffening the mixture into a gel, *right.* When the electric field is removed, the particles lose their rigid arrangement, and the fluid becomes runny, *far right.*

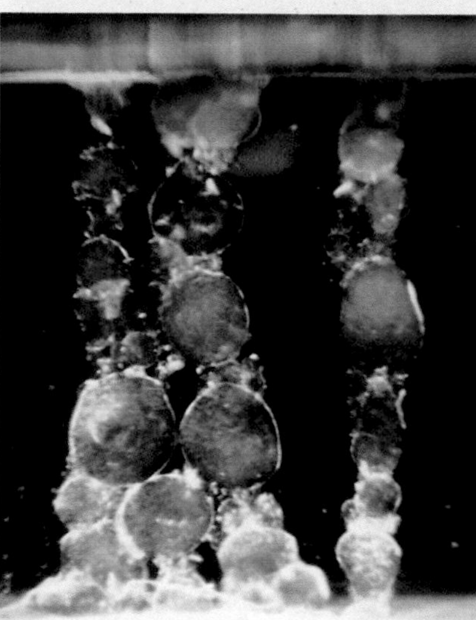

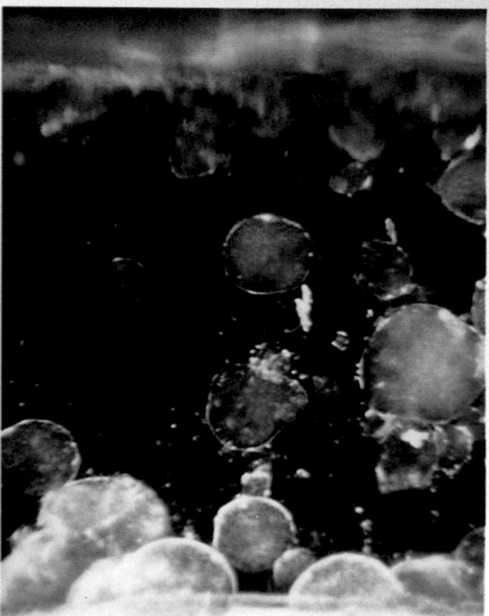

ing with the ground. As soon as the quake ended, the building would increase the electric field and change the fluid back into its stiff form, again anchoring the building to the ground.

Building the first smart structures

With a variety of sensors and actuators from which to choose, and with growing expertise in assembling these materials into complex structures, engineers are pursuing a wide range of projects. Air Force researchers were among the first to explore the field of smart structures, so it is no surprise that smart aircraft designs are among the most developed. One version could actually repair the type of defects that M.I.T.'s smart airplane wing detects. The wings would contain a network of optical fibers, and an on-board computer would monitor light traveling through them. If a crack forms in the wings, the computer detects it as a break in the beam of light from one of the fibers. The computer then sends electrical signals to actuators near the broken fiber. The actuators respond to the signals by pushing small sections of the wing together, sealing the crack. The computer maintains the electrical power to the actuators until human inspectors can make permanent repairs.

Aerospace engineers are also developing smart structures for use in space. Smart structures would be especially useful as satellites. As the weight of a satellite increases, so does the cost of its launch, making smart technology potentially cheaper than heavy reinforcements for

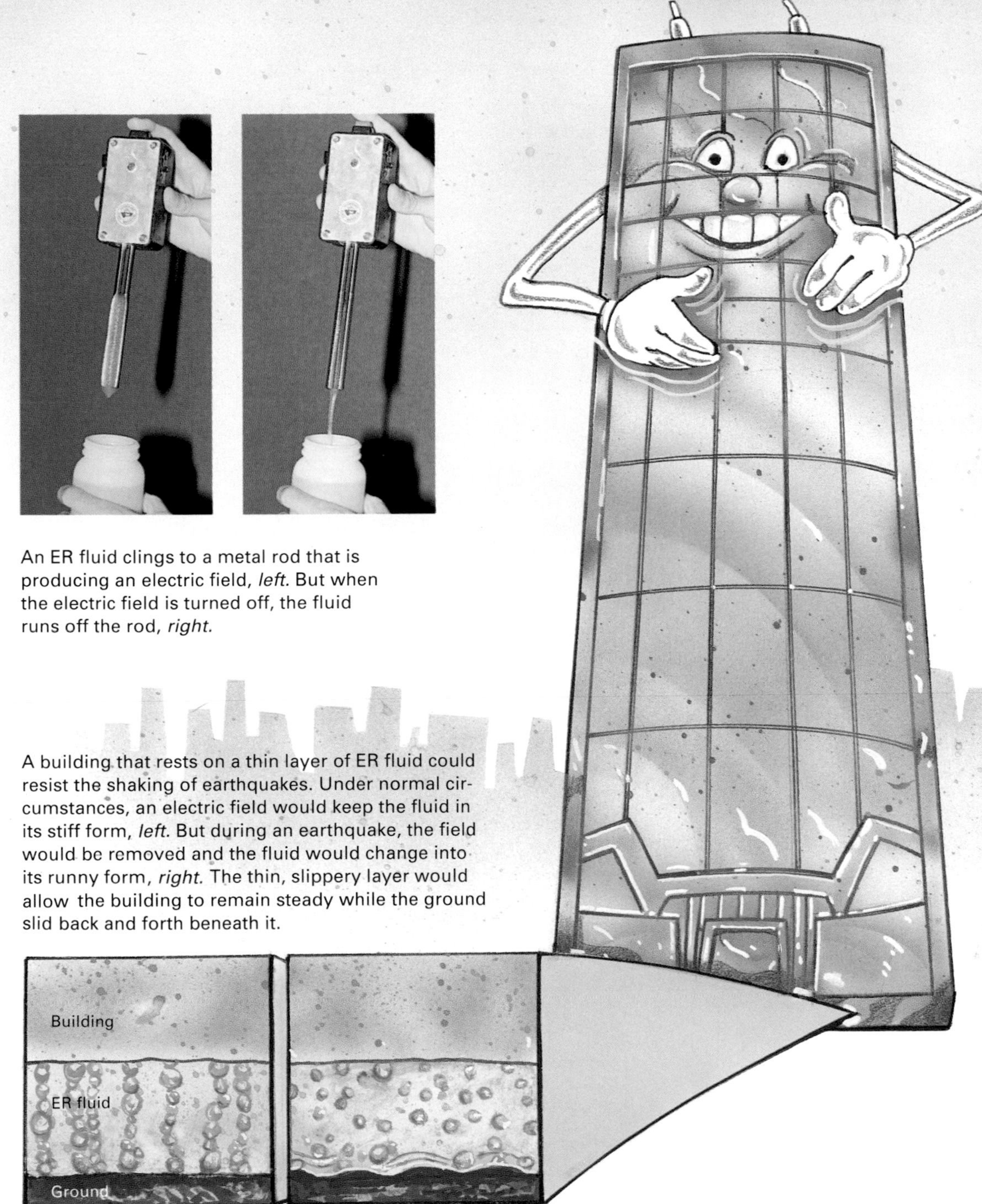

An ER fluid clings to a metal rod that is producing an electric field, *left.* But when the electric field is turned off, the fluid runs off the rod, *right.*

A building that rests on a thin layer of ER fluid could resist the shaking of earthquakes. Under normal circumstances, an electric field would keep the fluid in its stiff form, *left.* But during an earthquake, the field would be removed and the fluid would change into its runny form, *right.* The thin, slippery layer would allow the building to remain steady while the ground slid back and forth beneath it.

strength. And because repairing satellites after they are lifted into orbit is difficult and costly, smart satellites that can repair themselves would be invaluable.

Engineer Ben Wada of the Jet Propulsion Laboratory in Pasadena, Calif., and his colleagues are developing football-field-sized smart *trusses* (frameworks of beams) to be sent into orbit as multipurpose structures. As platforms, the trusses could hold telescopes or other scientific instruments, and as cranes, they could move materials to and from the cargo bays of space shuttles. The structures incorporate piezoelectric materials at regular intervals along the beams. Each piezoelectric segment can be made to independently expand, contract, or vibrate. A network of microprocessors will coordinate the piezoelectric activity among all the beams. The truss will therefore be able to adjust its size and configuration by sending electrical signals to various segments. Wada says such smart trusses will be able to counteract the expansions and contractions that occur as the truss's temperature rises and falls when it orbits in and out of sunlight. The structure will also be able to counteract mechanical vibrations that can blur the vision of orbiting telescopes or other instruments.

Uses for smart materials are also being developed for Earthbound structures. A typical example is the Stafford Building, which opened on the campus of the University of Vermont in 1993. Although it looks like many other buildings, the Stafford Building is smarter. Along its steel and concrete skeleton are 5 kilometers (3 miles) of optical fibers. One use of the fibers is to measure vibrations in the building, which could affect some of the scientific research going on inside. When the light in a fiber changes in a way that indicates the presence of vibration, a computer directs human technicians to the site to learn the cause of the problem. The optical fibers are also used to monitor the condition of the building's concrete. This information will help building administrators arrange for preventive maintenance and avoid unexpected repairs. The building lacks actuators, so human beings will be responsible for dampening vibrations or repairing cracks. But smart buildings of the future that contain actuators as well as sensors could take over these tasks themselves by making areas of the building more rigid or by moving together sections of walls to seal cracks.

Nevertheless, the Stafford Building serves as a model for aspiring architects of smart structures. Peter Fuhr and Huston Dryver, two engineers at the university who designed the Stafford Building's self-monitoring system, have already rigged other structures such as a bridge and a hydroelectric dam with similar networks of optical fibers.

A smart future

Structures such as smart submarine hulls and smart concrete are capable of performing specific tasks. However, researchers also envision a more general application of smart materials—what Craig Rogers calls a "birth-to-death health-care system" for structures such as buildings, planes, and bridges. The barest form of this idea calls for one set of sensors to monitor the condition of a structure for its entire lifetime. Sensors incorporated during manufacturing would relay information about factors such as temperature and chemical composition, helping engineers create the strongest possible components. The same sensors would continue tracking the condition of the components as they were assembled into the final structure. As the structure was put to its intended use, the sensors would indicate when components needed to be replaced. And the sensors would even detect when the structure had deteriorated beyond the point of repair.

Because the field of smart materials is so new, most of its challenges lie in the future. For example, researchers can embed optical fibers into construction materials such as concrete, but scientists are not yet sure how the fibers affect the strength and durability of the host material. Similar questions remain unanswered for the effects of incorporating piezoelectric compounds, shape-memory metals, and other smart materials into structures. Until researchers can be assured that smart construction materials are sturdy and reliable, engineers will be hesitant to use them in place of their dumber but more time-tested cousins.

Another problem for researchers is the difficulty of making smart structures small enough and light enough to be practical. In many prototypes of smart structures, such as the smart airplane wing, embedded sensors feed signals into computers outside of the host material. But no one will use a smart airplane if they have to haul around a cockpit full of weighty computers to operate it. To avoid that restriction, researchers have to develop ways to embed tiny electronic chips along with a vast network of wires into the structure without affecting its performance.

On an even more fundamental level, researchers are still learning precisely how to interpret signals such as the changes of light in optical fibers and the electrical signals from piezoelectric sensors. Scientists will not only have to understand what those signals indicate about conditions in the material, but they will also have to develop mathematical equations to express the relationship between the signal and, for example, the temperature or amount of vibration that provokes it. Only after engineers program such formulas into the "brains" of smart structures will the structures be able to respond to a range of environmental conditions.

As scientists and engineers continue fleshing out the future of smart structures, the boundary between the animate and inanimate worlds may start to blur a bit. With concrete that repairs its own cracks, buildings that call for help when they are damaged, and airplanes that alter the shapes of their wings as they fly, the constructed world will have abilities to adapt and respond that were previously reserved for the world of the living.

For further reading:

Amato, Ivan. "Animating the Material World." *Science,* Jan. 17, 1992, pp. 284-286.

Coghlan, Andy. "Smart Ways to Treat Materials." *New Scientist,* July 4, 1992, pp. 27-29.

Thompson, Brian S., and Gandhi, Mukesh V. *Smart Materials and Structures.* Chapman and Hall, 1992.

Science Studies

The Universe

BY MICHAEL S. TURNER

Scientists are making great progress in understanding the universe—how it is constructed on the smallest and largest scales, how it probably began, and how it may end.

Introduction

From earliest history, people's eyes have been drawn to the night sky. It contained much for the ancients to contemplate. The planets moved in irregular paths across the heavens. Comets appeared without warning, grew large, and then mysteriously departed. And the unchanging stars hinted at a perfect and eternal realm far above the Earth. But what were the stars, and how far away were they? Was the Earth the center of creation? Where did everything come from? Did the universe remain always the same, or did it change through time? Would it last forever or someday come to an end?

Philosophers devoted a lot of time trying to answer such questions. The ancient Greeks, who lived more than 2,000 years ago, believed that the universe was characterized by order and harmony and that it could be understood by the human mind. It is from *kósmos*—the Greek word for *order*—that we got the English word *cosmos,* the universe considered as a harmonious whole. The study of the cosmos is known as cosmology. But neither the Greeks, nor anyone who followed them during the next 1,500 years, made much headway in understanding the universe. That was because almost nothing was known about celestial objects or the laws of nature.

In recent centuries, however, the sciences of physics and astronomy have put cosmology on a firm footing. Today, we are adding to our knowledge of the universe at an astounding pace. High-technology telescopes and other advanced scientific tools are giving cosmologists the information they need to explain the universe on the largest and smallest scales from its infancy and into the future.

We now know that the universe is vast, perhaps infinite. We know also that it had a beginning and is constantly evolving. How—or if—it will end, however, is still uncertain. The overall scheme of things is coming into focus, but the details will surely be debated for years to come. And although we have learned a great deal about the universe, it is certain that questions will remain for the next generation of cosmologists to try to answer.

The author:
Michael S. Turner is professor of astrophysics at the University of Chicago and former deputy head of the NASA/Fermilab Astrophysics Center in Batavia, Ill.

The History of Cosmology

Human efforts to make sense of the universe undoubtedly began long before the dawn of civilization, but the speculations of the first stargazers can never be known. We do know that more than 2,000 years ago among the ancient Greeks, at least one person—the philosopher Aristarchus—believed that the Earth and other planets revolved around the sun.

Aristarchus and another Greek who came before him, Democritus, also pondered the stars, theorizing that they were other suns. Democritus was one of the boldest thinkers in ancient Greece. He proposed that everything is composed of atoms and that the hazy band of light across the sky formed by our own Milky Way galaxy was made up of stars too far away to be seen as individual pinpoints of light.

The ideas of Aristarchus and Democritus were revolutionary for their day. If their views had prevailed, the modern age of cosmology might have begun much sooner.

But most Greek scientists and philosophers, including the influential Aristotle and Plato, held other beliefs. They thought the Earth was at the center of everything, with the sun, stars, and planets circling about it affixed to transparent celestial spheres. As these vast spheres rotated, the heavenly bodies attached to them traced out circular orbits around the Earth.

According to Aristotle, who summed up the astronomical thinking of his day, the outermost sphere contained all the stars. But neither he nor any of the other ancient Greeks had any idea how huge the universe is. They estimated that the stars were a few thousand, or perhaps a few million, miles away. The nature of the stars was something they did not try to explain.

In the A.D. 100's, the basic scheme outlined by Aristotle was taken even further by the most famous of the early astronomers, Ptolemy, a Greek who lived in Alexandria, Egypt. Because the invisible celestial spheres helped explain the movement of heavenly bodies, Ptolemy retained them in his view of the universe, though he rejected the idea that they were actual, solid objects. But despite that simplification, Ptolemy's system added

Early views of the cosmos
A drawing from the 1500's, *above,* shows the view of the universe that most people had held since ancient times. The Earth is at the center of the universe with the planets and stars revolving about it on invisible spheres. A print, *right,* reflects the theory of the universe that had replaced the Earth-centered cosmos by the 1600's. In this scheme, proposed by Polish astronomer Nicolaus Copernicus, the Earth and planets revolve about the sun.

new complications to account for the motions of the planets across the sky.

Although it was cumbersome, the Ptolemaic system accounted quite well for the observed motions of the heavenly bodies, so it went unchallenged for nearly 1,500 years. But by the late Middle Ages, it had become evident to careful observers of the skies that Ptolemy's universe did not provide a truly accurate description of celestial phenomena. The time was ripe for a new view of the cosmos.

The influence of Copernicus

That new cosmic view arrived in 1543 with a book titled *On the Revolutions of the Heavenly Spheres,* by the Polish astronomer Nicolaus Copernicus. Like Aristarchus, whom he may have read about in the writings of the ancient Greek author Plutarch, Copernicus put the sun at the center of the solar system. Copernicus was motivated largely by a philosophical conviction that the structure of the universe must be simple and elegant. The Ptolemaic universe was anything but. Today, we date the modern era of astronomy and cosmology to the "Copernican Revolution" fostered by Copernicus' groundbreaking book.

But Copernicus' theory failed to gain immediate acceptance because many people were not ready to give up the idea that Earth was the center of the universe. Religious authorities, in particular, were committed to an Earth-centered cosmos. Realizing that he risked being denounced for contradicting the official position of the church, Copernicus had delayed publishing his book until he was on his deathbed.

Copernicus' theory, moreover, could not account precisely for the movements of the planets. Like the Greeks, Copernicus had assumed planetary movements to be circular—cross sections of great invisible spheres. The circle and sphere were perfect geometrical forms, and the heavens simply had to be the embodiment of perfection. But nature pays no attention to human ideas about how it should behave. The planets do not move in circular orbits, as the German mathematician and astronomer Johannes Kepler discovered in the 1600's.

Kepler was an avid supporter of the Copernican system, and he was determined to discover why it could not account for what astronomers observed. In 1600, Kepler became an associate of the Danish astronomer Tycho Brahe, who had spent years making detailed records of planetary motions. Brahe died in 1601, but Kepler continued to study Brahe's data, trying to make them fit the idea of circular orbits. Finally, Kepler realized that circular orbits were simply not correct, and it was then that he found the answer. The planets, he proclaimed in 1609, move in *elliptical* (oval) paths.

A contemporary of Kepler's, the Italian astronomer and physicist Galileo Galilei, advanced celestial observations with an important new tool: the telescope. Using a telescope to view the heavens, Galileo saw that the moon is covered with mountains and craters and that Venus, like the moon, goes through regular phases, changing in appearance from crescent-shaped to a round disk.

The stars were a particular surprise to Galileo. They not only apparently numbered in the millions but also varied tremendously in their range of brightness. Most stars were much too faint to be seen by the unaided eye. This finding indicated that some stars were considerably more distant than others, rather than all being equally far away, as most people had assumed since at least the time of Aristotle. The perceived universe was growing ever larger.

Galileo's observations greatly strengthened the Copernican hypothesis. For example, the fact that Venus goes through phases like those of the moon indicated that Venus circles the sun. Although absolute proof that the Earth was not the center of the universe was still lacking, Galileo became an outspoken critic of all who expressed doubts about the Copernican system. As a result, angry church authorities in Rome brought Galileo to trial in 1633 and forced him to renounce his belief in Copernicanism. Their victory, however, was nearly the last gasp in efforts to preserve the notion of an Earth-centered cosmos.

Newton, Herschel, and Einstein

In 1642, the year of Galileo's death, one of the greatest scientists in history, Isaac Newton, was born in England. Newton's interests and accomplishments were far-ranging, but it was his studies of gravity that were of most importance to the advance of astronomy. Newton discovered that the same gravitational force that causes an apple to drop to the ground from an apple tree keeps the planets in orbit about the sun and holds the universe together.

Newton used his theory of gravity to calculate how the planets should move around the sun. The answers to his equations revealed that the planets should move in elliptical orbits, as Kepler had found from his analysis of Brahe's observational data.

Another scientist in England, William Herschel, became the preeminent telescope builder and astronomer of the late 1700's. Using *reflecting telescopes* (telescopes that gather and focus light with mirrors rather than lenses) up to 122 centimeters (48 inches) in diameter, Herschel observed the panorama of the heavens in even greater detail.

Two giants of modern astronomy
Astronomer Edwin Hubble sits in the viewing cage of the 5-meter (200-inch) Hale telescope at the Mount Palomar Observatory in California in 1950. The telescope, which began operation in 1948, was the largest in the world for nearly 30 years. Hubble made major contributions to astronomy in the 1920's while working at the Mount Wilson Observatory, also in California. He discovered that the universe contains billions of galaxies and that they are all rushing away from each other. From the latter finding, scientists concluded that the universe is expanding.

Much of Herschel's time was spent studying glowing patches of light called nebulae. Astronomers earlier in the 1700's had begun finding these unusual objects, many of which they noted were disk-shaped. They assumed that the nebulae were part of the Milky Way, because the Milky Way was then thought to be the entire universe.

Other observations had suggested that the Milky Way itself was shaped like a disk. In a flash of insight, the German philosopher Immanuel Kant had proposed in 1755 that the disk-shaped nebulae were other galaxies like the Milky Way but far removed from it. "Island universes" he called them.

Seeking to resolve this question, Herschel scanned the heavens for nebulae and discovered about 2,000 new ones. He determined that many were huge clouds of gas or clusters of stars in the Milky Way. The disklike nebulae, however, appeared only as hazy smudges to his gaze.

Observations by later astronomers revealed that many of the disk-shaped nebulae had spiral shapes, but still no stars could be seen in them. So whether the nebulae lay within the Milky Way or were separate galaxies remained an open question for many years.

While astronomers surveyed the heavens, physicists were learning how the universe works. The greatest of these researchers, whose name would rank with Newton's, was Albert Einstein, who was born in 1879 in Germany.

In 1915, Einstein proposed a *general theory of relativity*, a new explanation for gravity. The theory described gravity in geometric terms, as a curvature of space. In this view, a planet or other large body warps the space around it, much as a bowling ball placed on a mattress would form a depression in it. The warping of space causes smaller objects to move in curved paths toward a massive object, just as marbles would roll toward the bowling ball because of the depression in the mattress.

Einstein also pondered the nature of the universe as a whole. The theory of relativity predicted that the universe could not exist in an unchanging condition, but rather had to be either expanding or contracting. But at that time, all the evidence pointed to a static universe. Rather than insisting that his theory was accurate, Einstein altered his equations by adding a factor he called the "cosmological constant," which counteracted gravity to produce a changeless universe.

A universe of galaxies

Einstein would soon have reason to discard the constant and call it the biggest error of his career. Discoveries made in the 1920's by the American astronomer Edwin Hubble of the Mount Wilson Observatory in California showed that Einstein's original conception of the universe had been correct after all. In addition, Hubble finally settled the question of whether the spiral nebulae are other galaxies.

Hubble studied stars called Cepheid variables,

which periodically become brighter, dimmer, and then brighter again. The time between peaks of brightness can be a few days or several months.

In 1912, Henrietta Swan Leavitt, an astronomer at the Harvard University observatory in Cambridge, Mass., had made a key finding about the Cepheids. She discovered that the longer the time period between a Cepheid's peaks of brightness, the greater the star's true brightness. Cepheids with the same period—the same length of time from peak brightness to peak brightness—always have the same maximum true brightness. (A star's true brightness, as opposed to its relative brightness as observed from Earth, is the actual amount of light it gives off. A faraway star might have a higher true brightness than a nearby star, but because of its distance, it has a lower relative brightness.)

Leavitt's finding meant that Cepheids could be used as cosmic measuring posts. Because the brightness of an object varies with its distance in a known way, it would be a simple matter to calculate the distance of a remote Cepheid. This could be done by comparing the star's relative brightness to a nearby Cepheid of the same period whose distance had already been calculated by other means.

In 1925, Hubble reported that he had found Cepheid variables in spiral-shaped nebulae and had calculated that the stars were much too far away to be within the Milky Way. Thus, the spiral nebulae were galaxies in their own right. Another class of nebulae, the ellipticals, with an oval shape but no spiral arms, were also soon recognized as being galaxies. The universe, astronomers now realized, consisted of billions of galaxies, each containing billions of stars.

Discovering the expanding universe

Beginning about 10 years earlier, an astronomer at the Lowell Observatory in Flagstaff, Ariz., Vesto M. Slipher, had begun obtaining data that would prove useful to Hubble in his next great discovery. Studying the light from spiral "nebulae"—their true nature as galaxies had not yet been determined by Hubble—Slipher found that in almost every case the light was shifted toward the red end of the spectrum.

This phenomenon, called the *red shift,* had been used by astronomers since the 1800's to measure the speeds of stars orbiting the center of the Milky Way. The red shift occurs because light waves emitted by a rapidly receding source become stretched out as they move away from the source. The stretching effect increases the wavelength of the light. Longer wavelengths fall toward the red end of the spectrum. In contrast, light waves from a source moving rapidly toward an observer are compressed and shifted toward the short-wavelength blue end of the spectrum.

Hubble was intrigued with Slipher's finding that the light from the spiral nebulae (spiral galaxies) was red-shifted. This could only mean, he said in a 1929 research paper, that all the other galaxies are rushing away from us at high speed. The universe, as Einstein's theory had predicted, was expanding.

The emergence of the big bang theory

But if the universe truly was expanding, why was it doing so? By the late 1940's, two theories had emerged to account for the expansion of the universe.

Three British astrophysicists—Hermann Bondi, Thomas Gold, and Fred Hoyle—developed the *steady state theory.* According to this theory, the universe has always existed and has always looked essentially the same, except that galaxies constantly move away from each other. In the growing empty regions between them, new stars and galaxies are formed from matter that somehow pops into existence from the vacuum of space.

A different scenario was presented by the *big bang theory,* whose most outspoken proponent was the Russian-American physicist George Gamow. The big bang theory held that the universe began in a huge explosion of matter and energy at some moment in the far-distant past. Matter created in this "primeval fireball" later came together to form the stars and galaxies.

In 1948, Gamow predicted that even though the universe had greatly expanded and cooled since the big bang, the explosion's leftover energy would still exist as very faint *microwaves* (high-frequency radio waves) throughout space. Detecting this *cosmic background radiation* would lend strong support to the big bang theory.

Gamow's prediction was not confirmed until the 1960's. In 1965, two researchers at Bell Telephone Laboratories in Holmdel, N.J.—Arno A. Penzias and Robert W. Wilson—were working with a large horn-shaped antenna designed for satellite communications. They detected a persistent hiss from wherever in the sky they pointed the antenna. That hiss was the sound of the cosmic background radiation.

With Penzias and Wilson's discovery, the big bang theory became widely accepted as the correct explanation for the origin of the universe. Although the theory has been refined, with details added by physicists studying matter on the smallest levels, the "big picture" of our cosmos being born in a moment of fiery energy appears to be correct. We live in a universe that has been here for a limited amount of time and that is evolving toward a future we have yet to determine.

The Universe on the Grand Scale

The nighttime sky is ablaze with light from objects near and far. With the naked eye we can see the moon, most of the planets of the solar system, nearby stars, and even some neighboring galaxies. With the most powerful telescopes, we can see remote galaxies in the far reaches of the universe.

The most important object in the heavens—to us—is the one we cannot see at night, the sun. The sun is a star of average size and brightness. It is the brightest object in the sky only because it is so close to Earth.

Orbiting the sun are the nine planets of our solar system: Mercury, Venus, Earth, Mars, Jupiter, Uranus, Neptune, and Pluto. The planets shine only because they reflect light from the sun. Most of the planets, including Earth, are themselves orbited by smaller bodies called moons. There are more than 40 of these natural satellites in the solar system, with the Earth's own moon ranking sixth in size.

In the grand scheme of things, the planets and their moons are a bit of an accident, having formed from the debris of the material that came together to form the sun about 4.6 billion years ago. Astronomers think it is likely that many stars have planets, though as of 1993 they had found little evidence of planetary systems around other stars.

Light bulbs of the universe

Stars are the light bulbs of the universe. They come in many sizes and brightnesses, but all of them are basically hot balls of gas that are powered by nuclear reactions at their cores.

Those reactions, in which a star consumes hydrogen to produce helium and energy, proceed at different rates, depending on the mass of the star. (Mass is the amount of matter something contains.) The most massive stars are also the brightest. That is because in these very large stars, the matter at the core of the star is squeezed together by gravity (which increases with mass) to a very high density, causing the nuclear reactions to occur at a more rapid rate. But the glory of such stars is short-lived. Giant blue-white stars typically exhaust their nuclear fuel in the relatively brief time of 10 million years or so.

The least massive stars, on the other hand, use their fuel extremely sparingly. Small orange or reddish stars may live for hundreds of billions of years. The sun, a yellow star of average mass, has been shining with its present radiance for about 4.5 billion years and will continue to do so for another 5 billion years before running out of fuel.

An estimated 80 to 90 percent of all visible stars are in the prime of life and are known as main-sequence stars. Stars that are past their prime include three types of exotic stars—white dwarfs, neutron stars, and black holes.

White dwarfs are small, dense stars that slowly cool and fade from view, like the dying embers in a fireplace. Average-sized stars like the sun end their days by first swelling into what are called red giant stars and then shrinking to become white dwarfs.

But if a star has more than about eight times as much mass as the sun, it dies a spectacular death. When the fuel of one of these large stars is used up, the star collapses and then blows itself apart in a titanic explosion known as a supernova. In most cases, gravity compresses the remaining core of a supernova to form a *neutron star,* an incredibly dense ball of tightly packed subatomic particles called neutrons. Some rapidly spinning neutron stars emit powerful beams of radio waves, resembling the rotating beams of a lighthouse. These beams are detected on Earth as pulses of radio waves. Thus, the stars are known as pulsars.

If the remnant of a supernova exceeds about three times the mass of the sun, its compression by gravity cannot be stopped. All the matter in the core is crushed together until it becomes a black hole, a body whose gravitational field is so powerful that not even light can escape its grasp. A black hole cannot be seen, but it can be detected by its effect on surrounding matter. As matter falls into a black hole, it releases huge quantities of energy into space.

The Milky Way and other galaxies

Stars do not exist in isolation. They are members of galaxies, vast collections of stars held together by gravity, the "glue" of the cosmos. Our own galaxy is known as the Milky Way because, from our vantage point on Earth, the more distant stars of the Galaxy appear not as distinct pinpoints of light but as a milky lane across the sky.

There are more than 100 billion galaxies in the universe. In only the nearest galaxies can astronomers make out individual stars. Most galaxies are so far away that their stars can be seen only as a haze of light.

The birth, life, and death of stars

Surrounded by glowing remnants of the clouds of gas and dust from which they were formed, the brightest stars of the Pleiades star cluster, *below,* shine with a brilliant bluish light. All stars are born, go through youth and middle age, and eventually die—though their life spans vary greatly. With an average age of about 100 million years, the Pleiades are cosmic youngsters.

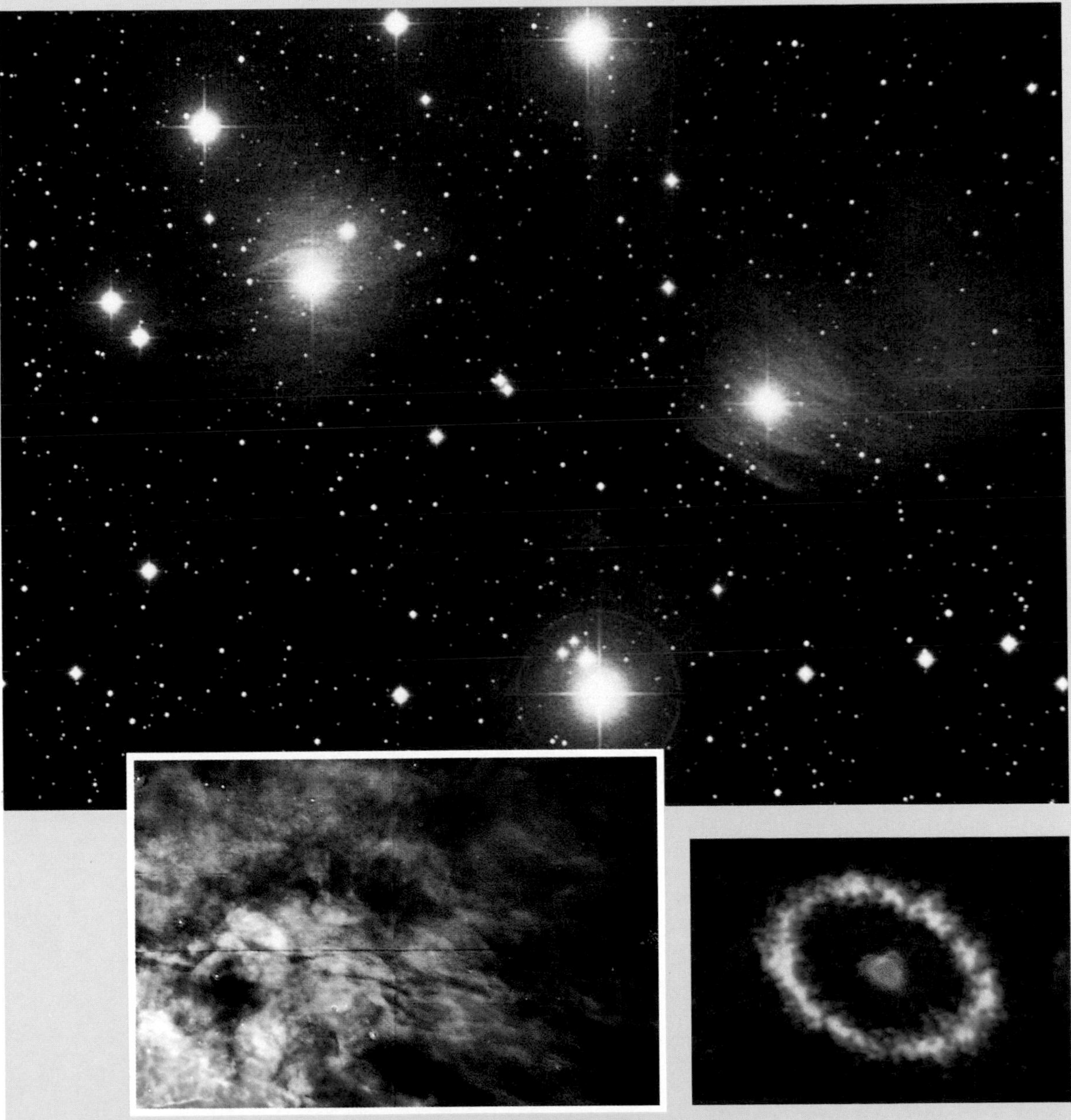

The Orion Nebula, *above left,* is one of the "maternity wards" of the Milky Way, where vast clouds of dust and gas are giving birth to many new stars. The largest stars spawned by Orion will end their days as *supernovae* (exploding stars). One such spectacular explosion, which occurred in a galaxy near the Milky Way, was seen in 1987. The gaseous outer layers of a star, blasted into space, now form an expanding ring around the collapsed core, *above right.*

Like stars, galaxies come in a variety of sizes and types. Although a typical galaxy contains about 100 billion stars, some have far less, and the smallest—called dwarf galaxies—have fewer than 1 million stars. The Milky Way is a relatively large galaxy, with about 200 billion stars.

Some galaxies have an irregular appearance, but most fall into one of two main categories. Many are *elliptical* (round or oval-shaped). Even more numerous are the spirals, which account for 70 to 80 percent of known galaxies.

Spiral galaxies are among the most beautiful objects in the heavens. Seen face on, they resemble great pinwheels in space, and from the side they look much like luminous flying saucers. The Milky Way and the neighboring Andromeda galaxy are both spirals. Our solar system is located in one of the spiral arms of the Milky Way, about halfway between the Galaxy's center and its outer fringes.

The most spectacular galaxies are quasars, which, despite being thousands of times smaller in size than large galaxies like the Milky Way, are hundreds of times brighter. Quasars are the most distant objects we can see in the universe, yet, because of their radiance, they appear in telescopes as bright, starlike objects. Most astronomers believe that quasars are powered by gigantic black holes at their centers. Quasars are thought to be a stage that many galaxies went through early in their evolution. The Milky Way, which seems to have a huge black hole at its center, may once have been a quasar.

Galaxy clusters, superclusters, and voids

Most galaxies have been pulled together by gravity into groups. The Milky Way and Andromeda, together with about 20 smaller galaxies, make up a collection of galaxies called the Local Group. Although most galaxies are found in small groups, about 10 percent of the galaxies in the universe are part of larger systems called clusters. Galaxy clusters may contain anywhere from hundreds of galaxies to thousands of galaxies. Astronomers have identified about 10,000 clusters. The Milky Way is located on the outskirts of a cluster called the Virgo Cluster, which consists of about 1,000 galaxies.

Clusters, in turn, are organized into even larger collections called superclusters. Both the Local Group and the Virgo Cluster are members of the Virgo Supercluster.

There are also vast regions of space, known as voids, that are nearly empty of galaxies. Astronomers' understanding of these areas is somewhat fuzzy. Voids could simply be regions of space that were left empty when gravity drew galaxies together to form clusters and superclusters. Or they could be expanses of space that were swept clean of matter by shock waves from supernovae or that never contained much matter to begin with. It is even possible that voids are not really empty. They may be teeming with galaxies that are just too faint to be seen.

The universe's vast distances

The distances between galaxies—and even between stars—are unimaginably huge. Astronomers measure cosmic distances not in kilometers but in *light-years.* One light-year, the distance light travels in one year at a speed of about 300,000 kilometers (186,000 miles) per second, is about 9.5 trillion kilometers (5.9 trillion miles). The nearest star outside the solar system, Proxima Centauri, is 4.3 light-years away. The farthest observable galaxies and quasars are up to 15 billion light-years from us—about the distance that light has had time to travel since the big bang. Although astronomers cannot see any farther than that, they are confident that the universe does not end there.

Because it takes time for light to travel across the vast distances in space, when we view faraway celestial objects, we are seeing the light that left them many years ago. Thus, by looking deep into space, we are also looking back in time. When astronomers view the surface of the sun, for example, they are seeing it as it existed 8.3 minutes earlier, because it takes sunlight that amount of time to travel the 150 million kilometers (93 million miles) to Earth.

The farther things are from us in space, the more removed they are in time. When we turn our telescopes on Proxima Centauri, we see it as it appeared 4.3 years ago, and in viewing the more distant Andromeda galaxy we are looking 2 million years into the past. Light from the most remote galaxies and quasars took 15 billion years to reach us, giving us a glimpse of the universe in its youth.

Mind-boggling sizes

The size of celestial structures is just as mind-boggling as the distances between them. Consider our own Galaxy. The starry spiral of the Milky Way spans a distance of about 100,000 light-years. If we could construct a scale model of the Milky Way the size of Chicago, the sun would be a speck visible only with a magnifying glass. The entire solar system would be no wider than a grain of salt.

The largest structure in the universe yet identified is known as the Great Wall. This is a chain of galaxy clusters about 550 million light-years long, 200 million light-years wide, and 10 million light-years thick.

A universe of galaxies

Stars are contained in galaxies, enormous collections of stars held together by gravity. A typical galaxy contains about 100 billion stars, and astronomers estimate that the universe contains more than 100 billion galaxies. Our own Milky Way is a larger-than-average spiral-shaped galaxy.

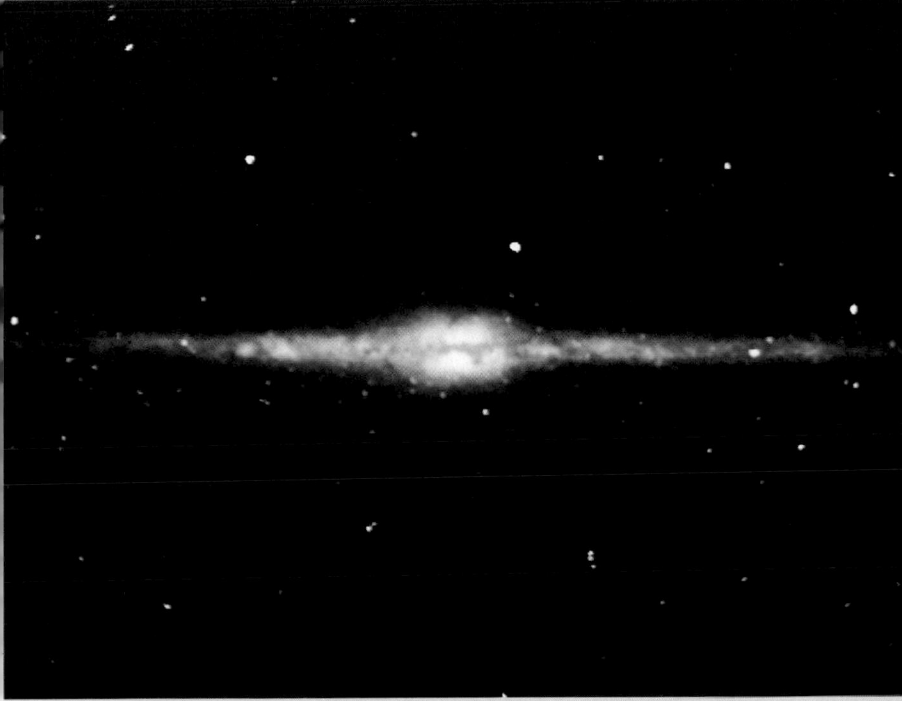

A panorama of the Milky Way, *left,* photographed in *infrared light* (long-wavelength light invisible to the unaided eye) by an Earth-orbiting satellite, is shaped like a disk with a bulge in the center. The Galaxy appears very distant because the solar system is far from the center of the galaxy.

Shining with the light of many billions of stars, a spiral galaxy in the constellation Cepheus resembles a great pinwheel in space, *above.* Spiral galaxies, which include our Milky Way, account for 70 to 80 percent of known galaxies in the universe.

Galaxies are drawn together by gravity into large groups called clusters, *above.* Clusters come together to form even larger collections known as superclusters.

Mapping the universe

Astronomers are engaged in a major effort to map many of the galaxies in the universe to gain an understanding of the universe's overall structure.

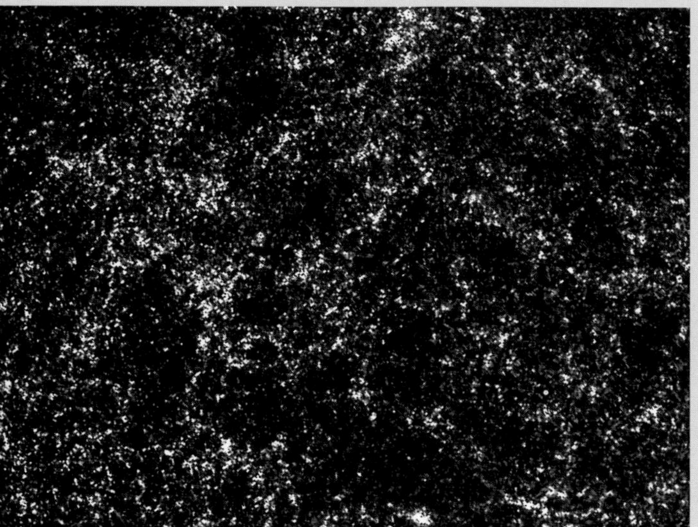

A map of a portion of the universe based on the positions of galaxies on the sky does not show the distances of the galaxies from Earth.

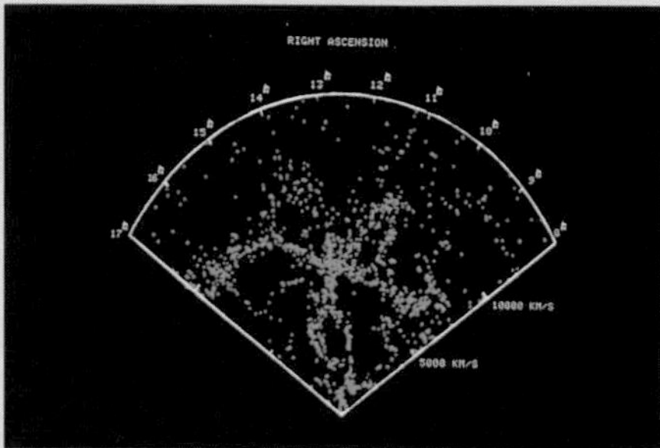

A three-dimensional map is created by piecing together "slices" of the universe. In one such slice, each dot represents a galaxy, the closest ones at the bottom point of the slice; the farthest ones, along the upper curve. The galaxies' distances from Earth are determined by the amount that each galaxy's light is shifted toward the red end of the spectrum. A galaxy's red-shift tells how fast the galaxy is moving away from us. That information gives the galaxy's distance because the galaxies that are farthest are moving fastest.

The mysterious Great Attractor

Perhaps the most mysterious object that we have yet come across in the universe is one discovered in the late 1980's by astronomers at the Carnegie Institution of Washington, D.C., Cambridge University in England, and several other U.S. and British institutions. They noted that the Milky Way and thousands of other galaxies within a radius of about 100 million light-years were moving in the direction of two large *constellations* (groups of stars) named Hydra and Centaurus.

The researchers concluded that the galaxies were being drawn in that direction by a large concentration of mass about 150 million light-years away. They dubbed the massive object the Great Attractor. By their estimate, the Great Attractor contains more than 10,000 times the mass of the Milky Way, causing it to exert an enormous gravitational pull on everything in its vicinity.

Another mystery: Dark matter

Astronomers have observed other gravitational effects in the universe that are even more mysterious than the Great Attractor. In the 1930's, the Swiss-American astronomer Fritz Zwicky calculated that galaxies in large clusters were moving so fast that the gravity provided by their visible stars was insufficient to hold the galaxies together. His work provided the first hint that there is much more matter in the universe than meets the eye. Researchers now estimate that this unseen matter—called *dark matter* because it emits no light—makes up at least 90 percent, and perhaps as much as 99 percent, of the universe's mass.

But what *is* dark matter? Astronomers have not yet been able to find the answer to that question. Dark matter could simply be stars of very low mass that are too faint to see, or it could be dead stars—neutron stars or black holes. Or dark matter could be something far more exotic, such as an unknown form of matter left over from the earliest moments of the universe.

Mapping the universe

While some astronomers contemplate this invisible portion of the universe, others are concentrating on mapping the parts we can see. This is a difficult task, and one that is far from complete. Our present understanding of the visible universe's "geography" is similar to Europeans' knowledge of the Earth's geography in the early 1500's, when world maps contained large areas of unknown territory.

Astronomers map the universe by charting the positions of galaxies as they appear across the "dome" of the sky and recording the positions in catalogs. Astronomers in the 1930's through the 1980's mapped the positions of more than 5 million galaxies.

Although the catalogs of galaxy positions are useful, they have one big drawback: They provide only a two-dimensional view of the universe, while the universe is three-dimensional. To appreciate the shortcomings of such maps, imagine being given the view of a 100-story building, looking upward from the basement, with the floor plans of all 100 floors overlying one another on a single sheet of paper. Such a map would be completely incomprehensible.

Only with a three-dimensional view that separates the layout of each floor from those of all the others would you be able to discern the overall design of the building. So it is with mapping the universe.

A valuable mapping tool—the red shift

Fortunately, astronomers have a tool at their disposal to produce a three-dimensional map of the universe. It is the red shift, the fact that light emitted by an object moving away from an observer is shifted toward the red end of the spectrum.

The red shift has played an important role in astronomy throughout the 1900's. In 1929, U.S. astronomer Edwin Hubble, after learning that the light from other galaxies is shifted toward the red end of the spectrum, concluded that the universe is expanding. Analyzing the red shift of galaxies, Hubble noted further that there is a consistent, proportional relationship between a galaxy's distance and the speed of its movement away from the Milky Way. The farther away a galaxy is, he discovered, the faster it is receding from us. (The Milky Way is also moving, of course. An observer in another galaxy would see our Galaxy receding at high speed.)

The relationship between a galaxy's velocity and its distance, which became known as Hubble's law, made it possible to figure the distance to any galaxy, no matter how far away, just by measuring its red shift. For example, a galaxy with a red shift indicating a velocity of 30,000 kilometers [18,600 miles] per second is about 2 billion light years away. A galaxy receding at twice that speed is twice as far away.

This discovery has been invaluable to astronomers, but analyzing the light from distant galaxies to obtain their red shifts is time-consuming work. Therefore, making three-dimensional maps of the universe—or red-shift surveys, as they are usually called—has proceeded at a much slower pace than earlier mapping efforts. By 1993, astronomers had measured the red shifts of about 40,000 galaxies.

The largest red-shift survey has been carried out by a research team led by astronomers Margaret Geller and John Huchra at the Harvard-Smithsonian Center for Astrophysics in Cambridge, Mass. By 1993, they had mapped the volume of space within about 500 million light-years of the Milky Way—only about 3 percent of the distance we can see.

Also in 1993, scientists at several U.S. institutions—the University of Chicago; the Fermilab accelerator laboratory in Batavia, Ill.; Johns Hopkins University in Baltimore; Princeton University in Princeton, N.J.; and the Institute for Advanced Study, also in Princeton—were working on an ambitious project to map the positions of about 1-million galaxies. Known as the Sloan Digital Sky Survey, the project will map 25 percent of the sky to a depth of 2.5 billion light-years from the Milky Way. The survey, which is expected to last 10 years, will be carried out with a specially designed 2.5-meter (100-inch) telescope being constructed in New Mexico. The telescope will be linked to an array of electronic recording instruments that will measure 600 red shifts at once.

Determining the structure of the universe

The researchers involved with this latest mapping effort hope their survey will be large enough to yield a representative sample of the universe. Astronomers have learned that on a very large scale, the universe is pretty much the same wherever you look. The New Mexico survey, which will be based on a sizable volume of space, should therefore contain enough information to enable astronomers to infer the structure of the entire universe.

This mapping effort can be likened to the voyages of discovery undertaken in the 1700's by the British naval explorer Captain James Cook. In three voyages in the 1760's and 1770's, Cook investigated and charted large areas of the Pacific Ocean, most of which had been unknown territory to Europeans.

Cook and his crews were the first Europeans to see Antarctica, New Zealand, and the Hawaiian Islands and to sail along the east coast of Australia. With Cook's expeditions, mapmakers could at last show the Earth as it really was. Although many blanks remained to be filled in, world maps would now contain all of the planet's major land masses, each shown its proper size and in its correct location. The Earth had become a more familiar place. So, too, as we map the galaxies, will we gain a clearer picture of how the universe looks on the grand scale.

A Subatomic View of the Universe

A microscopic view of the universe is very different from the view we get through a telescope. Planets, stars, and galaxies—along with trees, people, raindrops, and every other thing we know of—are all made of the same building blocks. These building blocks are the 92 naturally occurring elements. An element is a substance, such as gold or magnesium, that can be reduced to a single atom without losing its properties. Light elements such as hydrogen are made of small atoms, while heavy elements such as uranium are made of larger atoms.

Hydrogen is by far the most abundant element in the universe. This gas, the lightest of all the elements, makes up about 71 percent of the matter in a typical star such as the sun. Helium, the next-lightest element, accounts for another 27 percent. The other 2 percent of the atoms in a typical star include carbon, oxygen, and more than 20 other elements, the heaviest of which is iron. These heavier elements are formed by the nuclear reactions in the star's core, in which smaller atoms are fused together to form larger atoms.

Elements heavier than iron cannot be made in a normal star. Astronomers think that these heavier elements, such as uranium, were created in *supernovae*, the explosive deaths of massive stars. Supernovae produce the temperature and pressure necessary for creating the largest atoms.

From their study of matter at the atomic level, scientists have learned that atoms are not the smallest units of matter. Atoms are made of smaller particles called neutrons, protons, and electrons. Protons carry a positive electric charge, and electrons are negatively charged. Neutrons are so named because they have no charge.

Each atom has a tiny central core, called the *nucleus*, which contains its neutrons and protons.

The universe at the subatomic and atomic levels
All known matter in the universe is composed of fundamental particles, and four fundamental forces govern it. Quarks and electrons are the main types of fundamental particles. The strong force binds quarks together to make up protons and neutrons. The same force holds protons and neutrons together to form atomic nuclei. An atom consists of electrons orbiting a nucleus. The electromagnetic force holds the electrons in their orbit. The weak force is responsible for some types of radioactive decay. Gravity—the force that binds the solar system and galaxies together and dominates the universe at large—is very feeble at the atomic level.

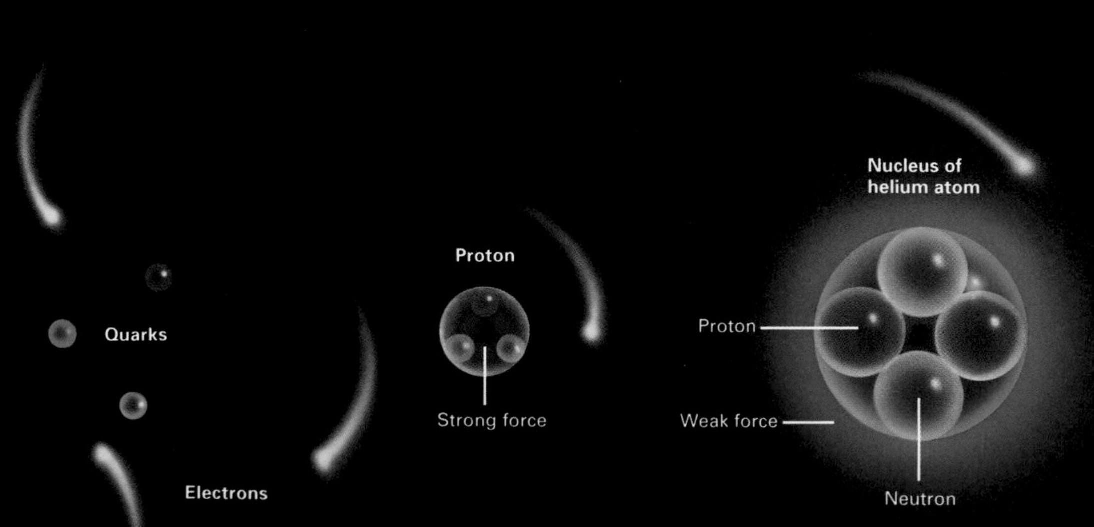

(The simplest nucleus, that of hydrogen, contains a single proton and no neutron.) The electrons—one for each of the atom's protons—orbit the nucleus at high speed. A complete atom contains an equal number of electrons and protons. Because the charges on those particles balance one another, the atom as a whole is electrically neutral.

The subatomic world does not end with protons, neutrons, and electrons, as physicists have learned by peering ever deeper into the structure of atoms. The "microscopes" they use for these investigations are *particle accelerators* (sometimes referred to as atom smashers), huge machines that boost protons or electrons to nearly the speed of light. These fast-moving particles then crash into a stationary target or into other particles speeding in the opposite direction. Such collisions result in a shower of subatomic particles, much like the debris scattered by an explosion. Scientists analyze the subatomic debris to learn how protons, neutrons, and electrons are constructed and how these particles interact with one another.

Research with particle accelerators has shown that protons and neutrons are composed of still smaller entities known as quarks. Physicists have designated one kind of quark the "up" quark and another the "down" quark. A proton is made of two up quarks and one down quark, while two down quarks and one up quark make a neutron.

Electrons, so far as physicists can tell, are fundamental particles that cannot be broken apart into smaller pieces. Electrons belong to a family of particles called leptons that also includes the uncharged neutrino, a bizarre little particle that is very light, and perhaps has no mass at all. Neutrinos interact so feebly with other matter that trillions of them, produced by nuclear reactions in stars, zip unimpeded through Earth every second.

Antimatter and dark matter

Physicists have learned that each atomic particle also has a counterpart that is the same in all respects except charge. For example, there is a particle called the positron that is exactly like the electron except that it carries a positive charge. Such particles are called antiparticles, and as a class they are known as antimatter.

Antiparticles have been formed in small amounts in particle accelerators and in violent processes in the cosmos, but there is no evidence for any significant quantities of antimatter anywhere in the universe. All the matter in our world, and most likely throughout the universe, is regular matter. That is a good thing, because when matter meets antimatter the two destroy each other in a burst of pure energy.

While quarks, leptons, their antimatter counterparts, and other related subatomic particles may seem of little relevance to the universe, that is not the case. During its infancy, the universe was just a hot soup of subatomic particles, and their interactions helped shape the universe we know today. In fact, many scientists believe that most of the matter in the universe is made up of subatomic particles unchanged since the earliest moments of the universe. This matter constitutes the mysterious dark matter that fills the universe.

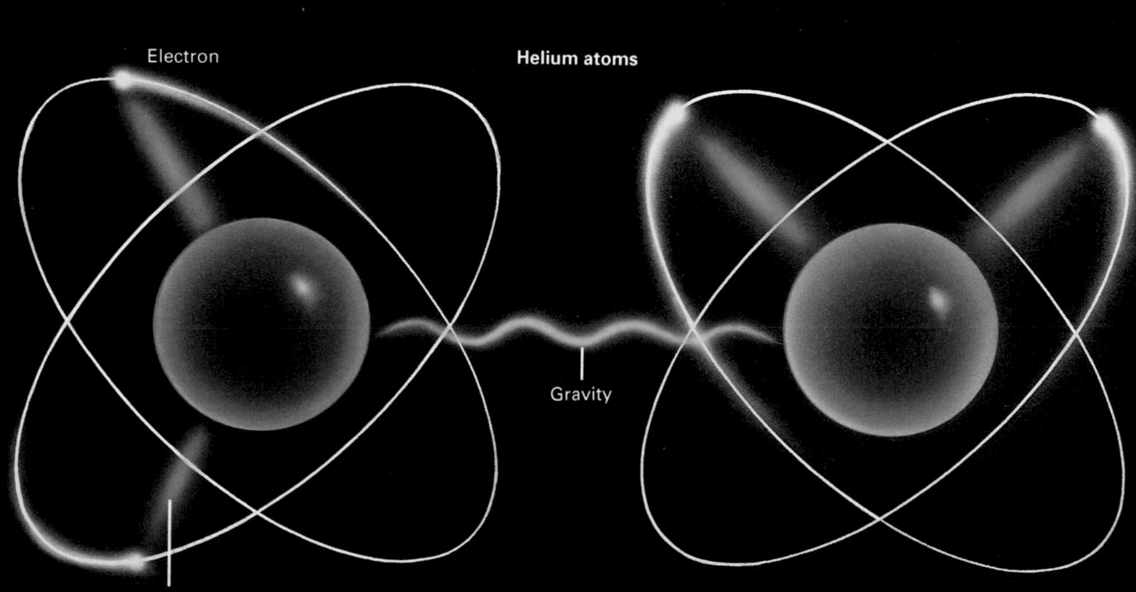

Exploring matter at the subatomic level
Curving green trails mark the paths of subatomic particles created by the collision of a proton and a particle called a neutrino. Experiments of this type are conducted with huge machines called particle accelerators, such as the one at Fermilab in Batavia, Ill. (inset). By smashing particles together at extremely high energies, such experiments reveal how the particles are constructed and how they interact. The experiments also simulate conditions in the early universe.

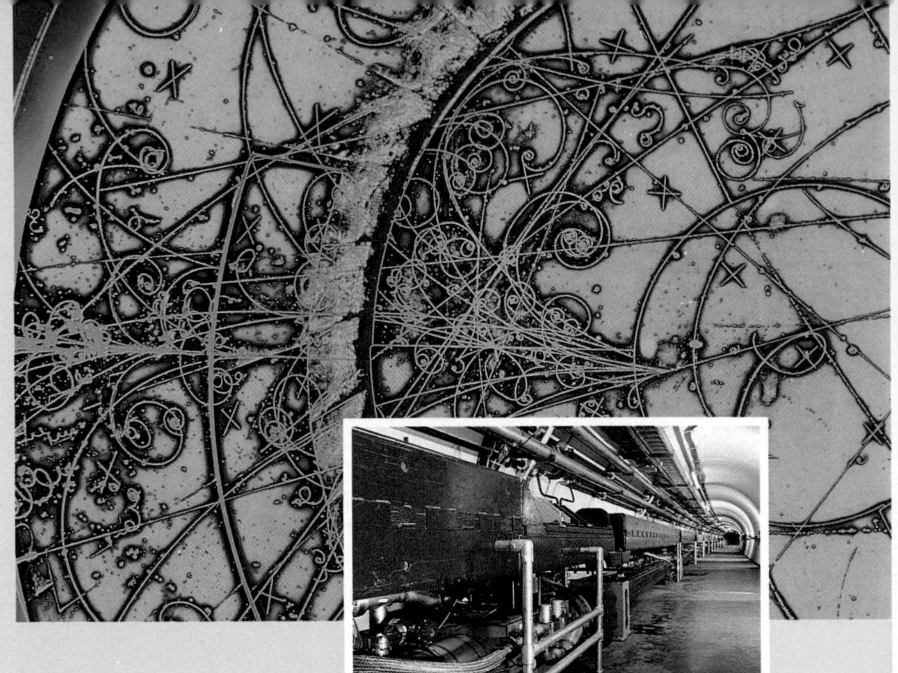

The four forces of nature

The fundamental building blocks of matter—six types of quarks, six types of leptons, and all of their antiparticles—are governed by four known fundamental forces of nature. These are the electromagnetic force, the strong and weak nuclear forces, and gravity. Electromagnetism and gravity are familiar to us in our daily lives; the strong and weak forces are evident only at the atomic level.

The electromagnetic force, transmitted by subatomic particles called *photons,* keeps electrons in orbit around the nucleus. It also enables atoms, through the sharing of electrons, to establish chemical bonds with one another to form molecules. We make constant use of the electromagnetic force in our world. Electric currents light our homes and offices and electromagnetic waves carry radio and television broadcasts through the atmosphere. Visible light is also a form of electromagnetic radiation.

The spectrum of electromagnetic radiation spans a vast range of energy levels and wavelengths. Radio waves have the lowest energy and longest wavelength. After that, in order of increasing energy and ever-shorter wavelength, come microwaves, infrared light, visible light, ultraviolet light, X rays, and gamma rays. All these forms of electromagnetic radiation travel at the speed of light.

The universe is filled with electromagnetic radiation emitted by stars, matter falling into black holes, and many other sources. Detecting that radiation with optical telescopes, radio telescopes, and other kinds of instruments enables astronomers to study these inhabitants of deep space. By far the most abundant kind of radiation we can detect is the microwave "echo" of the big bang, the moment when the universe exploded into being.

The electromagnetic force operates over an unlimited distance. In contrast, the strong and weak forces exert an influence only at subatomic distances, which is why their actions are not apparent in everyday life. The strong force, functioning like a kind of glue, binds three quarks together to form a proton or neutron. The weak force is responsible for some types of radioactive decay. The strong and weak forces are also transmitted by particles, known as gluons and W particles.

The fourth force, gravity, seems strong to us on Earth, but in fact it is the weakest of all the fundamental forces. The gravitational attraction between two atoms is all but nonexistent. But as the mass of objects increases, so does the gravity between them. On the scale of stars and planets, gravity is a powerful force. Like the electromagnetic force, gravity acts over unlimited distances. Galaxies millions—or even billions—of light-years apart exert a gravitational pull on one another, though not nearly as strong as the mutual attraction exerted by galaxies that are closer together.

When we view the galaxies, wheeling in their majesty amid the blackness of space, it seems incredible that they are made of the same basic stuff that we are. But such is the beauty and wonder of our universe that fewer than 100 kinds of atoms, together with four forces, can be used to construct a sea shell, a human being, or a spiral galaxy of 200 billion stars.

The Birth of the Universe

By the late 1970's, most cosmologists had accepted the idea that the universe began with an explosion of energy called the big bang some 15 billion years ago. The scientists then turned their attention to learning about the earliest moments of the universe, including the big bang itself.

Today, cosmologists believe that they can trace the history of the universe back to within 0.00001 second after the big bang, and they speculate about events that may have taken place even earlier. Many believe that what happened during the universe's first fraction of a second provided the foundation for all that followed and will explain some of the universe's most puzzling features.

"Time zero" for the universe was, according to the big bang theory, some 15 billion years ago. It is difficult even to speculate about this period. Nevertheless, some physicists have discussed the possibility that before time zero, space and time were churned together into a sort of foam or that there were six spatial dimensions in addition to the three—length, width, and height—that we have in our present-day universe. Those extra dimensions, the theorists say, would be too tiny for us to perceive today.

Other scientists think that perhaps there never was an actual time zero. They think the big bang may have been preceded by a "big crunch," in which a previous form of the universe came to a fiery end. In this view, the universe goes through endless expansions and contractions.

Such ideas are interesting but, for now at least, unprovable. We are on firmer ground when contemplating our present universe, beginning when it was 10^{-43} second old. When written out, that amount of time is a decimal point followed by 42 zeroes and a 1. It is a million trillion times shorter than the time it takes light to travel the diameter of a proton, one of the subatomic particles of the atomic nucleus.

To arrive at an understanding of the first fractions of a second, cosmologists—long concerned mostly with the universe as we now see it—joined forces with subatomic physicists, who study matter on the smallest scales. The physicists' research was aimed at discovering how matter behaves at extraordinarily high energies and densities—the very conditions that must have existed in the infant universe.

An important aspect of the physicists' work involved efforts to reach a deeper understanding of the four forces of nature—electromagnetism, gravity, and the strong and weak forces governing the atomic nucleus and subatomic particles.

Theorists believe that ultimately the laws of nature are simple and that all the forces of nature are just different aspects of a once-unified force. Likewise, they think that the great number of subatomic particles that physicists have detected in experiments are constructed of just a handful of fundamental particles. While the full simplicity of nature is not evident today, scientists think it would have been apparent in the earliest moments of the universe.

In the 1960's, three physicists—Sheldon Glashow and Steven Weinberg of the United States and Abdus Salam of Pakistan—formulated a theory showing how electromagnetism and the weak force are different aspects of a unified "electroweak force." By the early 1980's, experiments with particle accelerators, the huge machines that physicists use to study subatomic particles, had proved the theory correct.

During the 1980's, physicists went a step further and developed Grand Unified Theories (GUT's) that proposed how the strong force was once unified with the electroweak force. Although GUT's have continued to be an attractive idea to researchers, experiments have not yet proved them correct. Moreover, GUT's do not address the question of how gravity was unified with the other three forces.

Physicists' attention since the mid-1980's has shifted beyond GUT's to another unified theory called the superstring theory, which attempts to explain how gravity was once unified with the other three forces of nature. According to the superstring concept, subatomic particles of matter such as electrons and quarks, and force-carrying particles such as gluons, can be envisioned as being made from tiny loops of string. Previously, such particles had been thought of as points.

Fractions of a second after the big bang

So far, strings exist only in physicists' equations, and the theory may ultimately be disproved. Nonetheless, between superstring theory and GUT's, researchers think they know enough to describe events that might have occurred as early as 10^{-43} second after the big bang.

At that time, the universe was a formless "soup" of quarks, electrons, and other subatomic particles, many trillions of times hotter than the

A chronology of creation

Cosmologists can trace the history of the universe back to within a tiny fraction of a second after time zero—the moment, some 15 billion years ago, when the universe burst into existence in an explosion of energy called the big bang. Although they cannot yet describe the conditions at time zero, when all four forces of nature were unified, they can speculate about conditions as early as 10^{-43} second, when gravity became a separate force. Theorists think that in another fraction of a second, several other events occurred: The infant universe went through a momentary period of inflation, becoming many trillions of times bigger; the strong force split away to become a separate force; the universe became an extremely hot "soup" of subatomic particles; the remaining two unified forces separated into electromagnetism and the weak force; and quarks joined to form protons and neutrons. During the next three minutes, protons and neutrons formed atomic nuclei. Much later, nuclei joined with electrons to make complete atoms.

Time zero
All forces are unified.

10^{-43} second
Gravity becomes a separate force.

0.01 second to 3 minutes
Protons and neutrons join to form atomic nuclei.

300,000 years
Nuclei combine with electrons to form simple atoms.

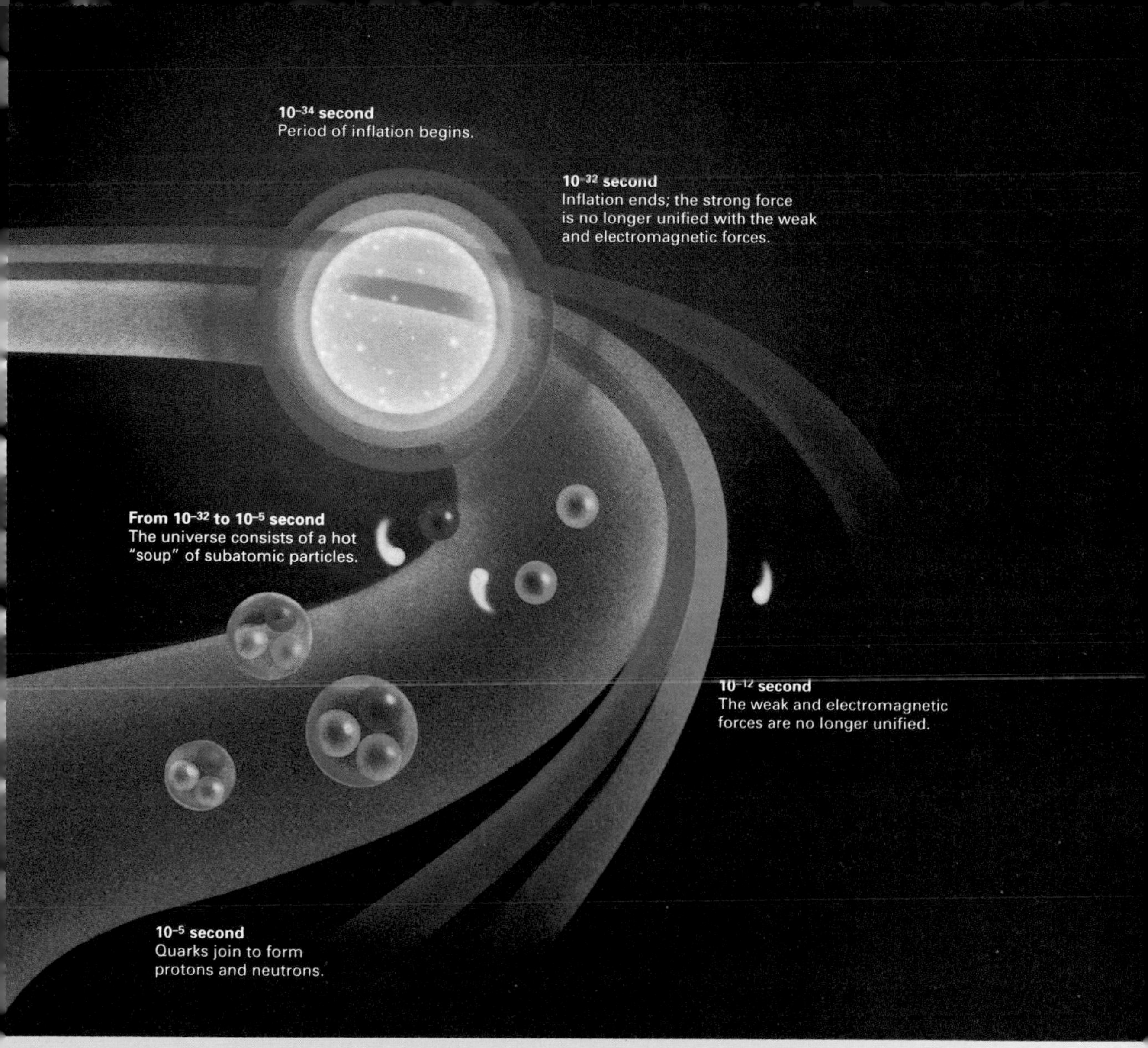

10^{-34} second
Period of inflation begins.

10^{-32} second
Inflation ends; the strong force is no longer unified with the weak and electromagnetic forces.

From 10^{-32} to 10^{-5} second
The universe consists of a hot "soup" of subatomic particles.

10^{-12} second
The weak and electromagnetic forces are no longer unified.

10^{-5} second
Quarks join to form protons and neutrons.

hottest star in today's universe. The portion of the universe that we can now see with telescopes was then less than the size of a proton. As the universe evolved from this almost unimaginable state, it expanded and cooled, with layer upon layer of structure developing, and with the single unified force blossoming into the four separate forces that we know today.

The first force to split off was gravity at 10^{-43} second. Then, at 10^{-34} second, the universe underwent an enormous and rapid expansion known as *inflation,* driven by an odd form of energy called vacuum energy. At about 10^{-32} second, the proton-sized universe increased in size by more than a million trillion trillion times, becoming as big as a grapefruit. In that moment of inflation, the universe expanded more, proportionately, than it would over the course of the next 15 billion years.

When the inflationary period ended, the energy that had fueled inflation was converted into a firestorm of subatomic particles. By this time, the strong force had split off to become a separate force. The universe now settled down to a slower rate of expansion.

During the period from 10^{-32} second to about 10^{-5} second, physicists think, several important things happened. In the final separating of the forces of nature, the electroweak force split apart to become electromagnetism and the weak force. And two events determined the amounts and kinds of matter that we have in the universe today.

The first of those events gave us a universe consisting almost entirely of ordinary matter rather than antimatter. Antimatter particles differ from particles of ordinary matter only in having an opposite electric charge. Physicists think the universe began with equal amounts of matter and antimatter. Matter and antimatter destroy each other in a burst of energy when they come in contact, so matter-antimatter annihilations went on at a furious pace in the first moments after the big bang, as did the creation of new particles and antiparticles.

Sometime between 10^{-32} second and 10^{-12} second, however, the universe developed a tiny excess of ordinary matter particles over antimatter particles—about a billion and one matter particles for every billion antimatter particles. At about 10^{-3} second, the universe had cooled to the point where the creation of new particles and antiparticles could no longer occur, though annihilations continued. When the last annihilations had taken place, the excess of regular matter remained. It was just the right amount of matter to come together to create the stars and galaxies. Were it not for the tiny excess of matter, there would be no matter left today.

The second event involves dark matter, the form of matter that cosmologists calculate makes up at least 90 percent of the universe's mass. Some theorists think dark matter is nothing more than dim stars and other large objects that are invisible to astronomers. Many other cosmologists, however, disagree with that view and theorize instead that dark matter is mostly subatomic particles created in the big bang.

By 10^{-5} second, the universe's temperature had dropped to about 1 trillion °C (1.8 trillion °F). Even though that was still tens of thousands of times hotter than the core of the sun, it was cool enough for quarks to combine to form neutrons and protons.

When the universe was about 0.01 second old, the temperature had fallen to 10 billion °C (18 billion °F). At that point, neutrons and protons began coming together to form the nuclei of the simplest atoms. By about 3 minutes after the big bang, nearly all the ordinary matter in the universe existed in the form of electrons and nuclei of hydrogen and helium. Nuclei of the metallic element lithium and of an alternative form of hydrogen and an alternate form of helium were also created, though only in tiny amounts.

The process by which all these nuclei were formed is referred to as *big bang nucleosynthesis,* and it provides an important verification of the big bang scenario. Theorists have calculated that big bang nucleosynthesis would have resulted in about 76 percent hydrogen nuclei and 24 percent helium nuclei, along with smaller amounts of lithium and the two rare forms of hydrogen and helium. The observed abundances of these five light elements in today's universe agree well with those calculations and provide a striking confirmation of the big bang theory all the way back to 0.01 second.

The matter era begins

When the universe reached the age of about 1,000 years, another important change occurred. Scientists call the first 1,000 years the "radiation era," because the energy of all the radiation in the universe was greater than the mass of all the matter. By 1,000 years, the expansion of the universe had decreased the energy of the radiation greatly, and the mass of the matter was now greater than that energy. The universe's "matter era" began. Although the radiation was still intense enough to prevent the formation of complete atoms, the gravitational force of the dark matter was now the most important influence in the universe. Regions of higher density in the universe's dark matter then began drawing together.

By the age of 300,000 years, the universe had expanded to about $\frac{1}{1,000}$ its present size and cooled to about 3000 °C (5400 °F). That temperature was finally low enough for nuclei and electrons to combine to form complete atoms of the elements hydrogen, helium, and lithium.

The formation of atoms was an important event in the history of the universe. Until then, electrons and nuclei had constantly collided with, absorbed, and reemitted photons of electromagnetic radiation. But hydrogen and helium atoms interact much less readily with photons than do electrons and nuclei, so the collisions all but ceased. The radiation was now able to travel freely, and the universe became transparent, much as if a fog had cleared. As soon as radiation went its own way, it stopped having any influence on matter. From then on, gravity alone would determine the development of the universe.

That liberated radiation, meanwhile, continued to stream through space, becoming increasingly stretched out and weakened as the universe expanded. Today, it is in the form of microwaves—low-energy radio waves—known as the *cosmic background radiation.* These radio waves allow us to view the universe at an age of 300,000 years.

The cosmic background radiation, which was predicted by the big bang theory, was discovered in 1964. The existence of the background radiation—together with other evidence, such as the relative abundances of hydrogen, helium, and lithium—has provided strong support for the big bang scenario. Few cosmologists doubt that the theory is the correct explanation of how our universe began, though the earliest history of the infant universe—the first fractions of a second—remains to be put on a firmer basis.

The Formation of Galaxies and Other Structures

Wherever in the sky astronomers point radio telescopes, they pick up a hiss of microwaves. What they are detecting is the cosmic background radiation, low-energy electromagnetic radiation that is the "fossil record" of the infant universe. By studying both that radiation record and the present universe, cosmologists have developed an explanation for how the universe's galaxies and other large structures must have formed.

The universe began as a hot, dense "soup" of subatomic particles. As the universe expanded and cooled, it evolved more and more structure, from quarks and electrons to atomic nuclei to atoms, and finally to *macrostructure*—stars, galaxies, clusters of galaxies, and superclusters. The early stages in the development of structure were driven by the strong and electroweak forces, which operate at subatomic levels. Later, gravity played the dominant role.

Researchers have been trying for nearly 30 years to understand how all this happened. In the 1960's, physicists P. J. E. Peebles of Princeton University in New Jersey and the late Yakov B. Zel'dovich of the Soviet Union theorized that the only thing required to get the ball rolling would have been a very slight amount of "lumpiness" in the early universe. By that, they meant small variations in the universe's density—some regions where there was slightly more matter than average and others where there was slightly less.

The regions of higher density would have had more gravitational pull than the regions of lower density and thus would have expanded less rapidly. The matter density in these regions would then have grown, relative to the rest of the universe, further increasing the regions' gravitational pull, slowing their expansion even more and increasing their density even further. In this way, gravity made the universe lumpier and lumpier, doubling the lumpiness every time the universe doubled in size. Ultimately, regions of higher density stopped expanding and formed bound clumps of matter that evolved into stars and other large structures.

Until the 1980's, physicists thought the density variations in the early universe would have had to be at least 0.1 percent for this series of events to get started. That is, if a given volume of space contained an average of 1,000 particles of matter, some regions would contain 1,001 particles and others only 999.

But that estimate was made before researchers became convinced that the great bulk of matter in the universe is unseen dark matter. Many cosmologists now suspect that dark matter is in the form of subatomic particles created in the big bang. If that is the case, then only a 0.001 percent variation in density in the infant universe—100,001 particles in some areas and 99,999 in others—would have been required for gravity to do its work. A smaller variation would be required because in a universe containing dark matter, the process of clumping began sooner.

If the universe contained just ordinary matter, gravity could not have begun amplifying the initial lumpiness until about 300,000 years after the big bang. Before then, electrons and nuclei were constantly colliding with photons of electromagnetic radiation. But atoms interact much less often with photons than electrons and nuclei do. So when atoms formed at 300,000 years after the big bang, matter and radiation "decoupled," freeing the matter so that gravity could begin its work.

But dark-matter particles—if indeed dark matter is in the form of subatomic particles—would not be affected by electromagnetic radiation. Thus, dark matter would have begun responding to gravity much earlier—as soon as about 1,000 years after the big bang. With a longer time for structure to grow, a smaller degree of lumpiness would be required to begin with.

A snapshot of the early universe

Until recently, cosmologists had little evidence to support their ideas about the development of structure in the universe other than the existence of the structure itself. That situation changed in April 1992 with the announcement that the Cosmic Background Explorer satellite (COBE)—looking far across the universe and far back in time—had detected the tiny density variations as differences in the intensity of the cosmic background radiation. This finding provided direct evidence for the lumpiness of the infant universe.

The COBE satellite, launched by NASA in November 1989, was designed specifically to study the cosmic background radiation, which gives us a "snapshot" of the universe at an age of 300,000 years. The COBE project is headed by John Mather of NASA's Goddard Space Flight Center in Greenbelt, Md.

COBE data revealed energy variations of 0.001 percent in the microwave radiation, corresponding to regions of lesser or higher density in the early universe. Gravity caused the energy variations. In regions of higher density in the early universe—areas where there was slightly more matter—gravity was a bit stronger, and the radiation from those areas lost more energy.

Cosmologists were ecstatic when the COBE findings were reported. Finally scientists had convincing evidence that there had indeed been the lumpiness in the early universe required to account for the development of all the structure we now see. And the amount of lumpiness was the amount predicted by the dark matter theories.

The next question to be answered was: Where did that lumpiness come from? Theorists have suggested a number of ideas to answer that question, almost all of which involve events that took place in the first moments of creation. The most promising of these ideas is the *theory of inflation,* which holds that the universe went through a brief period of very rapid expansion at about 10^{-34} second after the big bang. In 1980, physicist Alan Guth of the Massachusetts Institute of Technology proposed this theory, which was later modified by other physicists, including Andrei Linde of the

How gravity shaped the universe

One big question that cosmologists have tried to answer is how galaxies formed and how the universe developed its present structure. The universe was originally hot and smooth, but over time it has become increasingly cooler and more structured. The COBE image of the infant universe, *right,* has helped explain how this process got started and how it progressed. The COBE data showed that the universe, while still very hot, already contained areas of slightly higher density and thus stronger gravity (blue patches in image). The gravity of high-density regions pulled matter together, and those concentrations of matter eventually developed into star-filled galaxies. A type of matter known as *dark matter* played the dominant role in the formation of galaxies. Dark matter, which apparently makes up at least 90 percent of the mass of the universe, emits no light and so cannot be seen. Many cosmologists believe that dark matter is a still-undiscovered type of subatomic particle. If that is the case, then the universe developed its structure in the following way.

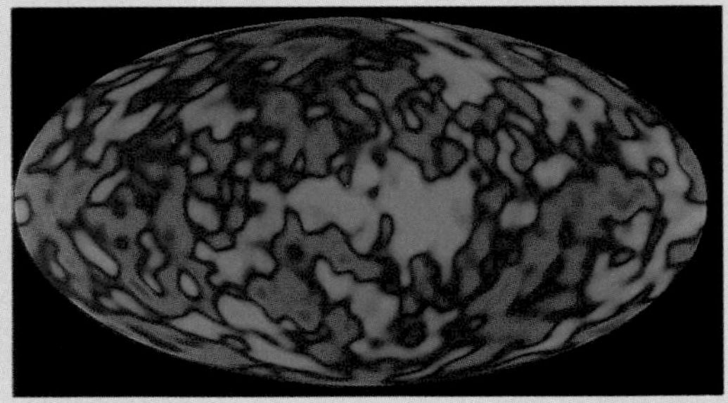

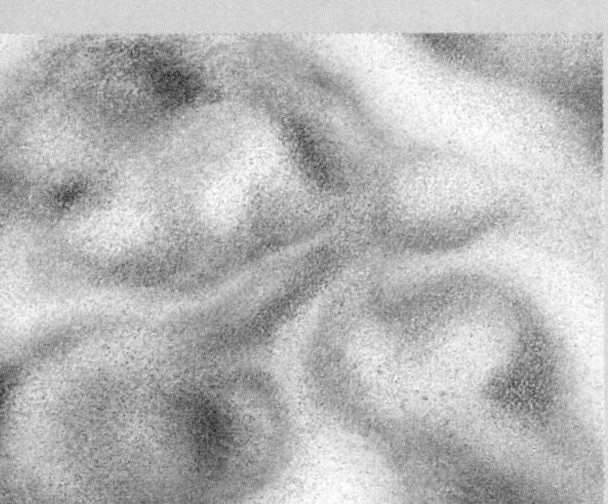

About 1,000 years after the big bang, the expanding universe was still an extremely hot "soup" of electrons, atomic nuclei, and dark matter particles. But already, regions of slightly higher density (purple areas) within the dark matter began to be drawn together by gravity.

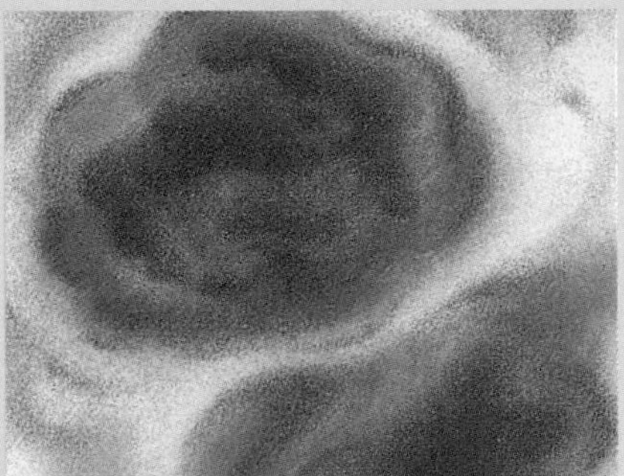

By about 300,000 years later, the universe had cooled enough for ordinary matter to form atoms, mostly of hydrogen. The high-density regions of dark matter then contracted further due to gravity and began to pull huge clouds of hydrogen gas (orange) toward them. Radiation left over from this period formed the COBE image.

After 1 billion years, the universe was dimmed of light and growing ever cooler. Large clumps of hydrogen gas and dark matter had drawn together.

former Soviet Union and Paul Steinhardt of the University of Pennsylvania. According to the theory, during inflation, slight variations in the universe's density developed. The theory also predicts that the universe's dark matter consists of subatomic particles left over from the big bang.

Hot or cold dark matter?

If these particles are truly the most abundant form of matter in the universe, how might they have guided the development of large-scale structure? Cosmologists have proposed two theories. One, known as the *hot dark matter theory*, says that dark matter particles are fast-moving neutrinos with a tiny amount of mass. The competing *cold dark matter theory* is based on hypothetical particles called axions or photinos. These particles are referred to as cold dark matter because they move very slowly.

Computer simulations have all but eliminated the hot dark matter theory. With hot dark matter, the simulations showed, superclusters of gas and neutrinos would have formed first and then fragmented to form galaxies and stars. The problem with this theory is that the oldest galaxies we can see today would not have had time to develop.

In simulations involving cold dark matter, structure forms from the bottom up—first galaxies, then galaxy clusters, and so on. In this view, gravity begins to amplify the tiny amount of lumpiness in the cold dark matter at about 1,000 years after the big bang. About 300,000 years later, ordinary matter is freed from the effects of radiation and is drawn by gravity to the lumps of dark matter. After about 2 billion years, the first *protogalaxies* (galaxies in formation) take shape. The protogalaxies contain many newly formed stars, though the details of how the stars would have formed are not well understood. Later, larger and larger structures form—fully developed galaxies, clusters of galaxies, superclusters, and even bigger structures in the future.

In these computer studies, cold dark matter seems to do much better than hot dark matter at reproducing the universe we see today. But simulations with cold dark matter have not worked perfectly. Consequently, some cosmologists have come to believe that the cold dark matter theory will have to be altered or even abandoned.

Some theorists have proposed, for example, that the early universe contained a mixture of both cold and hot dark matter. Others have suggested that the initial lumpiness was due not to inflation but rather to lumps or long strands of concentrated energy that were produced in the first moments of the universe and that served as gravitational "seeds" for the development of structure.

The debate shows that although our quest to understand the development of large-scale structure in the universe has come a long way, it is far from over. Cosmologists have a general understanding of how gravity shaped the universe but they are still uncertain about where the initial lumpiness came from and the details of how the universe's structure developed.

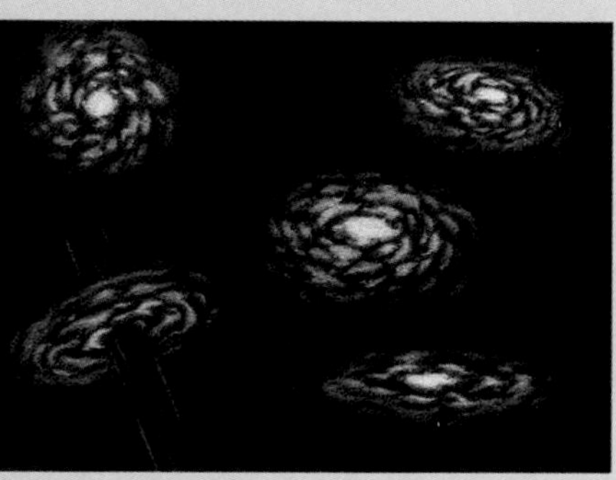

When the universe was about 2 billion years old, the clumps had contracted further, forming *protogalaxies* (galaxies in formation). Within them, the universe's first stars began to shine. Some protogalaxies became *quasars*, objects that for a while emit enormous amounts of energy.

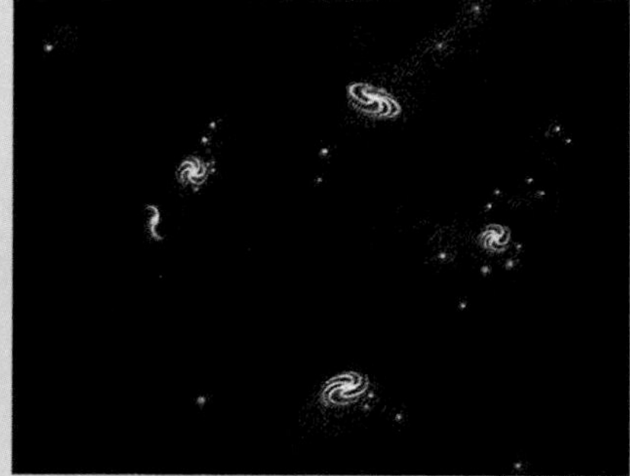

By 5 billion years after the big bang, the quasars and other protogalaxies had evolved into regular, star-filled galaxies with spiral or elliptical structures, and the galaxies began collecting into clusters. Within the galaxies, many stars had already finished their lives, but new stars were constantly being formed from the remaining clouds of hydrogen.

Today, an estimated 15 billion years after the big bang, the expanding universe continues to be dominated by the influence of gravity. The force has pulled galaxies into clusters and superclusters. Even larger structures will form in the future.

Future Challenges in Cosmology

Despite—and because of—all we have learned about the universe, many questions remain to be answered. As we answer old questions, new questions arise. What was the universe like in the very first instant of creation? What triggered the big bang? What is dark matter, and how much of it exists? Are the voids in the universe empty or do they hold unseen galaxies or other objects? How will the universe end? These are just a few of the issues that cosmologists are pondering. And there will undoubtedly be many other questions that no one has yet even thought to ask—questions that will pose a challenge to the next generation of cosmologists.

At the top of the list of questions that researchers are now trying to answer is the nature of dark matter, the mysterious stuff that accounts for most of the matter in the universe. A number of new experiments will help physicists determine whether dark matter is a previously unknown type of subatomic particle created in the big bang explosion that gave birth to the universe or whether it is ordinary matter in the form of *dark stars.* Dark stars are massive bodies—such as black holes, neutron stars, and Jupiter-sized planets—that give off no light. Two experiments that attempt to detect the dark matter particles that may pervade every part of our galaxy, including the solar system, were underway in 1993.

The Keck telescope, atop the Mauna Kea volcano in Hawaii, is the biggest optical telescope in the world. The Keck, completed in 1992, has a mirror 10 meters (400 inches) in diameter made of 36 hexagonal segments. With its great light-gathering power, the Keck produces images of celestial objects that are barely visible to smaller telescopes.

Determining the nature of the dark matter will help astronomers determine how much of it the universe contains. They might then be able to make a reliable estimate of the universe's matter content. That amount bears directly on one of the ultimate questions in cosmology: What is the fate of the universe?

Eternal expansion or "big crunch"?

If the matter content of the universe is less than a quantity that physicists refer to as the critical density, then the universe will continue to expand forever. That is because the gravitational pull of the galaxies on one another will not be enough to halt the expansion. If, on the other hand, the matter content exceeds the critical density, the expansion will, at some time in the far distant future, come to a halt. The galaxies—by then mostly burned-out debris—will then reverse direction, and billions of years later they will fall together into a "big crunch."

The current hunch among cosmologists—based in large part on the theory that the universe underwent a brief period of inflation in the first moments of the big bang—is that the universe is at the critical density. If so, the universe will expand at a slower and slower rate, but an infinite span of time would have to pass for it to come to a complete stop. In other words, there would be no big crunch.

Cosmologists do not yet have a reliable determination of the universe's matter content. They do know from their calculations of *big bang nucleosynthesis,* the process by which several kinds of atomic nuclei were created in the first minutes after the big bang, that ordinary matter contributes only 5 to 10 percent of the critical density. Thus, if the theory of inflation is correct, and if the universe is at the critical density, most of the matter must exist in the form of yet-to-be-discovered subatomic particles that were created in the moments after the big bang.

Two measurements reported by scientists in 1993 supported the idea that the universe is indeed at the critical density. The first measurement involved the cosmic background radiation, the hiss of microwaves that is the "echo" of the big bang. Data from the Cosmic Background Explorer satellite (COBE) and earlier measurements have shown that the microwave radiation is more intense in the direction of the constellations Hydra and Centaurus and less intense in the opposite direction. This finding can be likened to the fact that rain pelts harder on the windshield of a forward-moving car than on its rear window. It indicated that the Milky Way is moving in the direction of Hydra and Centaurus at a speed of about 620 kilometers (385 miles) a second.

The motion of the Galaxy is due to the gravitational pull of all the galaxies within a few hundred million light-years of us. The tug on the Milky Way due to any one galaxy depends upon that galaxy's mass, which is not known, and its location, which can be determined. In 1992, two research groups, led by astronomers Michael Rowan-Robinson in England and Michael Strauss of Princeton University in New Jersey, related the tugs of all the galaxies within a few hundred million light years of us to the COBE-measured velocity. From this, they inferred the matter content of this volume of space and estimated that the universe is near the critical density.

The second measurement reported in 1993 came from an astronomical satellite called ROSAT, which measures X-ray emissions from distant stars, galaxies, and clusters. ROSAT detected an enormous X-ray-emitting gas cloud surrounding three galaxies about 150 million light-years away. Astronomers calculated that the galaxies alone do not contain enough mass, in the form of visible stars, for their gravity to hold onto the gas cloud. They said the galaxies must have a much greater mass in the form of dark matter. If the mix of ordinary matter and dark matter in that group of galaxies is typical of the universe as a whole, then the universe would appear to be close to the critical density. Astronomers and physicists are hopeful that the dark matter question will be settled within a few years.

Seeking clues to creation

Astronomers also expect great progress in the coming years in mapping the large-scale structure of the universe. One mapping effort, the Sloan Digital Sky Survey, will measure the red shifts of more than 1 million galaxies and 100,000 quasars—information that will allow astronomers to construct a three-dimensional map of a significant portion of the universe. The survey will also answer questions about the nature of voids, regions of space that apparently contain no galaxies; superclusters; and, perhaps, the largest structures that exist. Once we better understand the structure of the universe today, we can put the theories about the evolution of structure in the early universe to the test. And then we can begin to unravel the mystery of the first moments of creation.

Workers in Texas stand in the enormous access shaft of a tunnel being excavated for the Superconducting Super Collider (SSC), the world's largest and most powerful particle accelerator. The SSC, with a circumference of 87 kilometers (54 miles), is designed to boost protons to extremely high energies. Collisions between those particles would simulate conditions that existed in the first moments of the universe.

The discovery by the COBE satellite in 1992 of small variations in the intensity of the cosmic background radiation, which provided the first evidence for the initial lumpiness that seeded the development of the universe's macrostructure, opened the door for the study of the lumpiness. The COBE data will be refined and extended through additional experiments, some with instruments carried aloft by balloons and others that will be conducted at the South Pole. The newer data will help scientists better understand the nature of the lumpiness and test theories of its origin in the earliest moments of the universe.

Other questions to be answered

There are a host of other questions that scientists expect to address in the late 1990's and in the early part of the next century. They hope, for ex-

A huge gas cloud surrounds a group of three galaxies in an image created by ROSAT, a satellite that detects X-ray emissions from celestial objects. Astronomers have calculated that the visible matter in the galaxies could not exert enough gravitational force to hold the gas cloud in place. Thus, they conclude that the group must contain large amounts of dark matter. If that group of galaxies is typical, the universe may contain enough matter to eventually halt its expansion. If so, the universe may someday contract into a fiery "big crunch."

ample, to see for the first time a galaxy in the process of forming, with its first stars lighting up. Researchers would also like to learn whether quasars are indeed powered by gigantic black holes, objects with such strong gravity that not even light can escape their gravitational pull.

Beyond those questions are the ones we are just beginning to ask. One such question is, why does the universe consist of regular matter, with almost no antimatter in evidence? The very existence of ordinary matter in the universe today stems from a slight excess of matter over antimatter in the moments after the big bang. But theorists want to learn precisely how and when that excess arose. Could it relate, they ask, to the slight preference that the laws of physics show to regular matter, a phenomenon discovered in 1964 by physicist James Cronin of the University of Chicago? Experiments with particle accelerators may help refine our understanding of the origin of the excess of matter over antimatter—and of matter itself.

There are many other questions that cosmologists are still a long way from answering. What happened before the big bang? Did the universe really undergo a period of rapid inflation in its earliest moments? Can we detect the neutrinos in today's universe that remain from just one second after the big bang? Does the space around us contain other dimensions, too small to perceive, in addition to the familiar dimensions of length, width, and height?

The role of technology in research

No doubt our ability to ask questions will always exceed our ability to answer them. Fortunately, however, we are living in a time of rapid technological progress, which has accelerated the pace at which we can build increasingly sophisticated instruments for investigating the universe.

Innovations in telescope technology are enabling astronomers to see out into space as never before. One major advance has been in the design of mirrors for telescopes. Before the 1980's, telescope builders could not make a reliable mirror much more than about 5 meters (200 inches) in diameter, but that limit has now been far surpassed. The largest telescope in the world is now the Keck telescope atop Mauna Kea, a volcanic peak in Hawaii. Completed in 1992, the Keck has a 10-meter (400-inch) mirror made up of 36 hexagonal segments. With such an enormous mirror, the Keck telescope has produced images of faint, faraway objects that smaller telescopes can barely make out.

The light-gathering power of the Keck and other new telescopes has been further enhanced by electronic detectors using charge-coupled devices (CCD's). A CCD captures light from the telescope mirror and converts it into electronic signals. The signals are stored in digital form, just like the images in a camcorder, and converted into an image on a computer screen. With the aid of specialized computer programs, the image can be enhanced, modified, and compared with other images. A CCD registers more than 70 percent of the light received by a telescope, compared with only a few percent for the best photographic plates.

Increasingly, astronomers are obtaining images from satellite observatories in orbit far above the Earth. The advantage of these space-based observatories is that they do not have to contend with Earth's atmosphere, which blocks the infrared, X-ray, and gamma-ray portions of the electromagnetic spectrum and blurs images in the visible part of the spectrum.

The National Aeronautics and Space Administration (NASA) has played the leading role in space-based astronomy. In April 1990, as the first event in its Great Observatories program, it launched the Hubble Space Telescope, named for American astronomer Edwin Hubble. The Space Telescope has been producing spectacular images despite an unexpected flaw in its primary mirror. (In the Special Reports section, see FIXING HUBBLE'S TROUBLES.)

The Hubble was followed into space in April 1991 by the Compton Gamma Ray Observatory, which surveys the universe in the highest-energy portion of the electromagnetic spectrum. NASA plans to launch two more observatories, the Advanced X-ray Astronomy Facility and the Space Infrared Telescope Facility, that will open other new windows on the universe.

Cosmologists are eying these projects with great interest. They are also looking forward to the construction of new research facilities that will give physicists their deepest look yet into the very heart of matter.

One of the most eagerly awaited projects is the Superconducting Super Collider (SSC), a huge particle accelerator under construction in Texas in 1993. With a circumference of 87 kilometers (54 miles), the SSC will dwarf all the accelerators now in existence.

The SSC will accelerate two beams of protons in opposite directions to higher energies than any so far achieved and smash them head-on into each other. The energy of the collisions will simulate the conditions that existed in the universe just 10^{-15} second after the big bang, helping cosmologists get a better understanding of the first moments of creation.

The costs and benefits of basic science

The SSC, when—and if—completed, would cost an estimated $8 billion. The Hubble telescope cost nearly $2 billion. The considerable expense of these and other large scientific projects has prompted many people in the United States—including some scientists—to question how much we should spend on research.

The quest for knowledge in all scientific fields is now limited by money, and one of the most important social questions of the 1990's is how much society can or should spend on the pursuit of knowledge. This question, relevant for all fields of science, is more difficult for research in subatomic physics and astronomy. Research in these fields offers few direct benefits for society and does little or nothing to directly address the life-and-death problems facing humanity and our planet.

Some scientists and business leaders argue that basic research almost always yields unexpected technological returns that far exceed the original investment. They cite many examples to support this view. For example, they point out, investigations into the structure and behavior of the atom led to the technology underlying virtually every electronic device in use today. And the space program has produced many high-technology "spinoffs," ranging from miniaturized electronics and high-speed computers to satellites that allow us to communicate around the world or look down on Earth to monitor the environment.

But some scientific endeavors produce benefits impossible to measure. The quest for understanding the universe involves not only professional scientists, but also students, who may help to discover something about the universe that no one has ever known before—like the precise mass of a subatomic particle or the existence of a very distant galaxy. The excitement of such a glimpse at one of nature's secrets can launch the career of a teacher, engineer, or doctor and ultimately benefit society as a whole.

And knowledge itself is a benefit, especially when it involves answering fundamental questions. Cosmologists are now learning the answers to questions that human beings have been asking since the dawn of history—how the universe came into existence, evolved to its present state, and will end.

For further reading:

Asimov, Isaac. *Atom.* Truman Talley Books, 1991.

Eames, Charles; Eames, Ray; and others. *Powers of Ten: About the Relative Size of Things in the Universe.* Scientific American Books, 1991.

Ferris, Timothy. *Coming of Age in the Milky Way.* William Morrow and Company, 1988.

Hawking, Stephen W. *A Brief History of Time.* Bantam, 1988.

Kaler, James B. *Stars.* Scientific American Books, 1992.

Lederman, Leon, and Teresi, Dick. *The God Particle.* Houghton Mifflin, 1992.

Overbye, Dennis. *Lonely Hearts of the Cosmos: The Story of the Scientific Quest for the Secret of the Universe.* HarperCollins Publications, 1991.

Riordan, Michael, and Schramm, David. *Shadows of Creation: Dark Matter & the Structure of the Universe.* W. H. Freeman & Company, 1991.

Voyage Through the Universe (series). Time-Life Books, 1989.

Weinberg, Steven. *The First Three Minutes: A Modern View of the Origin of the Universe.* Basic Books, 1988.

Science News Update

Science Year contributors report on the year's major developments in their respective fields. The articles in this section are arranged alphabetically.

Page 204

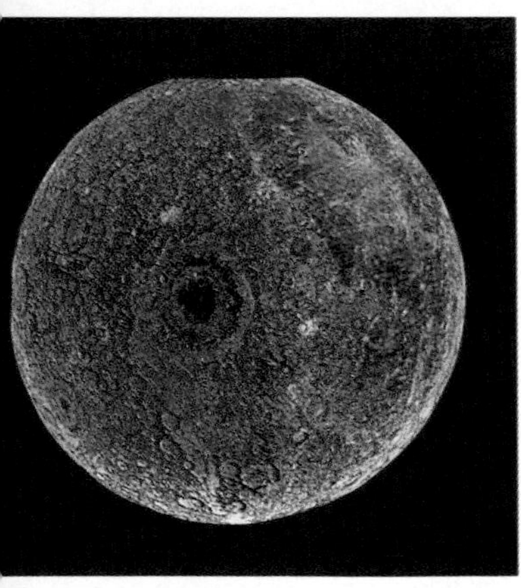

Page 221

Page 254

Page 290

A team of biologists reported in June 1992 that they had developed the first genetically engineered wheat plants. The researchers, at the University of Florida's Institute of Food and Agricultural Sciences in Gainesville and at Monsanto, Incorporated, of St. Louis, Mo., developed wheat plants resistant to a powerful herbicide that normally kills any plant that it touches.

Wheat is the world's most important food crop. Each year, growers worldwide produce about 550 million metric tons (600 million short tons) of wheat valued at $60 billion. But unlike rice and corn, which biologists have already altered genetically, wheat has proven difficult to genetically improve.

Research team leader Indra K. Vasil and his colleagues used a device called a gene gun to insert genes for resistance to the herbicide into wheat cells. The gun fired tiny gold pellets coated with genes into a petri dish containing wheat cells and nutrients. The cells were undifferentiated, very young cells that had not yet developed the specialized functions of the cells in a mature plant. Vasil then added the herbicide to the petri dish. The wheat cells without the gene died, but the cells that had taken up the gene from the gold pellets grew into plants. When the plants flowered, the biologists bred them with normal wheat plants. The resulting seeds produced plants that were resistant to the herbicide, indicating that the engineered parent had transmitted the desired gene. And the research team found that the resistant trait passed to many successive generations of wheat plants.

Electronic plant patrol. A new, 24-hour disease patrol was put to work in peanut fields in Texas for the first time in 1992. But the "inspector" worked quietly at a desk indoors and not among the plants outside.

Scientists from the Texas Agricultural Extension Service in July field-tested a computerized system for monitoring conditions that encourage sclerotium blight, a fungal disease. The blight attacks peanut plants, reducing peanut yields. It develops when the soil is wet and cooler than 27.8 °C (82 °F). Normally, a farmer detects the blight only

Researchers measure and record the amount of soil trapped uphill from a clump of switchgrass planted in a test channel designed to simulate the grade of a farm field. Soil-bearing water was sent down the channel the same way that rain water would flow from a field after a storm. In the laboratory and in field tests in summer 1992, switchgrass and other species of tall, stiff grass controlled soil erosion and conserved water better than did terracing using bulldozers.

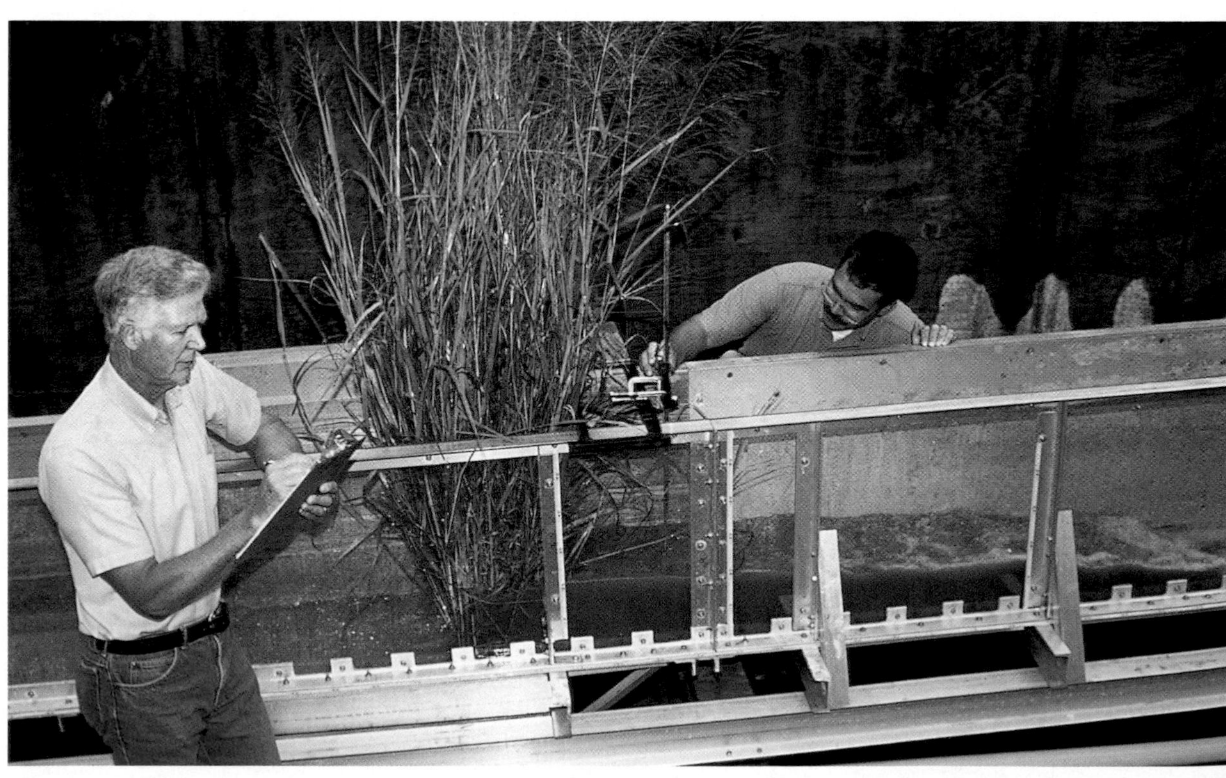

Using bees to prevent blight
In spring 1993, agricultural researchers used honey bees pollinating Utah test orchards to spread bacteria that prevent fire blight. Fire blight, a disease that can kill pear and apple trees, is caused by another type of bacteria that first infects tree blossoms.

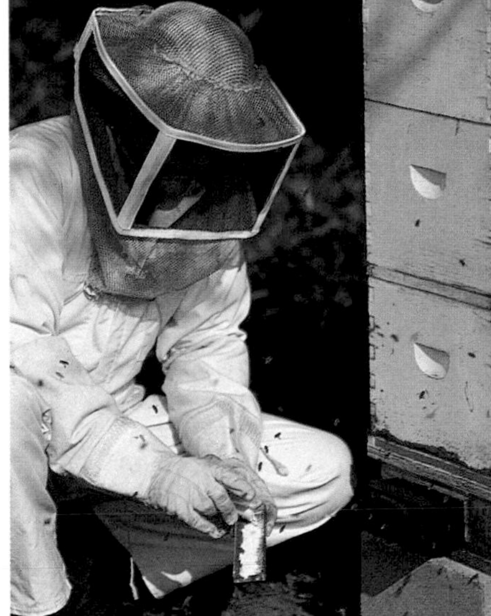

A scientist, *right,* fills a tray with a mixture of pollen and bacteria that keeps the disease-causing microbes in check. After the tray was placed in a beehive, bees entering and leaving the hive picked up the powdery substance on their body hairs. Whenever a bee visited an apple blossom, *far right,* it left beneficial bacteria behind.

after the plant leaves droop or other symptoms appear. But by this time, the blight has already damaged the plants so severely that fewer peanuts develop.

The Texas scientists devised a computer program that records and analyzes data gathered by two special sensors placed on a pole in the field. One sensor is aboveground to register air temperature, and the other is belowground to register the amount of moisture in the soil. The computer can be programmed to gather data as often as every 15 minutes.

The system detects conditions that invite an outbreak of blight some three weeks before the disease would be visible on the plants. In laboratory experiments, the system forecast Sclerotinia blight with 100 percent accuracy.

The Texas researchers said that using the computerized watchdog would enable a farmer to treat a threatened field with a fungicide before the disease took hold. The early-warning system also could prevent the application of chemicals as a precautionary measure when they were not needed.

Robot melon picker. A new robot locates melons, "decides" if they are ripe, and picks only the ones that are ready for market. Researchers from Purdue University in West Lafayette, Ind., and the Volcani Institute in Bet Dagan, Israel, devised the robot, which was tested in melon fields in Israel beginning in September 1992 and continuing into the 1993 growing season.

Purdue agricultural engineer Gaines Miles and his colleagues designed the robot, which resembles the skeleton of a large utility trailer. The robot's "eyes" are cameras mounted at two points on the trailer frame. As a tractor pulls the robot between the rows, a fan blows across the plants, moving the leaves and exposing the hidden fruit for the cameras to see.

A computer analyzes the camera images, looking for round, bright spots that may be melons. To help differentiate a melon from round but flat objects, a sensor projects a laser light beam onto the ground. When the laser strikes a spherical object at the same time the camera records a bright spot, the com-

Developing a robot that picks melons
A joint project by scientists in the United States and Israel developed a robot that successfully detected and picked ripe melons from a field in 1993. A scientist, *below,* tests the robot's electronic "sniffer," a sensor that judges a melon's degree of ripeness by measuring the amounts of aromatic gases given off by the fruit. The greater the levels of those gases, the riper the melon.

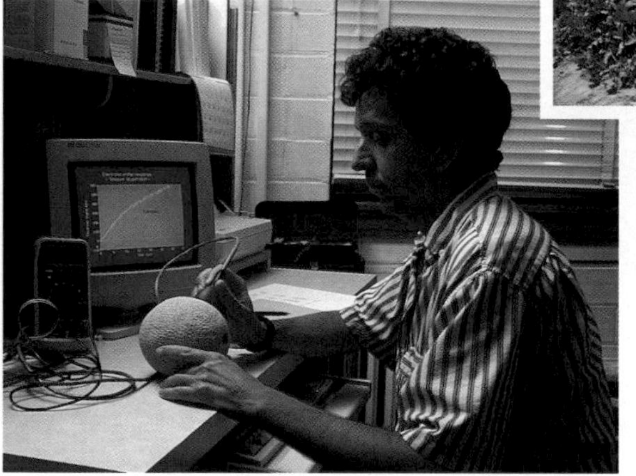

A technician drives a tractor pulling the robot in a melon field during tests in Israel, *above*. When cameras mounted on the frame recorded a bright image at the same time as a laser beam struck a spherical object, a computer determined that the object was a melon. Then, if the electronic sensor detected the proper degree of ripeness, gripper arms picked the fruit and sent it down a conveyor for packaging.

Agriculture Continued

puter determines that the object is a melon.

When the robot locates a melon, it determines its ripeness through another sensing device. That sensor measures the level of aromatic gases coming from the melon. The amount of gases the fruit gives off is directly related to the degree of its ripeness—the more gases, the riper the fruit. Tests showed that the sensor measured the time of ripeness accurately to within a day. Finally, the robot picks the ripe melons with one of two gripper arms.

A sticky solution to the problem of a highly destructive pest, the apple maggot, was reported by scientists at the University of Vermont in Burlington in 1992. Apple maggot infestations can ruin crops. Each female apple maggot fly lays as many as 15 eggs under the skin of one apple. After each maggot hatches, it burrows into the flesh of the fruit, leaving rusty streaks. The maggots feed inside the fruit from four to six weeks before dropping to the ground and entering the soil to change into pupae and eventually adult flies.

The Vermont researchers isolated a bacterium found in the digestive tract of the maggots that attracts adult flies of the species. The researchers found that if they put the bacterium on sticky traps (strips of paper similar to flypaper) in an apple orchard, the apple maggot flies sought out the bacterium, became stuck to the trap, and died before laying eggs.

Trapping adult apple maggot flies using bacteria is a method of biological pest control that may be effective for other orchard pests, according to entomologist George MacCollom, the research team leader. With a grant from the United States Environmental Protection Agency awarded in October 1992, MacCollom and his team started research to find a bacterium common to several other fly species that attack trees or developing fruit in orchards.

A shorter route to flour. A new compact flour mill suitable for use in developing nations, called the Kice Shortflow Unit Flour Mill, was designed in 1992. Agricultural engineer Steve Curran of Kansas State University in Manhattan and his associates at Kice Industries, Incorporated, of Wichita, Kans., developed the mill for use in regions where there is a need to grind grain quickly for immediate use. In the United States, on the other hand, mills are designed for storage as well as for grinding a variety of grain types. U.S. mills are often four to seven stories tall, contain sophisticated screening systems, and require more than a year to build.

The Shortflow mill is a compact 7.3 meters (24 feet) tall and only 4.7 meters (15.5 feet) wide. Its space-saving design calls for routing grain on a direct path to the grinding equipment. The mill can be shipped to a location, assembled, and bolted down to a concrete pad in less than a week. According to Curran, if set up in an existing building, the mill costs about one-tenth the cost of a traditional mill and requires only 40 percent of the electricity used to run a traditional mill. And, he said, one person can operate the mill with very little training.

Mice models for cows. Could a "herd" of genetically altered mice help agricultural scientists improve the production of meat and milk in cattle? Scientists at the University of Missouri in Columbia think so, according to their March 1993 announcement. By studying 600 transgenic mice, the researchers hoped to learn how farm animals can be genetically improved. (A transgenic animal has at least one gene from another species.) The mice, developed at Ohio University in Athens, had been given a cow gene that controls the production of a growth hormone.

The Missouri biologists, William Lamberson and Randall Pratheel Roberts, chose the mice to study because the animals are less expensive to care for than cattle, and they reproduce more quickly, allowing scientists to study more generations in a relatively short time. The scientists found that, in all cases, the transgenic mice were about 80 percent heavier than normal mice and that the weight differences were greater between the males of both types than between the females. The transgenic mice also used feed more efficiently than did the normal mice in the study.

[Steve Cain and Victor L. Lechtenberg]

In WORLD BOOK, see AGRICULTURE.

Anthropology

Three remarkably complete skulls found in a cave in Spain provide new clues to the origins of the Neanderthals, according to an April 1993 report by Spanish scientists. The Neanderthals were a type of prehistoric people who lived in Europe and western Asia from about 130,000 years ago to about 40,000 or 50,000 years ago.

The skulls were found by Spanish scientists excavating a cave called Sima de los Huesos (Pit of the Bones) in the Atapuerca Mountains of northern Spain. In the cave, scientists had previously found hundreds of human fossils. But most of these fossils were fragments, and scientists had determined only that the fossils belonged to a relatively primitive species of the genus *Homo.* The genus *Homo* includes all living people and their prehistoric human ancestors.

The three newly discovered skulls were those of two adults and one juvenile. Previous tests at the site helped date the fossils to at least 300,000 years ago—long before Neanderthals lived in the area. But the scientists reported that the fossils exhibited several traits commonly found in Neanderthal fossils. Thus, they appeared to represent an early stage in the evolution of the Neanderthals.

The discovery of the Atapuerca fossils provides support for the theory that the evolution of the Neanderthals occurred over a long period of time in Europe. The skulls also provide evidence that about 300,000 years ago, human beings evolving in Europe were different from those developing in Africa and Asia.

Ancient skulls in China. The discovery in China of two ancient skulls that may be from 200,000 to 300,000 years old has encouraged scientists who believe that modern human beings evolved simultaneously on more than one continent. Paleontologists Li Tianyuan of the Hubei Institute of Archaeology in China and Dennis A. Etler of the University of California at Berkeley announced the discovery of the skulls in June 1992.

Li and Etler reported that the skulls, which were crushed and broken, display a combination of primitive and more modern characteristics. For example, the skulls have a relatively massive, long, low, and sharply angled braincase (the part of the skull enclosing the brain), like that of *Homo erectus. H. erectus,* whose fossils have been found in Asia and Africa, first appeared about 1.8 million years ago. The Chinese skulls also have relatively short, broad, flat faces, somewhat like those of *Homo sapiens.*

Most anthropologists believe that *H. erectus* evolved into early *H. sapiens* about 500,000 years ago. Physically modern human beings, classified as *H. sapiens sapiens*—the subspecies to which all living people belong—appeared between 100,000 and 50,000 years ago.

Some scientists argued that the Chinese skulls provide support for the multiregional theory of human evolution, one of two major explanations for the origin of modern-looking human beings. According to this theory, physically modern human beings evolved from more primitive human beings simultaneously in Europe, Africa, and Asia. Scientists who support this theory argue that differences between modern racial groups, such as those in the face and skull, were apparent as early as several hundred thousand years ago.

The second explanation for the origin of modern-looking human beings is commonly called the "Out-of-Africa" theory. Advocates of this theory argue that modern human beings evolved in a relatively small region, probably Africa, and spread out to other continents beginning about 50,000 years ago, replacing all older groups of *H. sapiens.*

Early human beings. The idea that the evolution of the first human beings may have been more complex than most scientists had believed triggered continued debate in 1992 and 1993. The argument began in early 1992 when anthropologist Bernard Wood of the University of Liverpool in the United Kingdom published articles concluding that several fossils classified as *Homo habilis,* traditionally considered the oldest human species, actually represent two species of *Homo* that lived at the same time. Wood, a leading authority on early

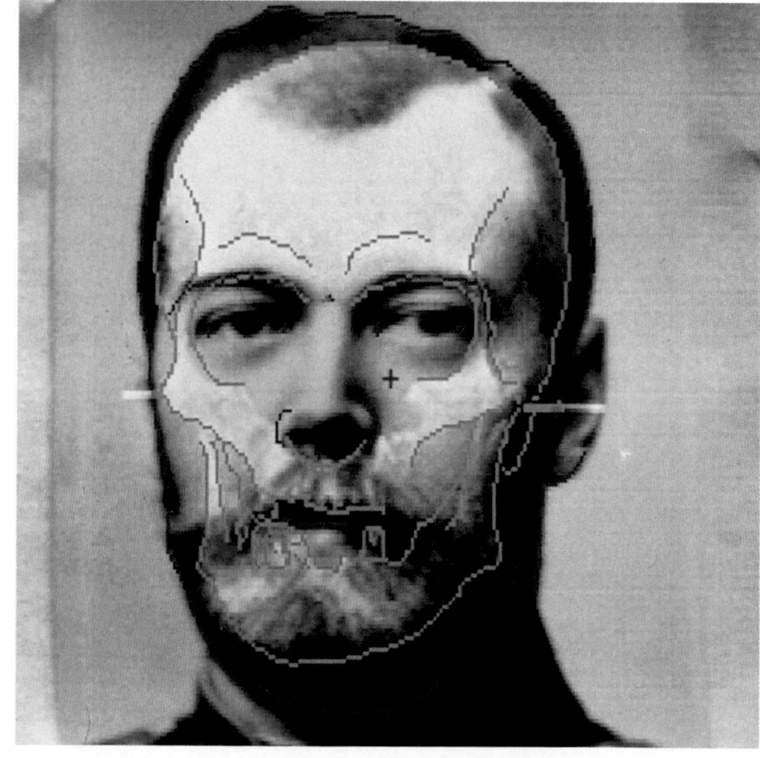

A computer model of a skull that Russian researchers identified in June 1992 as that of Russian Czar Nicholas II is superimposed on a photograph of the czar. An American anthropologist led a team of forensic experts who independently assessed the findings of the Russian researchers and found it "highly probable" that the skull and other skeletal remains were those of the czar and members of his family who were killed in 1918.

Transitional skulls?
Skulls discovered in China display characteristics of *Homo erectus* and *Homo sapiens* and may be from 200,000 to 300,000 years old, according to two scientists who analyzed the skulls and reported their findings in June 1992. The scientists said the discovery casts doubt on the theory that modern human beings originally evolved only in Africa.

human beings, based his conclusion on a reanalysis of several *H. habilis* fossils found in eastern Africa since 1964.

The most widely accepted theory of human evolution states that the relatively apelike *Australopithecus* evolved into *H. habilis,* believed to have been the first hominid to produce stone tools. (Hominids include human beings and their prehuman ancestors.) *H. habilis* evolved into *H. erectus,* a more advanced species, about 1.8 million years ago, and *H. erectus* evolved into *H. sapiens.*

In his articles, Wood contended that the *H. habilis* fossils actually represent two species—a previously identified species called *Homo rudolfensis* and a more narrowly defined *H. habilis.* Wood concluded that, compared to *H. habilis, H. rudolfensis* had a larger body and brain, and its legs and feet were more humanlike. In contrast, the face and teeth of *H. habilis* were more humanlike than those of *H. rudolfensis.*

Wood also theorized that neither *H. rudolfensis* nor *H. habilis* was the direct ancestor of modern human beings. He contended that this ancestor was a third species of *Homo,* also living in eastern Africa at that time. According to Wood, fossils that other scientists had classified as *H. erectus* actually belong to this species, called *Homo ergaster.*

Some scientists believe that Wood's *H. rudolfensis* and *H. habilis* are simply males and females of the same species. But Wood's research raised some fascinating questions, including how several human species may have managed to coexist and why all but one became extinct.

Ancient stone tools. Stone tools found in southern Ethiopia provide strong evidence that prehistoric people first began making hand axes about 1.4 million years ago. That conclusion was reported in December 1992 by a team of scientists from Ethiopia, Japan, and the United States.

The dating of the oldest previously known stone hand axes was based almost entirely on findings from a single site near Lake Natron in northern Tanzania. Those tools were also dated to 1.4 million years ago. The newly discovered hand axes were found at several sites known together as Konso-Gardula.

At Konso-Gardula, the scientists also unearthed fossils from the hominid that apparently made the tools. Anthropologists classify this hominid as either *H. erectus* or *H. ergaster.* Fossils from this hominid had been found in Tanzania as well.

Ancient tools are relatively abundant and easy to find, and they are useful for tracking changes in human behavior. The oldest known stone tools have been found in eastern Africa at sites that are roughly 2.3 million to 2 million years old. These tools, which are called Oldowan tools, consist mainly of sharp pieces of stone, called *flakes,* and the *cores* (large pebbles) from which the flakes were chipped. Oldowan pebble tools usually were chipped on only one surface and had a relatively short cutting edge.

Scientists are unsure exactly how these stone artifacts were used. However, experiments with modern replicas of Oldowan flakes suggest that they would have been useful for slicing through hides and meat. It also appears that the pebble tools would have been useful for breaking open bones to obtain marrow.

Hand axes and similar tools appeared after Oldowan tools. The newer tools are chipped on both surfaces and their cutting edge extends around the circumference of the stone.

Prehistoric people probably used hand axes for several purposes. Modern experiments with replicas indicate that hand axes were more useful than Oldowan tools for dismembering very large animal carcasses.

The improved ability to obtain meat may have helped hominids who were capable of making hand axes to colonize new regions, including the far southern and northern reaches of Africa and parts of Eurasia. Archaeologists have found hand axes dating from before 1 million years ago in the Jordan Valley of Israel. The oldest known hand axes found in Europe are about 500,000 years old. [Richard G. Klein]

In the Special Reports section, see UNCOVERING AFRICAN AMERICANS' BURIED PAST. In WORLD BOOK, see ANTHROPOLOGY; NEANDERTHALS; PREHISTORIC PEOPLE.

Archaeology, New World

Findings at the site of a Cherokee village established in the early 1700's in South Carolina indicate that, contrary to scholars' belief, Cherokee society changed slowly after these Indians' first contact with European colonists. In November 1992, archaeologist Gerald F. Schroedl of the University of Tennessee in Knoxville reported that his studies of the site indicate that the Cherokee used traditionally made objects and followed tribal customs for several decades after European colonists began moving into their territory.

By the early 1800's, the Cherokee were living in European-style houses and dressing in European-style clothing. In the 1820's, the Cherokee began using the Roman alphabet to write their language. For these reasons, scholars had assumed that the Cherokee had quickly adopted European customs.

The existence of the village, called Chattooga, was known from historical records. To guide their excavations there, Schroedl and his team relied on an instrument called a proton magnetometer. This device detects variations in the intensity and direction of magnetic fields below the ground. Such variations may be caused by buried structures and burnt material.

The main focus of the excavation was the remains of a Cherokee "townhouse," a structure used for community gatherings. The townhouse, which was about 15 meters (50 feet) in diameter, had burned down and had been rebuilt several times. In the burned debris from the most recent fire, archaeologists found fragments of stems from smoking pipes, which were dated by their distinctive shape to about 1735. The archaeologists concluded that the townhouse had been abandoned by the Cherokee shortly after that time.

Schroedl also excavated part of a dwelling in the village. This circular structure was about 8 meters (26 feet) in diameter and had a central hearth and four holes where large support posts once stood. According to Schroedl, the presence of the hearth indicated that this structure was used by the Cherokee during the winter months. In the summer months, the Cherokee cooked on

Forensic anthropologists in November 1992 examine the oldest and best-preserved skeletal remains of an early American colonist, one of three corpses entombed in lead coffins in the late 1600's and discovered in 1990 in a field near St. Mary's City, Md. Scientists said the find had "opened a window to the 17th century."

hearths located outside of their homes.

Schroedl and his team also found a variety and abundance of artifacts. Among these were traditionally made pots as well as stone tools and pipes made from rock available nearby. The archaeologists also unearthed a number of European-made artifacts, including glass trade beads, pipes, and bell-like ornaments. But they found only a few gun parts and no European-American knives, hoes, or axes.

Maps of the site made using the proton magnetometer showed that the village consisted of between 10 and 15 dwellings. The archaeologists estimated that about 90 people had lived in the village. Historical records about the village made by European settlers support this estimate.

Clovis engraved pebbles. Fifteen engraved limestone pebbles unearthed in central Texas and dated to about 11,200 years ago represent the largest collection of Clovis art ever found. The discovery of the pebbles was reported in October 1992 by archaeologists Thomas R. Hester, Michael B. Collins, and Pamela Headrick of the University of Texas at Austin. The Clovis people have been considered the first inhabitants of North America. Only one other engraved Clovis stone had been found previously.

The pebbles, discovered at an ancient Clovis campsite, range in length from 4.5 to 16 centimeters (2 to 6 inches). They bear a variety of designs executed in carefully engraved lines. One pebble has a grid pattern covering one side. Other pebbles are partially covered with grids. Still others are decorated with spirals and abstract symbols. One large flat pebble displays several plants and another displays what may be the profile of an animal.

According to Hester, the presence of what were apparently meaningful symbols on the pebbles indicates that the artifacts made by the Clovis people were more complex than archaeologists had believed. The purpose of the pebbles is unknown, but Hester and his colleagues theorized that they might have been used in rituals.

Olmec reinterpretations. Studies of the culture of the ancient Olmec Indians have yielded important new findings, according to a June 1992 report by archaeologist David Grove of the University of Illinois at Urbana-Champaign. The Olmec culture flourished on the tropical plain of the Mexican Gulf Coast between 1150 and 500 B.C. The new interpretations are based on recent findings by archaeologists excavating Olmec sites and on new readings of symbols and designs used by the Olmec.

Massive stone sculptures discovered in the 1930's and 1940's provided modern scholars with their first knowledge of Olmec civilization. These spectacular sculptures, which came to symbolize Olmec culture, included colossal stone heads weighing up to 22 metric tons (24 short tons). Since the heads were discovered, archaeologists have identified four major Olmec sites—La Venta, San Lorenzo, Tres Zapotes, and Laguna de Los Cerros, all in the Mexican states of Tabasco and Veracruz. At these sites, archaeologists found massive stone slabs weighing up to 33 metric tons (36 short tons) that they identified as altars, huge statues, and large *stelae* (upright stone slabs with carvings of rulers or priests).

Excavations and the study of Olmec symbols since the 1970's have led Grove and his colleagues to theorize that the colossal heads depict actual rulers. Similarly, they hypothesize that the "altars" served as kingly thrones.

In his 1992 report, Grove discussed the mutilated condition of numerous Olmec monuments. Many statues were missing the head and arms, for example, and large pieces of stone had broken off the thrones. Scholars once attributed this damage to later non-Olmec invaders. But Grove, citing excavation data from the Olmec site of San Lorenzo, believes that the Olmec themselves carried out the mutilation because the statues had been buried in Olmec times. The mutilations may have been part of sacred or ritual practices, perhaps at the death of an important ruler.

Findings at Olmec sites have also led scholars to revise their theories about the origin of the Olmec. Previously, they had believed that the monuments and settlements of Olmec civilization appeared suddenly after the ancestors of

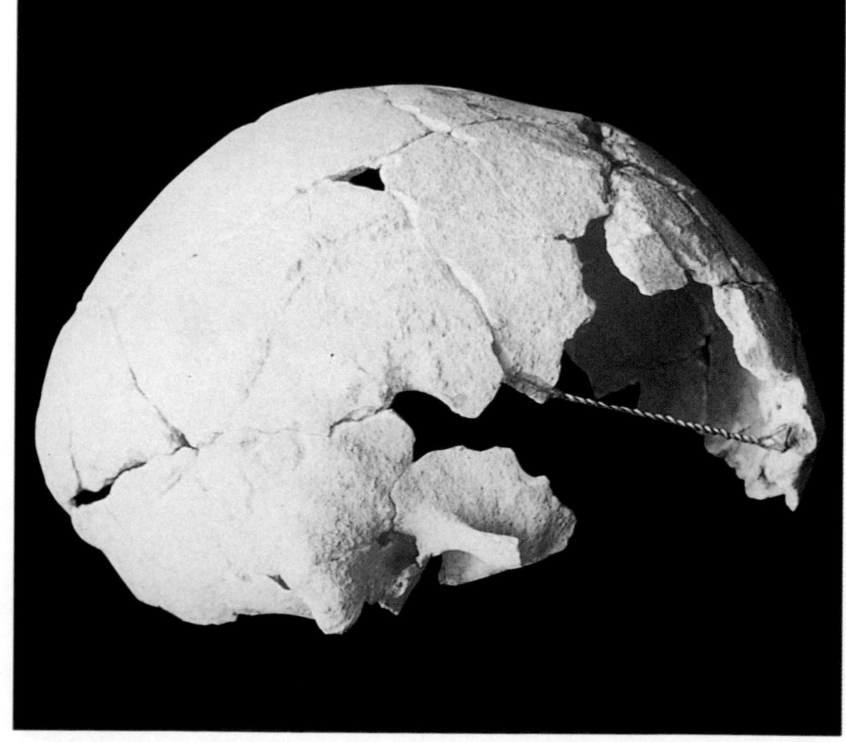

A skull fragment of Midland Woman, the name given to cranial and rib fragments discovered near Midland, Tex., was dated at 11,600 years old, making it the oldest human remains found in the New World. Archaeologist Curtis R. McKinney reported in October 1992 that he used sophisticated dating techniques to confirm the date.

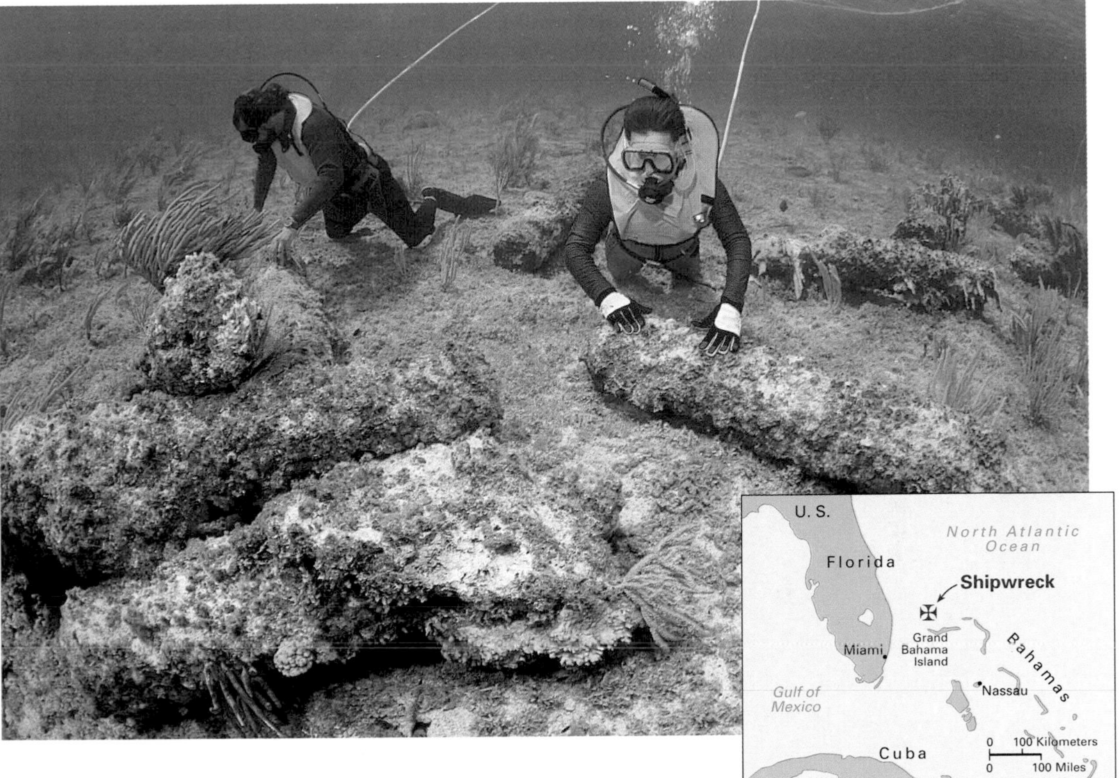

Divers search the wreckage of what may be the oldest shipwreck in the Americas, the well-preserved remnants of a Spanish sailing ship from the early 1500's. It was found off Grand Bahama Island (inset) in the West Indies. Archaeologists uncovered artifacts from the wreck, such as crossbows and guns, in summer 1992.

the Olmec migrated onto the coastal plain about 2000 B.C. Excavations by Grove and other archaeologists, however, revealed that the Olmec civilizations arose after a long period of cultural development, beginning with the establishment of agricultural villages in about 1500 B.C.

Another major change in scholars' thinking concerns the purpose of the Olmec sites. The presence of the massive stone monuments led scholars to conclude that the sites had been ceremonial centers occupied by small numbers of priests. Excavations by Grove and others confirmed that the sites were indeed the centers of political-religious power. But the archaeologists also discovered that thriving communities had once been associated with the centers.

The greatest change in the way scholars view Olmec culture concerns new interpretations of the symbols and designs on the Olmec's stone sculptures. These reinterpretations are based on advances in the 1980's made in deciphering Maya writing. Many of the symbols used by the Maya are also found on Olmec art.

For example, many Olmec monuments bear designs that resemble the facial features of a jaguar. Previously, scholars believed that the feline designs were linked to the concept of a jaguar deity. Now scholars believe that these Olmec symbols were not feline but instead represent the crocodile and a sharklike creature. These symbols reflected Olmec concepts of the human world and an underworld that was the domain of the supernatural. Another common Olmec symbol—the serpent—is now thought to represent a supernatural being affiliated with the elite or ruling class.

Olmec culture faded from the coastal plain after 500 B.C. Scholars believe the culture neither collapsed nor was destroyed. Rather, the Olmec were apparently absorbed by later Mexican cultures. [Thomas R. Hester]

In the Special Reports section, see UNCOVERING AFRICAN AMERICANS' BURIED PAST; SOLVING THE MYSTERY OF THE MOCHE SACRIFICES. In WORLD BOOK, see ARCHAEOLOGY; CHEROKEE INDIANS; INDIAN, AMERICAN; OLMEC.

The discovery of strands of silk in the hair of an Egyptian mummy dated to 1000 B.C. suggests that the ancient Egyptians began using silk about 1,300 years earlier than archaeologists had believed, according to a March 1993 report. That possibility in turn suggests that commerce along the Silk Road, the major trade route of the ancient world, may have begun about 900 years earlier than the date traditionally accepted by scholars.

The Silk Road was a network of caravan routes more than 6,440 kilometers (4,000 miles) long that began in China and passed through India and the Middle East. The routes ended in Rome, in Byzantium (now Istanbul, Turkey), and in other ports on the Mediterranean Sea. For about 1,500 years, traders traveled westward from China with silk, furs, and ceramics that they exchanged for gold, wool, ivory, and glass.

The silk strands were found in the mummy of a woman, 30 to 50 years old at the time of her death, who had been buried in a workers' cemetery at Thebes, a major religious center in ancient Egypt. Scientists at the University of Vienna determined that the strands were silk by analyzing the pattern of infrared light reflected by the strands. The researchers then dated the silk using a technique called amino-acid dating, in which scientists measure the amount of change that has occurred in the structure of certain amino acids (the building blocks of proteins) in a body since death.

The Austrian scientists concluded that although the use of silk did not become common in ancient Egypt until the A.D. 300's, its presence with the mummy suggests that silk existed in Egypt as early as 1000 B.C. If that is the case, then trade between China and Western countries did not begin in the late 100's B.C., as scholars have traditionally believed, but about 900 years earlier.

Ancient cave art. New explorations of a French cave decorated with some of the most beautiful prehistoric drawings and engravings ever found revealed many more ancient images in summer 1992. Among them were the first rock

A strip of land at the ancient seaport of Caesarea on the coast of present-day Israel holds the remains of a palace built by Herod the Great, a team of American archaeologists reported in December 1992. The archaeologists identified a mosaic-floored dining room and a huge pool at the site, which was once thought to be a public bath or a fish market.

paintings of penguins discovered in Western Europe. The French archaeologists who made the new finds also reported in summer 1992 that studies of charcoal found in the cave—believed to be from torches used by the artists to light their work—confirmed that some of the drawings are up to 27,000 years old.

The partially flooded cave, which was discovered in July 1991, lies deep within a cliff on the edge of the Mediterranean Sea, about 13 kilometers (8 miles) southwest of Marseille, France. A tunnel that slopes upward for about 140 meters (460 feet) connects the main chambers with the entrance, which now lies about 34 meters (112 feet) below the surface of the Mediterranean. At the time prehistoric people decorated the cave's walls, however, sea level was about 110 meters (360 feet) lower, and the cave entrance lay several kilometers inland.

Altogether, French archaeologists found more than 100 images in the cave, which may have been used for religious ceremonies or other rituals.

The archaeologists reported that the art may have been created during two separate periods. Art of the first period, which dates from about 27,000 years ago, consists of thousands of finger tracings, lines made by human fingers moving across the soft mineral coating on the rock surface, along with more than 45 stenciled outlines of human hands. Archaeologists believe that the hand stencils represent a kind of code because they are arranged in groups and many of the hands have one or more shortened fingers.

The second period of work dates from about 18,500 years ago to 19,000 years ago. It consists of drawings and engravings of penguins, seals, sea creatures that resemble jellyfish or squid, *ibex* (wild goat), red deer, horses, bison, and other animals that lived in the cool climate of this part of Europe at that time.

Oldest rock art? Engravings found on rock walls in south-central Australia and dated to more than 43,000 years ago may be the oldest known rock art. That conclusion was reported in August

A drawing of a penguin, one of three images of this cold-climate sea bird found on the walls of a French cave in summer 1992, was dated to about 18,500 years ago. The drawings—the first of penguins ever found among the prehistoric cave images of Western Europe—have survived from a time when the climate of this region was considerably cooler.

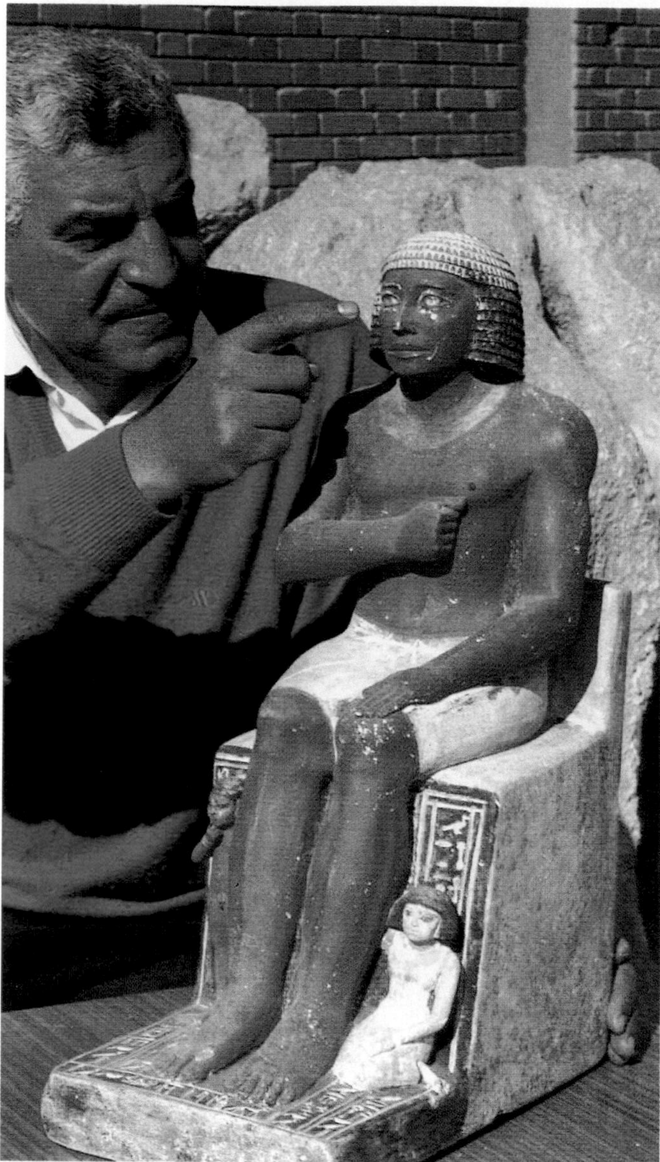

An Egyptian archaeologist points to a painted limestone statue of Kai, a high priest of ancient Egypt. The discovery of the 4,600-year-old statue was announced by Egyptian archaeologists in April 1993. It was found in an ancient cemetery for upper-class royal officials in the Egyptian city of Giza outside Cairo. Figures of a daughter and a son cling to the legs of the statue.

1992 by earth scientist Ronald Dorn of the University of Arizona at Tucson and archaeology student Margaret Nobbs of Flinders University of South Australia. The engravings consist of 19 images in the form of geometric designs; decorative lines; and animal tracks, especially those of emu and kangaroo.

According to Dorn and Nobbs, prehistoric Australian Aborigines were engraving complex designs on rocks about 20,000 years before prehistoric Europeans began decorating rock walls. The oldest previously known rock art, found in Spain and Germany, has been dated to 27,000 to 30,000 years ago. Other scientists expressed doubt about the reported age of the Australian engravings, however.

Caiaphas' bones? Bones discovered outside Jerusalem may be those of a man named Caiaphas, the Jewish high priest who handed Jesus over to the Romans for crucifixion, according to an August 1992 report. Caiaphas, who was the son of a man with the same name and who was also called Joseph, ruled as high priest in Jerusalem from A.D. 18 to 36. Israeli archaeologists found the bones in a tomb belonging to the family of Caiaphas. They identified the tomb by examining writings on the walls and artifacts found within.

In the tomb were 12 *ossuaries,* limestone boxes used to store bones after the flesh on them had decomposed. One of the ossuaries, which was decorated with an intricate pattern, was inscribed with the name "Joseph, son of Caiaphas." Israeli archaeologists believe the bones inside—those of a 60-year-old man—are those of Caiaphas. But they were unable to positively identify the bones as his.

Horse sculpture. A 4,300-year-old molded clay figurine found in northern Syria may be the oldest known sculpture of a domesticated horse. That conclusion was reported in December 1992 by a team of archaeologists headed by Thomas Holland of the Oriental Institute at the University of Chicago. The discovery suggests that the horse played an earlier, more significant role in the rise of ancient Middle Eastern empires than scholars had previously thought. Ancient Middle Eastern texts had indicated that people in this region were using horses by about 2300 B.C. The horse

A bonanza of ancient bronzes

More than 1,000 bronze heads, arms, legs, and other parts from disassembled statues—discovered off the southeastern coast of Italy in July 1992—are the most diverse collection of classical Greek and Roman bronzes ever found. Archaeologists believe the statue parts, which do not fit together and which span a period from the 300's B.C. to the A.D. 200's and 400's, represent an ancient cargo of scrap bronze destined for recycling.

A statue of a man in a toga, *above,* missing only part of his head, lies on the floor of the Adriatic Sea near the Italian city of Brindisi.

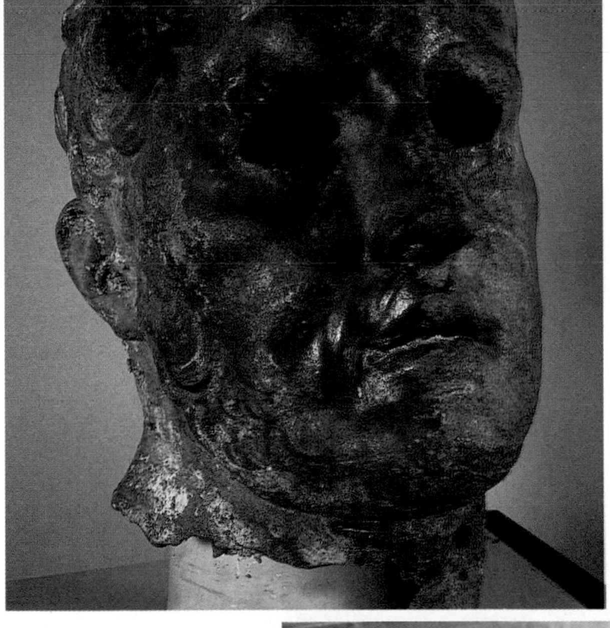

The head of a male statue, *above*—one of the best-preserved heads found at Brindisi—displays the wide eyes, open mouth, and puffy skin characteristic of an artistic style widely used about 200 B.C.

A bronze hand, *below,* is one of several pieces that are much larger than life-sized.

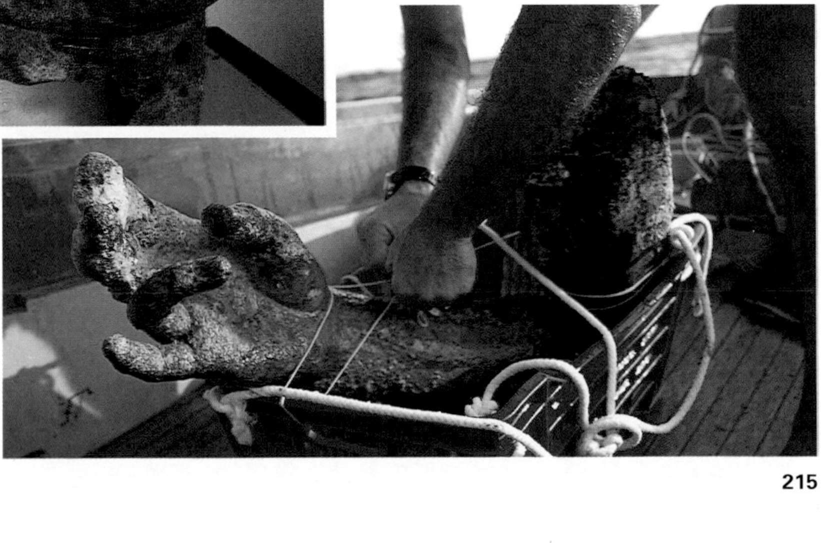

contributed to the development of the ancient empires by enabling rulers to move armies and equipment quickly and easily.

Holland and his colleagues believe the figurine represents a domesticated, rather than wild, horse because the muzzle of the figurine has a hole into which a ring for reins could have been inserted. In addition, the horse's mane, like those of domesticated horses, lies flat. The manes of wild horses stand upright.

Temple restoration. The completion of a six-year effort to restore the tomb of the ancient Egyptian queen Nefertari was announced in autumn 1992. Nefertari was the favorite wife of Ramses II, who ruled from about 1290 to 1224 B.C. The restoration of the tomb, which is located in Thebes, was organized by archaeologists from the Getty Conservation Institute in Malibu, Calif., and from the Egyptian Antiquities Organization.

Many scholars consider the wall paintings from Nefertari's tomb to be among the most beautiful in ancient Egypt. But over time, the paintings had deteriorated greatly. Some damage resulted from carelessness or vandalism, but most resulted from the separation of the paintings' plaster base from the limestone walls of the tomb.

The restoration process involved cleaning the paintings and applying mortar to spaces between the walls and the separated plaster base. Because water vapor from the breath of visitors promotes the separation of the plaster from the walls, access to the tomb will be restricted to a few scholars.

One unexpected bonus of the restoration project was the discovery of a tiny gold sheet, 2.5 by 2.5 centimeters (1 by 1 inch) that had been overlooked by scholars and grave robbers. The gold sheet, which is inscribed with hieroglyphs, may be a piece of a bracelet or armlet that once decorated Nefertari's mummy. [Robert J. Wenke and Nanette M. Pyne]

See also ARCHAEOLOGY, NEW WORLD. In WORLD BOOK, see ARCHAEOLOGY; EGYPT, ANCIENT.

Astronomy, Milky Way

A 20-year-old astronomical mystery was solved in 1992 with the identification of a strange source of cosmic radiation known as Geminga. Other findings about Geminga followed in early 1993.

Astronomers had been baffled by Geminga since 1972, when a satellite observatory called SAS-2 detected intense gamma rays coming from a featureless spot in the sky. Gamma rays are part of the electromagnetic spectrum, which also includes visible light, radio waves, and X rays. Gamma rays have the shortest wavelengths and highest energy of any form of electromagnetic radiation.

Studying Geminga. In May 1992, astronomers Jules Halpern of Columbia University in New York City and Stephen Holt of the National Aeronautics and Space Administration (NASA) Goddard Space Flight Center in Greenbelt, Md., reported that Geminga is a *pulsar.* A pulsar is a rapidly rotating collapsed star that sends out beams of radiation, much like the searchlight beams from a lighthouse. A beam that sweeps past the Earth can be detected as pulses of radio waves, gamma rays, or X rays, depending on which type of pulsar it is from.

Halpern and Holt made their discovery with the ROSAT satellite, which detects X rays from celestial objects. Data from the satellite showed that there is a pulse of X rays coming from Geminga every 0.237 seconds. A study of the data from previous studies showed that the object detected in 1972 by SAS-2 pulses with the same frequency in gamma rays. This finding confirmed that the source of the X rays and the source of the gamma rays are the same celestial object—Geminga—and that it is a pulsar.

A pulsar forms when a massive star runs out of nuclear fuel and its core collapses from gravity. The collapse results in a tremendous explosion called a *supernova.* In many cases, the remnant of a supernova is an extremely dense ball of *neutrons* (subatomic particles with no electric charge) called a neutron star. A pulsar is a rapidly spinning neutron star with an intense magnetic field that channels radiation into beams.

Astronomers Neil Gehrels and Wan Chen of the Goddard Space Flight Center set out to learn when the supernova that created Geminga occurred. The rotation of a pulsar slows down as the star ages, so the scientists examined the rate at which the pulsar's rotation speed is slowing. In February 1993, they reported that the pulsar formed about 330,000 years ago.

Also in February 1993, astronomers at the Institute of Cosmic Physics in Milan, Italy, reported that Geminga seems to be about 300 *light-years* from Earth. (A light-year is the distance light travels in one year—about 9.5 trillion kilometers [5.9 trillion miles].) They based that conclusion on studies of the intensity of Geminga's radiation and on careful observations of the faint point of visible light that is thought to be the pulsar.

Gehrels and Chen also investigated the possible relationship of Geminga to the Local Bubble, a huge pocket of thin gas surrounding our solar system. The Local Bubble has long been thought to be the result of a supernova explosion whose blast waves cleared a large area of space. Gas within the Local Bubble is much thinner and hotter than most gas clouds in the Milky Way.

Gehrels and Chen studied observations of Geminga's movement through the Milky Way and calculated the star's probable positions in the past. From their analysis, they concluded that at the time of the supernova explosion, the star would apparently have been at the center of what is now the Local Bubble. In other words, the explosion that created the Local Bubble seems to be the same one that gave birth to Geminga. If that is the case, the supernova occurred about 300 light-years from Earth, quite close by cosmic standards. It therefore may have had a significant effect on life on our planet 330,000 years ago.

Outward-bound pulsar. The discovery of the fastest-moving pulsar ever observed was reported in March 1993 by astronomer James M. Cordes of Cornell University in Ithaca, N.Y., and astrophysicist Roger W. Romani of Stanford University in California. The two researchers said the pulsar, designated PSR 2224+65, is moving about 1,000 kilometers (620 miles) a second. At that speed, an object could cross the United States in less than 5 seconds. Because it

A vertical purple band in an image created by the Hubble Space Telescope marks a shock wave from a *supernova* (exploding star) as it slams into huge clouds of thin gas, astronomers reported in January 1993. The shock wave compresses the gases, causing them to glow. The blast that created the shock wave and the disrupted gas clouds—a part of the Milky Way called the Cygnus Loop—occurred about 15,000 years ago.

is moving so fast, they said, the pulsar will eventually escape the gravitational grasp of the Milky Way and travel beyond the Galaxy.

Cordes and Romani studied the pulsar with the Arecibo radio telescope in Puerto Rico and the 5-meter (200-inch) Hale optical telescope at the Mount Palomar observatory in California. Their observations showed that the pulsar is about 16 kilometers (10 miles) in diameter, 6,000 light-years away, and 1,000 years old.

The scientists estimated that the pulsar will leave the Galaxy in about 20 million years. Because the pulsar is moving so fast, it is leaving a huge guitar-shaped shock wave in the cloud of dust and gas it moves through, much like the wake of a speedboat.

The pulsar's high speed probably resulted from the unsymmetrical explosion of the supernova from which it formed. Most of the original star's material was blown one way, and the core of the star—the newly born pulsar—was blown the other way with a tremendous amount of force. The scientists said the discovery of PSR 2224+65 raises the possibility that high-velocity pulsars and other neutron stars are common.

Giant molecules in space. Astronomers have long sought to learn the nature of particles in space that absorb a portion of the light from distant stars. In July and August 1992, astronomers Farid Salama and Louis Allamandola of NASA's Ames Research Center in Palo Alto, Calif., reported findings that may lead to a resolution of that question.

Astronomers first noticed the mystery when studying starlight that had been spread out to form a spectrum. Such spectra are made up of complex patterns of light and dark lines. By analyzing these lines, scientists can identify the atoms or molecules in the star or in clouds of gas or dust between Earth and the star. The bright *emission lines* are the signature of hot, thin gases, while the dark *absorption lines* reveal the presence of light-absorbing atoms and molecules in space beyond the star.

But astronomers have been unable to identify some of the absorption lines formed by the material in space. They

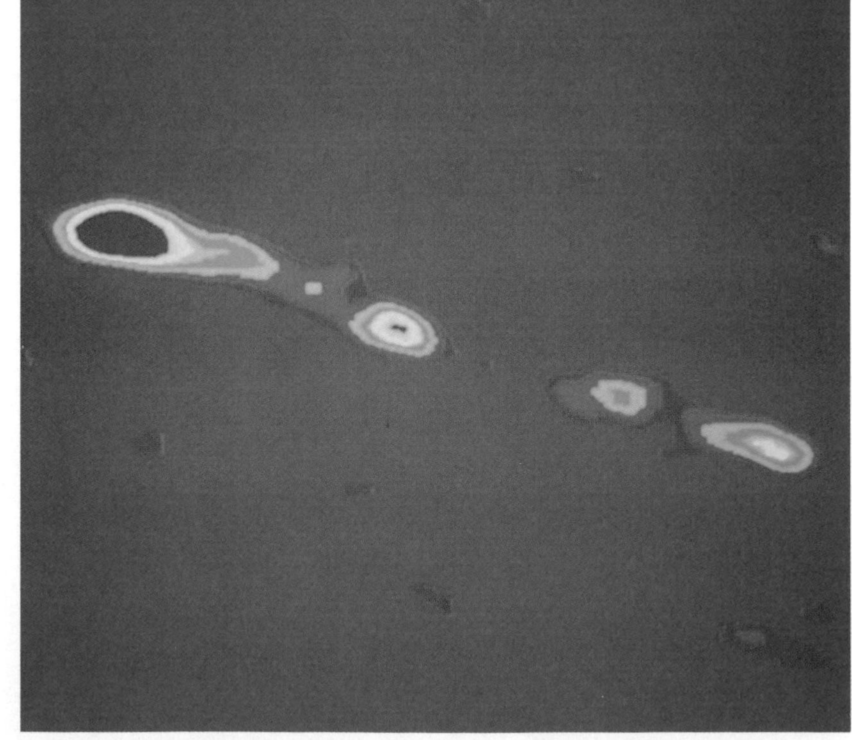

Two long jets visible on a radio wave image may be streams of electrons and *positrons* (the antimatter counterpart to electrons) gushing from a point near the center of the Milky Way known as the Great Annihilator, astronomers theorized in summer 1992. The Great Annihilator itself—located between the jets—is apparently a *black hole,* an object so dense that nothing can escape the gravity at its surface.

have tried to find out what is causing the lines by passing light through various molecules in the laboratory and comparing the resulting spectra against the spectra of the stars. But in none of these experiments did the absorption lines in the two spectra match.

Now, perhaps, astronomers may be on the verge of solving this puzzle. Beginning in the mid-1980's, astronomers began learning that another type of molecule may be abundant in space. These molecules, called polycyclic aromatic hydrocarbons (PAH's), consist of interlocking six-sided rings of carbon atoms with hydrogen, nitrogen, or other kinds of atoms attached to the carbon. Containing dozens of atoms, these molecules are far larger than any previously suspected of being common in the space between the stars.

In their normal form, PAH molecules do not produce absorption lines that match the mysterious lines formed in space. But Salama and Allamandola showed that the absorption lines created by *ionized* (electrically charged) PAH molecules are quite close in wavelength to the mysterious lines. A molecule becomes ionized when it loses one or more electrons. Starlight streaming through space carries enough energy to knock electrons off PAH molecules, thereby converting them into PAH ions.

Salama and Allamandola conducted laboratory experiments in which they embedded PAH ions in chemically inert materials and measured their spectra. They found some absorption lines formed by ions of pyrene and naphthalene (two kinds of PAH molecules) that closely match the lines in starlight.

The match, however, was not perfect, possibly because the material in which the ions were embedded distorted the wavelengths at which the ions absorbed light. The investigators were thus unable to conclude with certainty that PAH ions are the mystery molecules that astronomers have been searching for. By mid-1993, many astronomers were studying PAH ions in an attempt to settle this nagging question. [Theodore P. Snow]

In the Science Studies section, see The Universe. In World Book, see Milky Way.

Astronomy, Solar System

The discovery of frozen carbon dioxide (CO_2) and carbon monoxide (CO) on Triton, the largest of Neptune's eight known moons, was reported in May 1993. The finding was made by astronomer Dale P. Cruikshank and his colleagues at the National Aeronautics and Space Administration (NASA) Ames Research Center near Palo Alto, Calif.

Working with a British infrared telescope on the island of Hawaii, the investigators used a *spectrometer* to analyze sunlight reflected from Triton. A spectrometer separates light into the different wavelengths of the spectrum. Chemical elements and compounds on a planet or moon's surface absorb and reflect specific wavelengths of light. Using a spectrometer, an astronomer can analyze the light reflected from the object to determine which substances are present on its surface.

The spectrometer showed that Triton's surface contains ices of CO and CO_2. Studies in the 1970's and 1980's with less precise spectrometers had found frozen nitrogen and methane on Triton, but the 1993 findings marked the first time that CO and CO_2 had been detected there.

The frozen compounds and elements identified on Triton provide a chemical record of how the planets and moons of the outer solar system formed. Astronomers think the solar system formed from a large disk of gas and dust about 4.6 billion years ago. The outlying regions of the disk provided the raw materials for the outer planets and moons.

According to one theory, most of the CO and CO_2 in the outer disk was broken down to the elements carbon and oxygen before the planets and moons were created. But finding CO and CO_2 on Triton indicates that those compounds remained largely intact when the outer planets and moons formed.

Galileo's second Earth flyby. The United States space probe Galileo, launched in 1989, made its second and final pass by Earth on Dec. 7 and 8, 1992. During the flyby, Galileo passed just 304 kilometers (189 miles) above Earth. As the probe whipped around the Earth, the planet's gravity acted like a slingshot to increase the speed of the spacecraft as it travels to its destination, Jupiter. The probe is due to reach Jupiter in December 1995.

Increasing the speed of the spacecraft was the main purpose of the flyby. But the maneuver also enabled scientists at NASA's Jet Propulsion Laboratory (JPL) in Pasadena, Calif., and at participating institutions around the world, to conduct several measurements of the Earth and moon with Galileo's sophisticated instruments.

Galileo's path took it through the *magnetosphere,* the region of space around Earth in which the planet's magnetic field exerts a strong influence. The researchers mapped the magnetosphere from the side facing the *solar wind* (streams of charged particles from the sun), to the magnetosphere's tail, which extends for many tens of thousands of kilometers in the opposite direction. Data from these measurements were being analyzed in mid-1993.

Pointing Galileo's instruments toward the moon, scientists mapped the moon's north polar region with much greater accuracy than ever before. They also gathered information on the chemical composition of much of the lunar surface. Those data enabled them to identify several lunar regions, now heavily cratered by meteorite impacts, that were originally large, smooth plains made of hardened lava.

Asteroid close-up. The Galileo flyby also allowed engineers at JPL to recover all the data recorded by the space probe in October 1991 when it passed close to Gaspra, an asteroid 20 kilometers (12.4 miles) in length. The information had not been transmitted earlier because Galileo's main radio antenna had failed after launch, and the probe's small secondary antenna sends information at a very slow rate. Fortunately, the rate increases when the spacecraft is near Earth.

The images of the asteroid based on the complete data were much more detailed than the ones produced in 1991. Clark Chapman, a member of the Galileo imaging team, reported in March 1993 that the meteorite impact craters on Gaspra were fewer and smaller than would be expected for an aster-

A small icy body designated 1992 QB1 (arrow), discovered in August 1992, is the farthest known object in the solar system. It orbits the sun at a distance of about 6.3 billion kilometers (3.9 billion miles). 1992 QB1, estimated to be about 200 kilometers (124 miles) in diameter, is far smaller than any of the planets of the solar system. Astronomers think 1992 QB1 may be part of the theorized Kuiper belt, a large ring of icy debris that may have been left over from the creation of the solar system and from which many comets may originate.

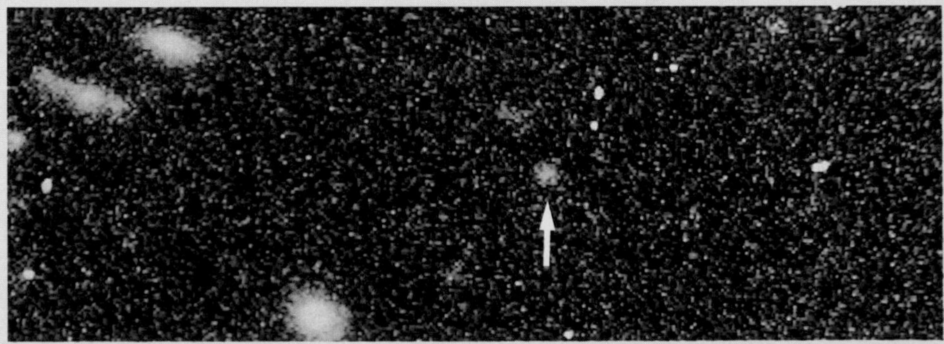

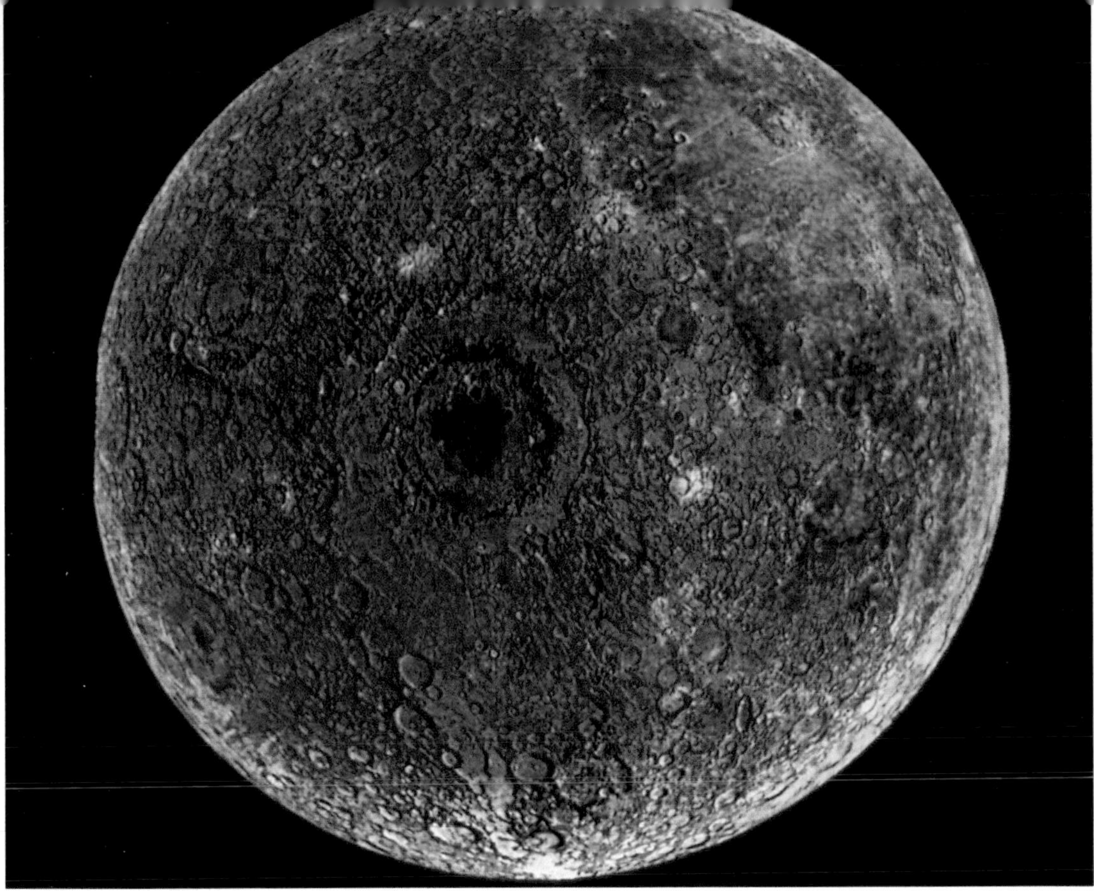

A false-color map of the moon's western edge and part of the side turned away from Earth helps reveal the chemical makeup of the surface. Regions that appear green and yellow, for example, indicate concentrations of iron and magnesium. The map, based on space probes' images of the far side of the moon in various wavelengths of light, was released in mid-1992 by the National Aeronautics and Space Administration.

oid that is presumably several billion years old.

That finding suggests that the asteroid's surface features were formed in just the past 200 million years. Chapman said Gaspra may have been hit by another asteroid 200 million years ago, a collision that might have shaken Gaspra's surface enough to erase all existing impact craters and create a "clean slate."

From some angles, Gaspra appears to be two objects fused together. This unusual shape suggests that collisions between asteroids may have been common events in the solar system. Asteroids could occasionally be split into pieces by collisions, and some of those pieces might later ram into other asteroids or chunks of asteroids, fusing together to form a larger object.

Earth-crossing asteroid. Another finding that suggests some asteroids may be two smaller objects joined together was reported in January 1993. JPL senior research scientist Steven Ostro said radar studies of an asteroid named Toutatis showed that it consists of two asteroid fragments stuck together.

Toutatis is one of about 150 known asteroids whose orbits cross the orbit of Earth. Scientists have found large impact craters on Earth indicating that such asteroids have slammed into our planet from time to time in the past.

Ostro and his colleagues made radar images of Toutatis on Dec. 8, 1992, as it passed within 3.5 million kilometers (2.2 million miles) of Earth. The images, the first close-up pictures ever made of an Earth-crossing asteroid, showed Toutatis to be roughly dumbbell-shaped. The two sections of the dumbbell are hunks of rock, one of them about 4 kilometers (2.5 miles) and the other about 2.6 kilometers (1.6 miles) across.

More clues to a warm, wet Mars. Warm, wet conditions may have prevailed on Mars billions of years ago, planetary geologists reported in February 1993. The researchers, Robert Craddock and Ted Maxwell of the Smithsonian Institution's National Air and Space Museum in Washington, D.C., came to that conclusion after analyzing images of Mars taken by the Viking spacecraft in the late 1970's.

Craddock and Maxwell examined a series of meteorite impact craters in the highland region of Mars's northern hemisphere. The craters were in various stages of erosion. The erosion patterns strongly indicated that the craters had been worn down by running water, the scientists said.

Evidence that Mars may once have had abundant surface water is not new. Since the 1970's, investigators studying images from Viking, as well as from the earlier Mariner 9 spacecraft, have identified what appear to be dried-up river channels and flood plains on Mars. But only since the late 1980's have scientists recognized that water may have dramatically reshaped the Martian terrain.

The Smithsonian scientists analyzed many craters of different sizes and ages. All the craters appeared to have been eroded by water, rather than by volcanic action or wind. This finding supported a theory that Mars had water over a large extent of its surface in the first 500 million to 1 billion years of its existence. During that early era, the planet apparently had many large lakes as well as rivers. For so much water to have existed, Mars must have had a much warmer climate and thicker atmosphere than it does today.

Whether life formed on Mars during that early warmer period is unknown. Biologists think that if there was life on Mars at that time, it would not have had time to evolve beyond simple forms such as bacteria.

Further revelations about Mars are expected from the Mars Observer spacecraft, launched by NASA in October 1992. The space probe, due to reach Mars in August 1993, will carry out a detailed two-year survey of the Martian atmosphere and surface.

The Mars Observer will search for traces of water on the Red Planet and radio back images of Mars's surface that will be much more detailed than any produced by Viking. Astronomers hope the mission will give them a better idea of how Mars evolved from a warm, water-filled world to the frozen desert we see today. [Jonathan I. Lunine]

See also SPACE TECHNOLOGY. In WORLD BOOK, see SOLAR SYSTEM.

Astronomy, Universe

A *supernova* (exploding star) in a nearby galaxy was reported on March 28, 1993, by Francisco Garcia, an amateur astronomer in Spain. Garcia discovered the supernova in a galaxy known as M81, which lies about 11 million light-years away. A light-year is the distance light travels in one year, about 9.5 trillion kilometers (5.9 trillion miles).

The new supernova, designated Supernova 1993J, is one of the closest and brightest exploding stars observed in the 1900's. Most supernovae are seen in galaxies tens or hundreds of millions of light-years from the Milky Way, though one recent supernova, in 1987, was just 150,000 light-years away.

Because Supernova 1993J was relatively close, it presented a new opportunity to study an exploding star in great detail. Astronomers were able to identify the star by looking at pictures of M81 taken before the blast.

Supernovae interest astronomers because they mark the death of massive stars containing 10 or more times as much matter as our sun. Stars shine by converting hydrogen to progressively heavier elements—including helium, carbon, nitrogen, and oxygen—in the process called nuclear fusion. When a massive star runs out of fuel, it collapses in on itself and explodes.

The tremendous heat and pressure of the explosion creates an array of even heavier elements, which are spewed into space along with the elements created by the original star. All this "debris" later condenses in space, where it can serve as the building blocks of new stars and perhaps planets. The material from which our own solar system was formed came from supernovae. Thus, coming to an understanding of supernovae is crucial to achieving a complete understanding of element formation.

Clues to the evolution of galaxies. One of the greatest mysteries in astronomy is how galaxies formed and developed. In 1992, by looking at groups of extremely distant galaxies, astronomers obtained some important clues that may help them understand that process.

It takes time for light to travel across the vast distances of space, so the light we see from the most distant galaxies

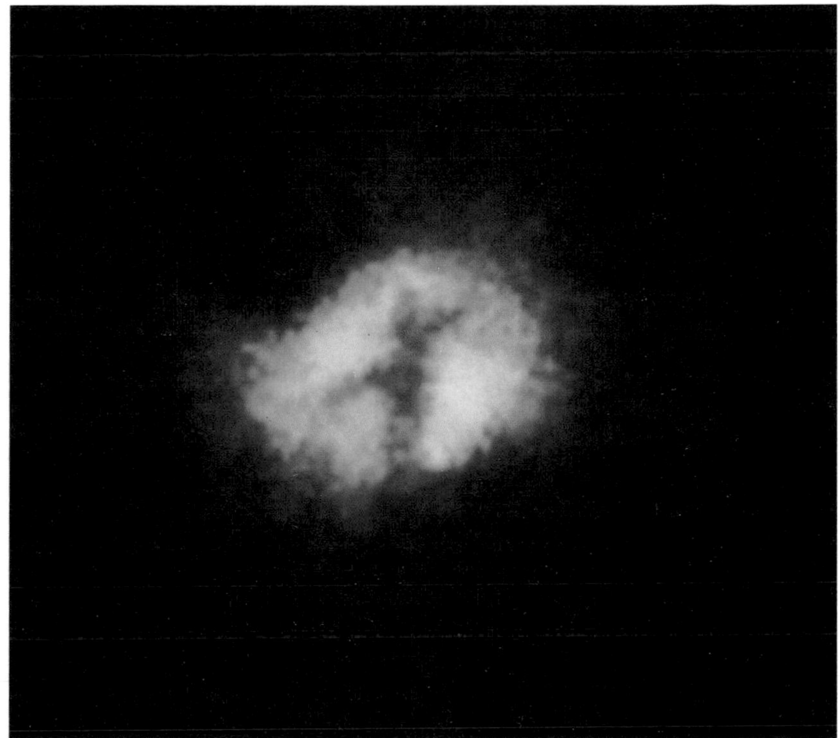

A dark X at the center of a galaxy known as the Whirlpool Galaxy marks the location of a probable black hole, in an image made by the Hubble Space Telescope in June 1992. Astronomers theorized that the wide bar making up one part of the X is an accretion disk, a flattened cloud of matter that gives off radiation as it circles into the hole. The intersecting bar is probably either a second accretion disk or some other concentration of matter.

was emitted billions of years ago. Because we are seeing those galaxies as they looked early in their development, they give us clues to how all galaxies may have changed over time.

Using the Hubble Space Telescope, an investigative team led by astronomer Alan Dressler of the Carnegie Institution of Washington, in Washington, D.C., observed two clusters of galaxies located 4 billion light-years away. The researchers reported in December 1992 that they saw the galaxies as they appeared when the universe was about 11 billion years old, approximately three-fourths its present age.

The Space Telescope images enabled the astronomers to identify different types of galaxies in the clusters by their shapes. The galaxies included *elliptical* (egg-shaped) and *spherical* (round) galaxies. They also included many pinwheel-shaped spiral galaxies—many more than they see in nearby clusters of galaxies. That finding indicated that spiral galaxies were more common several billion years ago than they are now.

If that is the case, what happened to the spirals to make them less numerous? The researchers offered two possible explanations. Perhaps, they said, many spiral galaxies collided with one another, merging to form elliptical galaxies. Another possibility is that spiral galaxies may once have contained many more bright, newly formed stars than they do now. With fewer such stars, the spiral arms—the regions where most new stars are formed—faded from view.

Extremely distant galaxies. Dressler and his colleagues also got a view of some of the most distant galaxies ever seen. One Hubble image revealed a cluster of 30 very faint galaxies at least 7 billion light-years away—the most remote cluster known.

Also using the Space Telescope, a team of astronomers headed by George Miley of Leiden Observatory in the Netherlands observed the farthest-known single galaxy. That galaxy, designated 4C41.17, is about 10 billion light-years away, the researchers reported in November 1992.

The image of 4C41.17 revealed a chain of fuzzy knots of light near the

Mirages in the Sky

One of the most exciting new branches of astronomy in the 1990's is the study of gravitational lenses. A gravitational lens is an object in space that bends light with its gravity just as a regular lens in a telescope bends it with curved surfaces of glass. These natural lenses in space may help astronomers find hidden matter in the universe and determine the universe's age.

The first gravitational lens was found accidentally in 1979 by a British radio astronomer, Dennis Walsh, and his colleagues. The researchers made their discovery while cataloging a number of astronomical *radio sources* (celestial objects that emit radio waves) and identifying them on photographs of the sky made with optical telescopes.

In one photograph, Walsh and his associates found a pair of bluish objects near a radio source. Analysis of the light from the two objects showed that they were both *quasars* (remote, starlike objects that give off huge amounts of energy). The astronomers then passed the two quasars' light through a spectrometer, a device that spreads out light into the spectrum. Usually, the spectrum of light from a star, quasar, or galaxy is as unique as a fingerprint. But the researchers were astonished to find that the quasars had identical spectra.

This finding suggested to them that they might be seeing not two different quasars but rather two images of the same quasar. Such mirror images, they theorized, could be produced by the intense gravitational field of a galaxy—one so faint that it had not been detected—that was lined up precisely between Earth and the quasar. The galaxy's gravity would bend light rays from the quasar and direct them toward Earth. Light passing on one side of the galaxy would bend to form an image of the quasar in one part of the sky, and light passing on the opposite side of the galaxy would produce a second image elsewhere.

Astronomers around the world immediately focused optical and radio telescopes on the twin quasars. Their observations showed that Walsh and his team were correct—there was indeed a galaxy between the two quasar images. Gravitational lenses were a real phenomenon.

The search was then on to find other gravitational lenses, and soon the "double quasar" image was joined by a "triple quasar." By 1993, several dozen other confirmed or suspected lenses had been identified.

Gravitational lenses were a significant discovery, but the notion that gravity can bend light was nothing new. The idea dates back about 300 years to the English scientist and mathematician Sir Isaac Newton, who formulated the first theory of gravity.

In more modern times, the German-American physicist Albert Einstein also predicted the bending of light by gravity. In his General Theory of Relativity, published in 1915, Einstein said that starlight passing near the sun would be very slightly bent by the sun's gravitational field. That effect, he calculated, would shift a star's apparent position in the sky by just under 2 seconds of arc, or about the diameter of a dime viewed from a distance of 2.4 kilometers (1.5 miles).

Astronomers realized this prediction could be tested during a total solar eclipse, when the moon blocks the bright light from the sun. While the sun was obscured, stars whose light was passing near the sun could be observed, and their measured positions during the eclipse could be compared with their true positions, known from observations at night.

This experiment was carried out in 1919 during a total eclipse in the Southern Hemisphere, when the sun was in line with a cluster of stars called the Hyades. Astronomers found that the positions of stars in the cluster were shifted by just the amount that Einstein had predicted. The outcome of this test was considered a triumph for Einstein's prediction that gravity could bend light.

Gravitational lenses provide an even more dramatic confirmation of the bending of light. The lens need not be a galaxy. It could be any object that is massive enough to have an intense gravitational field. It could be a star, a quasar, or a *black hole* (an object with such a strong gravitational field that light cannot escape its surface).

Depending on the shape and mass of the lens, the image we see of the distant light source can be distorted in many ways. It might be split up into several images, like the double and triple quasars. Or it might be stretched out into a long thin arc or even a complete circle. A lens can also make an image much brighter or much fainter than the actual object.

Astronomers find gravitational lenses interesting for several reasons. Aside from the pure scientific excitement the lenses have stirred, they promise to be useful as a tool that may help us answer some difficult questions.

One such question is, what are the distances to faraway galaxies? Astronomers have only been able to estimate how far galaxies are from us, and those estimates are derived through very complex measurements.

Establishing the exact distance to just one remote galaxy would be an important achievement. With that number in hand, astronomers could then figure the correct value of the *Hubble constant,* a factor that relates a galaxy's distance from

How gravity acts as a lens in space

When light from a celestial object passes near a strong gravitational field, it is bent, as though by a lens. This phenomenon produces multiple images of the object.

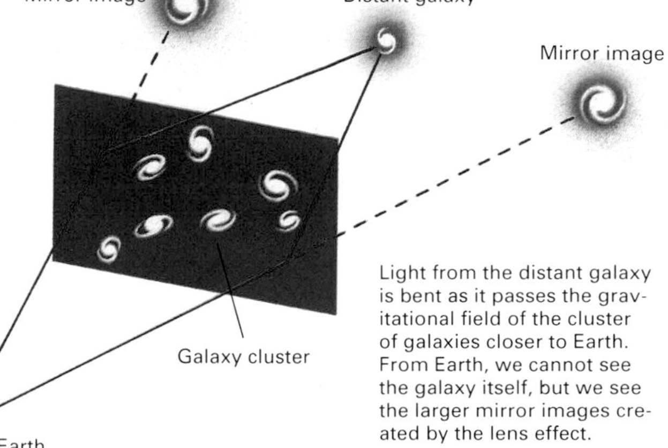

Two mirror images of a distant galaxy created by a gravitational lens appear as widely separated blobs of light in a photograph, *above,* made by the Hubble Space Telescope. The galaxy's true location lies directly beyond a cluster of closer galaxies, two of which can be seen between the mirror images.

Light from the distant galaxy is bent as it passes the gravitational field of the cluster of galaxies closer to Earth. From Earth, we cannot see the galaxy itself, but we see the larger mirror images created by the lens effect.

Earth with the speed at which it is moving away from us as the universe expands.

Knowing the correct value for the constant would in turn allow astronomers to calculate how long it took for all the galaxies to reach their observed positions since the big bang, the explosion that gave birth to the universe. And that information would give us the age of the universe.

Gravitational lenses offer the possibility of an elegant solution to these problems. The geometry of a lens system is very simple. The light rays from each of the images produced by a gravitational lens trace out two sides of a triangle in space. The two sides are the path of the light rays from the original object to the point where the light is bent by the gravitational lens and the path from there to Earth.

The size of a lens system can be calculated because light travels at a known speed. The light rays that produce each of the various images of a distant object follow slightly different paths, some of which are longer than others. Light rays following longer paths take longer to reach Earth than rays following shorter paths. Therefore, if the original object were suddenly to become brighter, the two or more images seen on Earth would become brighter at different times.

Using geometrical computations, astronomers can determine the relative lengths of the light paths of different images. They might find, for example, that a light path along the sides of one triangle is one-billionth longer than the light path creating a second image. This ratio, combined with the observed time differences in brightness changes and the known speed of light, would enable astronomers to determine the exact lengths of the light paths.

Researchers are trying to use this technique with the double quasar that Walsh found. The two images sometimes change in brightness. Although astronomers have not yet precisely established the time delay between two identical changes, it seems to be somewhere between 1 and 1.5 years. That time difference gives a value for the Hubble constant that agrees well with the current value being used by astronomers, which yields an age for the universe of about 15 billion years.

Another potential use of gravitational lenses is detecting black holes and other unseen concentrations of matter. Although we cannot see black holes and other dark objects directly, we might be able to detect them indirectly with the aid of gravitational lenses. Astronomers have found several apparent multiple quasar images that are not associated with any visible lens galaxy, suggesting that the images might have been made by enormous black holes.

At present, though, it is hard to predict the future course of gravitational lens research. Progress in science sometimes goes in unforeseen directions, and we can only guess how astronomers will use this valuable new tool. [Jerome Kristian] center of the galaxy. The astronomers said the knots might be large clusters of stars in formation. Each cluster would be about 1,500 light-years across and contain about 10 billion stars. It was also possible, they said, that the knots are not stars at all, but clumps of gas lit by radiation from matter surrounding a black hole. The black hole itself would be hidden in the galaxy's center. Further observations with the Space Telescope may help determine which explanation is correct.

New evidence for dark matter was discovered in a group of galaxies about 150 million light-years away, according to a report by NASA astronomers in January 1993. *Dark matter* is a mysterious form of matter, invisible to telescopes because it emits no light, that scientists believe exists in huge quantities in the universe.

Astronomers infer the presence of dark matter in galaxies from several kinds of evidence. One clue involves the fact that the galaxies are moving away from one another, carried along by the universe, which is constantly expanding. But galaxies seem to be moving apart too fast to be held together by only the stars and gas clouds we can see in them. The gravity needed for that cohesion must be provided by some sort of unseen matter. Moreover, if galaxies contained only the matter we can see, they would not have had sufficient gravity to clump together in the first place.

The amount of dark matter in the universe has been a subject of great debate among astronomers. According to most recent estimates, there is 2 to 4 times more dark matter than ordinary matter in galaxies, but there could be much more than that.

The NASA astronomers' announcement in January could help settle that question. The scientists conducted their research at NASA's Goddard Space Flight Center in Greenbelt, Md., using an instrument that detects X rays, a form of high-energy radiation emitted by some celestial objects. The X-ray telescope is aboard ROSAT, a satellite operated jointly by the United States, Germany, and Great Britain.

ROSAT images showed a gigantic gas cloud surrounding three galaxies that are called the NGC 2300 group. For the gas cloud to emit X rays, it must be very hot—about 10 million °C (18 million °F). But a gas cloud that hot should have dissipated into the surrounding space long ago. The gravitational force required to hold the hot cloud in place is far greater than would be exerted by the visible stars in the three galaxies.

The astronomers therefore concluded that a huge amount of dark matter is present in the NGC 2300 group. They estimated that there is about 25 times more dark matter than visible matter in and around the three galaxies.

If such a large amount of dark matter is present in other clusters of galaxies as well, it could have a profound effect on the future of the universe. The mutual gravitational pull of all the galaxies in the universe is slowing the universe's expansion, but the amount of visible matter in the galaxies is not sufficient to bring the expansion to a halt.

If the universe contains a great deal of dark matter, however, its gravity could eventually stop the expansion. The universe would then contract until all the galaxies crashed together in a "big crunch" sometime in the far future.

Evidence of large black holes in the center of each of two nearby galaxies was reported by astronomers in the United States in June and November 1992. The findings supported the theory that black holes may provide the energy that causes many galaxies, including our own Milky Way, to emit strong radio waves and other forms of electromagnetic radiation.

A black hole is an extremely dense, massive concentration of matter whose gravitational field is so strong that nothing, not even light, can escape it once in its grasp. Matter falling into a black hole first circles around it in a flattened cloud known as an *accretion disk.* As the matter spirals inward toward the hole, it heats up to millions of degrees Celsius, giving off intense radiation. Because black holes are themselves invisible, astronomers looking for black holes try to detect radiation emitted from the accretion disk or to see the disk itself.

Using the Hubble Space Telescope, a team of investigators led by astronomer Holland Ford of Johns Hopkins University in Baltimore studied the galaxy M51, known as the "Whirlpool Galaxy," which is about 20 million light-years away. The researchers reported in June 1992 that

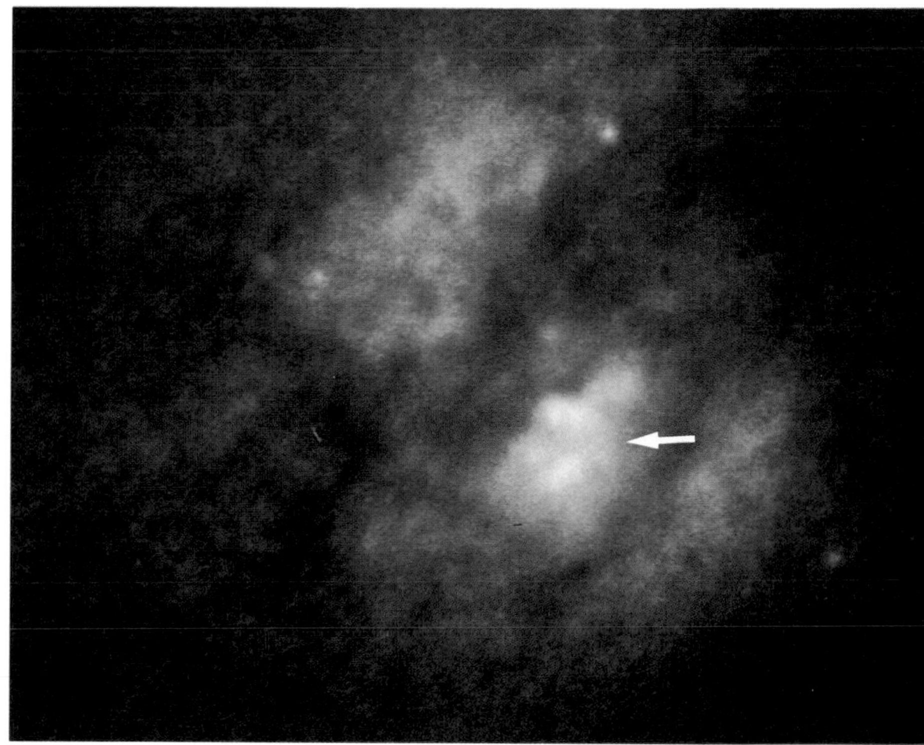

Several bright spots (arrow) in a galaxy called ARP 220 could be extremely bright clusters of young stars, the largest such clusters ever seen. Astronomers said in 1992 that ARP 220, a type of galaxy known as a starburst galaxy because it contains so many new stars, is apparently two spiral galaxies that have collided. ARP 220 is bisected by a dark band of dust.

they had observed a dark feature shaped like an X at the center of the galaxy.

They theorized that one bar of the X may be the silhouette of an accretion disk, seen edge-on, around a massive black hole. The intersecting bar, they said, could be either a second disk or some other concentration of matter.

In the center of another galaxy, called NGC 4261, about 45 million light-years distant, Ford and his colleagues observed a huge ring of gas and dust surrounding the center of the galaxy. In November 1992, the investigators identified the ring as another probable accretion disk around a black hole.

Big bang theory passes another test. Further research on the microwave radiation called the cosmic background radiation, reported in January 1993, bolstered the big bang theory that a huge explosion gave birth to the universe. The cosmic background radiation, often called the echo of the big bang, is a faint "glow" of high-frequency radio waves coming from every part of the sky.

Some astronomers have theorized that much of the energy in the background radiation was produced relatively recently in black holes, supernovae, and the disintegration of subatomic particles. If that were found to be so, the big bang theory, which holds that the background radiation originated almost entirely in the explosion that created the universe, would have to be revised.

The radiation has been extensively studied by the Cosmic Background Explorer Satellite (COBE), launched by NASA in 1989. Measurements from one of COBE's instruments, the Far Infrared Absolute Spectrophotometer (FIRAS), provided astronomers with strong support for the big bang scenario.

FIRAS data revealed that 99.97 percent of the energy in the cosmic background radiation was released within one year after the big bang. NASA astronomer John C. Mather said of the finding, "The big bang comes out a winner." [Laurence A. Marschall]

In the Special Reports section, see Fixing Hubble's Troubles. In the Science Studies section, see The Universe. In World Book, see Cosmology; Universe.

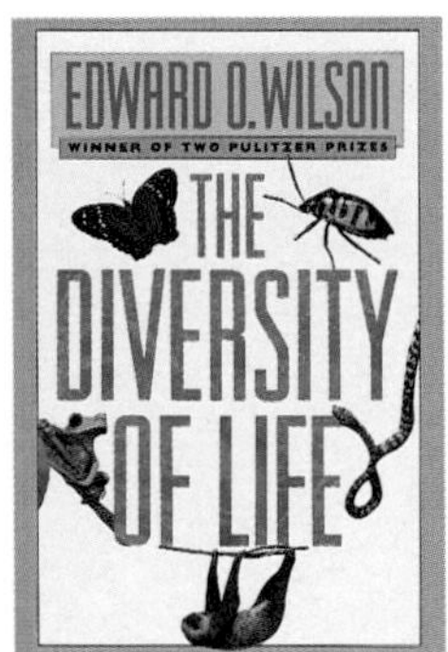

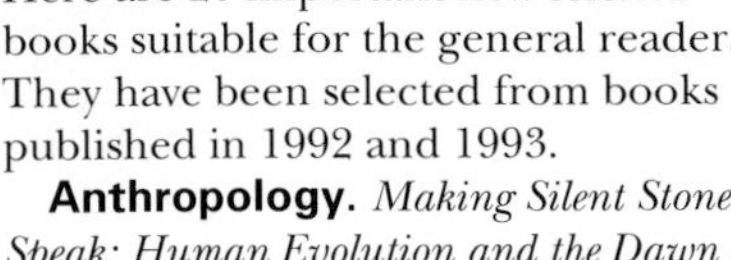

Here are 20 important new science books suitable for the general reader. They have been selected from books published in 1992 and 1993.

Anthropology. *Making Silent Stones Speak: Human Evolution and the Dawn of Technology* by Kathy Diane Schick and Nicholas Patrick Toth surveys prehistoric sites in east Africa and China and reconstructs the lives of our early toolmaking ancestors to show the means by which they made and used early stone tools. (Simon and Schuster, 1993. 352 pp. illus. $25)

Origins Reconsidered: In Search of What Makes Us Human by Richard Leakey and Roger Lewin is largely an account of the development in Leakey's thinking since his 1984 discovery of a nearly complete *Homo erectus* skeleton—the 1.5-million-year-old "Turkana Boy." Leakey's goal is to understand how our species came to possess the distinctly human characteristics of consciousness, creativity, and culture. (Doubleday, 1992. 375 pp. illus. $25)

Astronomy. *The Planets: Portraits of New Worlds* by Nigel Henbest is a collection of photographs and illustrations of the planets in our solar system and some of their moons. Most of the photographs were obtained by space probes, which dramatically extended our knowledge of these worlds. (Viking, 1992. 207 pp. illus. $35)

Biography. *Genius: The Life and Science of Richard Feynman* by James Gleick is the biography of the imaginative physicist who helped develop the theory of quantum electrodynamics and who shared the Nobel Prize in 1965 for his contributions to that theory. Feynman became a public figure toward the end of his life and became widely known as an investigator into the causes of the Challenger space shuttle disaster in 1986. (Pantheon, 1992. 532 pp. illus. $27.50)

Computer science. *Artificial Life: The Quest for a New Creation* by Steven Levy describes how scientists are using computers to simulate the metabolic and behavioral characteristics of living things. The purpose is to understand better how matter organizes itself into living systems and how those systems evolve. These investigations may yield a new understanding of what life is and how it is defined. (Pantheon, 1992. 390 pp. illus. $24.50)

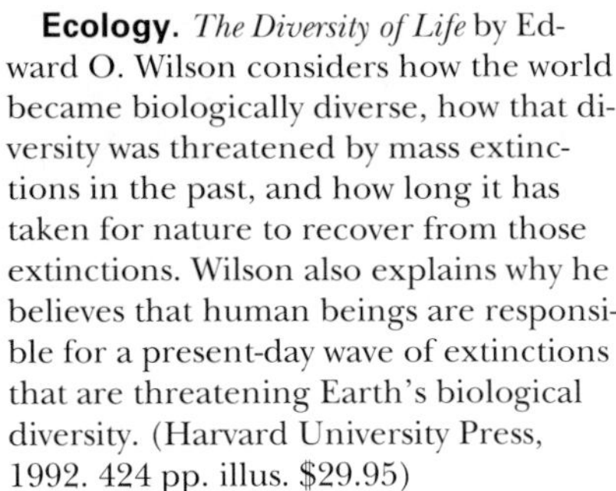

Ecology. *The Diversity of Life* by Edward O. Wilson considers how the world became biologically diverse, how that diversity was threatened by mass extinctions in the past, and how long it has taken for nature to recover from those extinctions. Wilson also explains why he believes that human beings are responsible for a present-day wave of extinctions that are threatening Earth's biological diversity. (Harvard University Press, 1992. 424 pp. illus. $29.95)

The Fate of the Elephant by Douglas Chadwick surveys the recent history and current situation of both the African and the Asian elephant and examines why the elephant is on the verge of extinction. Chadwick depicts the elephant's habitat and describes the consequences of its destruction. (Sierra Club Books, 1992. 492 pp. $25)

A Shadow and a Song: The Struggle to Save an Endangered Species by Mark Jerome Walters tells of the recent extinction of the dusky seaside sparrow, whose salt-marsh habitat in Florida was destroyed due to highway construction, the development of Cape Canaveral, and the use of pesticides to control mosquitoes. Walters also describes unsuccessful efforts by the United States Fish and Wildlife Service and Walt Disney World to save the sparrow from extinction. (Chelsea Green, 1992. 238 pp. illus. $24.95)

Environment. *Rubbish!: The Archaeology of Garbage* by William Rathje and Cullen Murphy describes a 20-year-old University of Arizona project to study garbage in landfills in the United States and Mexico. The surprising results of the study have significant implications for recycling policies and for the management of waste dumps. (HarperCollins, 1992. 250 pp. illus. $23)

Fossil studies. *Life in Amber* by George O. Poinar, Jr., is the story of amber, the fossils of resin that oozed chiefly from a now-extinct species of pine tree millions of years ago and trapped small life forms as it hardened. Poinar tells how and where amber was produced and what scientists have learned about the existence of life millions of years ago by studying insects and other fossils in amber. (Stanford University Press, 1992. 350 pp. illus. $55)

General science. *From Eros to Gaia* by Freeman Dyson is a collection of 35 essays and other writings by this noted physicist on a variety of science topics. These essays tell of his friendship with other famous scientists and give his views about the importance of such projects as the Superconducting Super Collider. In a concluding essay, Dyson speculates on the future of humankind. (Pantheon, 1992. 371 pp. $25)

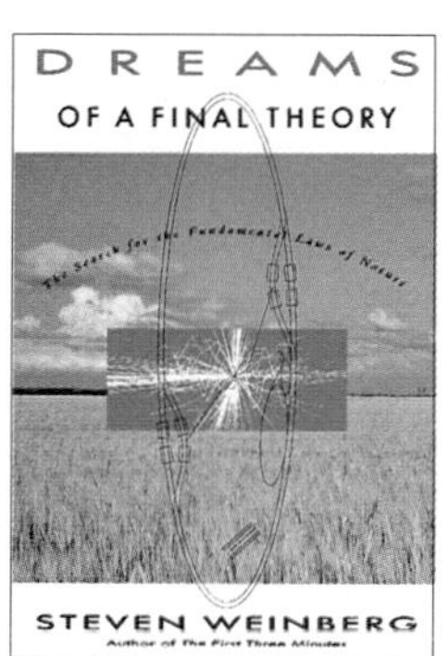

Geology. *Assembling California* by John McPhee is the final volume in the author's four-volume study of the theory of plate tectonics, which explains the forces that created most of Earth's surface features. McPhee describes the tectonic forces that led to the creation of present-day California, and he explains how these same forces are now slowly disassembling that region. (Farrar, Straus and Giroux, 1993. 304 pp. illus. $21)

Glaciers by Michael J. Hambrey and Jürg Alean examines the icy masses that cover one-tenth of Earth's land surface and are responsible for creating some of its most dramatic landscapes. This book explains how and where glaciers originate and the effects they have on wildlife, landscape, and climate. (Cambridge University Press, 1992. 208 pp. illus. $29.95)

Mathematics. *Pi in the Sky* by John D. Barrow introduces the reader to the usefulness and universality of mathematics. Barrow traces the earliest evidence of counting to people living in antiquity. He explores the origins of mathematical intuition and wonders if mathematics is a language like other languages, and why it has proved so useful for describing the physical world. (Oxford University Press, 1992. 317 pp. illus. $25)

Physics. *Dreams of a Final Theory* by Steven Weinberg, who shared the Nobel Prize for physics in 1979, addresses the search for a single theory that unites the four fundamental forces of nature. Weinberg summarizes the progress that has been made toward such a theory and speculates on how it might impact humankind. (Pantheon, 1992. 334 pp. $25)

The God Particle: If the Universe Is the Answer, What Is the Question? by Leon Lederman with Dick Teresi investigates a question first raised by the ancient Greeks: What are the fundamental building blocks of matter? Lederman, who shared the 1988 Nobel Prize in physics and is a former member of the *Science Year* Advisory Board, thinks scientists may be close to discovering a new particle that will enable them to reduce the laws of nature to a fundamental equation. Lederman expresses his belief that building the Superconducting Super Collider is important to aid in the search for the new particle. (Houghton Mifflin, 1993. 434 pp. illus. $24.95)

Technology. *From Sails to Satellites: The Origin and Development of Navigational Science* by J. E. D. Williams. Trade, conquest, and the lure of discovery have driven the development of navigational science. Williams explains how navigators have progressed from using simple geometry and calculations, dead reckoning, and the magnetic compass, to the use of satellite technology, which makes it possible to quickly determine locations at sea to within 30 meters (100 feet). (Oxford University Press, 1992. 310 pp. $35)

Zoology. *Beastly Behaviors: A Zoo Lover's Companion: What Makes Whales Whistle, Cranes Dance, Pandas Turn Somersaults, and Crocodiles Roar: A Watcher's Guide to How Animals Act and Why* by Janine M. Benyus describes activities common to mammals, birds, and reptiles. It then proceeds to detail some of the behaviors unique to different animals in the order a visitor to a zoo might encounter them. Sections on vital statistics give information on the animals' size, weight, longevity, and habitat. (Addison-Wesley, 1992. 366 pp. illus. $29.95)

Insects in Flight by John Brackenbury uses photographs to explain how insects have evolved the means of flying. The book examines the mechanics of insect flight, the design of their wings, and the hazards that accompany winged flight. Brackenbury also includes a section on insects that fly without the use of wings. (Blandford, 1992. 192 pp. illus. $35)

Landscape With Reptile: Rattlesnakes in an Urban World by Thomas Palmer is an account of a small surviving population of timber rattlesnakes in Blue Hills Reservation, a recreation area near Boston. Palmer shows how the rattler population reached its present reduced state and tells how the snakes survive in an environment surrounded by a metropolitan area. (Ticknor & Fields, 1992. 340 pp. $19.95)

[William Goodrich Jones]

Plants use electrical signals to send warnings from one part of the plant to another, according to a November 1992 report by a team of researchers from England and New Zealand. The electrical "warning system" is similar to an animal's nervous system.

Botanists have known of the existence of electrical signals in plants since the 1890's. But there has been ongoing controversy about the nature of the signals and their function. Scientific interest in studying the electrical signals plunged during the 1960's, after some *polygraph* (lie detector) operators hooked up plants to their machines and interpreted the readings as plant emotions.

But in 1992, plant physiologist David Wildon of the University of East Anglia in England and his colleagues studied what happened when a caterpillar damaged the *cotyledon* of a young tomato seedling. (A cotyledon is a structure that resembles a leaf and that emerges from the seed before the true leaves.)

The scientists found that the caterpillar's chewing caused the cotyledon to send an electrical signal to the plant's true leaves. The leaves responded by producing a protein that causes indigestion in caterpillars and other insects. In effect, when an insect chewed on one part of a plant, the rest of the plant released chemical defenses against insects.

Wildon's group then devised an experiment to distinguish between the action of chemical messengers such as hormones and the action of electric signals. The researchers found that by cooling the stem at the base of the damaged cotyledon they could stop the movement of chemicals out of the cotyledon. But the leaves continued to produce the substance that caused indigestion, indicating that electrical signals were responsible.

The electrical signals traveled slowly in a plant compared with nerve signals in an animal, but much faster than chemical messengers in plants. It took from two to four hours for the leaves to produce chemical defenses.

Fake flowers. When a rust fungus infects a species of mustard plant, the plant's development changes drastically. That was the March 1993 finding of plant ecologist Barbara A. Roy, currently at the University of California, Davis. Instead of producing normal pale blue flowers, mustard plants infected with rust fungus produce yellow structures that resemble buttercups. These yellow structures attract insects better than authentic flowers.

Roy studied the infected mustard plants while at the Rocky Mountain Biological Laboratory in Gothic, Colo. She found that the flowers of infected plants lack the structures in authentic flowers that produce pollen and seeds. The fake flowers were leaves covered with a yellow, sugar-coated mat of fungal sex cells.

Bees, butterflies, and especially flies spent more time on the fungal flowers than they did on genuine flowers, apparently because of the abundant sugary liquid. The insects picked up and distributed the sex cells, making possible the formation of *spores* (cells that enable an organism to reproduce). Spores formed only on the fake flowers that insects visited. These flowers then turned green, and they also stopped making the sugary liquid.

Glowing plants. Botanists have long known that touching a plant, bending it, or even blowing on it can alter its growth. Shaking a tomato plant for as little as 10 seconds each day, for example, can reduce its height by about 30 percent. In June 1992, scientists at the University of Edinburgh, Scotland, reported that they had developed a way to see the chemical reactions that occur within a plant when the wind blows on it. They made the plant glow.

The botanists speculated that stress such as wind might lead to a release of calcium ions from a plant cell's *organelles* (various specialized parts in a cell). The calcium ion leaving an organelle would enter the *cytosol,* the fluid that fills the cell. To find out if this happens, the scientists transferred a gene from jellyfish into tobacco plants. The jellyfish gene controls the formation of an *enzyme* (a protein that influences a chemical reaction). The enzyme converts another compound, called coelenterazine, into a form that is broken down by calcium ions. As the chemical is broken down, a blue light is given off.

Coelenterazine is not normally found in tobacco plants, so the researchers supplied it to the genetically engineered plants. In those plants, the enzyme reacted with the coelenterazine in the cytosol. The researchers then gave the

Largest organisms?

A lively debate began in 1992 when scientists in Michigan announced that mushrooms in a Michigan forest, *right,* could be considered part of the world's largest organism. The mushrooms sprout from rootlike cords called rhizomorphs, *inset,* that grow underground, and form a network covering 15.4 hectares (38 acres). The growth weighs an estimated 90 metric tons (100 short tons). Each mushroom is genetically identical, making the mass a natural *clone,* a group of individuals produced from a single organism asexually. But there may be breaks in the underground network, leading some scientists to insist that the fungus is not a single organism.

In November 1992, scientists reported a larger natural clone than the Michigan fungus: a stand of 47,000 aspen trees growing from a single root system covering 43 hectares (106 acres) in Utah. In autumn, such a clone is clearly visible, because the leaves on genetically identical trees turn golden at the same time, *below.* Scientists say other, larger clones are likely to exist as well.

In only 30 seconds, a few drops of water, *right*, transform dry star moss into a green plant, *far right*. In nature, star moss can withstand several years of drought, then spring to life when it rains. In July 1992, a Texas biologist reported identifying 74 proteins the moss makes as it speedily repairs dried cells. The scientist hopes to identify the genes that make the proteins, paving the way for the genes to be transferred into crop plants. Such plants could resist drought damage.

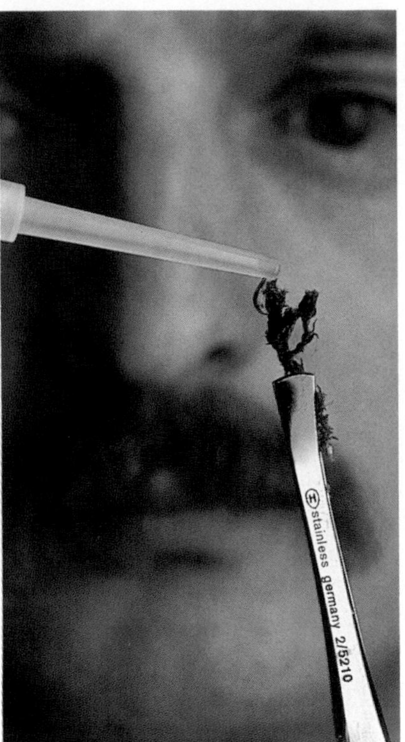

plants a brief burst of air. The plants glowed as long as they were stimulated by the wind, which showed that the stress caused calcium ions to be released into the cytosol. As the wind speed increased, the plants produced more light, indicating they were releasing more calcium ions.

A hormone breakthrough. A better understanding of the way the plant hormone ethylene affects plants was the result of work reported in February 1993 by molecular geneticist Joseph J. Kieber and his colleagues at the University of Pennsylvania in Philadelphia and the University of Arizona in Tucson. The scientists isolated and identified an enzyme that plays a role in plants' chemical responses to ethylene. Scientists have been studying plant hormones since the 1920's, but Kieber's work is the first to reveal one of the steps in the series of responses that a hormone causes.

In the 1960's, plant physiologists first realized that ethylene was a plant hormone and that it changes plant development in a variety of ways. For example, emerging seedlings respond to compacted soil by producing ethylene, which toughens cell walls, enabling the plant to break through the soil. Ethylene also affects the rate at which flower petals fade or fruit ripens.

Kieber's team examined more than a million seedlings of *Arabidopsis* (a tiny mustard plant that completes its life cycle in about 30 days) to find short plants with curled tips, an indication that the plants produced too much ethylene. Kieber gave those plants a substance that reduces ethylene production. Some then grew normally, but others remained stunted.

The scientists examined the genetic material of those stunted plants to determine how they were different from the plants that had a normal response. They found a *mutation* (change) in a gene that controls production of an enzyme. The enzyme was of a type known to be part of the hormone response in yeast, worms, fruit flies, and even human beings. This suggests that plants and animals may have similar responses to hormones. [Frank B. Salisbury]

In WORLD BOOK, see BOTANY.

A California research team in January 1993 reported developing a cheaper way to convert natural gas into *methanol,* a type of alcohol that can be useful as a cleaner-burning alternative to gasoline. The researchers said that the new technique may someday allow natural gas to replace petroleum as the source of the world's main transportation fuel.

The process, developed by chemist Roy A. Periana and his colleagues at Catalytica Incorporated, in Mountain View, Calif., is a highly efficient way to convert *methane*—a flammable gas that is the major component of natural gas—into methanol. Scientists estimate that the world's deposits of natural gas hold an amount of energy almost equal to that of the world's oil reserves. But transporting the explosive gas is difficult, so scientists have been looking for an inexpensive way to convert methane into methanol, which is a liquid.

Methane and methanol have similar chemical structures. Methane molecules consist of a carbon atom connected to four hydrogen atoms. Methanol molecules consist of a carbon atom connected to three hydrogen atoms and one *hydroxyl group* (an oxygen atom connected to a hydrogen atom). So the task for chemists was to devise a chemical reaction in which one of methane's hydrogen atoms would be replaced with a hydroxyl group.

The Catalytica team accomplished that feat by passing methane into a solution of water, sulfuric acid, and a compound containing the element mercury. The mercury compound serves as a *catalyst* (a substance that changes the rate of chemical reactions without itself being changed). It strips one hydrogen atom from methane, which then acquires an oxygen atom and a hydrogen atom from a water molecule to become methanol. The sulfuric acid holds the methane in an intermediate state in the reaction.

The researchers reported that their process converted 43 percent of the methane to methanol, compared with less than 5 percent by other methods. They also said their process was more energy-efficient than other methods, because it takes place at a temperature of 180 °C (356 °F) instead of the 700 °C (1300 °F) required by other processes.

Although methanol is less polluting than gasoline, it currently costs about 30 percent more than gasoline for an equivalent amount of energy. However, the Catalytica reaction may open the door to lower-cost methanol. Scientists must first determine whether the promising laboratory results can be duplicated on the scale of an industrial plant.

Fullerene news. Scientists in 1992 continued working with the hollow carbon compounds called fullerenes. New discoveries included tubes with fullerene walls and onionlike fullerene balls. See MATERIALS SCIENCE.

First antimatter-matter molecule. The world's first molecule containing a particle of *antimatter* was created—for a brief instant—by researchers in July 1992. An antimatter particle is identical to a particle of ordinary matter except that the electric charge is reversed. For example, the electron, a negatively charged atomic particle, has an antimatter counterpart known as the *positron,* which is the same in every respect except that it has a positive charge rather than a negative charge.

Antimatter particles can be produced in particle accelerators, but they usually do not survive long. When antimatter makes contact with ordinary matter, both are destroyed in a burst of energy.

But chemist David M. Schrader of Marquette University in Milwaukee, along with three scientists from Århus University in Denmark, created a molecule consisting of a hydrogen atom bound to an atom called positronium. Positronium consists of an electron and a positron orbiting each other. Scientists had first created positronium in 1949. But until 1992, no one had been able to bind the atom to ordinary atoms. Schrader's team created the molecule by striking methane molecules with a beam of positrons in a particle accelerator.

Not only was the new compound, called positronium hydride, the first known molecule containing antimatter on Earth, it existed for a comparatively long time—about 500 millionths of a second. The researchers believe it may be possible to make other such molecules. One goal is to create a hybrid form of water by substituting positronium for one of the two hydrogen atoms in a water molecule.

New tool for probing cells. Chemists in October 1992 reported constructing a new type of sensor for measuring the pH inside living cells. (A substance's pH indicates its acidity or alkalinity.) The new sensor, developed by chemist Raoul Kopelman and four co-workers at the University of Michigan in Ann Arbor, is a few millionths of an inch in diameter and can measure the pH of volumes as small as a few hundred trillionths of an ounce. Also, the sensor takes measurements very quickly, in about a thousandth of a second.

The Michigan researchers created the sensor using an *optical fiber*—a thin glass tube that carries light. First, they took an extremely thin optical fiber and narrowed it further using a pair of microscopic "pliers." They then coated the fiber with aluminum to prevent light from leaking out the sides. Finally, the scientists attached a special pH-sensitive dye onto the end of the fiber. When light strikes the dye, it glows. The characteristics of the light depend on the acidity of the fluid that surrounds it.

To measure the acidity of a cell, the researchers position the sensor so that the pH-sensitive dye touches the cell. Then the chemists send a pulse of light through the fiber. The light strikes the dye, making it glow. A computer analyzes the light emitted by the dye to determine the pH of the cell.

The researchers used their sensor to measure the acidity inside blood cells, frog cells, and rat embryos. They noted that other dyes sensitive to a variety of cellular conditions could be attached to the end of similar probes.

World's smallest battery. The world's smallest *galvanic cell*—a batterylike device that generates an electric current—was created by chemists at the University of California in Irvine, according to an August 1992 report. The flat, square cell, developed by a team led by chemist Reginald M. Penner, measures only a few millionths of an inch across.

The Irvine researchers used a tool called a scanning tunneling microscope (STM), which lets scientists see and manipulate individual atoms. With the STM, the researchers built two microscopic mounds, one of copper and the other of silver, on a flat carbon surface. Then they bathed the entire structure in an electrically conductive copper sulfate solution. Atoms in the copper mound immediately began dissolving into the solution, leaving behind electrons. The negatively charged electrons flowed through the carbon surface to the silver mound, where they rejoined positively charged copper atoms that had drifted over in the solution. The flow of electrons lasted about 45 minutes, until the silver mound was completely covered by copper atoms.

An electric current is the flow of electrons, so the galvanic cell could in principle have powered an electric device. The researchers estimated that the cell generated about 20 thousandths of a *volt* (the unit of measurement for the force of electricity). That voltage is far too low to power a flashlight or radio, but it might be sufficient to operate the microscopic machines under development by some scientists and engineers. However, the California researchers had not yet found a way to connect the tiny cell to a motor.

New sugar substitutes. A new group of sugar substitutes that contain no calories and that behave like natural sugars in cooking was described in October 1992. The new sweeteners have chemical structures and formulas similar to those of household sugars, said the compounds' discoverers, chemist Mansur Yalpani and his colleagues at Nutra-Sweet Corporation in Mount Prospect, Ill. But research indicated that the sugar substitutes are not recognized by the bacteria that break down sugar in the human stomach. So the compounds should pass through the digestive tract unchanged, without adding any calories to the diet.

In addition, the new compounds act like ordinary sugar in baked and cooked products, adding bulk and improving texture. As a result, foods made with the new sweeteners would probably taste more natural than diet foods made with sugar substitutes that do not respond to cooking as sugar does. However, because the new compounds are only 10 to 50 percent as sweet as table sugar, they would probably have to be mixed with sweeter artificial sugars to make the new compounds' taste acceptable.

The researchers had not yet tested their sugar substitutes on people. The compounds must be approved by the United States Food and Drug Administration before they can be sold in the United States. [Gordon Graff]

In WORLD BOOK, see CHEMISTRY.

Disastrous financial setbacks faced by the world's largest computer manufacturer dominated the computer industry during late 1992 and early 1993. International Business Machines Corporation (IBM), which is headquartered in Armonk, N.Y., announced in January 1993 that it had suffered a loss of more than $5 billion in 1992, one of the worst 12-month losses in the history of American business. The company laid off tens of thousands of workers in 1993, the first such action in its history. IBM's stock price shrank from about $100 a share in July 1992 to a low of $46 in early 1993. The stock had sold for a high of more than $125 in 1988.

IBM's troubles grew out of changes in the very industry the company had helped create. Much of IBM's business depended on mainframe computers, huge information-processing machines priced in the tens of millions of dollars. But since the mid-1980's, personal computers small enough to fit on a desktop, or even in a briefcase or pocket, have increased dramatically in power while dropping equally dramatically in price. IBM helped create a personal computer market with the introduction in 1981 of its first desktop computer, the PC. But many analysts said the company failed to anticipate the speed with which personal computer sales would reduce sales of mainframes.

Two new microprocessors were unveiled in 1993. Microprocessors are electronic devices that act as the "brains" of a personal computer, perfoming the calculations and other operations required by software programs.

The Intel Corporation of Santa Clara, Calif., introduced its latest generation of microprocessors in March 1993. Intel's new chip, called the Pentium, is one of the most powerful microprocessors ever built, with more than 3 million transistors. It is designed to replace the company's popular 486 chip. (The name *Pentium* broke with the series established by Intel with its four previous generations of chips, the 8086, 286, 386, and 486. In tel chose the name after a federal judge ruled in 1991 that the company could not be granted trademark protection for the "X86" name.)

New communicators
Personal digital assistants (PDA's), handheld devices that perform computing and communications tasks, attracted wide attention during late 1992 and early 1993. PDA's can read messages written on the screen with an electronic pen, then send faxes or communicate with other computers. The Personal Communicator 440 from EO and AT&T, *right,* is priced at $2,900. Apple Computer's Newton, *bottom,* was expected to sell for about $1,000.

The new chip doubles the computing speed of the 486. But the great advantage of the Pentium, analysts say, lies in the way it processes information. The Pentium uses a technology called RISC (*r*educed *i*nstruction-*s*et *c*omputing), which permits computers to process more than one instruction at a time. RISC technology is considered crucial to the next generation of computer software, which will employ more graphics, animation, and calculating instructions than current software.

Even as Intel began shipping the Pentium, its competitors were readying their own advanced microprocessors for introduction in mid-1993. In May, the Digital Equipment Corporation of Maynard, Mass., unveiled a personal computer powered by a new chip, the Alpha AXP, that operates at twice the speed of the Pentium. Like the Pentium, the Alpha AXP uses RISC technology. It is designed to make Digital's new computer—called the DEC PC AXP 150—the instrument of choice for users of new software from Microsoft Corporation of Redmond, Wash. The software, known as Windows NT, is intended for workstations, personal computers that can be linked in networks.

Personal digital assistants. As computers have grown more powerful, they have also become increasingly flexible, as well as smaller and more portable. The next step in this evolution is a device that is barely larger than a paperback book, yet capable of sophisticated data processing and communications tasks. Several of the devices, dubbed personal digital assistants (PDA's), were announced in 1992 and early 1993.

Apple Computer, Incorporated, of Cupertino, Calif., attracted the most attention as it moved ahead on plans to unveil a PDA called Newton. The final configuration of Newton, which was first announced in May 1992, had not been revealed by spring 1993. But Apple said Newton would be able to read messages written on the screen with an electronic pen and would include tools for scheduling and managing information. Connecting Newton to a facsimile (fax) machine or to a device called a modem would enable the product to transmit

Two computers in one

The Duo system, introduced by Apple Computer in October 1992, contains a portable computer that can be hooked up to a desktop system without the need to plug and unplug cables or adjust software files.

One of two special PowerBook Duo notebook computer models, *left,* is used with the system. The user slides the notebook computer into a docking module in the desktop system, *above,* whenever a large monitor or the other resources of a desktop computer are required.

data over telephone lines. Newton was expected to be priced at about $1,000.

In January 1993, the Tandy Corporation of Fort Worth, Tex., announced a PDA with the code name "Zoomer." The Zoomer was scheduled for a midyear release and was expected to be priced at about $600. Soon afterward, the American Telephone and Telegraph Company (AT&T) of Parsippany, N.J., and EO, Incorporated, of Mountain View, Calif., jointly unveiled their Personal Communicator 440. A version equipped with a cellular telephone and modem sells for $2,900.

A better mouse. Microsoft brought out a new mouse in May 1993. A mouse is used to guide a pointer or cursor that appears on a computer screen. The device represents the biggest change in mouse design since the mid-1980's.

The Microsoft Mouse 2.0 uses new software designed to make its tracking more accurate. The mouse's curved design is shaped to fit the human hand more comfortably than previous versions of the device. The mouse carries a suggested retail price of about $100.

Two computers in one. A docking system unveiled by Apple Computer in October 1992 enables users to combine the portability of a notebook with the increased resources of a desktop computer. The system, called Duo, eliminates the need to plug and unplug cables or change some software files every time the notebook is hooked up to the larger computer. Instead, it allows new versions of Apple's popular PowerBook notebook computer to simply be inserted into a docking module in a desktop system that includes a full-sized keyboard; color monitor; and optional CD-ROM (compact disc read-only memory) drive, which provides extensive storage capacity. The docking module alone is priced at about $1,080.

CD-ROM drives, and computer systems equipped with them, continued to show dramatic growth through early 1993. The growth coincided with the rising popularity of multimedia software, which employs animation, sound, and even video pictures. (See COMPUTER SOFTWARE.) [Keith Ferrell]

In WORLD BOOK, see COMPUTER.

Computer Software

Windows NT, a graphical user interface (GUI) intended for use on large systems of computers linked in networks, was introduced by Microsoft Corporation of Redmond, Wash., in May 1993. GUI's enable computer users to activate functions and programs by pointing a cursor at symbols called *icons* or at easy-to-read menus. Windows NT is a more powerful version of Microsoft's Windows, a GUI that remained the most popular software product for personal computers in 1992 and 1993.

The first GUI was introduced by Apple Computer, Incorporated, of Cupertino, Calif., for its Macintosh line of personal computers. Both Windows and Windows NT are designed for use in PC's—personal computers made by International Business Machines Corporation (IBM) of Armonk, N.Y., and IBM-compatible machines.

The new GUI, which analysts said was among the most complex pieces of software ever written, marked Microsoft's entry into the market for large corporate and institutional operations. It was priced at $400 apiece for each computer in a network in which it is used. Microsoft said it expected to sell 1 million copies of NT during the program's first year on the market. The basic Windows program sells about 1 million copies per month.

Also in May 1993, IBM introduced a new version of the company's own GUI, called OS/2. The new version, OS/2 2.1, offered a variety of features, including the ability to run more than one program simultaneously (a feature known as *multitasking*) as well as to work with CD-ROM (*c*ompact *d*isc *r*ead-*o*nly *m*emory) drives, which can store huge amounts of information. The program was well received, but analysts said it faced a struggle for acceptance in the Microsoft-dominated marketplace.

Operating system upgrades. In January 1993, Microsoft released MS-DOS 6.0, the sixth generation of its widely used DOS (*d*isk *o*perating *s*ystem) program for PC's. An operating system is the master program of a personal computer, controlling such operations as the transfer of data between disks and working memory. (Windows and some other GUI's are not true operating systems and require DOS or another operating system to run. The Macintosh GUI, in contrast, is part of the Macintosh operating system.)

MS-DOS 6.0 incorporates several new features, such as tools for recovering data that have been erased and software that detects and destroys computer viruses, programs that may damage data. Most notably, the new MS-DOS version features data compression. Data compression shrinks the amount of storage space required for large programs by *encrypting* (encoding) the programs through a kind of mathematical shorthand. As particular operations or data are called for, MS-DOS 6.0 decrypts their codes.

Apple also upgraded its operating system in 1993. The new version, System 7.1, extends the graphic and sound capabilities of Macintosh computers.

Multimedia software, which can incorporate text, animation, sound, photographs, and motion pictures, began to come of age in 1992 and 1993. New products included multimedia word processing programs from Microsoft and an interactive encyclopedia for Windows.

Microsoft released the first multimedia word processing program, a version of its popular Microsoft Word for Windows, in July 1992, and an expanded edition including video capabilities appeared in January 1993. The enhanced version of the program, priced at $595, comes on a single CD-ROM disc that also includes the entire user's manual, as well as optional reference tools such as a dictionary and an almanac. The program allows documents to carry sound, so that users can annotate documents by recording spoken comments. However, this feature requires a computer equipped with a microphone and built-in recording device.

In October 1992, Microsoft unveiled a new upgrade of Microsoft Word for the Macintosh. The upgrade, Microsoft Word 5.1, exploits the Mac's QuickTime technology, which enables users to insert and edit movies.

Another new multimedia product was Compton's Interactive Encyclopedia for

Zoo Keeper, from Davidson & Associates, was among the notable software offerings of 1992 and 1993. In the game, players learn about animals and their environments while protecting them from four Troublemakers, whose antics include stealing food and changing the climate.

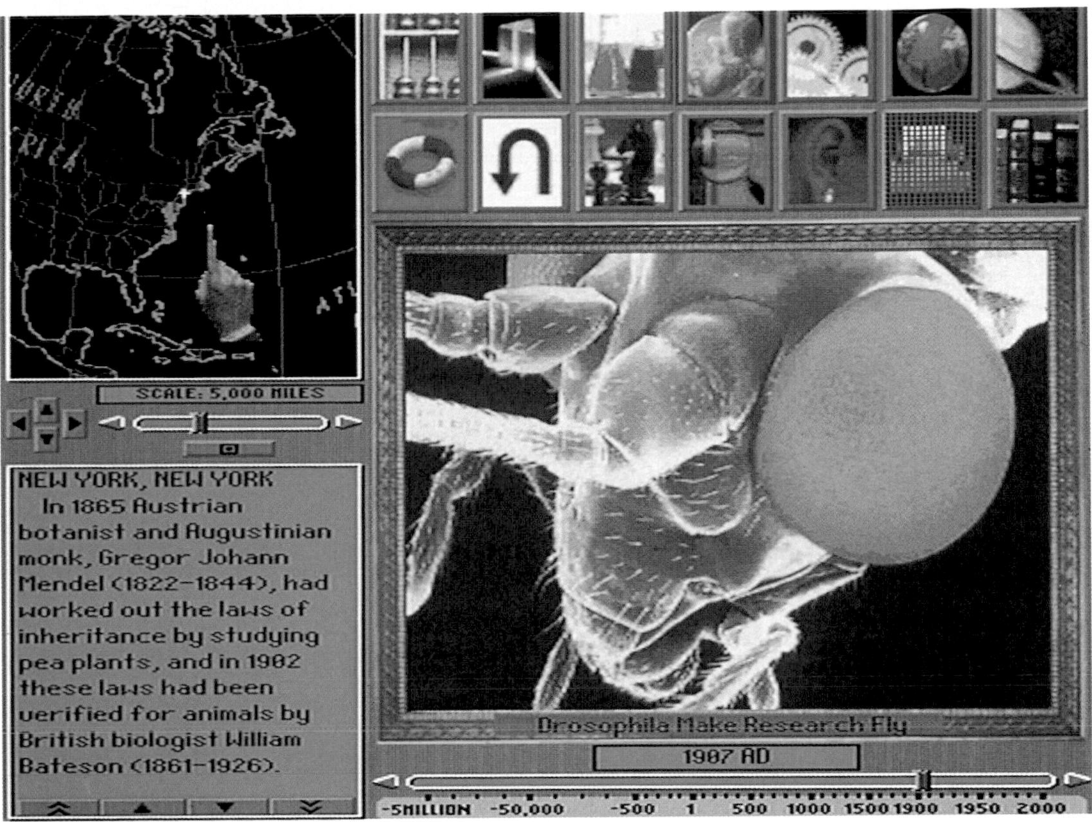

SCALE: 5,000 MILES
NEW YORK, NEW YORK
In 1865 Austrian botanist and Augustinian monk, Gregor Johann Mendel (1822-1844), had worked out the laws of inheritance by studying pea plants, and in 1982 these laws had been verified for animals by British biologist William Bateson (1861-1926).
Drosophila Make Research Fly
1987 AD
-5MILLION -50,000 -500 1 500 1000 1500 1900 1950 2000

Science Adventure, a new software package from Knowlege Adventure, offers sound, images, and text written by the late science writer Isaac Asimov. The software allows users to explore a wide range of subjects in biology, physics, chemistry, space, and other sciences.

Windows, released in November 1992 by Compton's NewMedia of San Diego, Calif. The encyclopedia, a new version of Compton's Multimedia Encyclopedia, is notable for its multitasking ability. For example, users can call up text, pictures, animation, and music to be displayed or played simultaneously. The product has a list price of $395.

In August 1992, Microsoft announced plans to release its own multimedia encyclopedia for Windows, to be called Encarta. Encarta was due for release in mid-1993.

Educational software. The increased computing power of most new personal computers, which permits such features as rapid, colorful animation, continued to benefit educational software developers in 1992 and 1993. Many new educational products have the look of animated films or video games.

One such program is Zoo Keeper, an educational game released by Davidson & Associates of Torrance, Calif., in October 1992. Zoo Keeper, which sells for about $60, challenges students to identify the threats, such as pollution or the destruction of habitats, that endanger animal species. Players then take steps to protect the species by restoring matters to their proper state.

Three new offerings in the Knowledge Adventure series from Knowledge Adventure of La Crescenta, Calif., were released in late 1992 and early 1993. The three sell for about $80 each and include Space Adventure, Dinosaur Adventure, and Science Adventure. (Science Adventure was written by Isaac Asimov, the noted science writer who died in 1992.) A typical Knowledge Adventure product allows students to explore various paths through the program's data and includes games designed to make learning more entertaining.

El Fish, introduced in April 1993 by Maxis, Incorporated, of Orinda, Calif., is among the most complex educational and entertainment packages ever released. El Fish (for "electronic fish") offers players the opportunity to design their own aquatic creatures, which then swim, eat, and even evolve as they adapt to their environment. El Fish is priced at about $60. [Keith Ferrell]

Notable scientists and engineers who died between June 1, 1992, and June 1, 1993, are listed below. Those listed were Americans unless otherwise indicated.

Balassa, Leslie L. (1903-July 2, 1992), Hungarian-born chemist who patented a number of pain remedies and headed several companies engaged in pharmaceutical research.

Bay, Zoltan L. (1900-Oct. 4, 1992), Hungarian-born physicist who played a pivotal role in helping found the field of radio astronomy by using radio signals to study the moon's surface in 1946.

Bohm, David J. (1919?-Oct. 27, 1992), physicist who wrote several books on quantum mechanics.

Braus, Harry (1915?-March 1, 1993), chemist who held patents on more than 50 products and developed a method for identifying and measuring pollutants in air and water.

Buchsbaum, Solomon J. (1929-March 8, 1993), physicist who served as a science adviser to United States Presidents Richard M. Nixon, Jimmy Carter, Ronald Reagan, and George Bush on a variety of science councils and science advisory boards.

Denis P. Burkitt

Burkitt, Denis P. (1911-March 23, 1993), British medical researcher who was among the first to explore a possible connection between viruses and certain cancers. Burkitt's finding that a high-fiber diet helps protect against colon cancer changed the eating habits of millions of people throughout the world.

Feld, Bernard T. (1919-Feb. 19, 1993), physicist who helped develop the atomic bomb and then became a leading advocate of nuclear disarmament as editor in chief of the *Bulletin of Atomic Scientists.*

Robert W. Holley

Glasse, Robert M. (1929-Jan. 1, 1993), anthropologist who specialized in the study of the Huli people of Papua New Guinea and who helped uncover the cause of a fatal neurological disease among New Guinea natives.

Gorenstein, Daniel (1923-Aug. 26, 1992), theoretical mathematician renowned for his mastery of the Theory of Groups, a complex branch of algebra used in particle physics.

Polykarp Kusch

Harrington, Robert S. (1942-Jan. 23, 1993), astronomer known for his contributions to the field of positional astronomy and his research on the possible existence of a 10th planet.

Harrington, William J. (1923-Sept. 5, 1992), medical researcher who became internationally known for his studies of blood diseases and who made the first documented discovery of an autoimmune disorder.

Haury, Emil M. (1904-Dec. 5, 1992), anthropologist and archaeologist noted for his investigations of the Hohokam culture, which flourished for about 1,000 years, until A.D. 1400, in what is now central and southern Arizona.

Holley, Robert W. (1922-Feb. 11, 1993), biologist who shared the 1968 Nobel Prize for physiology or medicine for helping unravel the structure of ribonucleic acid (RNA), the chemical that carries out the genetic instructions of DNA (deoxyribonucleic acid), the basic unit of heredity. Holley's findings gave scientists important new insights into how genes determine the function of cells.

Holtfreter, Johannes F. K. (1901-Nov. 13, 1992), German-born zoologist who helped pioneer the study of embryo development and developed several new techniques used in embryology, including the invention of a solution for growing embryonic cells and tissues in test tubes.

Horn, Daniel (1916?-Oct. 7, 1992), psychologist and public health official who coauthored several important studies in the 1950's that established a strong statistical association between cigarette smoking and lung cancer.

Huntington, Hillard B. (1910-July 17, 1992), physicist who helped develop integrated circuits and computer chips by discovering the phenomenon known as *electromigration,* the movement of atoms in metals. Huntington's research at Rensselaer Polytechnic Institute in Troy, N.Y., contributed to creating the field of solid-state physics.

Jordan, Joseph (1919-Aug. 14, 1992), chemist whose early research showed how hemoglobin substitutes could be manufactured to transport oxygen in the blood and who later developed an instrument that aids diabetics in monitoring their own blood-glucose levels.

Kroon, Reinout P. (1907-Aug. 4, 1992), engineer who helped design and develop the first turbojet engine produced in America.

Kusch, Polykarp (1911-March 20, 1993), German-born physicist who shared the Nobel Prize in physics in 1955 for experiments that determined how an electron responds to an external magnetic field, often called the magnetic moment of the electron. The finding helped confirm the theory of quantum electrodynamics, which concerns the interaction of electrons and electromagnetic radiation.

Laubenstein, Linda J. (1947?-Aug. 15, 1992), physician who documented some of the earliest AIDS cases in the United States.

Barbara McClintock

London, Perry (1931-June 19, 1992), psychologist known for his studies of *altruism* (selflessness), particularly among people who risked their lives to rescue Jews during World War II (1939-1945).

Mayall, Nicholas U. (1906-Jan. 5, 1993), astronomer and former director of the Kitt Peak Observatory in Arizona whose work aided scientists in estimating the age and size of the universe.

McClintock, Barbara (1902-Sept. 2, 1992), geneticist who is often cited as one of the most important figures in the history of genetics because of her many pathbreaking findings. She won the 1983 Nobel Prize for physiology or medicine for her discovery of transposable genetic elements—often called jumping genes—which showed that gene fragments can move and thereby change the way genes control cell growth and development.

Jan H. Oort

Newell, Allen (1927-July 19, 1992), computer scientist and pioneer in the field of artificial intelligence.

Nolan, Thomas B. (1901-Aug. 2, 1992), geologist and former director of the United States Geological Survey who supervised the first photographic mapping of the moon.

Oort, Jan H. (1900-Nov. 5, 1992), Dutch astronomer credited with making several major discoveries about the Milky Way Galaxy, including its rotation and the location of our solar system within it. He was also known for his research on comets and his theory that comets originated in the outskirts of the solar system in a region that came to be called the Oort Cloud.

Albert B. Sabin

Roe, Arthur (1913-April 27, 1993), chemist and science educator who became the first director of the National Science Foundation's international program, working on joint projects with the former Soviet Union and other nations.

Sabin, Albert B. (1906-March 3, 1993), medical researcher who developed the oral polio vaccine. The oral vaccine was licensed in 1961, six years after the injected vaccine, but it became the most widely used polio vaccine around the world. During World War II, Sabin isolated the viruses that caused sand fly fever and created vaccines against dengue fever and Japanese encephalitis, illnesses that plagued American troops fighting in the Pacific.

Schulman, LaDonne H. (1936-Aug. 12, 1992), geneticist known for her findings regarding transfer RNA (tRNA), molecules that carry specific amino acids to complete the process of forming particular proteins. Schulman was recognized for work showing a link between enzymes and tRNA.

Storms, Harrison A., Jr. (1916-July 11, 1992), aeronautical engineer who played a key role in developing military aircraft during World War II and later in developing spacecraft for the Apollo Project, the U.S. program to land astronauts on the moon.

Warkany, Josef (1902-June 22, 1992), Austrian-born medical researcher who focused attention on the importance of prenatal care, showing that diet deficiencies during pregnancy can result in birth defects.

Watson, William W. (1899-Aug. 3, 1992), physicist who was part of the Manhattan Project, the secret United States effort to develop the first atomic bomb.

Zorn, Max A. (1907?-March 9, 1993), German-born mathematician who developed a fundamental premise of algebraic theory.

Zuckerman, Lord (Solly Zuckerman) (1904-April 1, 1993), South African-born British scientist who began his career as a zoologist but later became a science adviser on military strategy to several British prime ministers. Zuckerman established his scientific reputation as an expert on primate behavior, but, during World War II, he became interested in explosives and acted as a counselor on military strategy to Prime Minister Winston Churchill. He later warned against the stockpiling of nuclear arms and the development of the Strategic Defense Initiative, a space-based missile defense system better known as "Star Wars." [Rod Such]

From the time a child's first baby tooth emerges until the child has all 20 baby teeth may be the best time for preventive treatments against the major bacterium that causes tooth decay, *Streptococcus mutans*. Researchers at the University of Alabama at Birmingham reported this finding in January 1993. They called the period a "window of infectivity."

The researchers tracked the amount of *S. mutans* in the mouths of 46 infants at birth and at various times thereafter until each child reached 6 years of age. For 38 of the children, the age of first getting the bacteria ranged between 19 and 31 months. The appearance of the bacteria coincided with the eruption of the 20 baby teeth. The remaining 8 children in the study were free of the bacteria until well past four years of age.

The researchers found that once the last baby tooth had emerged, the window of infectivity closed until the child's first permanent tooth erupted, at about 6 years of age. Therefore, they suggested that researchers attempting to develop a vaccine against *S. mutans* or some other preventive should focus on treatments suitable for teething babies.

Vaccine against gum disease. A vaccine that prevented periodontal disease in rats was developed by researchers at the University of Copenhagen in Denmark and at the State University of New York at Buffalo and at Stony Brook, according to a July 1992 report. Periodontal disease destroys the bone that anchors teeth in the jaw and causes the teeth to loosen and fall out.

The research team had previously immunized rats against periodontal disease using whole, killed *Porphyromonas gingivalis (P. gingivalis)*, the most common bacterium that causes the disease. However, using whole bacteria in a vaccine increases the risk of producing harmful side effects. Thus, the scientists searched for components of the bacterium that would nevertheless cause the rat to produce *antibodies*, disease-fighting molecules that the body produces in response to a particular germ or other foreign invader. Either infection or vaccination can cause the body to produce a specific type of antibody that will attack that germ if it ever invades again.

The scientists tested two preparations made from two components of *P. gingivalis*. One component was found on a hairlike structure on the bacteria called a *fimbria*. The other was part of the cell surface. The scientists tested the preparations in four-week old germfree rats. Some rats were fed live *P. gingivalis*, and others were injected with one of the two vaccine preparations and then given the live bacteria. Forty-two days later, blood and saliva samples showed that all the rats had developed antibodies to the bacteria. The researchers then used a microscope and X rays to view and measure the amount of bone loss around the rats' teeth.

The scientists found that the rats vaccinated with the fimbria preparation had lost almost no bone around the teeth. The cell-surface vaccine was ineffective, despite the fact that tests showed it had caused the rats to produce antibodies in the blood. The researchers speculated that the difference in effectiveness may have been related to the fact that the fimbria vaccine caused antibody production in saliva as well as in blood. The cell-surface vaccine produced only a few antibodies in saliva.

The scientists suggested that the fimbria is a critical part of the bacterium needed to produce bacterial infections. The fimbria binds to mucous membranes in the mouth, enabling the bacteria to form colonies. The *toxins* (poisons) they produce then lead to events that result in destruction of bone around the teeth. Antibodies in saliva may prevent such colonization by binding to the fimbriae, preventing them from sticking to mouth surfaces.

Repairing teeth with gum. Toothpaste and chewing gum capable of building up the mineral surface of teeth may eventually be offered to consumers, according to an October 1992 report. Ming S. Tung at the United States National Institute of Standards and Technology in Gaithersburg, Md., developed the technology behind such products, in which noncrystalline calcium compounds or carbonate solutions are incorporated into toothpaste and chewing gum. Using the products would cause the compounds to crystallize and form hydroxyapatite, the major mineral in teeth. This could fill in at an early stage decalcified areas of tooth enamel that might otherwise progress to form actual cavities. [Paul Goldhaber]

In World Book, see Dentistry.

In 1992 and 1993, doctors reevaluated established therapies to treat AIDS, stomach ulcers, and high blood pressure. New reports suggested that the established drugs for these conditions were not as beneficial as once thought.

AIDS drug doubts. The results of a three-year study of the AIDS drug AZT (zidovudine) suggested that early treatment with the drug provides no benefit, according to an April 1993 report by British and French scientists. Since 1987, doctors have prescribed AZT to treat patients with early stages of infection with *human immunodeficiency virus* (HIV), the virus that causes AIDS. Previous research had shown that AZT slowed the progression to AIDS and slightly extended the life span of some patients with fully developed AIDS.

The British and French study involved 1,749 patients infected with HIV but not yet having AIDS symptoms. One group was given AZT immediately, whereas people in the second group were given the drug only when they began to develop AIDS symptoms. The patients were followed on average for three years. The study found that patients in both groups developed AIDS at about the same rate and died at the same rate. Because AZT can have serious side effects, the findings of this study may lead to changes in the approach of many doctors who have routinely given AZT to individuals with early-stage HIV infections.

Bacteria and ulcers. Antibiotics can successfully treat the most common kind of stomach ulcer, according to a February 1993 report by Austrian physicians. The researchers claimed the antibiotic treatment offered an alternative to long-term drug therapy or surgery.

The doctors' research reflected a new understanding of the cause of stomach ulcers, also called peptic ulcers. Physicians have long believed that stomach acid was the primary cause. They treated ulcer patients by giving them drugs that would either decrease the production of acid or increase the resistance of the stomach lining to the effects of the acid. Then, in the early 1990's, some studies showed that a bacterium called *Helicobacter pylori* was associated with the occurrence of peptic ulcers.

In the new study, doctors at the University of Vienna School of Medicine in Austria successfully treated 48 out of 52 patients with ulcer disease by giving them antibiotics that kill *H. pylori*. The doctors reported that treating the bacterial infection enabled the ulcers to heal quickly. As many as 85 percent of patients with ulcers caused by *H. pylori* have their ulcers return. But only four patients (8 percent) in the group treated with antibiotics had a recurrence.

Blood pressure drug debate. In January 1993, a committee of the National Institutes of Health (NIH) sparked controversy by issuing a report recommending that doctors switch from prescribing the newest drugs to treat high blood pressure to prescribing drugs approved in the 1970's. Treatment of high blood pressure has been a major health concern for more than 20 years. Improved treatment has contributed to a steady decrease in strokes and heart attacks.

According to the NIH report, years of experience with older drugs, such as hydrochlorothiazide and propranolol, has proved them to be effective, inexpensive, and easy for most patients to take. The report noted that as increasing numbers of drugs have been developed to treat high blood pressure, doctors have tended to use the newer drugs more and the older drugs less. But, the NIH committee said, the newer drugs' benefits and drawbacks are less well understood. Also, some are 10 or 20 times more expensive than the older drugs.

Some specialists called the report a step backward in the treatment of high blood pressure. They claimed that the committee ignored important scientific evidence that some of the new drugs might be better than the older ones. These experts said that the newer drugs such as angiotensin-converting enzyme (ACE) inhibitors and calcium channel blockers are more effective in preventing some consequences of high blood pressure, such as heart attacks and kidney disease. Observers said the debate may not be settled without additional scientific study.

New migraine drug. In December 1992, the U.S. Food and Drug Administration (FDA) approved sumatriptan for treating migraine headaches. The drug is injected under the skin. In clinical trials of the drug, 75 percent of the more than 1,000 patients participating said they had relief of headache pain within one hour of the injection, and 80 percent had relief after two hours. In a few cases, relief came within 30 minutes.

Taxol approved. Also in December, the FDA approved taxol for the treatment of advanced ovarian cancer. The approval came in a record five months from the time the application was first submitted to the FDA. In studies, the drug had reduced the size of cancerous tumors in 20 to 30 percent of women.

Taxol is made from the bark of a rare, slow-growing tree, the Pacific yew. Environmental concerns surrounded the discovery of the drug because the company that makes taxol, Bristol-Meyers Squibb Company, has a unique agreement with the U.S. Forest Service to harvest trees on public lands in the Pacific Northwest. But in January 1993, Bristol-Meyers Squibb announced that progress in creating synthetic taxol in the laboratory had proceeded so rapidly that no Pacific yew trees would need to be harvested from public lands in 1993.

Sickle cell anemia drugs. Researchers reported significant advances in the treatment of sickle cell anemia in 1993. Sickle cell anemia is an inherited disease that affects the red blood cells. The disease causes the cells to form abnormal, sicklelike shapes. Such cells cannot pass through small blood vessels. The cells clog the blood vessels and interfere with blood flow, thus depriving tissue of oxygen. Left untreated, the illness can affect almost all parts of the body. Although the disease continues to be fatal, most individuals live into their 20's, and some live well into middle age.

Studies reported in January 1993 suggested two new approaches to treating sickle cell anemia. In the first report, researchers said that a medicine called butyrate helped patients produce more of the normal red blood cells and fewer sickled cells. In the second study, researchers at the National Institutes of Health reported that they had found that a combination of two drugs—erythropoietin and hydroxyurea—also produced more of the normal red blood cells. Further studies of all three agents will be needed before the new treatments are widely available. (See also MEDICAL RESEARCH.) [B. Robert Meyer]

In WORLD BOOK, see DRUG.

Ecology

Ecologists focused much attention during 1992 and 1993 on how *fragmenting* (splitting apart) natural habitats affects plant and animal species. In one major study, a group of ecologists headed by Michael S. Gaines, then of the University of Kansas in Lawrence, and Robert D. Holt of the same university reported in July 1992 that different characteristics of an ecosystem react differently to such fragmentation.

To perform their study, the researchers divided a large field into many study areas of several sizes. Six large patches measured 50 meters (165 feet) by 100 meters (330 feet); 18 medium patches were 12 meters (40 feet) by 24 meters (80 feet); and 82 small patches measured 4 meters (13 feet) by 8 meters (26 feet). Mowed strips of land from 5 to 10 meters (16 to 33 feet) wide separated the patches. For six years, the investigators studied how the numbers, types, and sizes of plants and animals in each type of study area changed.

The researchers found that many broad characteristics of the ecosystem remained unaffected by the fragmentation. Fragmentation did not change the amount of minerals in the soil, for instance, and it did not affect *plant succession* (the orderly changes that take place over time in the kinds of plant species in an area). Nor did the diversity of plant or animal species within the patches change.

But the populations of some plant and animal species did change. The study showed that plants that reproduce *asexually*, in which a new organism develops from parts of a single parent organism, were less likely to survive in the smaller patches. Patch size also influenced the populations of *vertebrates* (animals with backbones). For example, snakes and rodents were markedly affected by patch size. The snake population was more dense in the larger patches than in the smaller ones. For rodents, the effects depended on the species' body size. Large rodents, such as cotton rats, were proportionally more numerous in the larger patches. Populations of smaller rodents, such as deer mice, were most dense in the smaller patches.

Do clear-cut forests recover?
A greater variety of plants exists in a North Carolina forest that has never been cleared, *top*, than in a forest in the same area that has grown for about 50 years after clearing, *bottom,* according to ecologist David Duffy of the State University of New York at Stony Brook. In a June 1992 report, Duffy said his studies indicated that clearing forests results in a loss of nutrients in soil, limiting future plant growth.

Benefits of monogamy. Mice that behave monogamously (having only one mate) have more surviving offspring compared with mice who mate with multiple partners. Ecologist David O. Ribble, then at the University of California at Berkeley, used a technique called DNA fingerprinting to make this finding, which he reported in June 1992.

Ribble studied a mouse called the California mouse, which ecologists believe is one of the few completely monogamous mammals. The first step in Ribble's study was determining the relationships of parents to offspring in the population of mice. In most species, it is impossible to use observation alone to determine with certainty which animals are another's parents. An observer often cannot be absolutely certain that only one male fertilized the eggs of a female, for instance. Observers also cannot be certain that all of the youngsters a female is raising are her offspring.

To overcome these problems, Ribble used DNA fingerprinting to determine relatedness with certainty. In this technique, scientists analyze genetic material from the animal's cells. By comparing the genetic makeup of different mice, Ribble was able to determine how they were related.

Ribble found that the average California mouse couple produced 4.7 offspring that survived beyond nursing. He compared this number with estimates of surviving offspring of similar species of mice that are not monogamous. This comparison showed that more California mice survived past nursing. Monogamous behavior may have led to a higher survival rate because the fathers may have given their offspring increased care. Such behavior is not often found in species that are not monogamous, Ribble said.

Mast cycles in oaks. Several factors influence cycles of *mast* (nut) production in oak trees, ecologists reported in early 1993. In certain years, trees of the same species produce far more seeds than usual. Despite the widespread occurrence of this phenomenon, ecologists had not known whether masting was a response to weather changes or a a natural cycle in a given species.

Victoria L. Sork and her colleagues at the University of Missouri and Washington University, both in St. Louis, studied the mast production of black, white, and red oak trees in Missouri over an eight-year period. The researchers measured the number of acorns that the trees produced and the number of blossoms on the trees, as well as the area's air temperature and rainfall.

Their study showed that the black, white, and red oak species experienced mast cycles that lasted two, three, and four years, respectively. Warm weather and plentiful rain resulted in greater acorn production, but the amount of time elapsed since a tree's last mast production was also important.

The researchers theorized that the trees needed time to build up energy resources to produce another abundant masting. The cycles of mast production in oak trees may vary because different species use their stored energy and replenish it at different rates, according to the researchers.

Call for a catalog of species. How many species of living things are on the Earth? Two well-known biologists issued a 50-year plan in November 1992 for answering that question. Peter H. Raven, director of the Missouri Botanical Garden in St. Louis, and Edward O. Wilson, curator in entomology at Harvard University in Cambridge, Mass., said that the plan was necessary because science knows little about the diversity of species on Earth, and human activity threatens vast numbers of species each year. As a result, as many as 50,000 species become extinct each year, they said.

Raven and Wilson estimated that science has identified approximately 1.4 million species of living things on Earth. However, this represents only 2 to 15 percent of all living species, they said. The authors advocated sampling all of Earth's habitats with the hope of cataloging as many species as possible while there is still time. [Robert H. Tamarin]

See also ZOOLOGY (Close-Up). In the Special Reports section, see BEWARE THE EXOTIC INVADERS; THE CASE OF THE DISAPPEARING SONGBIRDS; and CHAMPION OF THE DIVERSITY OF LIFE. In WORLD BOOK, see ECOLOGY.

Electronics

The United States consumer electronics industry suffered the effects of a poor economic climate in 1992 and 1993. Nevertheless, the industry continued to introduce new products, many of them employing digital technology to bring computerlike capabilities to traditional consumer electronics applications.

Two new audio products aimed at attracting a new generation of customers were unveiled in late 1992. The last important new audio product to take hold of the market was the compact disc (CD), introduced in 1982. Like CD's, the two new products—tiny tape cassettes and optical discs—store sounds as digital code, a series of 0's and 1's. But unlike CD's, they are also recordable.

Digital compact cassettes (DCC's), developed by N. V. Philips of the Netherlands, were unveiled in October 1992. DCC's are small magnetic-tape cassettes. DCC systems, priced at about $800, can play both DCC's and traditional audio cassettes and can record DCC's.

Japan's Sony Corporation began selling its MiniDisc (MD) system in December 1992. An MD player/recorder employs lasers to record sound on a compact disc only 6.4 centimeters (2.5 inches) in diameter. A portable MD unit capable of recording is priced at $750. A playback-only version sells for $550.

Several hundred prerecorded tapes and MD's accompanied the introduction of the DCC and MD systems. Traditionally, the availability of prerecorded music has been crucial to the success of new audio products.

TV goggles. Although many people wear glasses to watch television, a new product places the television inside the glasses. In January 1993, Virtual Vision of Redmond, Wash., introduced the Virtual Vision Sport, a headset with a built-in TV screen. The "screen" is a small mirror located in the lower portion of one lens, an arrangement designed to give the effect of watching a large-screen TV from a distance of about 4.5 meters (15 feet). The image blocks only a small portion of the field of vision, so wearers of the goggles can also perform other activities. The product, priced at about $900, includes a rechargeable battery pack, as well as an antenna that can pick

Digital recording
Two products unveiled in autumn 1992 enable consumers to record sound digitally. Sony Corporation's portable player/recorder, *right,* records MiniDiscs, CD's measuring only 6.4 centimeters (2.5 inches) in diameter. The recorder sells for about $750. A unit from N. V. Philips, *below,* plays and records digital compact cassettes (DCC's), audiotapes with the sound quality of CD's. It is priced at about $800.

up television signals from up to about 30.5 meters (100 feet).

Applications for the technology that are not geared to the consumer market are also in the wings. For example, a surgeon might use the glasses to view images recorded by a tiny video camera placed inside a patient's body during an operation.

Videophones. A long-awaited technology became reality as two companies began selling videophones in 1992 and 1993. In September 1992, American Telephone & Telegraph (AT&T) of Parsippany, N.J., brought out its VideoPhone, which had been announced in January 1992. MCI Communications Corporation of Washington, D.C., unveiled its own VideoPhone in March 1993. Both devices, which plug into standard phone outlets, enable people to see one another on a small screen as they speak. The caller and the receiver must have a videophone made by the same company for the phones to work.

The AT&T videophone, originally priced at $1,500, was reduced to $1,000 in January 1993. The MCI product sells for $750. Both products transmit video images at a rate of about 10 frames per second, in contrast to the 30-frame-per-second rate of TV images. Their slow transmission rate makes videophone images appear slightly jerky. For this reason, as well as the phones' relatively high price, videophones were expected initially to be purchased chiefly for uses in business and education.

New game machine. The January 1993 announcement of a new video game machine by the 3DO Company of San Mateo, Calif., generated excitement in the consumer electronics industry. The machine promised to raise the stakes in the video game industry—and perhaps throughout the electronics market—by using digital technology, compact discs, and computer power to deliver interactive entertainment and educational products that look and sound more like TV or motion pictures than like traditional video games. 3DO units were expected to be available in late 1993. [Keith Ferrell]

In World Book, see Compact disc; Electronics; Tape recorder.

United States President George Bush in October 1992 signed the Energy Policy Act, legislation outlining the country's goals for energy production and use through the year 2010. The act includes provisions to cut oil imports by a third and boost the use of renewable energy sources, such as wind and solar power, by 20 percent. The act also eases licensing regulations for nuclear power plants and mandates efficiency standards for lighting equipment, motors, and heating and cooling equipment.

The new laws, along with a 1993 proposal by President Bill Clinton to levy a federal energy tax, highlighted national efforts to increase energy efficiency and reduce pollution. Other such efforts include advances in the efficiency of consumer appliances and vehicles as well as in the larger arena of commercial and utility energy production.

A more efficient motor. An improved motor for home air conditioners and heating appliances was put into production at the Fort Wayne, Ind., General Electric Motors Division in July 1992. Called the ECM, for "electronically commutated motor," the new motor has a built-in computer that varies the motor's speed according to the appliance's needs. This feature allows the motor to run efficiently over a wide range of speeds.

Motors in conventional home air conditioners and furnaces do not run efficiently at low speed, and so they are designed to simply turn on and off when heating or cooling is necessary. The ECM's speed control allows the motor to run at a reduced speed when heating or cooling needs drop, saving as much as 20 percent in energy costs compared with a motor of similar size with no speed controls.

The motor's extra efficiency may help make up for some of the expected slight loss of efficiency in the next generation of consumer cooling appliances, which will use replacement refrigerants for those that contain chlorofluorocarbons (CFC's). Because CFC's destroy the protective ozone layer in Earth's upper atmosphere, CFC manufacturers in developed countries must stop making the chemicals by 1996. Manufacturers claim that refrigerants designed to replace CFC's may be 5 to 10 percent less efficient than CFC's.

Solar energy storage. A material that efficiently stores solar energy as electrical energy was developed by chemists at Princeton University in Princeton, N.J. The lack of an efficient way to store solar energy has been an obstacle to the increased use of solar energy. But in August 1992, chemists Lori Vermeulen and Mark Thompson reported making a storage device using a film made of two compounds—zirconium phosphonate and an organic material called viologen-halide—sandwiched together. The zirconium phosphonate absorbed some of the energy of sunlight, causing a chemical reaction that freed subatomic particles called electrons from the atoms in the layer. The freed electrons moved to the viologen layer but were trapped there. This constitutes stored electrical energy, because the electrons could be dislodged later to produce useful electricity. Researchers familiar with the Princeton work reported that scientists still had not found a way to easily dislodge the electrons and thereby recover the trapped energy from the viologen layer, however.

Battery advance. A nickel-hydride battery shows promise as a device to power electric vehicles. That was the determination of the Advanced Battery Consortium, a partnership of major U.S. automakers and the U.S. Department of Energy, who awarded an $18.5-million contract in June 1992 to Ovonic Battery Corporation of Troy, Mich., to develop the battery. The automakers plan to test the batteries in electric vehicles. If the battery performs well, the consortium will collaborate on building an electric car using it.

The nickel-hydride battery provides about twice as much energy for its weight as does its main competitor, the lead acid battery currently used in gasoline-powered cars. An average 180-kilogram (400-pound) nickel-hydride battery would allow a compact car to travel about 320 kilometers (200 miles) compared with 195 kilometers (120 miles) for the lead acid battery.

A drawback to the nickel-hydride battery is a relatively short life span. It needs to be replaced after about 500 charges compared with 1,000 charges for a typical lead-acid battery. Technology development may improve this performance.

An array of 9,600 photovoltaic solar panels began supplying power to homes in Davis, Calif., in January 1993. The array of panels is the largest ever to successfully supply power to an electric utility. The panels, manufactured by Advanced Photovoltaic Systems Incorporated of Princeton, N.J., can power 150 homes.

Tires to energy. A power plant in Montebello, Calif., that uses old tires as fuel began operation in December 1992. The new plant, owned by Texaco Incorporated of White Plains, N.Y., employs a form of energy production called *gasification,* in which liquid or solid fuel is turned into gas. In this plant, the tires become a mixture of gases. Some of the gas can be separated and sold, and the remaining gases are burned to power a gas turbine that runs a generator.

Texaco hopes the new plant will help solve the environmental problems old tires cause. The company estimates that U.S. consumers discard about 250 million tires per year, most of them ending up with the 2 billion to 5 billion tires already in landfills. Energy engineers at Texaco also say that the Montebello plant's emissions of sulfur dioxide—a major pollutant from many power plants—are 10 times lower than current Environmental Protection Agency limits for electric power plants.

In the Montebello plant, operators mix shredded tires with waste oil to form a heavy *slurry* (liquid with solid particles). Operators pressurize the slurry, inject oxygen, and heat the mixture to 370 °C (700 °F). The heat and pressure cause molecules in the mixture to break down into *synthesis gas*—a mixture that includes hydrogen, carbon monoxide, methane, and hydrogen sulfide.

The combustion of synthesis gas can power a gas turbine to produce electricity. Filters must separate sulfur gas from the synthesis gas to avoid creating sulfur dioxide during combustion. The filtered sulfur can be sold as fertilizer. Methane in the synthesis gas may also be broken down into ammonia and methanol, both of which can be used as industrial chemicals.

Gasification and fuel cells. A 20-kilowatt fuel cell system was commissioned at a coal gasification plant in Plaquemine, La., in February 1993. The Electric Power Research Institute (EPRI) in Palo Alto, Calif., and Energy Research Corporation (ERC) in Danbury, Conn., sponsored the demonstration project. Coal gasification turns coal into synthesis gas in a process similar to that used by Texaco's tire-fueled gasification plant in California.

Dan Rastler, manager of fuel cell systems at EPRI, said that the project is designed to determine how contaminants in gas obtained from a coal gasification plant affect the performance of fuel cells. Destech Energy Incorporated, a subsidiary of Dow Chemical Company Incorporated of Midland, Mich., will use its Plaquemine plant to provide gas to ERC's carbonate fuel cells.

Unlike conventional power plants, which burn *fossil fuels* (coal, oil, or natural gas), fuel cells produce electricity by electrochemical reactions that transform the energy in hydrogen or natural gas into electricity. A fuel cell power plant works in a manner similar to a battery. Pumps inject natural gas or hydrogen along with pressurized steam into one electrode (the anode). Oxygen and carbon dioxide enter at the other electrode (the cathode).

Chemical reactions change the natural gas into hydrogen and carbon dioxide, which generate electrochemical reactions with the oxygen and carbon dioxide. These reactions create a flow of charged molecules across the fuel cell and liberate electrons through the external circuit to create an electric current. Waste products include carbon dioxide, heat, and water.

The environmental benefits of fuel cells are impressive. They emit significantly fewer pollutants than do coal-fired power plants. Conventional power plants emit nitrogen oxides and sulfur dioxide, which cause acid rain. But fuel cells emit as much as 250 times less nitrogen oxide and no sulfur dioxide. Fuel cells are currently three to five times more expensive than conventional power plants, but the expense of fuel cell power production should fall as the technology advances and more plants are produced.

ERC's carbonate fuel cell will use hydrogen from the synthesis gas derived from Destech's gasification plant. The fuel cell was scheduled to begin producing electricity by June 1993.

[Pasquale M. Sforza]

In WORLD BOOK, see ENERGY SUPPLY; SOLAR ENERGY.

Environment

Two major oil spills created environmental damage during late 1992 and early 1993. On Dec. 3, 1992, an oil tanker containing 90 million liters (24 million gallons) of crude oil ran aground on rocks at the entrance of La Coruña Harbor in northern Spain. Heavy seas split the Greek tanker, the *Aegean Sea,* in two. Approximately 82 million liters (22 million gallons) of oil spilled into the ocean, nearly double the amount that spilled when the *Exxon Valdez* ran aground in Alaska's Prince William Sound in 1989. The oil that washed ashore contaminated 200 kilometers (125 miles) of Spain's coastline.

The oil contaminated many shellfish in the area, and first-year losses to mussel harvesters were estimated at $15 million. But experts said the environmental damage could have been worse. The fact that the cargo in the *Aegean Sea* was *volatile* (quickly evaporating) light oil lessened the environmental impact of the spill.

On Jan. 5, 1993, the United States-owned oil tanker *Braer* ran aground on the rocky coastline of the Shetland Islands off the northeast coast of Scotland. The tanker had lost power, and fierce winds pushed it into a steep cliff where the ship began to spill its cargo of 97 million liters (26 million gallons) of crude oil. The oil threatened nesting sites for endangered birds and other marine life such as seals, otters, and porpoises.

Fewer U.S. oil spills. Researchers in July 1992 reported a sharp decline in oil spills from tankers in United States waters in 1991. Researchers at *Golob's Oil Pollution Bulletin* in Cambridge, Mass., reported that tankers spilled 209,000 liters (55,000 gallons) of oil in U.S. waters in 1991, well under the 12.9-million liter (3.4-million gallon) yearly average for 1978 through 1991.

Lasting effects of *Valdez*. Oil from the *Exxon Valdez* spill caused lasting harm to many marine organisms in Prince William Sound, according to a panel of biologists and ecologists who reported on the effects of the spill in February 1993. One of the panelists, marine biologist D. Michael Fry of the University of California, Davis, estimated

Two oil spills

The United States tanker *Braer* sinks off the southern coast of the Shetland Islands, *right,* after hitting the rocks on Jan. 5, 1993. The *Braer* spilled 97 million liters (26 million gallons) of oil, causing widespread damage to marine life. The Greek tanker *Aegean Sea* burns after running aground near La Coruña Harbor, Spain, on Dec. 3, 1992. The tanker spilled 82 million liters (22 million gallons) of oil into sensitive marine environments.

AEGEAN SEA

that the spill killed 500,000 birds, 10 times more than any oil spill in U.S. history.

Not only did the spill kill many birds, but it also hampered their breeding. According to marine biologist Dennis Heinemann, then with the Manomet Bird Observatory in Manomet, Mass., one-third of the adult common murres (a diving bird that resembles a small penguin) died as a result of the spill. Breeding in the remaining population had nearly halted, he reported. Most of the Harlequin ducks that survived the spill had also failed to breed, he said.

Fish suffered breeding problems as well. Biologist Evelyn D. Biggs of the Alaska Department of Fish and Game reported that year-old herring living in oil-polluted waters in 1989 produced half as many healthy young in 1992 compared with herring living in unaffected waters.

Many plants and animals were recovering from the *Exxon Valdez* spill, however. Although 880 of 8,000 bald eagles in Prince William Sound were believed to have been killed by the spill, the eagle population had returned to normal, according to the Alaska Department of Fish and Game.

Plants and global warming. Plants and soils could release large amounts of stored carbon if Earth's surface temperature begins to warm significantly, two ecologists at the University of Virginia in Charlottesville reported in February 1993. The finding, which was contrary to previous research, was important because of scientific and public concern over *global warming,* a potential rise in Earth's average temperature due to an excessive buildup of carbon dioxide and other gases in the atmosphere. These gases cause warming by trapping energy from the sun. Human activities, such as burning *fossil fuels* (coal, gasoline, and natural gas), also release carbon dioxide into the atmosphere.

Plants and soils affect the level of carbon dioxide in the atmosphere. Plants take in carbon dioxide from the air during *photosynthesis* (the process by which plants convert carbon dioxide and water to sugar and oxygen). They release carbon dioxide through respiration. Soils contain carbon from dead and decaying plants. This carbon is released as carbon dioxide by the organisms that cause the decomposition of plant matter.

Scientists Thomas M. Smith and Herman H. Shugart used computer models of global climate to predict climate changes that would occur if carbon dioxide levels in the atmosphere doubled. They used these predictions to estimate how a warmer Earth would affect vegetation and soils, and how those effects would alter the amount of stored carbon.

From their computer simulations, Smith and Shugart concluded that a warmer climate would create drier conditions in some regions more quickly than many plants could adapt. Their study suggested that these plants would die because of the rapid climate change. After the plants died, decomposing organisms would release the carbon in the plants as carbon dioxide gas. The scientists also suggested that drier conditions would lead to forest and brush fires, which also release carbon dioxide.

Carbon dioxide from tundra. Smith and Shugart's claim was supported by another report in February 1993. Biologist Walter C. Oechel and his colleagues at San Diego State University studied levels of carbon dioxide released from *peat* (partially decayed moss and other plants) in a 200-kilometer (125-mile) stretch of *tundra* (a vast, treeless plain) in northern Alaska.

The scientists measured changing concentrations of gases in sealed plastic enclosures about 5 meters (16 feet) by 2 meters (6.5 feet) on the peat for five summers beginning in 1983. They compared their readings with ones taken in the same region in the 1970's and determined that the peat emitted more carbon dioxide in the 1980's than in the 1970's.

The team related the carbon dioxide releases to increased temperatures in the region. According to Oechel, the average summer temperature in the region had risen by between 1.4 and 2.4 Celsius degrees (2.5 and 4.3 Fahrenheit degrees). The scientists theorized that the increased temperatures were stimulating the growth of bacteria, which help decompose the plant material and release carbon dioxide in the process.

Russia's radiation hazards. A tank of radioactive waste exploded at a chemical plant in the Russian city of Tomsk in March 1993, contaminating the plant and areas around it. The accident highlighted severe problems with radioactive waste disposal in the former Soviet Union that surfaced in late 1992.

Russian authorities revealed in November 1992 that they had dumped radioactive waste from nuclear power plants into the Berents Sea since the 1960's. The Russians also said that nuclear reactors from four sunken Russian nuclear submarines remained in the Arctic Sea in 1993.

Another record ozone loss was detected above Antarctica in 1992. Late September measurements taken with a National Aeronautics and Space Administration (NASA) satellite showed that the annual destruction of the protective ozone layer was about 30 percent greater than ever before. (Ozone is a molecule made of three oxygen atoms. In Earth's upper atmosphere, it absorbs damaging ultraviolet radiation from the sun.)

The ozone layer above Antarctica thins every year between August and October, when distinctive wind patterns trap large amounts of artificially produced chemicals called *chlorofluorocarbons* (CFC's). CFC's are used in air conditioners, to make foam insulation, and to clean metals in industry. In the atmosphere above Antarctica, CFC molecules collect on ice particles in extremely cold clouds. When sunlight strikes the CFC molecules, they break down into highly reactive chlorine atoms, which react with ozone molecules and destroy them. This destruction creates a "hole" in the ozone layer.

Scientists have recorded a widening ozone hole every year since they began to measure ozone levels above Antarctica in 1985. But the depth of the hole also increased in 1992. In late September and early October 1992, balloon measurements showed that all the ozone had been destroyed between 14 to 18 kilometers (8.7 to 11.2 miles) above Earth's surface. This 4-kilometer (2.5-mile) band exceeded the band of complete ozone loss 1 to 2 kilometers (0.6 to 1.2 miles) deep that was observed in previous years.

Arctic ozone levels. Scientists also reported unprecedented ozone loss over regions of the Northern Hemisphere in early 1993. In April 1993, atmospheric scientists with the Jet Propulsion Laboratory in Pasadena, Calif., reported on satellite measurements of the upper atmosphere above the Arctic and large regions of Russia, Canada, Scandinavia, and Europe. The data showed that ozone concentrations in those areas were 10 to 20 percent lower from mid-February through March 1993 than during the same period in 1992. The scientists said that the ozone loss was twice the normal year-to-year variations in ozone concentrations.

The researchers attributed the ozone loss to record levels of the chemical chlorine monoxide in the upper atmosphere above the Arctic regions combined with colder than normal winter temperatures. Chlorine atoms produced by the breakdown of CFC's react with chlorine monoxide. It, in turn, can react with ozone molecules, destroying them. Cold temperatures promote this destruction by helping ice particles form. The ice particles provide a surface for the chemical reactions that destroy ozone.

Pinatubo and ozone loss. Atmospheric scientist David J. Hofman of the National Oceanic and Atmospheric Administration in Boulder, Colo., reported in September 1992 that sulfuric acid from the eruption of Mount Pinatubo in the Philippines helped reduce ozone levels above Antarctica to these record lows. He reported that sulfuric acid droplets suspended in the upper atmosphere provided a surface upon which the chemical reactions that break apart CFC's molecules could occur.

Firmer ozone treaty. In November 1992, representatives from 74 countries agreed in Copenhagen to a quicker phaseout of CFC's than called for in the Montreal Protocol, an existing international agreement to ban CFC's. The revised agreement bans CFC's by January 1996, four years earlier than the old agreement. The revised agreement also regulates the production of a type of CFC that had not been controlled. These chemicals, called hydrochlorofluorocarbons (HCFC's), are used for many of the same purposes as are CFC's. HCFC's are 20 times less damaging to the ozone layer than are CFC's, but HCFC's still present a slight threat of ozone destruction. [Daniel D. Chiras]

In WORLD BOOK, see ENVIRONMENTAL POLLUTION.

Important discoveries of two new dinosaur species were reported by paleontologists in 1992 and 1993.

Earliest dinosaur. A fossil of the most primitive dinosaur yet known was described in January 1993 by paleontologist Paul C. Sereno of the University of Chicago. Sereno and a team of paleontologists had found the fossil, named *Eoraptor* ("dawn raptor"), in 1991 in northwestern Argentina. The fossil was dated to about 225 million years ago, the mid-Triassic Period. Paleontologists believe that dinosaurs arose about 240 million years ago.

Eoraptor measured 1 meter (3 feet) in length and weighed about 11 kilograms (25 pounds). Paleontologists said it was a primitive member of the group of upright-walking dinosaurs known as saurischians, which included *Tyrannosaurus. Eoraptor* possessed short arms with three-fingered hands adapted for gripping and slashing prey. However, the dinosaur lacked a sliding joint in its lower jaw, a trait of later saurischians.

Ferocious raptor. Another new species of dinosaur—and one of the most ferocious predators that has ever lived—was reported in 1992 by paleontologists Donald Burge of the College of Eastern Utah Prehistoric Museum in Price and James I. Kirkland of the Dinamation International Society of San Juan Capistrano, Calif. The creature, named *Utahraptor* ("Utah's raptor"), was dated to 125 million years ago, the early Cretaceous Period. The fossil had been unearthed in a quarry in eastern Utah in 1991.

Utahraptor was about 6 meters (20 feet) long and weighed about 1 short ton (0.9 metric ton). It was armed with bladelike claws on its fingers and toes and an upright 38-centimeter (15-inch) sicklelike claw on each foot. Paleontologists believe that *Utahraptor* used the claws to slash and disembowel prey.

Utahraptor belonged to a group of relatively small, agile carnivorous dinosaurs called dromaeosaurids. The raptors in Michael Crichton's 1990 novel *Jurassic Park* were patterned after a closely related dromaeosaurid called *Deinonychus.*

Fossil microorganism. The discovery of one of the oldest fossil ciliates was reported in January 1993 by paleontolo-

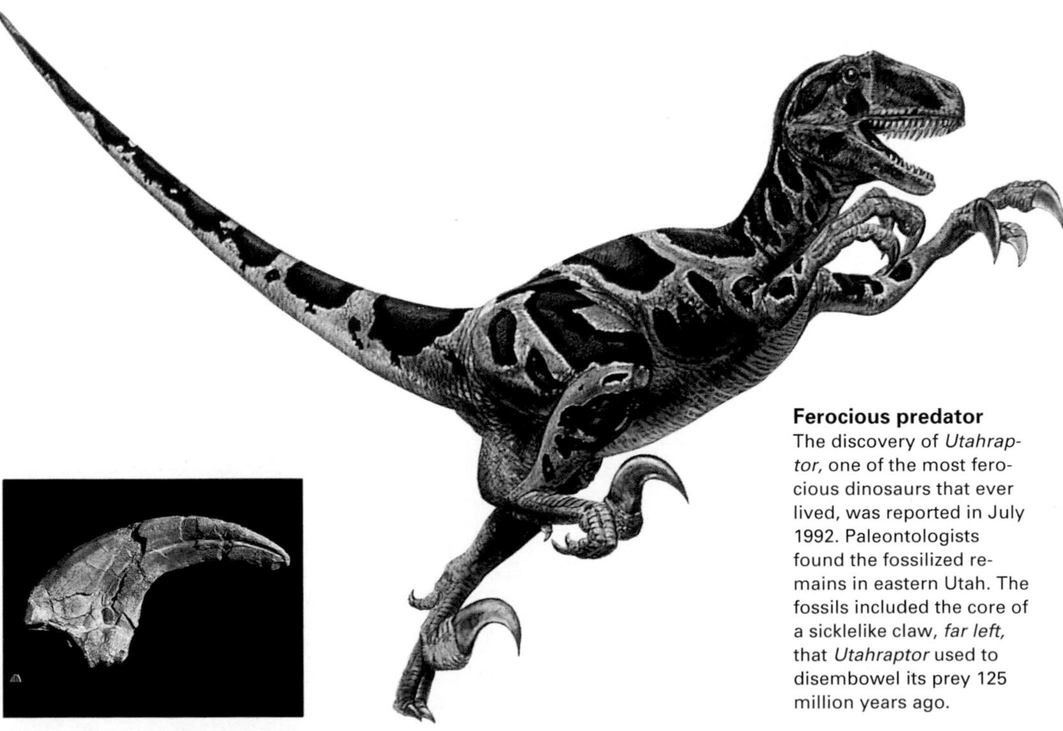

Ferocious predator
The discovery of *Utahraptor,* one of the most ferocious dinosaurs that ever lived, was reported in July 1992. Paleontologists found the fossilized remains in eastern Utah. The fossils included the core of a sicklelike claw, *far left,* that *Utahraptor* used to disembowel its prey 125 million years ago.

Most primitive dinosaur
An *Eoraptor,* the most primitive dinosaur known, chases a small mammallike reptile known as a cynodont, *below.* The fossilized skull of an *Eoraptor* fits neatly in a human hand in an X-ray image, *right.* Paleontologists found the dinosaur in northwestern Argentina and reported their findings in January 1993. *Eoraptor* lived about 225 million years ago.

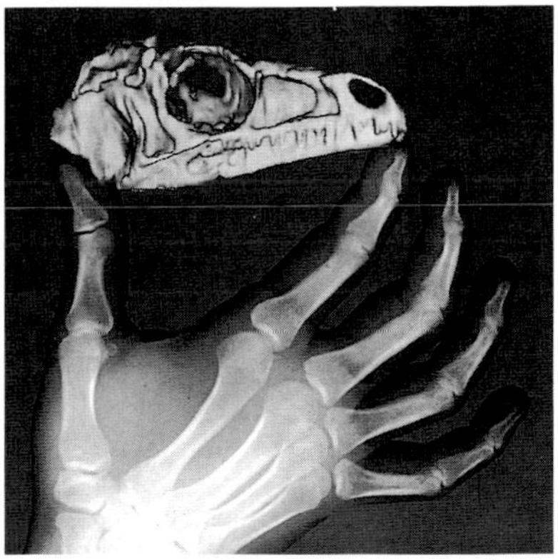

gists George D. Poinar, Jr., and Benjamin Waggoner of the University of California at Berkeley. Ciliates are *protozoans* (one-celled organisms) that are covered with tiny hairlike projections called cilia. The researchers found the ancient ciliates trapped inside pieces of *amber* (fossilized tree resin) from Bavaria in southern Germany. They dated the ciliates to the Triassic Period, about 220 million years ago.

The fossil ciliate possessed virtually the same form as a modern ciliate called *Nassula.* And among the ciliates in the amber was one caught in the act of feeding on colonies of bacteria. Today, *Nassula* feed on similar bacterial colonies. So *Nassula* appears to have changed very little after more than 200 million years.

Fossil insect DNA. Discoveries of other fossil organisms trapped in amber were announced in September 1992 by two groups of researchers. In both cases, the fossils were so well preserved that the scientists were able to extract and analyze fragments of DNA (deoxyribonucleic acid, the molecule of which genes are made). The DNA fragments were dated to some 30 million years ago.

Entomologists with the American Museum of Natural History in New York City and Yale University in New Haven, Conn., examined fossils of an extinct group of termites previously thought to be an evolutionary link between modern cockroaches and termites. The fossils were located in amber from the Dominican Republic. By comparing the fossil DNA to DNA from modern cockroaches and termites, the researchers showed that the extinct termites were more closely related to modern termites.

Scientists from California Polytechnic State University in San Luis Obispo and the University of California at Berkeley conducted similar analyses on DNA extracted from fossil stingless bees, also preserved in Dominican amber.

A June 1993 report announced the finding of the oldest fossil genetic material yet—DNA from 120-million-year-old extinct weevils found in amber from Lebanon. (See also GENETICS.)

Saber-toothed tiger DNA. Fossil DNA was also obtained from saber-toothed cats (also called saber-toothed tigers) by a team of researchers led by geneticist Stephen J. O'Brien of the National Cancer Institute in Atlanta, Ga., according to an October 1992 report. The DNA was extracted from the bones of saber-toothed cats that had been trapped in the La Brea tar pits in what is now Los Angeles. The bodies of the cats were entombed in the tar about 14,000 years ago. By analyzing the DNA, the scientists showed that the saber-toothed cats were closely related to modern large cats, such as lions and tigers.

Flightless bird? The discovery of an unusual turkey-sized fossil was reported in April 1993 by paleontologist Mark Norell and his colleagues at the American Museum of Natural History and the Mongolian Academy of Science. The researchers claimed that the 75-million-year-old fossil, named *Mononychus* ("one claw"), represented a strange flightless bird. But other scientists argued that the fossil was only a dinosaur that looked like a bird.

The scientists studied two specimens of *Mononychus* that had been found in the Gobi Desert in Mongolia in 1987 and 1992. The fossils possessed several traits of dinosaurs, such as sharp, strong teeth and a long neck and tail. However, the fossils also had fused wristbones and a breastbone shaped like the keel of a ship—all characteristics of a bird. Yet the animal had only 7.5-centimeter (3-inch) arms, not wings, and so could not fly.

Another odd feature of the fossils was the large single claw at the end of each arm. Some scientists speculated that the claws were used for digging.

The New York researchers argued that *Mononychus* is more closely related to modern birds than is the earliest known bird, *Archaeopteryx,* which lived 147 million years ago. But other paleontologists pointed out that *Mononychus* seemed too primitive for a bird that appeared tens of millions of years after *Archaeopteryx*. Some suggested that *Mononychus* came from a line of animals that had descended from *Archaeopteryx* but had then lost the ability to fly.

[Carlton E. Brett]

In WORLD BOOK, see DINOSAUR; PALEONTOLOGY; PROTOZOAN.

Genetics

A genetic defect that identifies people who have an extremely high chance of developing some form of cancer—often colon cancer—was reported by geneticists in May 1993. The defect also sheds light on a previously unknown mechanism of cancer development.

New cancer gene. Researchers led by Bert Vogelstein at Johns Hopkins University in Baltimore and by Albert de la Chapelle at the University of Helsinki in Finland examined the DNA of members of families with a history of colon cancer. The researchers discovered a genetic defect that was inherited by family members with the disease but not by their healthy relatives. The researchers then found the same defect in dozens of other people with a family history of colon cancer.

The researchers were not able to identify the gene, but they tracked it to a very small area on chromosome 2. They said a large percentage of people with the gene will develop colon cancer. An additional group of people with the gene will develop other cancers.

Scientists had previously known about defective genes that cause cancer by creating uncontrolled cell division. But the researchers said the gene appears to produce cancer by creating mutations in many other genes. They speculated that the defective version of the gene produces a protein that creates mistakes in DNA or that prevents mistakes from being corrected during DNA replication. Cells with the defective gene continually develop mutations as they divide. After many years, the accumulated mutations may cause the cells to turn cancerous.

The researchers estimated that the gene is carried by 1 in every 200 people. They said that a genetic test for the defect in people with a family history of colon cancer may be ready within a year.

Cancer researchers focused on other genes as well. In June 1993, for example, researchers reported developing compounds that block a cancer-causing gene called the ras oncogene in laboratory tests. (See also Close-Up.)

ALS gene. A gene that causes a form of amyotrophic lateral sclerosis (ALS), often called Lou Gehrig's disease, was identified in March 1993 by a team of 32

A mouse with cystic fibrosis
A researcher holds a mouse, *right,* genetically engineered to produce the abnormally thick mucus that causes the human disease cystic fibrosis. As in a person with cystic fibrosis, the mouse develops a severe mucus blockage in its colon, *below.* Mice containing the abnormal gene responsible for cystic fibrosis were unveiled in August 1992 and will help scientists study the disease and develop treatments.

researchers led by Robert H. Brown, Jr., of Massachusetts General Hospital in Boston and Teepu Siddique of Northwestern University Medical School in Chicago. ALS, one of the most devastating diseases of the nervous system, destroys *motor neurons,* nerve cells in the brain and spinal cord that control the muscles. The muscles gradually wither as a result of lack of use, and paralysis and death follow.

About 10 percent of ALS cases are hereditary, caused by genes transmitted from parent to child. To uncover the responsible genes, the researchers examined the DNA (deoxyribonucleic acid, the molecule of which genes are made) of families with hereditary ALS. The researchers discovered that 13 of the families they studied possessed a *mutation* (change) in the DNA of one particular gene. That gene normally produces a protein called superoxide dismutase, which helps cells neutralize destructive molecules called free radicals.

The researchers theorized that the family members with hereditary ALS produce a defective form of superoxide dismutase that allows free radicals to build up in motor neurons and destroy them, thereby causing the disease. The researchers said more work needs to be done to confirm the roles of superoxide dismutase and free radicals in ALS. But if free radicals do cause ALS, scientists said it might someday be possible to treat the disease with drugs that target those molecules.

Genetically altered mice. The creation of mice that showed symptoms of the human disease cystic fibrosis was announced in August 1992 by researchers at the University of North Carolina in Chapel Hill. The mice will help scientists study cystic fibrosis, a genetic disease in which certain cells of the body produce large amounts of thick mucus. The mucus clogs the passageways of the lungs and other organs. About 30,000 Americans are afflicted with the disease.

Geneticists in 1989 had identified the gene whose mutation is responsible for cystic fibrosis in human beings. The normal human gene produces a molecule called CFTR. Mice also have this gene.

To develop the modified mice, the researchers used a new technique called *gene targeting*. First, the scientists created a defective form of the mouse CFTR gene and mixed copies of it with cells taken from a mouse embryo. In some of the cells, the altered gene was incorporated into the mouse DNA and replaced the normal gene. The researchers injected these cells into the mouse embryos and implanted the embryos into the wombs of female mice.

The resulting newborn mice carried the defective gene in some of their cells. By mating selected pairs of mice, the researchers were able to breed mice that carried two copies of the defective CFTR gene—just as human beings with cystic fibrosis do.

Geneticists called the animals "knockout mice" because the researchers had "knocked out" their normal genes and replaced them with defective versions. The knockout mice developed symptoms similar to those in human beings, including mucus blockages in their intestines. However, the mice did not have the mucus-clogged lungs characteristic of human cystic fibrosis patients.

Despite that difference, researchers said the mice should help them understand how the defective CFTR gene leads to cystic fibrosis in human beings. The existence of the mice should also speed the development and testing of treatments.

Two chromosomes mapped. The first detailed maps of human *chromosomes*, the threadlike structures in a cell that carry the genes, were unveiled in October 1992 by two teams of scientists. The maps of chromosome 21 and the Y chromosome were the first milestones in the Human Genome Project, a long-term, international scientific mission to locate and identify the approximately 75,000 genes that reside on the human chromosomes.

A team of 36 researchers from France, Spain, Japan, and the United States mapped chromosome 21. Scientists believe that it carries the genes involved in several well-known genetic disorders. An extra copy of this chromosome causes Down syndrome.

Researchers at the Whitehead Institute for Biomedical Research in Cambridge, Mass., mapped the Y chromosome. The Y chromosome is usually found only in male human beings.

The two groups did not discover the locations of all of the genes on the chromosomes. Rather, they created what geneticists called *physical maps*—maps of "signposts" along the chromosomes that will enable other geneticists to locate newly identified genes more easily. Previously, geneticists who determined that a gene resided on chromosome 21, for example, had to examine the entire chromosome to locate the gene. But the researchers broke up the chromosomes into pieces and then identified unique DNA sequences called *markers* on each piece. By learning which of the markers lies near a new gene, geneticists can limit their search to the part of the chromosome containing that marker.

Huntington's disease gene. A 10-year search for the gene that causes Huntington's disease, an incurable, fatal disorder of the nervous system, ended in March 1993. Researchers in 1983 had tracked the gene to chromosome 4. But the gene is located near the tip of the chromosome, a region of DNA that is particularly difficult to study, and geneticists had been continually thwarted in their search. The scientists involved in the discovery came from research centers in the United States and Europe—the Hereditary Disease Foundation in Santa Monica, Calif.; the Massachusetts General Hospital in Boston; Boston University Medical School; the Massachusetts Institute of Technology in Cambridge; the University of California at Irvine; the California Institute of Technology in Pasadena; the University of Michigan in Ann Arbor; the University of Wales; and the Imperial Cancer Research Fund in London. The scientists had shared the results of their work for several years to lead to the finding.

Huntington's disease involves the destruction of brain cells responsible for controlling movement. Patients experience a loss of control over body movements, decreasing mental abilities, and eventually death. The disease afflicts about 30,000 Americans.

After locating the Huntington's gene, the researchers learned that the mutation in the defective gene involves a sequence of three *nucleotides* (the genetic subunits of DNA). The normal gene usually contains a row of 11 to 24 copies of the sequence. But the defective gene contains up to 86 or more of these repe-

Cancer's Common Genetic Thread?

For years, cancer researchers have been seeking a common thread among the various types of cancer that could provide them with a single means of diagnosing and treating cancer, no matter which organ it affects. By 1993, a growing body of evidence suggested that a protein called p53 may turn out to be such a thread.

The p53 protein is found in virtually all cells, but researchers have found that many cancer cells possess an abnormal form of the protein. All proteins are produced by *genes* (the blueprint for all cell functions), and the abnormal form of p53 is caused by a *mutated* (changed) gene. The mutated p53 gene has been found in patients with many types of cancer, including 75 percent of lung cancers, 70 percent of colon cancers, and 30 to 50 percent of breast cancers. Scientists therefore believe that p53 plays an important part in the process by which cells become cancerous.

Researchers discovered p53 in 1979. But it was not until 1992 that scientists began to determine in a detailed way how p53 functions in human cells. It acts as a brake on the growth of damaged cells, and uncontrolled growth characterizes all cells that have become cancerous.

This insight was reported by a team of scientists led by geneticist Michael Kastan at the Johns Hopkins University Medical School in Baltimore. The researchers had exposed noncancerous human cells to a form of electromagnetic radiation called ultraviolet (UV) radiation, which can damage DNA (deoxyribonucleic acid, the molecule of which genes are made). Such damage can lead to cancer. For example, skin cancer can be caused by exposure to UV rays from the sun. The scientists observed that the cells exposed to UV rays churned out p53. Soon afterward, the cells ceased growing and dividing. After about 15 hours, the levels of p53 dropped and the cells resumed their normal growth.

Upon further examination, the researchers discovered that p53 had stopped cell growth just before the cells replicated their damaged DNA. The researchers concluded that by producing p53, the cells gave themselves time to repair the damage to their DNA that had been caused by the ultraviolet radiation. After the damaged DNA was repaired, the cells shut down production of p53 and resumed normal growth.

The researchers hypothesized that the situation in cells with mutated p53 genes would be much different. Like normal cells, such cells would respond to DNA damage by producing p53. But the abnormal protein would be unable to stop cell division. The cells would keep growing and dividing, despite the damage to the DNA that could lead to cancer.

Scientists are still investigating exactly how p53 halts cell division. Studies indicate that p53 somehow recognizes when DNA has become damaged and binds to certain parts of the DNA molecule, thereby activating the genes responsible for stopping the cell's reproductive machinery. Several teams of researchers in 1992 identified stretches of DNA where p53 binds to DNA. However, scientists are only beginning to understand the full chain of events that start with the production of the p53 protein.

Another 1992 discovery was that normal p53 can be implicated in cancer. A second group of researchers at Johns Hopkins University Medical School led by geneticist Bert Vogelstein reported finding an abnormal version of a protein called MDM2 that attaches itself to normal p53. When that happens, the p53 cannot bind to the cell's DNA when damage occurs, and cell growth proceeds unchecked. Types of cancer called sarcomas may result. Scientists at the National Institutes of Health found another abnormal protein called E6AP that acts in a similar way.

While some scientists work to clarify the role of p53 in cancer development, others are using the existing knowledge of p53 to improve cancer diagnosis and treatment. For example, geneticist Stephen Friend and his colleagues at the Massachusetts General Hospital in Boston announced in 1992 that they had developed a genetic test for determining whether people have inherited the defective p53 gene from their parents. Such a test could alert physicians to some patients with a high risk of cancer. The test still requires years of study before it can be approved.

Other scientists are attempting to fine-tune traditional cancer *chemotherapy*—treatment with anticancer drugs—to take into account the activity of p53. In one proposed treatment, a doctor would give a cancer patient a drug that stimulates the production of p53 and thus halts the growth of normal cells. Cells with abnormal forms of p53, however, would keep growing. The physician would next administer a drug that kills only the body's dividing cells. Such a combined drug treatment could prove to be a major weapon against cancer.

Cancer is a complex disease influenced by a number of genetic and environmental factors. Nevertheless, many genetic scientists view p53 as a key player in the disease, and they predict greater advances as the gene gives up more of its secrets.

[Steven Dickman]

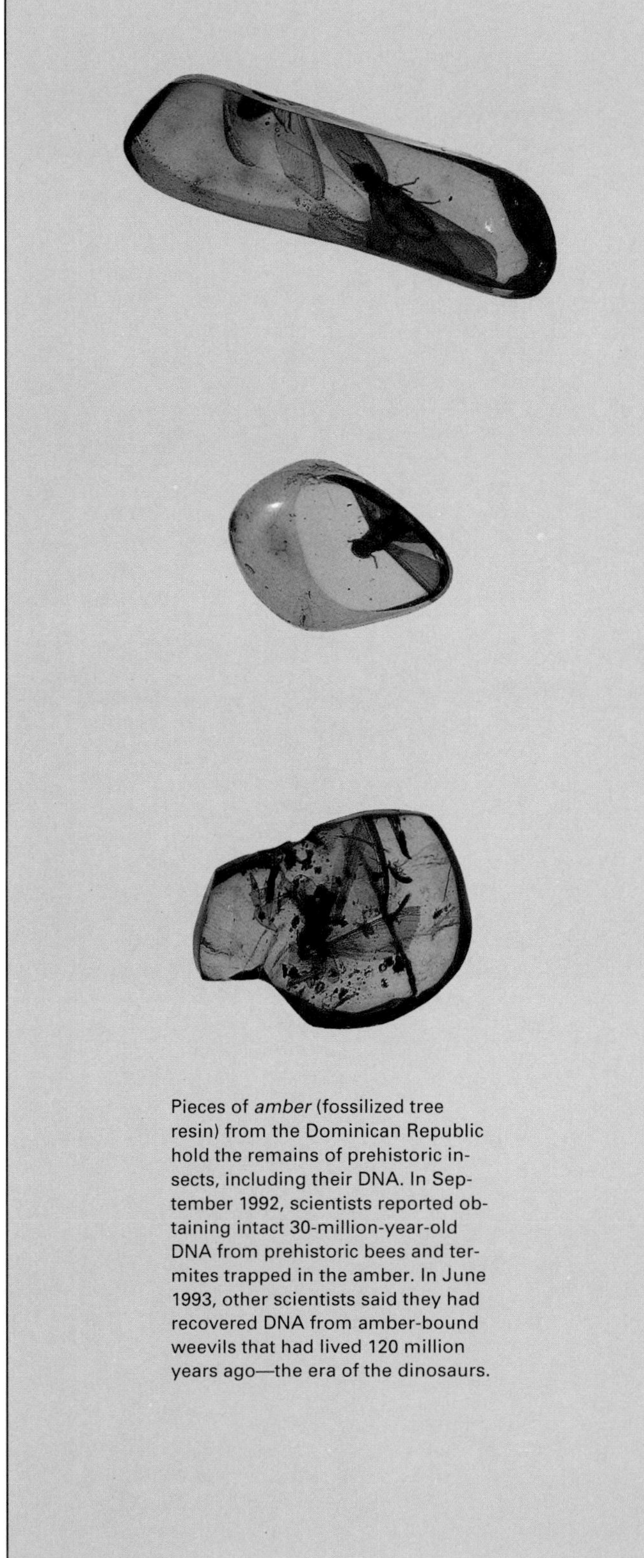

Pieces of *amber* (fossilized tree resin) from the Dominican Republic hold the remains of prehistoric insects, including their DNA. In September 1992, scientists reported obtaining intact 30-million-year-old DNA from prehistoric bees and termites trapped in the amber. In June 1993, other scientists said they had recovered DNA from amber-bound weevils that had lived 120 million years ago—the era of the dinosaurs.

titions. The additional DNA apparently prevents the gene from producing the proper protein.

Scientists said the next step in their research was to determine how the defective protein causes brain cells to die. Scientists can also begin developing an improved genetic test for the disease, which typically does not reveal itself until middle age.

Gene therapy controversies. Two major controversies erupted in the scientific community in 1992 and 1993 over experiments involving *gene therapy.* Gene therapy is an approach to fighting genetic diseases in which defective genes are replaced with normal ones.

The first controversy arose in December 1992 when the director of the National Institutes of Health, Bernadine P. Healy, approved a new type of gene therapy on a human patient with fatal brain cancer. In the therapy, doctors inject the patient with brain cancer cells that have been genetically altered to stimulate the immune system. This approach had not been tried before in human beings and so normally would have undergone a review process lasting several months. But doctors said the patient did not have months to live, and Healy granted them permission for the experiment. In doing so, she bypassed the Recombinant DNA Advisory Committee (RAC), the NIH group charged with overseeing all human gene therapy experiments.

The move outraged some members of the RAC, who argued that the treatment was unproven, but Healy defended her act on the grounds of compassion and the need for quick action. In the end, Healy and the RAC agreed on revised guidelines that permitted the early, compassionate use of gene therapies when the RAC cannot evaluate them quickly enough.

The second controversy emerged in February 1993 when an NIH advisory committee significantly cut the funding for an experiment by geneticist Steven A. Rosenberg, who is considered a pioneer of gene therapy. The committee had looked at an ongoing study in which Rosenberg injected genetically modified immune system cells into seriously ill cancer patients. The members of the committee said they were concerned that the experiment, which had been underway for more than two years, had not yet shown signs of success. At a hearing, Rosenberg could not produce enough data to convince the members that the experiment was working, so the committee voted to delete $225,000 in funding on the project.

Some scientists contended that it was too early to cast judgment on an experiment as unusual as Rosenberg's. But other scientists viewed Rosenberg's shortage of supporting data as confirmation that he was moving too quickly from animal studies of gene therapy into work with human patients.

Abnormal genes in leukemia. New evidence of the role of abnormal genes in the development of leukemia was presented in 1992 by a number of research teams, including groups at the University of Tennessee in Memphis and at the University of Chicago. Leukemia is a cancer of the blood in which the body produces immature and therefore ineffective white blood cells.

The researchers studied several abnormal genes linked to leukemia. All were of a type known as fused genes. A fused gene is a mutation created when a piece of DNA breaks off from one chromosome and attaches to another.

The researchers discovered that the fused genes dictate the structure of proteins that may be the cause of the cancer. For example, one fused gene being investigated by the scientists makes a protein known as a transcription factor. A transcription factor binds to particular stretches of a cell's chromosomes and determines how and when other genes are activated. The scientists theorized that the transcription factor from the fused gene improperly regulates the genes responsible for cell growth, thereby causing white blood cells to be created at the wrong stage of development.

Scientists hope to create a test to detect the products of fused genes in a person's blood. Such a test could be used to diagnose leukemia at its earliest stages. The test could also be given to leukemia patients undergoing chemotherapy to learn when the last blood cells containing fused genes have been destroyed. [David S. Haymer]

In WORLD BOOK, see CELL; GENETICS.

Geology

About 1 billion years ago, the formation of the Earth's ocean crust underwent a major change, according to a theory proposed by geologist Eldridge M. Moores of the University of California, Davis, in January 1993. Moores theorized that the ocean crust that formed more than 1 billion years ago was much thicker than the crust being formed today, and that the thinning of the ocean crust had far-reaching effects on the planet.

At present, ocean crust forms by volcanic activity along a vast undersea mountain range. The range extends from the Arctic Ocean down the center of the Atlantic. It also extends into the Indian and South Pacific oceans and the Gulf of California and appears off the coast of northern California, Oregon, and Washington.

The center of this range is called a spreading center, because over millions of years the ocean crust that forms there spreads out from the crest of the range, creating the floor of the ocean basins. As new ocean crust moves out, old ocean crust is *subducted* (pushed down) beneath the continents or beneath other slabs of ocean crust along deep-sea trenches.

Moores argued that throughout most of the Proterozoic Eon (2.5 billion to 570 million years ago) until the middle of the Neoproterozoic Era (1 billion to 570 million years ago), the ocean crust being formed at the spreading centers was as much as two to three times thicker than that forming the ocean floor today.

Evidence for this idea can be found in *ophiolites*, pieces of ancient ocean crust that were shoved up onto the continents, where they can be studied today. The ancient crust in ophiolites dating to the early Neoproterozoic Era or before is more than 10 kilometers (6 miles) thick. Today, the ocean crust is about 6.5 kilometers (4 miles) thick.

The average depth of the ocean basins today is about 3.8 kilometers (2.4 miles), and so a thicker crust would have made the ocean basins much shallower. Moores speculated that sea level about 1 billion years ago

Evidence of a New California Fault

On June 28, 1992, a strong earthquake struck the Mojave Desert in southeastern California, near the small town of Landers. This shock, which registered 7.6 on the Richter scale, attracted a great deal of public attention because it was the largest tremor to hit California since 1952. For many geologists, however, the main focus of interest was not the force of the quake but its surprising behavior. In fact, the Landers quake was unlike any other ever observed. Attempts to explain it have fueled an ongoing debate over whether a major new *fault* (break in Earth's crust) is forming in southern California. Some scientists argue that this proposed new fault will eventually replace the deadly San Andreas Fault as the most active earthquake zone in the United States.

Earthquakes in California occur along the boundary between the North American and the Pacific plates, two of the huge tectonic plates that make up Earth's outer layers. The North American Plate carries most of that continent's land mass, whereas the Pacific Plate contains a large portion of the Pacific sea floor. At the boundary between the two plates, the Pacific Plate is creeping to the northwest past the North American Plate at the rate of about 50 millimeters (2 inches) per year. The boundaries between many tectonic plates are not narrow corridors marked by a single fault but regions hundreds or even thousands of kilometers wide made up of several seismic zones, areas riddled with faults along which clusters of earthquakes have occurred.

Geologists have identified at least four seismic zones in the boundary between the North American and Pacific plates. The first zone extends along the eastern side of the Sierra Nevada, a huge mountain range in eastern California. The second extends north from the Mojave Desert through west-central Nevada. The third extends north-northeast from the Mojave Desert through central Utah to Yellowstone Park in Wyoming.

The fourth—and most active—seismic zone is made up of the San Andreas Fault and the faults that branch from it. The San Andreas Fault itself extends more than 1,200 kilometers (750 miles) from off the coast of northwestern California to the southeastern part of the state near the Mexican border. The most famous California earthquakes have occurred along or very near the San Andreas. These tremors have included the 1906 jolt that devastated San Francisco and the 1989 Loma Prieta quake that killed at least 62 people and caused an estimated $6 billion in damage. Because the San Andreas seismic zone is the most active zone in the plate boundary, geologists consider it the main dividing line between the North American and Pacific plates.

Earthquakes in the North American-Pacific plate boundary occur because two sections of rock sliding past each other along a fault become "stuck." The sides of such a fault segment may be squeezed together by the weight of the rock, or jagged sections of a fault may become locked together. As the plates try to move, pressure builds up in these sections. Eventually, the pressure becomes too great for the rocks to resist and the stuck sections move suddenly, producing an earthquake.

The Landers earthquake happened in basically this way. But the quake surprised many geologists for three reasons. First, the movement associated with this tremor cut across several different previously mapped faults in a zone extending 70 kilometers (44 miles) from north to south. The motion began near the southern end of the fault zone and moved northward, jumping from fault to fault. Previously, geologists had thought that earthquakes occurred along only a single section of a single fault.

The second surprise was that the Landers quake and the quakes that occurred just before and after it fell on a line that bends to the west about 150 kilometers (100 miles) east and slightly north of Los Angeles. Previously, geologists had thought that the movement that takes place along a fault during an earthquake and its aftershocks occurred only along straight segments of the fault and would be stopped by a sharp bend.

The third surprise concerning the Landers earthquake was that a large aftershock took place about 30 kilometers (19 miles) west of Landers on a different fault, called the Pinto Mountain Fault, which extends to the northeast. In general, aftershocks occur in faults that run in the same direction as does the fault of the main earthquake.

The behavior of the Landers quake was not quite so surprising to one group of geologists, however. In 1989, geologists led by Amos Nur of Stanford University in Palo Alto, Calif., had proposed that a new fault could be forming in the Mojave Desert and that this fault would eventually replace the San Andreas Fault as the main dividing line between the North American and Pacific plates.

Nur and his colleagues based their theory on the occurrence since 1947 of several quakes north and east of a bend in the San Andreas. Throughout much of its length, the San Andreas extends to the northwest, roughly parallel to the direction in which the Pacific Plate is moving. North and

Hints of a new fault

A crack in the floor of the Mojave Desert (dark line, *right*) appeared after the June 28, 1992, earthquake near Landers, Calif. Several unusual characteristics of the quake seemed to support a theory that a new fault is forming in the Mojave Desert. For example, the Landers quake occurred on a line with a 1992 quake near Joshua Tree, Calif., as well as earthquakes that struck in 1908 and 1947 (starred areas, *below*). That line may mark the presence of a new fault that extends north, according to the theory.

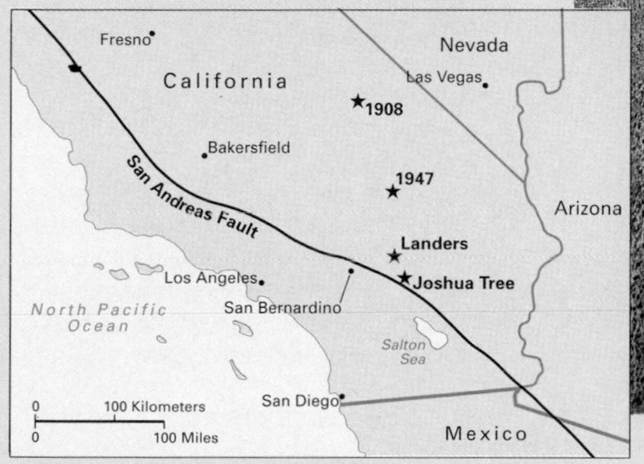

east of Los Angeles, however, the fault makes a wide turn to the west-northwest for about 100 kilometers (62 miles). Then the fault resumes its path to the northwest.

The crust just north and south of the bend is being squeezed together because the two sections of rock there cannot move smoothly past each other. This squeezing action has resulted in the rise of the steep San Bernadino, San Jacinto, and San Gabriel mountains.

Nur and his colleagues proposed that the faults branching from the San Andreas are also being bent. As a result, the San Andreas Fault system in this area cannot absorb the motion caused by the movement of the plates as easily as it absorbed it in the past. According to the geologists, the motion in the plate boundary is being transferred to a new fault. They believe this new fault extends north from the bend in the San Andreas Fault past the eastern side of the Sierra Nevada into western Nevada. The Landers quake and a magnitude 6.3 quake in Eureka Valley on April 22 fell along this line.

Other geologists studying the Landers earthquake dispute Nur's ideas. They argue that the Landers earthquake actually occurred along an existing fault in an area where faults are not well mapped. These geologists also argue that the Landers earthquake and its aftershocks—especially a quake that struck near Big Bear Lake on the Pinto Mountain Fault on June 28—were only to be expected considering the tectonic activity taking place in the Mojave Desert. Because of plate movement, the desert is being squeezed between the San Andreas Fault and the southern end of the Sierra Nevada.

Still, Nur's theory poses several interesting questions. How do new faults form? And how would we recognize such an event if it did occur? Many geologists believe that much of the San Andreas Fault is less than 5 million years old. Before then, seismic activity in the North American-Pacific plate boundary seems to have been heaviest along faults off the coast of California. When the San Andreas developed, all of what is now California west of the fault began moving as part of the Pacific Plate rather than as the North American Plate. According to Nur's theory, the Sierra Nevada will be the next region to be added to the Pacific Plate. At the rate the plates are moving, however, it may take geologists about 10 million years to confirm his theory. [Eldridge M. Moores] may have been up to 2 kilometers (1.2 miles) higher than it is today. With such a high sea level, most of the continents would have been flooded, and only high mountain ranges would have projected out of the sea.

During the Neoproterozoic Era, the continents drifted together to form the supercontinent scientists call Rodinia, which existed from about 1 billion to 750 million years ago. As Rodinia broke up to form smaller continents, Moores speculated, the ocean crust thinned and the ocean basins deepened. The continents emerged from the sea, and for the first time in the history of the Earth, large areas became dry land.

Moores theorized that the emergence of the continents started a chain of events that led to the development of many types of land animals. First, weathering and erosion of the rocks exposed on the land would have washed nutrients such as phosphate into the sea. The nutrients fertilized the sea, allowing *phytoplankton* (tiny floating plants) and other marine plants to multiply in profusion. The green plants engaged in *photosynthesis* (the process by which plants harness energy from the sun). In photosynthesis, plants remove carbon dioxide from the air and convert it to sugars and living tissue. As a by-product, plants release oxygen into the atmosphere.

Ordinarily, when plants decompose, this process occurs in reverse. Decomposing organisms consume oxygen from the atmosphere as they convert the organic matter into carbon dioxide and water. Moores theorized, however, that in the late Neoproterozoic much of the organic matter produced by photosynthesis was buried in sediments and therefore did not decay. The resulting increased amount of atmospheric oxygen then led to the diversity of animal life forms—oxygen-breathing creatures—known to have appeared at the beginning of the Phanerozoic Era 570 million years ago.

Modeling the Earth. An advanced computer model of the transfer of heat in the Earth indicates that there is an important transitional boundary at 670 kilometers (400 miles) below the surface. That was the February 1993 finding of geophysicists Paul J. Takley and David J. Stevenson of the California Institute of Technology in Pasadena; Gary A. Glatzmaier of Los Alamos National Laboratory in New Mexico; and Gerald Schubert of the University of California, Los Angeles.

The existence of a 670-kilometer boundary has been known for many years, because earthquake vibrations travel at a higher velocity below this zone than above it. In their computer model, the scientists assumed that the rock in that region is undergoing a change related to its temperature. Lighter, less dense minerals would be in the mantle above 670 kilometers, whereas heavier, more dense minerals are in the region below. At the boundary itself, the lighter materials absorb heat and become heavier and denser, a transition much like ice melting in a pool of water.

The scientists found that their computer model accurately simulated many features of plate tectonics, the theory that Earth's outer layers behave as a set of rigid plates. The model showed that material would rise in cylindrical plumes from the base of the *mantle*, the region between Earth's core and its outer shell. Those plumes correspond to "hot spots," regions of volcanic activity at such places as Hawaii and Iceland. The model also showed that material enters the mantle along linear zones, which correspond to the known subduction zones, where one plate is being shoved under another.

According to the computer model, descending slabs of cooler surface material tend to accumulate on top of the 670-kilometer boundary before they absorb enough heat to change into the higher density material found below. The model suggests that the change can occur rapidly, with accumulated material cascading like an avalanche into the lower mantle. The scientists suggested that this feature may account for why studies of earthquake vibrations moving through Earth's interior do not always show what is expected from observations of surface features.

Cataclysmic floods. Evidence of the largest known flood was reported in January 1993. The authors of the new report were paleohydrologists (scientists who study the record of rivers in the geologic past) Victor R. Baker and

Gerardo Benito of the University of Arizona in Tucson and glacial geologist Alexey N. Rudoy of Tomsk State Pedagogical Institute in Tomsk, Siberia, Russia.

The scientists described the results of their study of a region in the Altai Mountains in southern Siberia just west of the Mongolian border. They found that tremendous floods resulted from the collapse of ice dams in the Altai Mountains during the late Pleistocene Epoch, about 10,000 years ago. According to the researchers, floodwaters 400 meters (1,300 feet) high poured into the Ob River.

At the flood's peak, more than 1 billion cubic meters (35 billion cubic feet) of water cascaded into the river every second. That volume would make the peak flow of the Altai floods a thousand times greater than the combined flow of all the world's rivers today. The scientists concluded that the Altai flood was the largest for which a geological record exists.

A report by glacial geologists Derlad G. Smith and Timothy G. Fisher of the University of Calgary, in Alberta, Canada, in January 1993 described another great flood. This flood cascaded from one of the glacial lakes that existed on what are now the Canadian prairies. After studying boulder and gravel deposits in the Clearwater and Athabasca river valleys of Alberta, the scientists proposed that glacial Lake Agassiz in the province of Saskatchewan drained catastrophically about 9,900 years ago. They theorized that a natural dam was breached, and that the surface of the lake dropped 46 meters (150 feet) during a period of about 78 days. The resulting flow was about 2.4 million cubic meters (85 million cubic feet) per second, much less than that of the Altai flood but still of enormous proportions.

What was most significant about the flood was that the water flowed north into the Mackenzie River and then into the Arctic Ocean. The scientists theorized that the large influx of fresh water into the surface waters of the Arctic may have melted sea ice and contributed to a warming of the climate in the Arctic region known to have occurred at that time.

Interestingly, the Altai floodwaters also flowed into the Arctic Ocean. Their effect on the climate has not yet been determined, because the exact time of the Altai floods is not known.

Iceberg outbreaks in the North Atlantic during the last ice age may have been caused by the surges of an ice stream. That was the December 1992 finding of glacial geologists John T. Andrews and Kathy Tedesco of the University of Colorado in Boulder.

Scientists learned about these icebergs from a 1988 study of *cores* (samples) of sediment from the North Atlantic. The cores corresponding to the last ice age contained ice-rafted debris—layers of sand and pebbles transported by icebergs. These layers appear to represent two episodes in which thousands of icebergs drifted into the North Atlantic. The episodes were termed *Heinrich Events,* after Hans Heinrich, the German geologist who discovered them. The first event occurred between 21,400 and 18,900 years ago and the second between 16,400 and 13,400 years ago.

While examining cores of sediment from the bottom of the northwestern Labrador Sea, Andrews and Tedesco found concentrations of grains of limestone of the same age as the Heinrich Events in the North Atlantic. Limestone, a sedimentary rock, is usually eroded by being dissolved and washed to the sea in rivers. In the cold climate of a glaciated area, however, the limestone does not dissolve but is crushed into small particles. The geologists thus assumed that the limestone grains had been transported by glaciers.

Andrews and Tedesco theorized that the layers containing limestone particles were brought to the sea by a glacial ice stream that extended from the Hudson Bay region to the edge of the northwestern Labrador Sea. Andrews and Tedesco theorized that because of the increasing depth of the bottom of the Hudson Strait, an ice stream flowing through it would have been unstable and prone to surging. Glaciers surge forward when they become detached from the surface on which they rest, which often happens when a glacier enters the sea. The ice is buoyed up by the water and, with no friction to hold it back, advances rapidly, spawning masses of icebergs. [William W. Hay]

In WORLD BOOK, see GEOLOGY.

Scientists in 1992 and 1993 continued to discover new forms of the hollow carbon molecules called fullerenes. The most famous fullerene is *buckminsterfullerene,* which consists of 60 carbon atoms in the shape of a soccer ball.

Buckytube manufacturing. In July 1992, materials scientists Thomas W. Ebbesen and P. M. Ajayan at NEC Corporation in Japan reported a simple recipe for making *buckytubes*—hollow fullerene tubes. Buckytubes had been discovered in 1991, but scientists had not known how to mass-produce them.

The NEC scientists made the tubes by shooting an electric current across two carbon rods in a special chamber. The current caused carbon from one rod to build up on the other rod in the form of 0.6-centimeter-wide (0.25-inch-wide) cylinders. When the researchers examined the rods, they found that the small cylinders were full of tiny buckytubes.

In January 1993, other NEC scientists reported that they had filled buckytubes with lead. The researchers heated buckytubes in the presence of lead at a temperature high enough for the lead to melt. Afterward, the researchers found that molten lead had been drawn into some of the tubes, where it had hardened into wires a few atoms across.

Filling the fullerenes. Fullerenes constructed like onions, with ever-smaller fullerenes nested inside of one another, were unveiled in October 1992 by microscope specialist Daniel Ugarte of the Federal Polytechnic School of Lausanne in Switzerland. Ugarte made the structures by shooting a beam of subatomic particles called electrons at carbon. The force of the beam caused the carbon particles to form microscopic spheres with up to 70 nested shells.

In March 1993, chemist Martin Saunders and his colleagues at Yale University in New Haven, Conn., and at the University of Rochester in New York reported that helium atoms can become trapped inside fullerenes as the cagelike molecules are formed. Scientists typically create fullerenes in a chamber containing helium atoms. When Saunders and his co-workers heated a sample of fullerenes made in such a chamber, they detected the release of helium atoms.

Tiny transparent magnets were unveiled in July 1992 by chemists at the Xerox Webster Research Center in Webster, N.Y. The magnets, which measure only a few hundred millionths of an inch across, consist of a compound of iron and oxygen. Their creators said they could someday be used in data storage devices for computers.

Colorful silicon portraits
The faces of United States President George Washington, *left,* and singer Elvis Presley, *right,* were "painted" without the use of pigments on wafers of silicon. Researchers in June 1992 reported making each image by projecting a black-and-white photograph onto silicon that was being etched in acid. Afterward, when the silicon was exposed to light, the image reappeared in various colors.

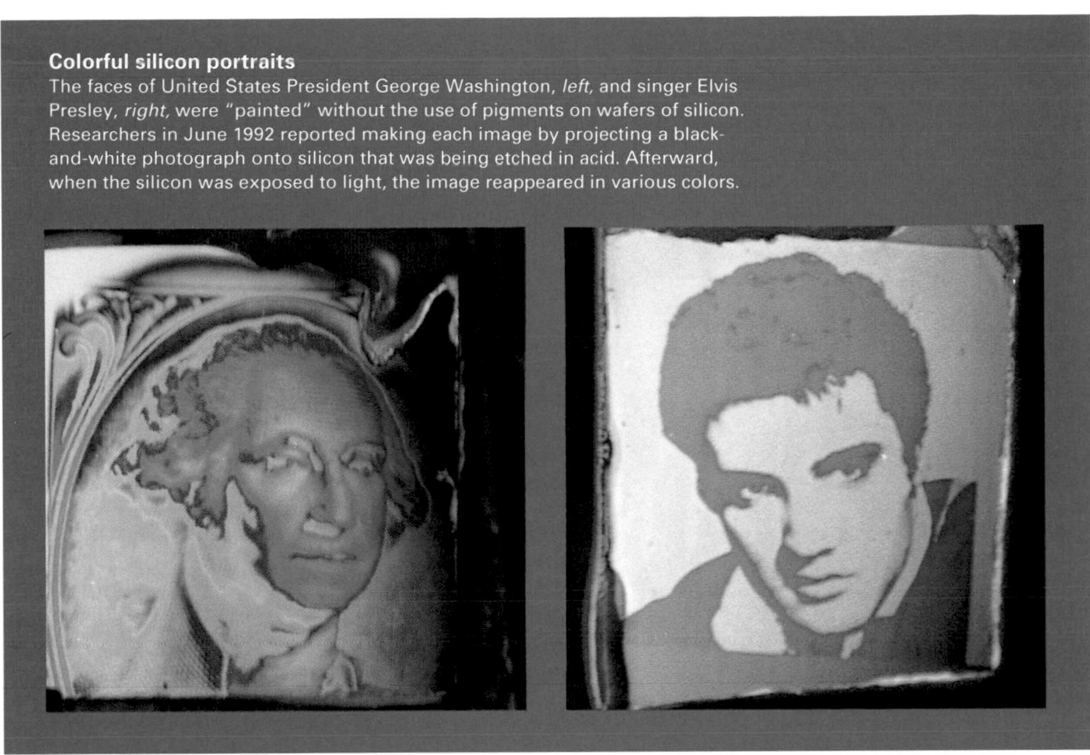

Further investigation indicated that the helium atoms had come from inside the fullerenes. The researchers theorized that when they heated the fullerenes, bonds between some carbon atoms in the cagelike molecules broke open, making "doorways" through which the helium atoms could escape.

Fullerene relative. In December 1992, Reshef Tenne and a team of materials scientists at the Weizmann Institute of Science in Israel reported that they had created a new kind of molecular cage that contains no carbon at all. Instead, the structure is composed of atoms of tungsten and sulfur. The researchers made fullerenelike balls and tubes with the elements. But because the new structures contain no carbon, scientists expect the properties to be much different from those of fullerenes.

Polymer sheets. A new technique that makes it possible to create sheetlike polymers was the January 1993 finding of materials scientist Samuel I. Stupp and his colleagues at the University of Illinois at Urbana-Champaign. A polymer is a molecule made of smaller molecules that are identical to one another.

The Illinois researchers started with many short, identical polymer strands. Each strand contained reactive atoms at two sites—one at the top and the other in the middle. Given the chance, reactive atoms are likely to form bonds with other atoms. The researchers placed the strands in water, where they bunched together in two layers, like the bristled ends of two toothbrushes facing each other. When the researchers heated the strands, the two layers bonded together because of the reactive atoms at the ends. The strands in each layer also bonded to their neighbors because of the reactive atoms at the middle of each strand. The result was a single polymer molecule measuring 10 nanometers thick but several square micrometers in surface area. The sheets last longer and withstand heat better than polymer chains do, the researchers said.

Magnetic writing surface. A special surface can be repeatedly written on and erased using a magnet, according to a November 1992 announcement by three Japanese companies. The key component of the new writing surface is *liquid crystal*—the substance in the liquid-crystal displays (LCD's) used in calculators and digital watches. The color of liquid crystal changes depending on the arrangement of its molecules. When liquid-crystal molecules are aligned, the liquid crystal has a certain color. But when the molecules are disturbed, such as by a magnetic field, the liquid crystal may change color.

The new writing surface consists of a layer of liquid crystal laid on a base material and covered with transparent plastic. Users write words with a device that applies a magnetic field to specific parts of the surface, creating gray letters on a light background. Passing the entire surface over a magnet realigns the liquid crystal and erases the message.

Pictures on silicon. Using light and chemicals, scientists can create colorful pictures on silicon. Chemists Michael J. Sailor and Vincent V. Doan at the University of California in San Diego reported this finding in June 1992. The new report was the latest in a series of discoveries about the optical properties of the element silicon. Scientists in 1990 had discovered that silicon that has been *etched* (eaten away by acid) glows when exposed to light. The etched substance is called *porous silicon* because the acid creates tiny *pores* (holes) in the surface of the silicon. The arrangement and sizes of the pores seem to determine the color of light reflected by the silicon, though scientists do not yet know how.

The California researchers created their color pictures by a simple process. First, they dipped a silicon wafer in acid and projected a black-and-white image onto the wafer. The light from the picture interacted with the acid to create a pattern of shallow and deep pores on the silicon that matched the light and dark areas of the picture. Afterward, when the wafer was exposed to light, the image reappeared. The researchers say the technique could be used to store color images without pigments or to create simple *holograms* (images that can be viewed from three dimensions).

[Elizabeth J. Pennisi]

In WORLD BOOK, see MATERIALS SCIENCE; POLYMER.

Medical Research

The first attempt to use gene therapy to treat cystic fibrosis, the most common lethal hereditary disease among whites in the United States, began on April 17, 1993, at the National Heart, Lung and Blood Institute in Bethesda, Md. Gene therapy is an experimental form of medical treatment that involves introducing new *genes* (the units of heredity and the blueprint for all cell functions) into a patient's body to correct a disease-causing genetic defect.

Cystic fibrosis affects some 30,000 people in the United States. People with the disease, which has no cure, are plagued by excessive mucus in the lungs and other organs, which leads to recurrent infections and breathing difficulties. Most patients die before age 30.

An abnormal gene lies at the root of the disorder. In healthy people, a gene called the cystic fibrosis transmembrane conductance regulator (CFTR) gene controls production of a protein that regulates the flow of chloride through the body's cells. People who have cystic fibrosis have flawed CFTR genes that do not provide the correct instructions for chloride regulation. This leads to thick mucus secretions and other symptoms of cystic fibrosis, such as excessive salt in sweat. People who have inherited one normal CFTR gene from one parent and an abnormal gene from the other parent do not develop cystic fibrosis. But people with two flawed genes—one from each parent—will have the disease.

The early phases of the gene therapy experiment took place in the laboratory as the researchers inserted working CFTR genes inside microbes called adenoviruses. Normally, adenoviruses invade the cells that line the air passages of the nose and lungs, where they multiply and cause coldlike symptoms. However, researchers had altered the adenoviruses they used in the experiment so that the viruses could not multiply, though they could still enter the airway cells. The researchers hoped that some of the viruses would deliver normal CFTR genes to the patients' cells, and that the genes would then instruct the body's cells to make the protein that controls chloride levels. Animal studies had shown that only 10 percent of the

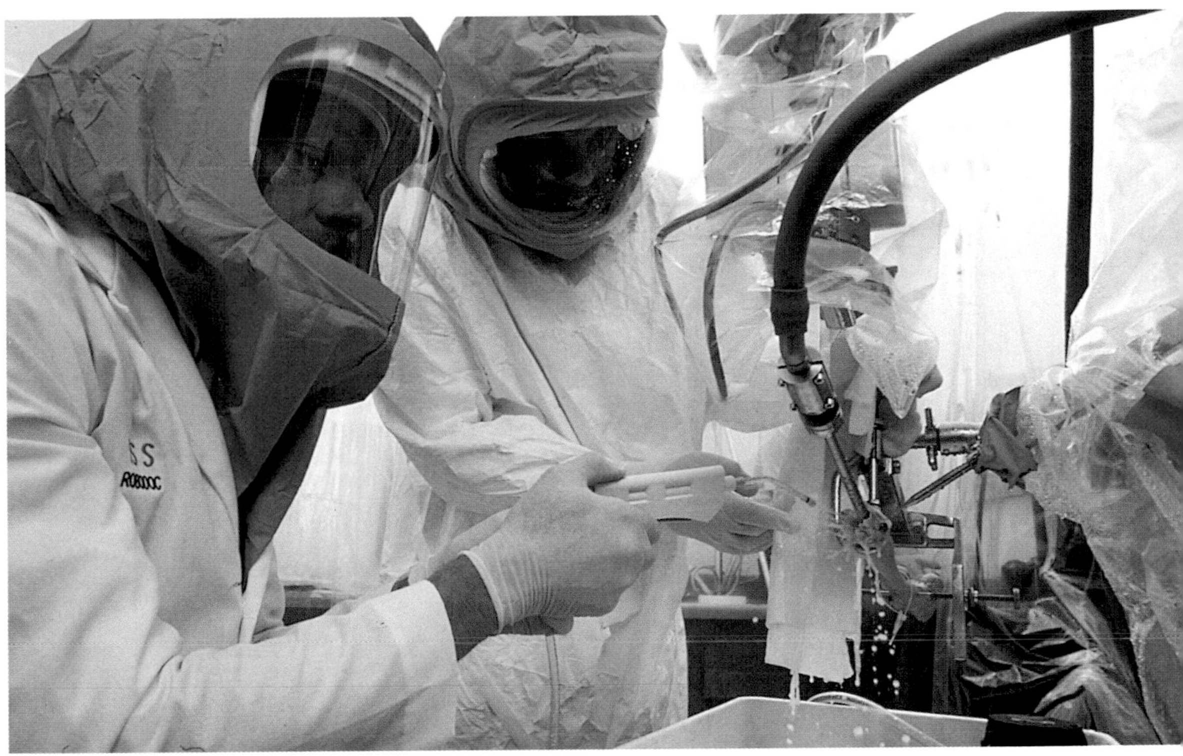

In preparation for a November 1992 hip-replacement surgery, surgeons practice using a new robotic device that drills a hole into a thighbone. Using instructions from a computer, the robot, nicknamed Robodoc, precisely carves a space into which the surgeon can snugly fit an artificial hip joint.

airway cells needed to accept and activate the gene to relieve the symptoms.

The researchers administered the therapy in two steps. First, they sprayed adenoviruses containing the CFTR gene into the nasal passages of a 23-year-old man with cystic fibrosis. The next day, they threaded a thin tube down the patient's trachea (windpipe) and trickled droplets containing the altered viruses directly into his left lung.

During the next phase of the experiment, the researchers planned to determine whether CFTR genes had entered the patient's cells and whether the genes were making the all-important CFTR protein to reverse the build-up of mucus in the airways.

Embryo test for cystic fibrosis. A new method of genetic testing can determine—before a pregnancy begins—whether the genes that cause cystic fibrosis are present in an embryo. Researchers in England and the United States in September 1992 reported on their use of this technique.

In the study, three couples underwent *in vitro fertilization,* in which eggs are removed from a woman's ovaries, placed in a laboratory dish, and combined with sperm obtained from a man to create one or more fertilized eggs. The egg becomes an embryo as it divides into two cells, and then four, and so on. The embryo may be implanted in the woman's uterus, where it can grow into a fetus.

In the study, the researchers tested four-celled and eight-celled embryos before implantation. They began by using extremely small instruments to immobilize each embryo and puncture a hole in its outer layer. The researchers then gently *aspirated* (sucked) one cell out of the embryo and into a tiny tube called a micropipette.

To more easily study the genes contained in the embryo cell, the researchers made multiple copies of its genes using a chemical technique called polymerase chain reaction. The researchers then analyzed the genes to determine whether the abnormal cystic fibrosis gene was present. Two of the three couples in the study had healthy embryos, which were implanted into the respective mothers. One of these embryos developed, resulting in the birth of a baby girl free of cystic fibrosis.

The new testing technique, which is accomplished within three days of fertilization, is much faster than conventional genetic testing for cystic fibrosis that is done after a pregnancy has been established for at least nine weeks. However, experts pointed out that more studies are necessary to determine the method's safety as well as its practicality, since it is likely to be very costly. The researchers noted that this approach could also be used to diagnose other inherited diseases for which a flawed gene has been identified. (See also GENETICS.)

Treatment for inherited anemias. Two new treatments benefited people with certain hereditary *anemias* (conditions in which the blood lacks oxygen-carrying red blood cells), according to two independent studies reported in January 1993. Each of the new treatments used a drug to stimulate the body's production of an alternate form of *hemoglobin,* a protein in red blood cells that plays a crucial role in transporting oxygen.

One of the diseases treated was sickle cell anemia, a condition in which a defective gene causes the body to make an abnormal form of hemoglobin. The abnormal hemoglobin can still carry oxygen, but it causes red blood cells to become rigid and sickle-shaped. The sickled cells are easily destroyed, leading to anemia. The abnormal cells also become stuck in tiny blood vessels, causing numerous health problems.

The other anemia treated was beta thalassemia, also called Cooley's anemia. Like sickle cell anemia, it is caused by a defective hemoglobin gene.

The new treatments were based on earlier research showing that human beings have two genes that control hemoglobin production. One, called the fetal gene, causes a fetus to produce a type of hemoglobin suited to its needs in the womb. About two months after a baby is born, this gene "turns off," and a different gene instructs the body to make adult hemoglobin, which is better suited to life outside the womb. The amount of fetal hemoglobin in the body then drops to less than 3 percent of the total.

Because sickle cell disease does not occur until after the body has started producing adult hemoglobin, scientists reasoned that if they could coax the dormant fetal gene to dramatically boost fetal hemoglobin production, patients with sickle cell anemia would improve. Although fetal hemoglobin is not as effective as adult hemoglobin in transporting oxygen after birth, it does not cause sickled cells.

In one study, hematologist Susan P. Perrine of Children's Hospital Oakland Research Institute in Oakland, Calif., gave six patients arginine butyrate, a natural fatty acid used as a food additive. Three of the patients had sickle cell anemia and three had beta thalassemia. The patients received a continuous intravenous dose of butyrate for two to three weeks. When the treatment ended, the patients had suffered no serious side effects, and their fetal hemoglobin levels were 6 percent to 45 percent higher than their pretreatment levels.

In another study, scientists from the National Institutes of Health in Bethesda, Johns Hopkins Medical School in Baltimore, and Nippon Medical School in Tokyo reported similar results using different drugs to treat four patients. Previous studies had demonstrated that a cancer drug called hydroxyurea could increase fetal hemoglobin levels. But the drug did not work well for all patients, and it had side effects that limited its use.

In the study, the researchers gave the patients lower doses of hydroxyurea along with iron supplements and a synthetic version of erythropoietin, a natural substance that helps stimulate the production of blood cells. This approach appeared to be more effective and better tolerated by patients than treatment with hydroxyurea alone.

Robot performs surgery. For the first time in history, surgeons used a robot to perform a crucial step in a hip-replacement operation. William Barger, an orthopedic surgeon in Sacramento, Calif., employed a robotic arm during surgery in November 1992. The computer-driven device, dubbed "Robodoc," is not designed to automate hip surgery. But it helps the surgeon by performing one step in the operation with greater precision than the most skillful surgeon. As of mid-1993, the California researchers had used Robodoc to perform 10 hip-replacement operations.

Test detects cystic fibrosis gene in embryos
A cell from a human embryo is suctioned into a tiny instrument, *below,* in a new testing procedure that detects—before a pregnancy begins—the gene that causes cystic fibrosis. Researchers reported in September 1992 that the test was performed on four- and eight-celled embryos developed from fertilized eggs in the laboratory. After testing, embryos without the defect were implanted in the mother's womb.

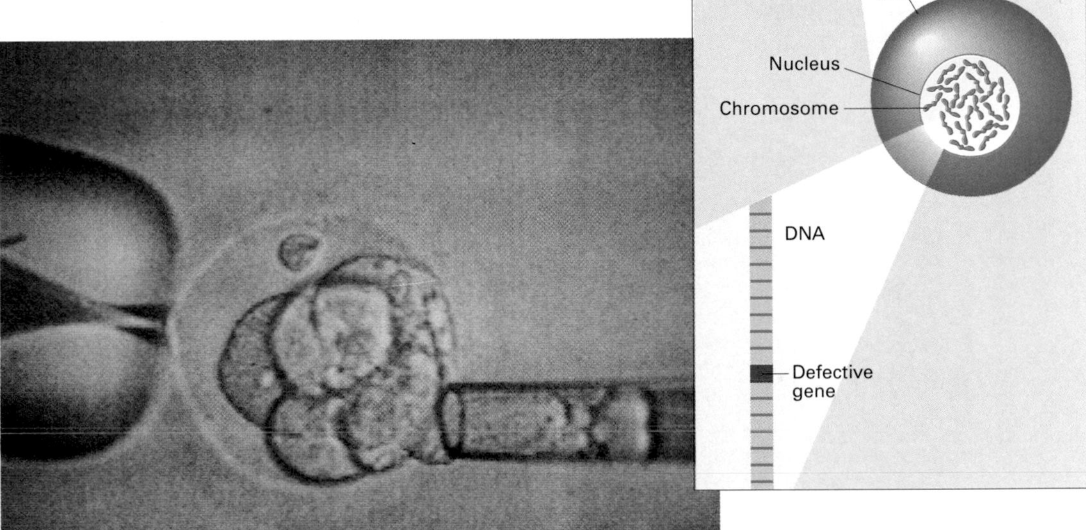

In hip-replacement surgery, the surgeon implants a cup-shaped socket into the pelvic bone and a metal shaft into the *femur* (thighbone). The shaft has a ball on one end that protrudes from the femur and fits into the socket, creating an artificial joint that resembles the natural ball-and-socket hip joint that is being replaced.

To insert the shaft into the femur, surgeons must create a 15- to 25-centimeter (6- to 10-inch) cavity down the middle of the femur. To make the hole, surgeons have traditionally used a mallet to pound a spike down into the bone. Unfortunately, this method does not yield a hole that can offer the shaft a smooth, snug fit within the surrounding bone. This poor fit reduces the implant's durability and thwarts a successful outcome for the patient.

The researchers designed Robodoc to drill a hole in the femur that would achieve at least 90 percent contact between the implanted shaft and the surrounding bone. This accuracy is accomplished not only by the robotic "arm" that drills the hole, but also through the use of a computerized tomographic (CT) scanner, which takes multiple X rays to generate a detailed cross section of the femur. A computer analyzes the CT images and the size and shape of the implant, and it helps the surgeon plan a strategy for creating an ideal cavity in the femur. Then, during surgery, the surgeon guides the robotic arm to the bone, where it uses a high-speed drill to mill a cavity according to the computer's instructions.

AIDS-like illness. At the Eighth International AIDS Conference held in Amsterdam in July, reports of cases of an AIDS-like illness not caused by human immunodeficiency virus (HIV) created an uproar among health experts and anxiety among the public. Several researchers at the meeting described patients who had some signs of AIDS, most importantly a shortage of white blood cells called CD4+ T-lymphocytes, which play a key role in orchestrating the body's immune defenses. But none of the patients were infected with HIV, which causes AIDS. Nor had they been exposed to other factors, such as chemotherapy, that are known to severely deplete CD4+ T-lymphocytes. The researchers called the ailment idiopathic CD4+ T-lymphocytopenia (ICL), which means "a shortage of CD4 cells due to an unknown cause."

The news ignited speculation that an unidentified virus might cause the AIDS-like malady, and it prompted meetings by public-health officials to consider the evidence. In February 1993, four separate research teams published reports concluding that ICL did not represent a deadly new epidemic. The studies suggested that the ailment is a collection of rare diseases that may have existed for years. Researchers at the Centers for Disease Control and Prevention (CDC) in Atlanta, Ga., found just two people who fit the criteria for ICL after the researchers reviewed more than 230,000 cases.

The studies also uncovered reassuring evidence that ICL is not contagious and is not spread through blood transfusions. Researchers found no sign of low CD4 counts in spouses or other sex partners, children, or roommates of people with ICL, and the illness did not occur more often in patients who had previously had blood transfusions.

None of the researchers could find an infectious agent or any other single cause for the ICL in the patients they studied. Furthermore, the condition appeared to vary considerably from person to person. Some people with ICL died; in others, the condition remained stable; and still others showed an unexplained improvement in their CD4 levels and overall health. Such findings led researchers to conclude that ICL may actually be several different diseases having a variety of causes.

One study also found that ICL has existed at least since 1983. This discovery led health experts to suggest that ICL cases were likely to have existed before 1983 but were not identified until tests for measuring CD4 cells became routine in the early to mid-1980's, after AIDS was recognized.

Help for weak immune systems. Researchers at the Fred Hutchinson Cancer Research Center in Seattle reported in July 1992 on a new technique that may help patients with seriously weakened immune systems resist potentially life-threatening viral infections. Immunosuppressed individuals include people with AIDS, people who take certain cancer medications, and patients who receive drugs to prevent rejection of transplanted organs. Because few medications are effective against viral infections, these individuals are particularly threatened by viruses.

The study involved three leukemia patients who had undergone a bone marrow transplant, a procedure that involves destroying a patient's own bone marrow with radiation and replacing it with bone marrow cells from a healthy donor. The bone marrow is destroyed because it produces cancerous cells. Because marrow is also the source of infection-fighting cells, destroying it leaves the patient dangerously vulnerable to infections for 30 to 100 days after the procedure, until the transplanted marrow becomes established and begins making new infection-fighting immune cells. About half of all transplant patients are infected by cytomegalovirus (CMV), a common virus that is usually harmless to healthy people but which can cause a fatal pneumonia in transplant recipients during this vulnerable period.

To shore up the patients' immune defenses, the researchers gave them an infusion of white blood cells called "killer" T cells. The investigators obtained the cells from each of the three bone marrow donors, who had previously been exposed to CMV. This prior exposure enables the T cells to recognize and attack CMV cells. About one month after the patients' bone marrow transplant, the researchers gave them huge doses of the killer T cells in a series of four weekly treatments.

The results were encouraging. The CMV-fighting cells persisted in the transplant patients for up to 12 weeks and caused no toxic reactions. More important, none of the patients developed CMV infections. But because the experiment involved only three people, more studies are needed to prove that the technique works.

Gene fix for high cholesterol. The first use of gene therapy to treat a patient with extremely high blood-cholesterol levels was undertaken in July 1992 by researchers at the University of Michigan Medical Center in Ann Arbor. The patient's disease, called familial hypercholesterolemia (FH), is caused by a flawed gene that produces an abnormal form of a liver protein called the low-density-lipoprotein (LDL) receptor. The normal form of the LDL receptor enables the liver to remove from the blood the type of cholesterol that leads to clogged arteries.

There are two forms of FH, which differ in severity. About 1 in 500 Americans have only one abnormal LDL gene, and they have a mild form of FH, which can be treated with a combination of diet and cholesterol-lowering drugs.

However, these measures offer little help to people with the lethal form of FH, which affects about 1 in 1 million Americans. People with severe FH have two abnormal genes. As a result, the liver has a total or near-total inability to remove LDL from the blood, leading to life-threatening cholesterol levels very early in life. One-third of all children with severe FH develop chest pains and heart attacks by age 10, and most die by age 20.

The patient who underwent the gene therapy was a 29-year-old woman with severe FH. The researchers removed a small piece of the woman's liver, cut it up into tiny pieces, and treated the tissue with a cell-releasing enzyme. The cells were then placed in a nourishing solution. Later, the researchers exposed the cells to harmless viruses that contained copies of the normal LDL receptor gene. The scientists expected the viruses to ferry copies of the normal gene into the liver cells, giving them the ability to produce the normal LDL receptor. The liver cells were then injected back into the patient through a vein leading to the liver.

If the treatment works as hoped, the altered liver cells will produce the normal LDL receptor protein in the patient's liver, allowing it to remove harmful cholesterol from the blood. In the months following the procedure, the researchers planned to monitor the patient's cholesterol levels to see if the altered liver cells were producing the desired effect. [Joan Stephenson]

In World Book, see AIDS; Cholesterol; Cystic fibrosis; Genetic engineering; Immunity; Sickle cell anemia.

Meteorology

The strongest winter storm in many years battered most of eastern North America with heavy rains, deep snow, and hurricane-force winds in mid-March 1993. This strong, fast-moving low-pressure system—termed an extratropical cyclone because it occurred outside the tropics—sprang up in the Gulf of Mexico late in the evening of March 12.

The center of low pressure moved across Georgia the next morning and across the Mid-Atlantic states on March 13. Severe thunderstorms associated with the intensifying low-pressure system swept across Florida early on March 13, spawning an outbreak of 27 tornadoes. Early on March 14, the center of lowest pressure crossed the Gulf of Maine. The center moved across Maine that afternoon and finally moved offshore and headed toward Greenland early on March 15.

Damage from the storm was estimated at more than $1 billion. The death toll was estimated at 243, with another 48 people missing at sea and presumed dead. The toll could have been much higher if people had not been given warning of the storm's approach. The National Weather Service had forecast the likelihood of a major winter storm earlier in the week.

Heavy, wet snow and high winds caused most of the damage during the March storm. Snow blanketed the area from Mississippi to Maine, with the heaviest snowfall along the Appalachian Mountains. More than 60 centimeters (2 feet) fell in many areas from North Carolina through New England. The storm also produced winds of about 180 kilometers (110 miles) per hour in Franklin County, Florida, and atop Mount Washington, New Hampshire.

Hurricane Andrew. A rare Category 5 hurricane hit Florida just south of Miami at about 5 a.m. on Aug. 24, 1992. The hurricane, named Andrew, created sustained winds of 230 to 240 kilometers (145 to 150 miles) per hour, with gusts to 280 kilometers (175 miles) per hour. The storm passed over the Florida peninsula and picked up speed again in the Gulf of Mexico, continuing west until it struck Louisiana on August 26.

Andrew was the third strongest hurricane to strike the United States in the 1900's, surpassed in intensity only by Hurricane Camille in 1969 and the Labor Day Hurricane of 1935. Andrew became the most devastating and most costly natural disaster in U.S. history, inflicting between $20 billion and $25-billion in damage in south Florida and in Louisiana, according to the National Hurricane Center in Miami.

Reports of deaths from Andrew varied, depending on the manner in which the statistics were compiled. The National Hurricane Center reported in March 1993 that 43 people lost their lives from causes directly related to Andrew in Florida, and another 15 died in Louisiana. The Centers for Disease Control and Prevention in Atlanta, Ga., reported that 32 people died in Florida, and 17 died in Louisiana.

Andrew began in an area of low atmospheric surface pressure called an easterly wave off the coast of west Africa. Early on the morning of August 18, the storm strengthened into a *tropical storm* (a class of storm with winds up to 116 kilometers [72 miles] per hour), and meteorologists gave the storm its name.

Andrew's winds reached 120 kilometers (75 miles) per hour on August 22, which classified the storm as a hurricane. Explosive growth during the next 36 hours raised the central winds to 220 kilometers (135 miles) per hour, giving the storm a force seen only a few times since 1980.

Andrew damaged 80,000 homes and left 180,000 people homeless in Florida. In a few hours, the storm passed into the Gulf of Mexico and continued westward with winds of 225 kilometers (140 miles) per hour.

Gradually, Andrew curved northwest and made its second landfall early on the morning of August 26 between Morgan City, La., and New Orleans. About 25,000 people were left homeless in Louisiana. As Andrew moved inland, it spawned 13 tornadoes. On August 26, Andrew lost its tropical storm characteristics amid torrential downpours across Louisiana and Mississippi.

Powerful whirlwinds were Andrew's most devastating winds, according to a February 1993 report. Senior severe weather researcher Theodore T. Fujita of the University of Chicago found evidence that Andrew contained destructive whirlwinds called suction vortexes that form around a hurricane's eye.

Fujita examined photos of the damage from Hurricane Andrew in south Florida and discovered that the most intense damage appeared in narrow streaks that followed the path of the hurricane's eye. Most meteorologists had assumed that the most severe damage from hurricanes resulted from gusts of wind. But instead of forming straight lines of destruction that gusts of wind would produce, Fujita found that the most intensely damaged areas appeared along swirling paths.

Fujita speculated that powerful swirling winds that formed at the edge of Andrew's eye were the culprits. These whirlwinds, which resemble those that occur in tornadoes, are about 0.8 kilometer (0.5 mile) wide and bring down columns of air at speeds as high as 320 kilometers (200 miles) per hour. Such blasts gave rise to the swaths of intense damage within the overall pattern of damage from the storm, said Fujita.

Hurricane Iniki, with winds of 210 kilometers (130 miles) per hour and gusts of up to 260 kilometers (160 miles) per hour, struck the Hawaiian island of Kauai late on the afternoon of Sept. 11, 1992. Iniki was the strongest storm to hit Hawaii in the 1900's and became the third most costly natural disaster in U.S. history, surpassed only by Andrew and Hurricane Hugo, which struck South Carolina in 1989. Iniki damaged 14,350 homes on Kauai, completely destroying 1,421 of these and severely damaging another 5,152. Three deaths were attributed to the storm.

Cool year. Atmospheric scientists reported in January 1993 that Earth's average temperature in 1992 was 0.19 Celsius degree (0.34 Fahrenheit degree) cooler than in 1991. Scientists from England's Meteorological Office in Berkshire used measurements of temperatures at Earth's surface to determine the average temperature. Researchers believe that the 1992 decline in global temperature resulted from a thin veil of volcanic ash in Earth's *stratosphere* (upper atmosphere) that partially blocked sunlight.

The eruption of Mount Pinatubo in the Philippines in June 1991 and of Mount Hudson in southern Chile in August 1991 blasted between 23 million

Deadly hurricanes

Two severe hurricanes battered the United States in 1992. In a satellite image superimposed on a map, *below right,* Hurricane Andrew (swirling orange and red area) sweeps across the Bahamas and heads for Florida on Aug. 23, 1992. Andrew's sustained winds reached 240 kilometers (150 miles) per hour, making the storm one of the most severe of the century. An area of Miami lies in ruins after the hurricane, *right.*

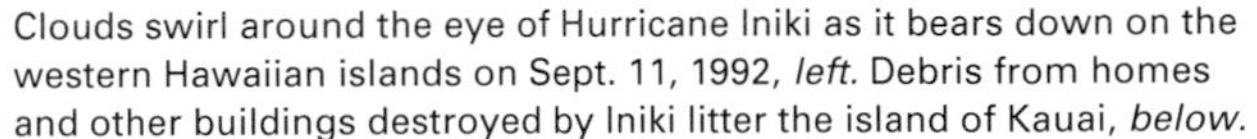

Clouds swirl around the eye of Hurricane Iniki as it bears down on the western Hawaiian islands on Sept. 11, 1992, *left.* Debris from homes and other buildings destroyed by Iniki litter the island of Kauai, *below.*

and 27 million metric tons (25 million and 30 million short tons) of sulfur dioxide into the stratosphere. There, the sulfur dioxide reacted with water vapor to produce a cloud of sulfuric acid droplets. High-level winds carried volcanic ash around the world. The cloud of particles reached maximum thickness in the Southern Hemisphere in November 1991 and in the Northern Hemisphere in April 1992.

A report confirming the belief that Mount Pinatubo's eruption helped cool Earth appeared in March 1993. Atmospheric scientists at the National Aeronautics and Space Administration (NASA) and the National Oceanic and Atmospheric Administration (NOAA) based their report on satellite observations of radiation levels reflected from Earth.

The team of NASA and NOAA scientists reported that the volcanic cloud of ash from Mount Pinatubo resulted in a 3.8 percent increase in the amount of solar radiation reflected from Earth's atmosphere in the months following the eruption. The increased reflection of sunlight reduces the amount of heat energy reaching Earth and results in lower temperatures. The team determined the amount of solar radiation reflected from Earth's atmosphere and compared the readings with the average from 1985 through 1989.

Although scientists had long suspected that volcanic debris caused such reflection, the NASA and NOAA team's work was the first direct evidence of it. The team also predicted that most of the volcanic ash would wash out of the atmosphere by early 1995.

Many scientists regarded Mount Pinatubo's cooling effect in 1992 as a temporary change from the increases in worldwide average temperatures of the 1980's through 1991. The warmest year on record was 1991, and the average worldwide temperature during the 1980's was 0.2 Celsius degree (0.36 Fahrenheit degree) warmer than the average for 1951 through 1980.

El Niño. The cooling in mid-1992 was made all the more dramatic by the fact that an unusually strong El Niño had made temperatures higher earlier in the year. The El Niño, a complex weather system linked to a warm current in the eastern Pacific, persisted until early 1993. An El Niño usually causes warmer than normal conditions in many regions of the world, including large parts of Australia and southern Africa. (In the Special Reports section, see El Niño and Its Effects.)

Ozone levels drop. The ozone hole in the stratosphere above Antarctica deepened in 1992, according to measurements taken from a NASA satellite. Scientists also reported unprecedented declines above Arctic regions in early 1993. Ozone is a molecule made up of three oxygen atoms. In Earth's upper atmosphere, it absorbs damaging ultraviolet radiation from the sun.

Atmospheric scientists reported in September 1992 that the ozone hole above Antarctica was 30 percent wider and deeper than ever before. Above the Arctic, ozone levels fell as much as 20 percent compared with 1992 levels. (See Environment.)

California drought ends. California's six-year drought ended in early 1993. Rain and snowstorms throughout California and much of the West during late 1992 and early 1993 filled reservoirs to 90 percent of capacity.

California's drought left reservoirs at near-record lows. In many areas, cities instituted water rationing, and the agricultural industry suffered millions of dollars of losses due to damaged crops. Seven straight years of drought would have been unprecedented in the region's 400 years of recorded history.

Heavy rains that began in December 1992 caused flooding, mud slides, and huge snowdrifts throughout California. In Los Angeles, heavy rains caused flooding that killed four people in February 1993. Snowstorms in February shut down Interstate 80, a major link between California cities, and mud slides also occurred in the Santa Cruz Mountains south of San Francisco.

The California Department of Water Resources said that the amounts of water from various natural sources were at or above normal in 1993. Rainfall was 60 percent higher than normal, runoff was average, and water in snowpack was estimated to be 50 percent higher than normal. In contrast, the rainfall for 1987 through 1992 had been as much as 80 percent below normal. [John T. Snow]

In World Book, see Meteorology; Ozone.

Researchers in 1992 and 1993 made several discoveries about Alzheimer's disease. Alzheimer's disease is the gradual deterioration of some of a person's brain cells, resulting in an increasing loss of memory and mental functioning.

Alzheimer's disease protein. In September and October 1992, four teams of researchers reported that *beta amyloid,* a protein that builds up in the brains of Alzheimer's patients, is also present in healthy cells throughout the human body. The role of beta amyloid in the disease is unknown, though the protein can kill brain cells in laboratory tests, and *plaques* (deposits) of beta amyloid have been found in the brains of many deceased Alzheimer's patients.

In the new research, the scientists found beta amyloid in other types of cells, such as blood vessel cells, and in the fluid that surrounds the brain and spinal cord. This finding suggests that beta amyloid has a normal function in cells.

Brain chemical link. The first evidence of a link between the build-up of beta amyloid and another common characteristic of Alzheimer's patients—an abnormally low level in the brain of a chemical called *acetylcholine*—was reported in October 1992 by neuroscientists Roger M. Nitsch and Richard J. Wurtman and their colleagues at the Massachusetts Institute of Technology in Cambridge and at Massachusetts General Hospital in Boston. Acetylcholine is an important *neurotransmitter,* a chemical that relays information between *neurons* (nerve cells).

The researchers studied human cells that produce *amyloid protein precursor (APP),* a molecule that cells break down to create beta amyloid. The scientists placed the cells in solutions containing various levels of acetylcholine and compared the resulting cellular activity. The researchers found that the amount of acetylcholine in the solution determined what the cells did with APP. When acetylcholine levels were normal, the cells broke down APP into a harmless protein that does not build up into plaques. But when acetylcholine levels were low, the cells turned APP into beta amyloid.

A possible role for beta amyloid in the destruction of brain cells was revealed in January 1993 by scientists at the National Institute of Diabetes and Digestive and Kidney Diseases. The researchers learned that beta amyloid molecules collect in a cell's *membrane* (outer layer) and create a channel from the outside of the cell to the inside. The molecules then pump a large quantity of calcium atoms from the fluid outside the cell into the cell. Calcium regulates cellular activity, and too much calcium in a cell creates disorder and can kill the cell. So beta amyloid may kill brain cells by flooding them with calcium. The beta amyloid plaques might then be left over when the cells die.

Neuron protector. A naturally occurring neurochemical prevents the death of *motor neurons* (nerve cells that control movement) after injury, three teams of neuroscientists reported in December 1992. The chemical, called brain-derived neurotrophic factor (BDNF), is found in the brains and spinal cords of mammals.

One research team, led by neuroscientist Qiao Yan of Amgen, Incorporated, in Thousand Oaks, Calif., and including scientists from Washington University in St. Louis, Mo., applied BDNF to the motor nerve fibers of newborn rats after cutting some of the fibers, which extend from the bodies of neurons in the spinal cord and brain. Neurons in early developing mammalian nervous systems usually die when their fibers are cut, but applying BDNF caused most of the neurons to survive. Similar experiments at the Max Planck Institute for Psychiatry in Germany and at the Bowman-Gray School of Medicine in Winston-Salem, N.C., also found that BDNF has a protective effect on motor neurons with damaged nerve fibers.

Researchers are not sure how BDNF works. Nevertheless, experts say these findings may open the way for the development of a treatment for nerve cell injuries in human beings.

A gaseous neurotransmitter. The discovery of a new neurotransmitter that takes the form of a gas was reported in January 1993 by neuroscientists Ajay Varma and Solomon H. Snyder and their colleagues at Johns Hopkins University Medical School in Baltimore. The researchers announced that carbon monoxide (CO), a gas that is poisonous in large amounts, is a neurotransmitter.

The same laboratory in 1989 had

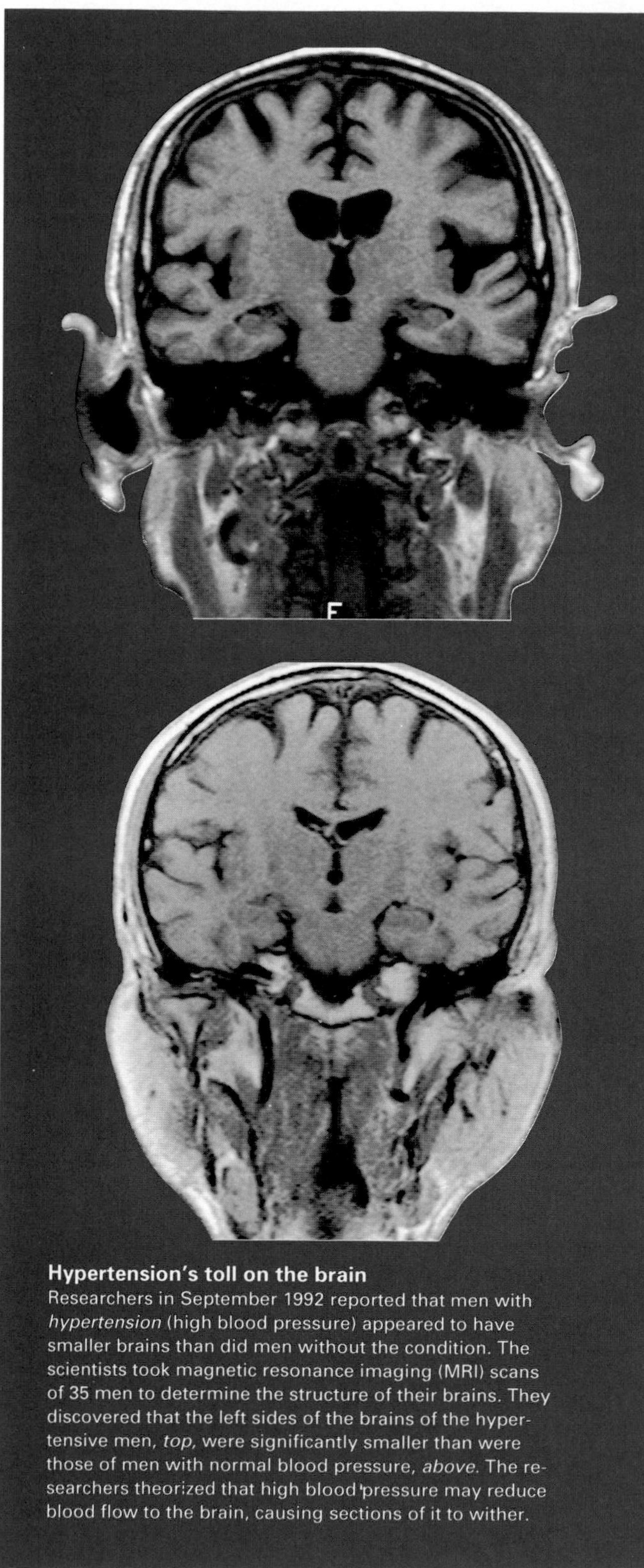

Hypertension's toll on the brain
Researchers in September 1992 reported that men with *hypertension* (high blood pressure) appeared to have smaller brains than did men without the condition. The scientists took magnetic resonance imaging (MRI) scans of 35 men to determine the structure of their brains. They discovered that the left sides of the brains of the hypertensive men, *top,* were significantly smaller than were those of men with normal blood pressure, *above.* The researchers theorized that high blood pressure may reduce blood flow to the brain, causing sections of it to wither.

found the first and only other known gaseous neurotransmitter, nitric oxide (NO). Scientists had discovered that compounds chemically similar to a known neurotransmitter are likely to be a neurotransmitter as well, so the Baltimore researchers had decided to see if gases similar to NO might also be neurotransmitters. They examined CO because the brain uses the gas for other purposes, such as activating proteins, and molecules often play several roles in the body.

The scientists probed neurons in the brains of rats for the molecule that produces CO in cells. The researchers found significant amounts of the molecule in the part of the brain that processes smells and in the *hippocampus,* a structure involved in long-term memory. When the researchers applied substances that blocked the formation of CO, the neurons in those areas could not communicate with one another. These experiments showed that CO was a neurotransmitter.

The discovery of gaseous neurotransmitters opened a new world of study for neuroscientists. Researchers are intrigued by the idea that gases may exist throughout the brain, flowing into and out of neurons and helping to regulate the brain's activity.

Another neurotransmitter finding was reported in 1992 by neuroscientists at the Oregon Health Science University in Portland and at University College in London. The researchers showed that a molecule called adenosine triphosphate (ATP) can act as a neurotransmitter.

Scientists had already known that ATP is a vitally important biochemical that provides the energy for virtually all living cells. In the 1970's, some scientists suggested that ATP might also be a neurotransmitter. However, researchers could not test this idea because there existed no substance that blocked the workings of ATP. Because neuroscientists could not observe the brain's behavior in the absence of ATP, they could not determine ATP's role in the brain. In early 1992, researchers at last identified a reliable blocking agent, and they were able to establish that some neurons communicate with ATP.

Brain-mapping technique. A new method for observing the brain at work was described in August 1992 by neurosurgeons George A. Ojemann and Michael M. Haglund and computer scientist Daryl W. Hochman of the University of Washington Medical School in Seattle. The researchers detected neural activity by shining a special type of light on the surfaces of exposed brains.

Ojemann and his colleagues were performing brain surgery on patients with *epilepsy,* a disorder in which uncontrolled electrical discharges in the brain lead to sudden seizures. The patients can remain conscious and unanesthetized during the operation because the brain cannot feel pain.

The researchers removed part of each patient's skull, then shined near-infrared light—invisible rays of light—on the exposed brain. They then asked the patients to move their tongues and to name objects shown on a screen. As the patients performed each task, a special camera recorded the near-infrared light reflected off the brain. By analyzing the images, the researchers identified areas of the brain's surface that changed in reflectivity during the two tasks. They concluded that those regions of the brain must be necessary for the tasks. (In the Special Reports section, see MAPPING THE HUMAN BRAIN.)

Chemical craving. A brain protein called galanin may be at the root of some people's craving for fatty foods, researchers suggested in autumn 1992. Neuroscientist Sarah F. Leibowitz and her colleagues at Rockefeller University in New York City found a relationship between rats' fat intake and the amount of galanin in the rats' brains. In all mammals, galanin is found in the *hypothalamus,* a brain structure known to be involved in regulating hunger, emotion, and other unconscious body functions.

The researchers gave rats a choice of three types of food: protein, in the form of milk protein; carbohydrates, in the form of a sugar-cornstarch mix; and fat, in the form of lard. Some of the rats lapped up the lard before turning to the other foods. When the researchers examined the brains of those rats, they found higher concentrations of galanin in their hypothalami than were present in the hypothalami of rats that did not eat as much lard. And when the scientists injected galanin directly into the hypothalami of rats, the rats ate much more lard than did unaltered rats.

The researchers' work on galanin may someday help people with such disorders as obesity, diabetes, and atherosclerosis, all of which may be related to the overconsumption of fats. In 1992, the researchers were already testing a drug that seemed to block the activity of galanin in rats and shut off their craving for fats.

Ear cell regeneration. Reports in March 1993 from two collaborating research teams may offer new hope for people with some types of hearing loss. For the first time, the researchers observed in adult mammals the regrowth of damaged ear cells critical for hearing. Scientists had previously believed that after such cells in adult mammals are damaged, they cannot regenerate. The research teams were led by cell biologist Andrew Forge of University College in London and by neuroscientist Mark E. Warchol of the University of Virginia School of Medicine in Charlottesville.

The researchers looked at *hair cells,* which are located deep in the ear. Part of the job of hair cells is to detect vibrations caused by sound waves and to relay appropriate signals to the brain. Hair cells can become damaged as a result of physical injury, drugs, high noise levels, or simply old age.

The London team studied hair cells in live guinea pigs. First, they gave the animals a drug that selectively kills hair cells. When the researchers examined ear tissue taken from the animals four weeks later, they saw young hair cells growing in areas where the drug had wiped out earlier hair cells.

The Virginia team also examined ear tissue grown in the laboratory. They first removed ear tissue from guinea pigs and human beings and grew it in test tubes. Then the scientists applied a drug that killed the hair cells. After one week, the scientists found new hair cells growing in both the guinea pig and human tissue.

The researchers said they needed to perform further tests to learn whether the new hair cells function properly. They also reported that they did not know what triggered the hair cells to regenerate. Discovering a chemical signal for hair cell growth could someday lead to a cure for some types of deafness.

[George Adelman]

In WORLD BOOK, see BRAIN.

Contributions to high-energy physics, improvements in the understanding of complex chemical reactions, and the discovery of a regulatory mechanism affecting most cells were the bases for Nobel Prize awards in physics, chemistry, and physiology or medicine in October 1992. Each prize was worth $1.2 million.

The chemistry prize was awarded to Canadian-born theoretical chemist Rudolph A. Marcus of the California Institute of Technology in Pasadena for his mathematical analysis of how electrons move from one molecule to another in a solution. Marcus accomplished most of the work on which the award was based in the late 1950's and early 1960's. Other chemists greeted the work skeptically at the time, however. It was not until the 1980's that scientists verified Marcus' theoretical predictions by experimentation.

Marcus' early work resulted in mathematical formulas for describing the rate of oxidation-reduction processes, also known as *redox reactions*, in which one molecule in a solution gains an electron while another molecule gives up an electron. This electron transfer changes the structures of the two molecules as well as those of nearby molecules in the solution. Marcus' later research in electron transfer concerned more complex chemical reactions, such as *photosynthesis,* the energy-converting activity of green plants. The Nobel Committee of the Royal Swedish Academy of Sciences in Stockholm, Sweden, noted that Marcus' work "has greatly stimulated experimental developments in chemistry."

The physics prize went to physicist George Charpak, a Polish-born French citizen who has worked since 1959 at the European Organization for Nuclear Research (CERN), a research center for the study of subatomic particles, located near Geneva, Switzerland. Charpak was recognized especially for his invention in 1968 of the multiwire proportional chamber, an electronic detector used in *particle accelerators* (devices that hurl subatomic particles together at extremely high energies). During particle collisions, Charpak's detector could record the paths of millions of subatomic particles each second. The detector made it

French physicist George Charpak of the European Organization for Nuclear Research, a research facility near Geneva, Switzerland, won the Nobel Prize for physics in October 1992 for his work in detecting subatomic particles.

American biochemists Edmond H. Fischer, left, and Edwin G. Krebs work in their lab at the University of Washington in Seattle. The two men shared the Nobel Prize in physiology or medicine in October 1992 for their discovery of an important process affecting the way most cells function.

possible for physicists to discover several new particles.

The invention advanced a technique originally used as early as 1908. Charpak placed a closely spaced array of positively charged wires in a chamber filled with gas molecules. The wires acted as sensors that detected the paths of subatomic particles as they ripped negatively charged electrons from the gas molecules in the chamber. The sensors in turn were connected to computers that recorded the millions of instantaneous events.

The physiology or medicine prize was shared by biochemists Edmond H. Fischer and Edwin G. Krebs, professors emeritus at the University of Washington in Seattle. The scientists were honored for a discovery they made about cell proteins in the mid-1950's, which is now known to "concern almost all processes important to life," according to the Nobel Assembly of the Karolinska Institute. Working together, Fischer and Krebs discovered a process known as *protein phosphorylation* while studying muscle contractions. They were following up work done by American biochemists Carl and Gerty Cori, who won a Nobel Prize in 1947 for their discovery of a muscle enzyme called phosphorylase that releases stored energy and causes a muscle to contract. Fischer and Krebs sought to learn what causes phosphorylase to switch on or off as needed.

They found that phosphorylase is switched on (its stored energy is released) by another enzyme, one of a group of proteins known as protein kinases. This enzyme carries certain phosphate groups to inactive phosphorylase molecules. The two researchers learned that the process also works in reverse. Phosphorylase can be switched off when another kind of enzyme strips away phosphate groups.

The Nobel Assembly noted that Fischer and Krebs's findings had "initiated a research area which today is one of the most active and wide-ranging. . . . Step by step, it has become evident that protein phosphorylation constitutes a fundamental mechanism, influencing all cellular functions." [Rod Such]

In WORLD BOOK, see NOBEL PRIZES.

Iron and Heart Disease

Iron—the mineral most people have thought of as a source of strength and the answer to "tired" blood—may be a culprit in processes that can lead to heart attack, according to a study published in September 1992 by Finnish researchers. The study was the first to provide objective evidence for a link between iron and cardiovascular disease in human beings. The idea that iron may promote heart disease was first proposed in 1981 by Jerome L. Sullivan, a pathologist at Veterans Affairs Medical Center in Charleston, S.C.

Epidemiologists at the University of Kuopio, Finland, randomly selected 1,931 men between the ages of 42 and 60 who had no symptoms of heart disease. The men answered questions regarding diet, smoking, and living habits. The men also underwent a physical examination that included tests to measure levels of *cholesterol* (a fatlike substance) and *ferritin* (the protein that carries iron in the blood). Because researchers cannot easily measure iron stored in body tissues, they used ferritin levels as an indicator of the total amount of iron in the body.

The researchers monitored the men for an average of three years. During that period, 51 of the men had heart attacks. The investigators compared the ferritin levels of the men who had heart attacks with the ferritin levels of the men who did not have heart attacks. They found that men who had ferritin levels above 200 micrograms per liter of blood were more than twice as likely to suffer a heart attack as were men with lower ferritin levels.

The researchers also found that a high level of low-density lipoproteins (LDL's), long considered a major villain in heart disease, was not a significant risk factor by itself. (LDL's carry cholesterol in the blood.) The scientists discovered, however, that men who had higher than average levels of both LDL-cholesterol and ferritin had the highest risk of all for heart attack.

The investigators calculated that ferritin levels were associated more closely with heart attacks than were several other risk factors, including high blood pressure, diabetes, and low levels of high-density lipoproteins (molecules thought to remove cholesterol from the blood). Smoking a pack of cigarettes a day for several years was the only risk factor more closely linked to heart attacks than was ferritin.

How might iron lead to heart attacks? Iron reacts with oxygen in the bloodstream to release *free radicals,* unstable molecules that readily combine with other molecules in a process called oxidation. Oxidation of LDL-cholesterol molecules appears to be a necessary step in the development of *atherosclerosis,* a condition in which arteries be-

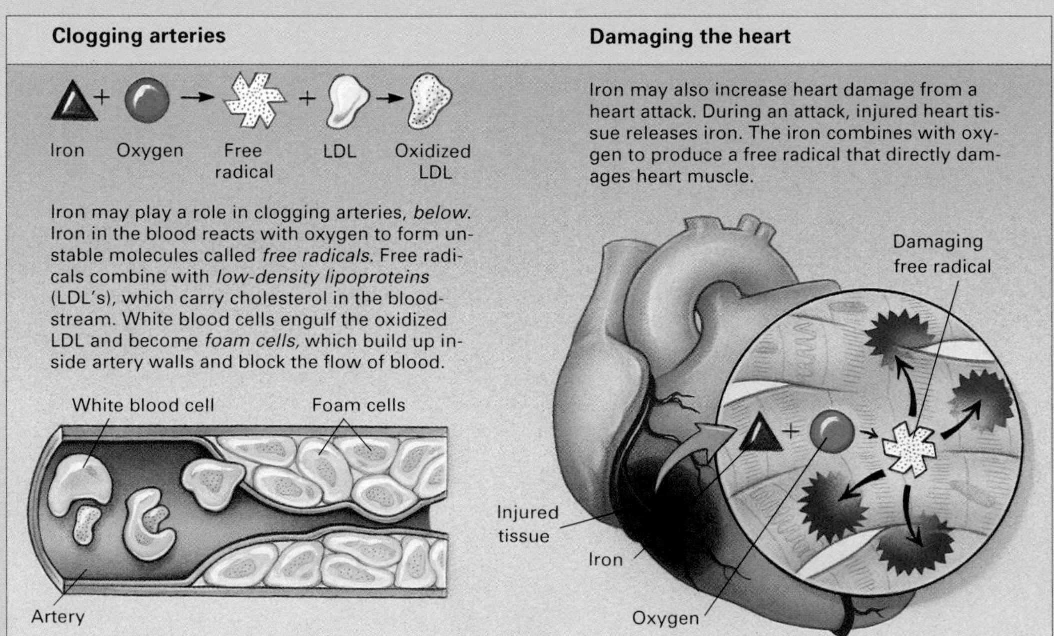

Iron may play a role in clogging arteries, *below.* Iron in the blood reacts with oxygen to form unstable molecules called *free radicals.* Free radicals combine with *low-density lipoproteins* (LDL's), which carry cholesterol in the bloodstream. White blood cells engulf the oxidized LDL and become *foam cells,* which build up inside artery walls and block the flow of blood.

Iron may also increase heart damage from a heart attack. During an attack, injured heart tissue releases iron. The iron combines with oxygen to produce a free radical that directly damages heart muscle.

come clogged and hardened. This process occurs as *macrophages,* a type of infection-fighting white blood cell, target oxidized LDL-cholesterol as a foreign or undesirable substance. The macrophages engulf the cholesterol molecules, become embedded in the artery, and turn into "foam" cells—so named for their foamy appearance. As these foam cells accumulate, they produce fatty streaks that have a rough, craggy surface, which traps dead cells, calcium, and other debris. Eventually, this build-up hardens into plaque and clogs the artery, reducing or stopping the flow of blood to the area fed by the artery. If an artery feeding the heart becomes completely blocked, a heart attack may result.

During a heart attack, iron may further damage the heart, according to researchers. In a heart attack, part of the heart is unable to function because its nourishing blood supply has been cut off, usually by either atherosclerosis or a blood clot. This heart tissue releases iron, which joins with oxygen to form a free radical. The free radical may then directly damage more heart tissue.

Although there seems to be a substantial case for the link between iron and heart attacks, controversy still exists. In February 1993, investigators from Harvard Medical School in Boston reported on their study of ferritin levels and heart attack rates in 476 American men. They found that ferritin levels above 200 micrograms per liter increased heart attack risk by only 10 percent, not 120 percent, as in the Finnish study. In a second study, also reported in February, other Harvard researchers found that the type of iron found in red meat increased the risk of heart attack by 43 percent.

Scientists have noted that some groups of people at low risk for heart attack share characteristics that lead to reductions of stored iron. Researchers suggest, for example, that women who have not undergone menopause (the time of life when menstrual periods end) have a very low rate of heart attack because they lose a large amount of iron in menstrual blood. However, no studies have yet researched heart disease in premenopausal women or other people who have low levels of stored iron.

More iron studies, including some involving women, are planned. Meanwhile, many medical experts believe that the evidence for iron's role in heart disease is strong enough to warrant precautionary advice regarding iron supplements. Although extremely low iron levels can lead to anemia and other health problems, most Americans obtain enough iron in their daily diet to avoid such iron deficiencies, experts say. They caution consumers to be wary of taking iron supplements unless their physician prescribes them.

[Beverly Merz]

Nutrition

Men and women who take vitamin E supplements may have a lower risk of heart disease than do people who take no such supplements. Researchers at the Harvard School of Public Health and the Brigham and Women's Hospital, both in Boston, reported this finding in May 1993.

In one study, researchers reviewed questionnaires completed by 87,245 women aged 34 to 59 participating in the Nurses Health Study, an ongoing research project. The questionnaires contained information about the nurses' health, dietary habits, and use of vitamin supplements over an eight-year period. During this time, 552 heart attacks occurred among the women.

Among women who took at least 100 units of vitamin E per day for two years or more, the rate of heart attack was 41 percent lower than it was among women who took no supplements. In making their calculation, the researchers had taken into account other factors related to heart disease—such as age, weight, and smoking—as well as the possible effects of other vitamins.

In another project, a four-year study of men, researchers analyzed questionnaires from 39,910 male health professionals aged 40 to 75. In this study, 667 men developed heart disease. Among men who took at least 100 units of vitamin E daily for two or more years, the rate of heart attack was 37 percent lower than it was among men who took no such supplements.

Laboratory studies had previously shown that vitamin E is an antioxidant, a chemical that prevents molecules from *oxidizing* (reacting with oxygen). In this capacity, vitamin E could, theoretically, help prevent low-density lipoprotein (LDL), a main carrier of cholesterol in the bloodstream, from becoming oxidized. Oxidized LDL can form deposits in arteries that reduce blood flow, and this may lead to heart attack or stroke.

Despite their findings, the Boston researchers said that recommendations regarding vitamin E use should await controlled studies to prove its safety and usefulness.

Folic acid and birth defects. Before and during pregnancy, a woman should consume 0.4 milligrams of folic acid per day to reduce the risk of giving birth to a baby having a neural tube defect. The

United States Public Health Service (PHS) offered this advice in September 1992.

Defects in the neural tube, which is the tissue from which a fetus's brain and spinal cord develop, occur in the first two months after conception. About 2,500 babies with neural tube defects are born in the United States annually. The most common results of these defects include *anencephaly,* a fatal condition in which major parts of the brain are missing, and *spina bifida,* a dangerously crippling condition in which the spinal column does not completely encase the spinal cord.

The recommendation from the PHS was prompted by several studies, including one reported in December by Hungarian geneticists Andrew Czeizel and Istvan Dudas. The researchers selected 7,540 Hungarian women who were under age 35 and planning to become pregnant. The scientists randomly assigned half of this group to take a daily multivitamin supplement containing 0.08 milligrams of folic acid. The remaining women took a supplement containing only a small amount of minerals. Both groups took the supplements during the time they were trying to become pregnant. In all, 4,753 of the women became pregnant during the study, and these participants continued to take their supplements at least two months after becoming pregnant.

The researchers received information on the outcome of the pregnancies of 4,156 women. Of these, 2,104 women had taken folic acid supplements and all delivered babies who had no neural tube defects. Among the 2,052 women who had not taken folic acid supplements, 6 delivered babies with neural tube defects.

At least seven other studies reported since 1981 have shown that consuming various amounts of folic acid dramatically decreases the risk of neural tube defects. An excessive amount of folic acid, however, may mask the symptoms of vitamin B_{12} deficiency, which could cause serious health problems. Thus, the PHS advised women to keep their total folic acid consumption to less than 1 milligram per day unless they are closely supervised by a physician.

Milk and diabetes. A protein in cows' milk may trigger an immune response in infants that leads to Type I diabetes, according to a study published in July 1992. Type I diabetes is a disorder in which the immune system destroys cells in the pancreas, called islet cells, that make insulin, a hormone that regulates the body's use of sugar.

Type I diabetes, which can seriously damage the body's organs, affects 1 million Americans and usually requires a lifetime of daily insulin shots. Many scientists think that both hereditary and environmental factors play some role in the development of Type I diabetes.

In the study, a team of Canadian and Finnish researchers analyzed blood samples from 142 children in Finland who had Type I diabetes and from 79 children who did not have the disease. They found that only the diabetic children had an increased number of antibodies that target a protein called albumin in cows' milk. The antibodies attack this protein as a foreign invader. The cows' milk protein is chemically similar to a protein called p69, which is contained in human islet cells. Because of this similarity, the researchers hypothesized that during the first several months of life, the immune systems of some children may mistake p69 cells for cows' milk albumin and destroy them and the islet cells.

Margarine's down side. Margarine and vegetable shortening could contribute to coronary artery disease by increasing LDL levels, according to a March 1993 research report. The researchers studied the results of questionnaires from 85,095 women in the Nurses Health Study. They found that women who ate four or more teaspoons of margarine per day had a 50 percent greater chance of developing heart disease than did women who ate margarine only once a month. These effects remained even after the researchers adjusted for the women's age, total energy intake, and other known risk factors. The researchers said that the health effects may be linked to margarine's manufacturing process. Manufacturers *hydrogenate* (combine with hydrogen) vegetable and other oils, making them solid or semisolid. The process creates trans fatty acids, which behave much like saturated fat in the body, the researchers said. [Johanna T. Dwyer]

In World Book, see Nutrition.

Scientists aboard the drill ship *JOIDES Resolution* in October 1992 reported finding methane hydrate deposits below the ocean floor off the coasts of Oregon and Vancouver Island, Canada. The deposits may offer vast sources of natural gas. Geophysicists Satish C. Singh and Timothy A. Minshull of the University of Cambridge in England and George D. Spence of the University of Victoria in Canada reported the finding.

The scientists discovered the deposits after the ship drilled into the ocean floor at the intersection of two *tectonic plates* (gigantic plates that make up Earth's crust) to retrieve samples of rock and sediment. There, the smaller Juan de Fuca plate is slipping beneath the larger North American plate. The pressure crushes rocks and sediments and squeezes out dissolved chemicals. Over millions of years, this force has converted *organic* (carbon-containing) material in the sediments into methane (natural gas), then into hydrocarbons, compounds found in petroleum.

Due to the low temperatures below the ocean floor, the methane freezes. The result is methane hydrate, a frozen white substance.

Studies using sound waves to probe the ocean floor indicate that methane hydrate may exist under the ocean floor along the coasts of many continents. When scientists bounce sound waves off the ocean floor, the unique density and makeup of the methane hydrate produces a distinctive wave reflection. Scientists have discovered these distinctive patterns a few hundred meters beneath the ocean floor along many continents, and they appear to reflect the surface of the sea floor, regardless of the geological jumble caused by the crust's bending and twisting.

The study aboard the *JOIDES Resolution* confirmed many geologists' belief that a layer of methane hydrate exists wherever this pattern appears. Such regions could potentially store some of the largest reservoirs of natural gas in the world.

The discovery also suggests that if the ocean warms, a vast amount of methane could be released into the atmosphere. This could occur as a result of *global*

Rough seas near Antarctica are revealed by a map of global wave heights made by the TOPEX/Poseidon satellite in autumn 1992. The satellite gathered the data by bouncing radio waves off the surface of the ocean and analyzing the reflected signals. The highest waves— between 5.5 meters (18 feet) and 8 meters (26 feet)—appear as yellow and red areas in the satellite image.

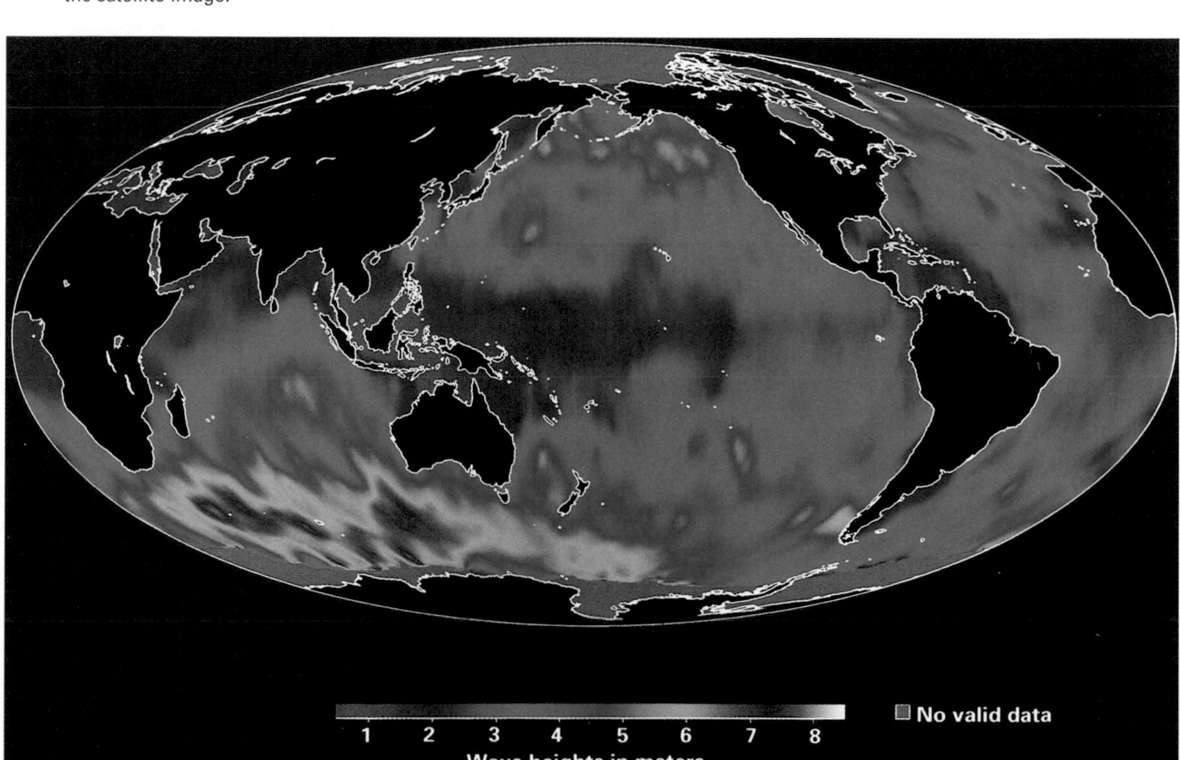

warming, the potential rise in Earth's average temperature due to the accumulation of excess *greenhouse* (heat-trapping) gases in the atmosphere. Greenhouse gases occur naturally and help keep Earth's atmosphere warm enough to support life. But human activities, mainly burning *fossil fuels* (coal, oil, and gas), also produce significant amounts of greenhouse gases.

Carbon dioxide is the main greenhouse gas, but the methane in methane hydrate is capable of trapping the sun's heat even more effectively than carbon dioxide. Some scientists speculate that if the methane hydrate below the ocean floor melted, releasing the methane into the atmosphere, any greenhouse warming would accelerate.

Tiny killers. A type of *phytoplankton* (plantlike, one-celled marine organisms) that kills fish with a powerful toxin in order to feed on them was described by biologists at North Carolina State University in Raleigh in July 1992. The scientific team discovered the phytoplankton, a type called a dinoflagellate, in two *estuaries* (broad river mouths) off the coast of North Carolina.

Dinoflagellates have been known to cause massive fish kills in occurrences called red tides. During red tides, dinoflagellates multiply rapidly, and, in the process, release toxins that kill fish. Scientists had previously believed that dinoflagellates released toxins strictly for defensive purposes, however. The discovery in North Carolina is the first in which phytoplankton have been discovered killing fish in order to feed upon them.

The scientists determined that the single-celled killer lies dormant in a protective membrane on the seabed until live fish approach. Then, the organism breaks out of its protective membrane and multiplies, releasing a powerful poison that kills fish, sometimes on a massive scale. As flesh falls off the fish, the phytoplankton feeds by grabbing pieces with a tonguelike feature. After several hours of feeding, each dinoflagellate covers itself again in a membrane, sinks into sediment, and awaits more fish.

Laboratory and field studies confirmed that the toxin is lethal to 11 species of fish, including commercially valuable striped bass, southern flounder, and eel. It is deadly at temperatures ranging from 4 °C to 28 °C (40 °F to 82 °F). The researchers had not discovered the toxin's chemical makeup or mapped its distribution beyond the two estuaries, the locations of repeated fish kills. Nevertheless, the researchers predict that the organism is widespread and is the cause of many other mysterious coastal fish kills.

Whale family tree. Toothed whales and toothless whales may be more closely related than previously thought. That was the January 1993 finding of zoologist Michel C. Milinkovitch of the Campus Erasume Université Libre de Bruxelles in Brussels, Belgium, and ecologist Axel Meyer of the State University of New York at Stony Brook. The scientists compared DNA (deoxyribonucleic acid, the molecule of which genes are made) from 16 whale species to reach their conclusion.

Scientists have traditionally divided whales into two distinct branches, which were believed to have had long, independent histories. The toothed whales include the sperm whale and various dolphins and porpoises. These animals find food and navigate by *echolocation* (sending out sound waves and listening for echoes). Toothless whales, also called baleen whales, obtain food by filtering large amounts of water through a comblike feature in their jaws called baleen. They cannot echolocate. Baleen whales include humpbacks and blue whales.

Scientists have long believed that the two branches split from a group of toothed whales some 40 million to 45 million years ago. But when researchers compared DNA from 16 species of living whales, they found that the sperm whale DNA more closely resembled the DNA of baleen whales than that of dolphins and other toothed whales. The scientists said that the split between the toothed and toothless whales could have occurred as recently as 13 million years ago. The scientists also compared whale DNA with that of a cow, a sloth, and a human being. That analysis provided support for the theory that all whales descended from hoofed, plant-eating land animals called ungulates, which are related to cows. [Lauriston R. King]

See also GEOLOGY. In WORLD BOOK, see OCEAN; WHALE.

In January 1993, construction began on the tunnel for the Superconducting Super Collider (SSC), a controversial new particle accelerator being built near Waxahachie, Tex., about 32 kilometers (20 miles) south of Dallas. The SSC, the world's largest and most expensive instrument designed for purely scientific purposes, was expected to help physicists study highly unstable particles not ordinarily seen in nature. But as of June 1993, the future of this mammoth project remained in doubt.

A particle "race track." Particle accelerators are devices that speed up subatomic particles such as electrons or protons to nearly the speed of light, giving them enormous amounts of energy. Magnetic fields guide the paths of the particles, forcing them into a fixed orbit. In accelerators of the type to which the SSC belongs, the particles (in this case, protons) are guided in two beams moving in opposite directions around a track. At certain points, the beams cross, and a small number of the protons collide. When this happens, some of the protons' energy is converted to mass, creating a variety of other particles.

The SSC plans called for an oval ring about 87 kilometers (54 miles) in circumference, located in a tunnel carved out 46 meters (150 feet) underground. The oval's curved ends will together hold more than 10,000 superconducting magnets, each about 16 meters (52 feet) long and cooled to only a few degrees above *absolute zero* (–273.15 °C or –459.67 °F), the lowest possible temperature. These magnets will guide two beams of protons traveling in narrow pipes within the main ring. At each of two collision points on the straight sections of the oval, a particle detector the size of a large building and equipped with sensitive instruments will track the paths of the emerging particles.

Each proton in the two beams will carry as many as 20 trillion electronvolts (20 TeV) of energy. (An electronvolt is a unit of energy equal to the amount of energy gained by an electron when it moves across an electric potential of 1 volt.) At the world's most advanced existing accelerator, the Tevatron at the Fermi National Accelerator Laboratory (Fermilab) near Batavia, Ill., particles are accelerated to less than 1 TeV. Scientists hope the enormous energy generated when protons collide in the SSC will help them answer such fundamental questions as why matter has mass. Physicists believe mass may be conferred by an extremely heavy particle known as the Higgs boson, whose existence has not yet been proven. The Higgs boson is thought to be too heavy to be detected by existing accelerators.

Milestones reached. The SSC passed its first milestone on Aug. 14, 1992, when scientists successfully tested six superconducting magnets designed for use in the accelerator. The six magnets make up a basic unit known as a half-cell. Each half-cell includes five magnets designed to force the speeding protons to follow a curved path. The sixth magnet brings the beams of particles to a focus, much as a lens focuses a beam of light. The completed accelerator will need 1,720 of these half-cells.

The SSC passed a second milestone on Jan. 14, 1993, when workers began digging the collider's main tunnel. In preparation, workers had first dug a vertical shaft, which they then widened near the bottom into a large cavern. A huge digging machine, or "mole," was brought down the shaft in pieces and assembled on the cavern floor. The mole bores through layers of soil and rock, laying a concrete lining behind it as it moves. Eventually, up to six boring machines may be at work constructing the main tunnel and smaller tunnels that will house accelerators designed to boost the protons to their proper speed.

Uncertainty over funding dogged the project all year, however. Since the SSC was announced in 1988, many members of the Congress of the United States had balked at its $8-billion price tag—as had some scientists, who feared that the collider might divert funding from other projects. In June 1992, the House of Representatives voted to kill the SSC. In August, however, the Senate—at the urging of the Administration of President George Bush—voted to maintain funding, though at a lower level than the SSC's directors had originally requested. After negotiations, the House went along with the Senate action.

The Administration of President Bill Clinton, who took office in January 1993, was said to support the SSC, though officials planned to slow down construction of the collider to help reduce the federal budget deficit. In June 1993, however, the House of Representatives again voted to kill the project. The Senate was expected to vote on the issue by autumn 1993.

One factor that disturbed particle physicists and politicians was the failure of the SSC to draw the expected support from foreign nations. The government of Japan said it favored the project, but it was reluctant to make a definite commitment of funds. Meanwhile, the 14-nation European Organization for Nuclear Research (CERN) moved ahead with plans for a European proton collider, to be constructed at the CERN laboratories near Geneva, Switzerland.

Whatever the SSC's fate, it seemed unlikely in June 1993 that it would be completed as planned in 1999. At probable funding levels, observers said, the opening was likely to be delayed at least until the year 2003.

Closing in on the t quark. Excitement grew at Fermilab's Tevatron in spring 1993, as researchers began to close in on a particle believed to be the twelfth, and final, fundamental building block of matter to be observed. The particle, known as the *t quark*, has eluded physicists since the 1970's, when the theory that predicts its existence first gained wide acceptance.

According to the theory, called the Standard Model, there are two families of fundamental building blocks. One family, known as *leptons*, includes electrons, as well as muons, taus, and three kinds of neutrinos. The other family, called *quarks*, are the fundamental particles that make up protons and neutrons, the two types of particles in the atomic nucleus. The Standard Model also includes a third family of particles, the bosons. Bosons transmit forces between particles, providing the "glue" that holds matter together.

The Standard Model suggests that there must be one kind of quark for every kind of lepton. There is strong evidence that the lepton family includes only the six leptons already discovered. But there are only five known quarks: the u, d, c, s, and b quarks. (The letters stand for *u*p; *d*own; *c*harmed; *s*trange; and *b*ottom or *b*eauty—fanciful names that do not describe qualities of the quarks themselves. The *t* of the t quark is derived from *t*op or *t*ruth.)

Only 4 of the 12 fundamental particles—the u quark, d quark, electron, and electron neutrino—exist under ordinary circumstances. But scientists have discovered 7 of the 8 others through particle accelerator experiments. Physicists believe that the t quark remains undiscovered only because it is too heavy to be produced in any accelerator less powerful than the Tevatron. The t quark is thought to be as heavy as an entire atom of silver, which itself consists of hundreds of u and d quarks.

The difficult search. In the Tevatron, beams of protons and antiprotons travel in opposite directions around a ring 6.5 kilometers (4 miles) long. (An antiproton is the antimatter counterpart of a proton. Antimatter consists of fundamental particles with the same mass as the fundamental particles of ordinary matter, but with the opposite charge.) The Tevatron speeds each proton and antiproton to an energy level of about 0.9 TeV. When a proton and antiproton collide, they release about 1.8 TeV of energy, some of which is converted to mass in the form of numerous particles.

These collisions are capable of creating t quarks. But a fundamental rule of physics dictates that whenever a particle is produced, its antiparticle must be produced at the same time. Physicists estimate that creating the combined mass of a t quark and its antiparticle would require about 0.3 TeV of energy. Although this falls well within the capability of the Tevatron, it is very rare for such a large share of the collision energy—about 15 per cent—to be converted to just one pair of particles.

The major obstacle to discovering the t quark is not so much producing it as recognizing it when it is produced. Scientists believe that the t quark is an extremely unstable particle likely to *decay* (disintegrate) in less than one-billionth of one-billionth of a second. So it is impossible to detect directly. Instead, it must be recognized through the pattern of lighter particles it forms as it breaks up. But these too are unstable particles, observable only through the products of their own decay.

The t quark hunter thus starts at least two steps away from his prey, trying to make sense of up to 20 or more particles that result from this chain of disintegrations, plus a few stray particles

The development of a flexible light-emitting diode (LED) made of plastic was announced by scientists at Uniax Corporation in Santa Barbara, Calif., in June 1992. LED's—devices that give off light when an electric current runs through them—are widely used in electronic displays. Experts said the flexible LED's might find uses in billboards or other large displays.

from other sources to confuse the issue. Fortunately, scientists have identified a small number of distinctive patterns that provide a strong indication of a t quark-antiquark pair. The one likely to occur most often includes, among other particles, two easily identified leptons—an electron and a muon. If the t quark exists, and if it has its expected mass, this pattern should be observed about once in every 10 billion collisions.

Tevatron detectors had seen four examples of this telltale pattern by June 1993. While these results were promising, the experimenters were not ready to claim the discovery of the t quark. It will take many more sightings to settle this matter. Fortunately, the Tevatron was scheduled to run nearly continuously until spring 1994, allowing Fermilab physicists to sample as many as 500 billion collisions—about 10 times as many as they had searched by June 1993.

More neutrinos from the sun. Findings published in summer 1992 added a new twist to the mystery of the "missing" solar neutrinos. These fundamental particles are electron neutrinos produced by nuclear reactions inside the core of the sun and other stars. These particles pass through the dense inner layers of the sun and travel into space in endless streams. Based on calculations of the rate at which the sun generates energy, scientists can estimate the number of solar neutrinos that should be detectable on Earth. But before 1992, neutrino detectors had recorded less than half the predicted number.

The new findings came from the Gallium Experiment, or GALLEX, a neutrino detector established by American, French, German, Israeli, and Italian scientists beneath Italy's Gran Sasso Mountain. According to research published in July 1992, GALLEX—which began collecting data in May 1991—recorded many more neutrinos than had earlier detectors, though still not as many as expected. The findings added to scientists' puzzlement over the nature of solar neutrinos.

The sun produces energy through a chain of nuclear reactions that convert four hydrogen atoms into one helium atom. The nucleus of most hydrogen atoms contains a single proton and no neutrons, while a helium nucleus consists of two protons and two neutrons. When four hydrogen nuclei fuse to form helium, two protons must be transformed into neutrons, a process that always produces an electron neutrino.

Scientists estimate that 70 billion solar neutrinos pass through each square centimeter of Earth every second. Most of these neutrinos travel right through the Earth without leaving a sign of their presence.

Detecting neutrinos. Like other electrically neutral particles, neutrinos can be observed only when they interact with a negatively charged electron or a positively charged nucleus—something that happens very rarely. Of all the neutrinos passing through 100 metric tons (110 short tons) of material over one week, only a few will interact with an electron or a nucleus.

Besides GALLEX, three neutrino detectors are in operation. Scientists set up the first one in South Dakota's Homestake gold mine in 1968. The Homestake detector sees only 30 percent of the expected number of neutrinos. A second detector, the Kamiokande, operating since 1987 in a Japanese lead mine, sees only 40 percent as many neutrinos as expected.

The Homestake and Kamiokande detectors are sensitive only to high-energy electron neutrinos produced by a type of nuclear reaction that generates only a small fraction of the sun's energy. Most solar neutrinos—more than 98 percent—are emitted through a type of reaction called a *proton-proton (pp) reaction,* in which two protons collide. Neutrinos produced by pp reactions have energy levels of only 0.42 million electronvolts (0.42 MeV) or less.

A third detector, one sensitive to the low-energy neutrinos produced by pp reactions, was set up in 1990 in southern Russia. In this detector, known as SAGE (for Soviet-American Gallium Experiment), neutrinos pass through a tank of the rare element gallium. Neutrinos with energy levels as low as 0.23 MeV can convert a gallium nucleus to a radioactive germanium nucleus, which can be detected with a radiation coun-

Building a particle race track
The Superconducting Super Collider (SSC), a particle accelerator being built in Texas, passed two milestones in 1992 and 1993, despite uncertainty over the project's continued funding. In August 1992, scientists successfully tested a group of 16-meter (52-foot) magnets, *left,* designed to guide and focus subatomic particles. In January 1993, workers began digging the collider's main tunnel using a huge boring machine, *below.*

ter. However, problems in estimating the effects of background radiation have prevented the SAGE physicists from coming up with a reliable measurement of the detected neutrinos.

The GALLEX detector, like SAGE, uses gallium to detect low-energy neutrinos. The data published in July 1992 revealed that GALLEX has found about 60 percent of the expected number of neutrinos. GALLEX thus comes closer to the predicted number than do the other detectors, but it fails to account for all the missing neutrinos.

Scientists have proposed two explanations for these puzzling results. One theory holds that the center of the sun is about 10 percent cooler than the current estimate of about 15,000,000 °C (27,000,000 °F). If that is the case, the sun would produce fewer of the higher-energy neutrinos observable by all the detectors, while leaving the pp neutrinos recorded only by GALLEX unaffected. But if this explanation is correct, scientists' understanding of what goes on inside the sun and other stars will require major revision.

The other proposed explanation is that the sun's neutrinos are being converted to another type of neutrino before they reach Earth. There are two other types of neutrinos, the mu and tau neutrinos, both of which are currently difficult to detect. Scientists believe that such a reaction, if indeed it occurs, would affect primarily high-energy neutrinos. That would explain why the GALLEX data come closer to the predicted number of neutrinos than do data from the Homestake and Kamiokande detectors.

Particle physicists find the second explanation more interesting, because it would require neutrinos to have a tiny amount of mass. So far, there is no evidence that neutrinos have any mass at all, but neither is there any reason to believe that their masses are exactly zero. Further data from GALLEX or SAGE—or from more sensitive detectors now under construction or on the drawing boards—may help solve the mystery.

[Robert H. March]

In WORLD BOOK, see NEUTRINO; PARTICLE PHYSICS; QUARK.

Psychology

In several nations, the number of people who experience major depression has risen in succeeding generations since 1915, according to a December 1992 report. Major depression is a psychological disorder involving recurring periods during which a person feels helpless, hopeless, apathetic, intensely sad, and, frequently, suicidal.

Psychiatrist Myrna M. Weissman of Columbia University in New York City organized the study, which compared rates of major depression previously uncovered in nine independent surveys of 43,000 people. The study participants lived in Canada, France, Germany, Italy, Lebanon, New Zealand, Taiwan, Puerto Rico, and the United States. Weissman found that some populations have undergone "peaks and valleys" in the rate of depression that may reflect the influence of war, economic hardship, and other local conditions, the report noted.

Suicide risk in adolescents. Suicidal behavior is not uncommon among teenagers. And those who exhibit highly aggressive behaviors are more likely to think about and attempt suicide than are their less aggressive peers. Epidemiologist Carol Z. Garrison of the University of South Carolina in Columbia and her colleagues reported this finding in February 1993 after analyzing questionnaire responses of 3,764 public high school students in South Carolina.

Students in grades 9 through 12 answered 70 questions regarding their behavior and possible suicidal thoughts during the year prior to the survey. Seventy-five percent of the students said they had not thought of or attempted suicide. Eleven percent reported having had serious suicidal thoughts, and 7.5 percent acknowledged making a suicide attempt. Teens who engaged in the most aggressive behavior, such as carrying weapons and regularly getting into fights, most often thought about, planned for, and attempted suicide.

Babies sum it up. Babies as young as 5 months of age may be able to add and subtract small numbers of items, according to a study reported in August 1992. Psychologist Karen Wynn of the University of Arizona in Tucson said her research with infants indicates that "humans are innately endowed with arithmetical skills."

Wynn's experiment was based on the established knowledge that babies usually gaze longer at new or unexpected objects than at familiar objects. In the study, 32 normal 5-month-old babies were divided into two groups. One group of babies saw a researcher place a rubber Mickey Mouse doll on a table and put a screen in front of the toy. The babies then saw another doll placed behind the screen. Then the screen was removed to reveal either one doll or two dolls. The babies who saw one doll looked at it considerably longer than did the babies who saw two dolls.

The researchers showed the other group of babies a similar sequence of events, but items were removed from the table. The scientists noticed the same reaction in the babies. Another set of trials found that babies also looked longer when the final number of dolls in the first experiment was three rather than two.

The infants' prolonged gaze at the unexpected number of dolls indicated that they had anticipated a different outcome, and thus, had already added or subtracted on some simple, unconscious level, according to Wynn. But infants' counting ability probably does not extend much beyond the amounts tested in these studies, she said.

Anxiety over surgery. People who feel anxious before undergoing surgery maintain stable levels of two key stress hormones after their operation. In contrast, surgical patients who use relaxation techniques to get rid of their presurgery anxiety later display a surge of the same hormones that may interfere with recovery from their operation, according to a June 1992 report.

Psychologist Anne Manyande of University College in London and her colleagues reported that their findings suggest that worrying may sometimes provide a protective mechanism when people must deal with a stressful event. And distracting oneself from such mental preparations for a stressful experience may, in fact, backfire.

The researchers measured in 40 patients levels of the hormones adrenaline

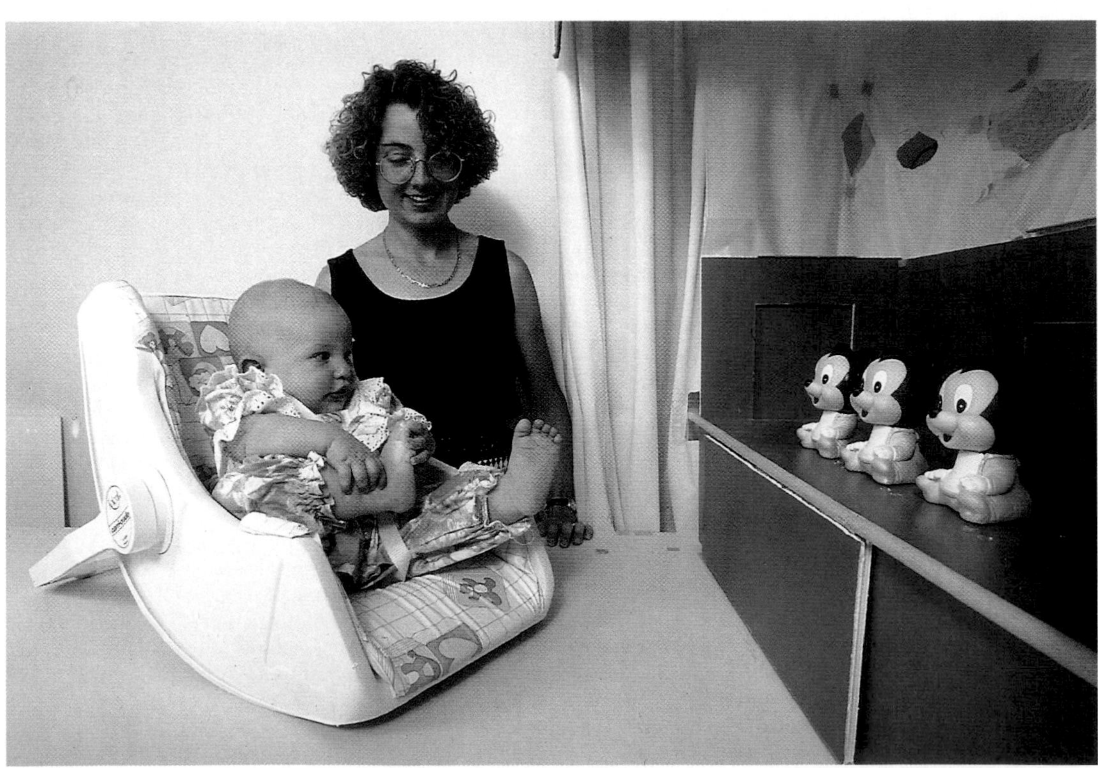

A baby taking a long look at three dolls used in an experiment may indicate that the infant possesses rudimentary arithmetical skills, according to an August 1992 report from a researcher at the University of Arizona in Tucson. In the study, infants were shown dolls being placed behind a screen. When the screen was removed, the babies tended to stare at the dolls longer—indicating surprise—if the researchers had added or removed a doll out of the babies' sight.

and cortisol, which the body releases in response to stress and danger. On the day before their operations, such as ulcer repair, 21 patients listened to a cassette tape describing mental strategies to reduce muscle tension. Another 19 surgical patients heard tapes about the hospital and its staff. The researchers measured the level of hormones before and after the patients listened to the tapes, just before and just after surgery, and two days following surgery.

Levels of adrenaline and cortisol increased substantially during surgery and in the following two days among only those who heard relaxation tapes. Previous studies found that this biological reaction fosters weight loss, fatigue, and a weakened immune system. Such consequences may offset the advantages noted by the relaxation group who, in comparison with the other patients, had less anxiety before and after surgery, needed fewer painkillers, and had lower heart rate and blood pressure following the operation.

Schizophrenia and the brain. The failure of brain cells to settle in appropriate layers during fetal development may lie at the root of schizophrenia, a severe mental disease characterized by disruption of thought and emotion. Researchers reported this finding in March 1993 after studying the distribution of a particular class of nerve cells in the brains of seven schizophrenic adults and seven nonschizophrenic people shortly after their deaths. Normally, these nerve cells move up from deeper layers in the brain to areas near the surface of the cortex (the outer portion of most of the brain) as a fetus develops in the womb. But in the schizophrenics' brains, these cells were found to have congregated in the deeper layers.

The cells, which may help establish important connections between various brain structures, either failed to move to their final destinations or underwent excessive pruning during fetal development, according to the researchers. These abnormal outcomes may result from hereditary factors combined with fetal exposure to viral infections, the investigators said. [Bruce Bower]

In WORLD BOOK, see PSYCHOLOGY.

Public Health

During the first three months of 1993, the number of newly reported AIDS cases increased by about 200 percent, according to the United States Centers for Disease Control and Prevention (CDC) in Atlanta, Ga. The jump largely reflected a revision in the official definition of AIDS that became effective January 1.

The CDC's expanded definition of AIDS takes into account the number of CD4+ T-lymphocyte cells in the blood. These infection-fighting white blood cells are destroyed by the human immunodeficiency virus (HIV), which causes AIDS. Levels indicate how well a person's immune system is functioning. According to the new definition, a person who is infected with HIV and has no more than 200 CD4 cells per microliter of blood would be considered to have AIDS.

The definition also added three conditions that, along with HIV infection, would warrant a diagnosis of AIDS. Those conditions are pulmonary tuberculosis, recurrent pneumonia, and invasive cervical cancer.

CDC officials predicted that because of the new AIDS definition, the number of newly reported cases would increase from about 47,000 in 1992 to at least 90,000 in 1993. Such increases would probably not continue beyond 1993, after the preexisting cases meeting the new criteria had been reported, public health experts noted.

Contaminated food. U.S. President Bill Clinton in March 1993 proposed an overhaul of the nation's meat inspection system following four deaths and more than 500 illnesses caused by eating tainted meat. Public health investigators said the meat was contaminated with a strain of *Escherichia coli* bacteria that can cause diarrhea, colon inflammation and hemorrhaging, kidney failure, and death.

Officials from the CDC and the Department of Agriculture traced the cause of the illness to hamburgers served at fast-food restaurants in four Western states. CDC officials said that the ground beef used in the hamburgers was contaminated before it reached the restaurants. However, thorough cooking normally kills *E. coli*.

A study reported in May 1993 caused

further concern about food safety. In this study, CDC researchers found that 23 people had been sickened by *E. coli* in fresh apple cider made at a small cider mill that did not wash its apples. Since *E. coli* cannot be detected by taste, smell, or food appearance, health experts called for a national plan for *E. coli* testing. They advised consumers to thoroughly cook ground beef and not to drink fresh, unpreserved apple cider made from unwashed apples.

More children need vaccinations. Although 97 to 98 percent of U.S. children are immunized before or shortly after entering grade school, only 10 to 42 percent of 2-year-olds have received all their recommended vaccinations, said an April 1993 CDC report. The lack of vaccinations puts young children at risk for developing and transmitting such diseases as measles, polio, mumps, rubella, pertussis, and diphtheria.

Only 52 to 71 percent of 2-year-olds have received measles inoculations, the CDC noted. And the measles epidemic of the early 1990's signals the need for changes in the health care system's methods of delivering immunizations.

War's effect on public health. Infants and children in war-torn nations are at particularly high risk of death, according to public health reports in 1992. The United Nations Children's Fund reported in December 1992 that in Baidoa, Somalia, a city plagued by drought and by a civil war that has raged since 1990, children under age 5 face a greater risk of dying than do adults.

Between April 3, 1992, and Nov. 21, 1992, 17 in every 10,000 people of all ages died each day in Baidoa. But for children under the age of 5, the number of deaths was 30 per 10,000 people. Fifty-six percent of the children's deaths were caused by diarrhea, 23 percent by measles, and the remainder by other causes. By November, only 9 percent of Baidoa's population was under 5 years of age. In contrast, children in this age group comprise 20 percent to 25 percent of populations in most developing nations.

In September 1992, researchers led by epidemiologists at the Harvard School of Public Health in Boston reported

A researcher at the University of California at Irvine displays a photograph in July 1992 of previously unidentified viral particles he linked to AIDS-like illnesses in two patients. Other doctors also reported AIDS-like symptoms in patients not infected with HIV (the virus that causes AIDS). The announcements created widespread concern until a scientific review of the cases published in February 1993 found no evidence for a new AIDS virus.

An upswing in the number of new cases of AIDS reported in the United States was expected after public health officials broadened their definition of the disease in 1993. AIDS experts predicted that about twice as many new AIDS cases would be reported in 1993 as had been recorded for each of the previous several years.

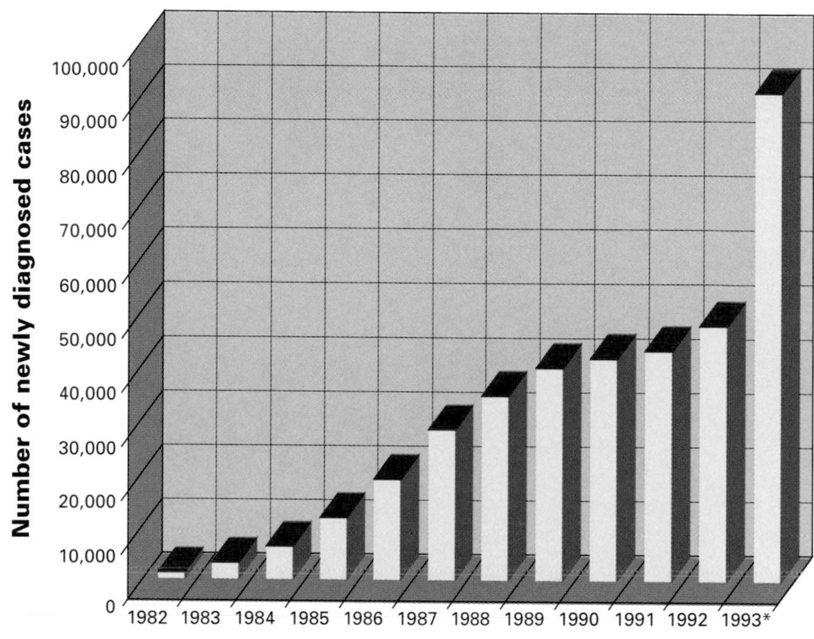

that the Persian Gulf War (1991) and the trade sanctions imposed on Iraq caused a threefold increase in deaths among Iraqi children under 5 years of age. The researchers estimated that between January and August 1991, the deaths of more than 46,900 children were related to the war and its aftermath. The increased rates of disease and death were probably due to the loss of electricity that powers sewerage, refrigeration, and medical facilities, as well as to the decreased access to health care and food supplies, the researchers said.

Children in Armenia have also been hard hit, health officials say. A "drastic reduction" in food, fuel, electricity, and health care resources have occurred in Armenia since the 1988 imposition of an economic blockade, according to a February 1993 report from epidemiologists at the Armenian Institute of Public Health. And during 1992, 5.3 percent of Armenian infants and children had signs of "wasting," visible as significantly lower-than-normal height and weight for their age.

Hurricane Andrew's toll. With fierce winds, Hurricane Andrew hit the southeastern coast of Florida on Aug. 24, 1992. As the storm moved across the state, 2.5 million Florida residents lost electric power, and 56,000 homes were destroyed or severely damaged, according to a September 1992 report from the CDC. Medical examiners in eight Florida counties reported 32 deaths related to the storm.

Some of the people who died had refused to evacuate their homes. Fourteen deaths were directly associated with the hurricane, mainly the result of blunt trauma or suffocation when dwellings collapsed. The other 18 deaths were indirectly related to the hurricane. Eleven of those were from natural causes, such as heart attacks brought on by stress, and 5 deaths resulted from accidents during clean-up efforts or falls from damaged buildings. The causes of 2 deaths were unknown. (See also METEOROLOGY.) [Richard A. Goodman and Deborah Kowal]

In WORLD BOOK, see AIDS; PUBLIC HEALTH; IMMUNIZATION.

Winners in the 52nd annual Westinghouse Science Talent Search were announced on March 8, 1993, and winners of the 44th annual International Science and Engineering Fair were named on May 14. Science Service, a nonprofit organization in Washington, D.C., conducted both competitions. Other science student competitions included international olympiads in chemistry, mathematics, and physics, all held in July 1992.

Science Talent Search. Forty finalists in the Westinghouse competition were chosen from 1,662 entrants from high schools throughout the United States. The top 10 finalists received scholarships totaling $175,000 from the Westinghouse Electric Corporation of Pittsburgh, Pa.

First place and a $40,000 scholarship went to Elizabeth Michele Pine of the Illinois Mathematics and Science Academy in Aurora. For her project, Pine examined the relationship between two groups of fungi, mushrooms and false-truffles, which appear to be unrelated because of their physical dissimilarities. She compared microscopic *spores* (the cells that enable fungi to reproduce) from a false-truffle with spores from a genus of mushrooms by studying their DNA (deoxyribonucleic acid), the basic unit of heredity. She found that false-truffles were closely related to the mushroom genus and should probably be reclassified as belonging to that genus.

Second place and a $30,000 scholarship went to Xanthi M. Merlo of Washington Park High School in Racine, Wis. In a series of experiments, Merlo demonstrated that a recently discovered protein known as protein Z plays a role in blood clotting.

Third place and a $20,000 scholarship were awarded to Lenhard Lee Ng of Chapel Hill (N.C.) High School. Ng used mathematics to study the probability of accurately determining a grocery bill by rounding off the price of each food item to the nearest dollar.

Fourth place and a $15,000 scholarship went to Constance Chen of La Jolla (Calif.) High School.

Fifth place and a $15,000 scholarship were awarded to Ryan Egeland of Wayzata (Minn.) Senior High School.

Sixth place and a $15,000 scholarship were awarded to Wei-Hwa Huang of Montgomery Blair High School in Silver Spring, Md.

Seventh place and a $10,000 scholarship went to Mahesh Mahanthappa of Fairview High School in Boulder, Colo.

Eighth place and a $10,000 scholarship went to Steve Chien of Montgomery Blair High School.

Ninth place and a $10,000 scholarship were awarded to Elizabeth Mann of Montgomery Blair High School.

Tenth place and a $10,000 scholarship went to Zachary Freyberg of Midwood High School at Brooklyn College in New York City.

Science fair. The 44th annual International Science and Engineering Fair took place in Mississippi Beach, Miss., from May 9 to 15, 1993. The 826 participants were selected from finalists at high school science fairs throughout the United States, Puerto Rico, Guam, and several foreign countries. The four top awards recognized the most outstanding scientific accomplishments from among the fair's First Award winners.

Two of those top awards were all-expense-paid trips to the Nobel Prize ceremonies in Stockholm, Sweden, in December 1993. The other two top awards were all-expense-paid trips to the European Community Contest for Young Scientists in Berlin, Germany, in September 1993.

Selected for the Glenn T. Seaborg Nobel Prize Visit Award were Lana Israel of North Miami Beach (Fla.) Senior High School, and Mahesh Mahanthappa of Fairview High School in Boulder. Selected to represent the United States as guest observers at the European Community Contest were Robert Knorr of Sawyer (N.D.) Public School, and Lea Potts of Mount Vernon High School in Alexandria, Va.

The First Award winners of $500 each in the 13 disciplinary categories were:

Behavioral and social sciences. Ivan Retzignac of Atlantic Community High School in Delray Beach, Fla., and Lana Israel of North Miami Beach Senior High School.

Biochemistry. Kristofer Thiessen of Athens (Ala.) High School, and Ravi Kamath of Shawnee Mission East High School in Prairie Village, Kans.

Botany. Robert Knorr of Sawyer (N.D.) Public School; Justin Thornton of Gold Beach (Ore.) High School; and

Dianella Howarth of Moanalua High School in Honolulu, Hawaii.

Chemistry. Laura Becvar of Coronado High School in El Paso, and Lea Potts of Mount Vernon High School in Alexandria, Va.

Computer science. Sarita James of Homestead High School in Fort Wayne, Ind.

Earth and space sciences. Paige Provenzano of Chamberlain High School in Tampa, Fla.

Engineering. Stephen Barton of Central High School in Brooksville, Fla.; Christopher Monaghan of Winter Park (Fla.) High School; Leland Piveral of Central High School in St. Joseph, Mo.; and Jeff Smith of Sylvania (Ohio) Southview High School.

Environmental sciences. Aaron Aycock of Lake Gibson High School in Lakeland, Fla.; Karin Wilhelm of Stuart-Menlo Community School in Stuart, Iowa; Dotty Hammersley of Tuscarawas Valley High School in Zoarville, Ohio; and George Lee of Mission San Jose High School in Fremont, Calif.

Mathematics. Jason Sanders of Ely High School in Pompano Beach, Fla., and Mahesh Mahanthappa of Fairview High School in Boulder.

Medicine and health. Michael Weiner of Tom Clark High School in San Antonio; Christopher Nelson, Jr., of Lakewood High School in St. Petersburg, Fla.; and Jyothi Vinnakota of Nicolet High School in Glendale, Wis.

Microbiology. Carlo Contreras of Jefferson High School in Lafayette, Ind., and Jennifer Bullard of Lakeside High School in Atlanta, Ga.

Physics. Fred Niell of Lausanne Collegiate School in Memphis, and Daniel Stevenson of Hudson (Ohio) High School.

Zoology. Kimberly Hanisak of Vero Beach (Fla.) Senior High School; Nathan Terracio of Heathwood Hall Episcopal School in Columbia, S.C.; and Freddy Medina-Colon of Carvin School in Carolina, Puerto Rico.

Chemistry Olympiad. The 24th International Chemistry Olympiad was held for the first time in the United States. The competition was conducted in Pittsburgh. Teams from 34 nations

Elizabeth Michele Pine of Chicago, center, winner of the top prize in the Westinghouse Science Talent Search in March 1993, celebrates her scholarship award with second- and third-place runners-up Xanthi M. Merlo of Racine, Wis., right, and Lenhard Lee Ng of Chapel Hill, N.C.

competed by taking a five-hour written examination and a five-hour test of practical laboratory skills. China placed first, Hungary second, Poland third, and the United States fourth.

The top-scoring American on the four-member U.S. team was Swaine Chen of O'Fallon (Ill.) Township High School, who won a gold medal and placed 15th among the 136 contestants. Silver medals went to Jeffrey Chuang of Bellaire High School in Houston, and Logan McCarty of Amherst (N.Y.) Central High School. A bronze medal was awarded to Christopher Herzog of Highland Park (N.J.) High School.

Math Olympiad. The 33rd annual International Mathematical Olympiad was held in Moscow. Teams from 52 nations competed by working on six challenging math problems. First, second, and third place went to China, the United States, and Romania, respectively.

Every member of the six-member team from the United States won a medal. Gold medals went to Kiran Kedlaya of Georgetown Day High School in Washington, D.C.; Robert Kleinberg of Iroquois Central High School in Elma, N.Y.; and Lenhard Ng of Chapel Hill (N.C.) High School. Silver medals were awarded to Wei-Hwa Huang of Montgomery Blair High School in Silver Spring, Md.; Sergey Levin of Classical High School in Providence, R.I.; and Andrew Schultz of Evanston (Ill.) High School.

Physics Olympiad. The 23rd annual International Physics Olympiad was held in Helsinki, Finland, and drew participants from 37 nations. Five-member teams competed by testing their experimental skills and understanding of theoretical physics. Every member of the U.S. team won an award. Eric Miller of San Francisco University High School won a gold medal, and he placed 5th among the 177 competitors. Szymon Rusinkiewicz of Bellaire High School in Houston also won a gold medal. A silver medal went to Michael Schulz of Baldwin (N.Y.) High School. Carwil James of Hawken School in East Cleveland, Ohio, and Dean Jens of Ankeny (Iowa) High School received honorable mention awards. [Rod Such]

Space Technology

The world's space agencies continued to struggle with the enormous costs of space projects in 1992 and 1993. The end of the Cold War and the breakup of the Soviet Union in 1991 offered these agencies both challenges and opportunities. No longer would nations sponsor expensive projects solely to demonstrate their technological prowess. But at the same time, the barriers to cooperation between former rivals had come down.

Space station redesigned. The space station planned by the United States National Aeronautics and Space Administration (NASA), to be called Freedom, came under fire in early 1993. The station, which carried a $31-billion price tag for construction costs alone, was first proposed in 1984. In March 1993, the Administration of President Bill Clinton ordered NASA to redesign Freedom, making it smaller and less costly to build and operate.

In early June, the Administration accepted a design for a scaled-down Freedom that would use many components already being developed for the station. Officials said the redesigned station would cost $18 billion through the year 2000 and could be in operation by then.

On June 23, the station narrowly survived a crucial vote in the U.S. House of Representatives. Members rejected a bid to cancel the project by a single vote.

Joint projects increasing. In June 1992, U.S. President George Bush and Russian President Boris Yeltsin signed an accord expanding the two nations' cooperation in space. Under the accord, a Russian cosmonaut would take part in a U.S. shuttle flight in autumn 1993, and an astronaut would serve aboard Russia's Mir space station in 1995.

The international nature of space programs was illustrated dramatically during 1992. In August, space travelers born in six different nations were in orbit simultaneously, with astronauts representing Costa Rica, Italy, Switzerland, and the United States aboard Atlantis and cosmonauts representing Russia and France on Mir. Soon afterward, a Swedish spacecraft carrying Canadian, German, and American instruments was placed in orbit by a Chinese launch vehicle, while a U.S. military experiment was carried aboard a Russian rocket.

Tether trouble. An experiment with a space tether—one of the most intriguing U.S. space shuttle missions ever—ended in only partial success in August 1992. The mission began when the shuttle Atlantis went aloft on July 31, 1992, carrying the Italian-built Tethered Satellite System (TSS), a spherical satellite that was to have been unreeled at the end of a thin cable 20 kilometers (12.4 miles) long. The exercise, a joint project of NASA and the Italian Space Agency, was to have tested the applications of tethers in space.

The mission experienced problems from the start. Before tackling the tether, Atlantis' seven-member crew released a European Space Agency satellite called Eureca, which failed to enter its intended orbit. Ground controllers in Darmstadt, Germany, eventually boosted the satellite to the proper orbit.

After releasing Eureca, the Atlantis crew turned its attention to the TSS. But the tether, a copper and fiber cable only 2.5 millimeters (0.1 inch) thick, jammed repeatedly before the astronauts gave up trying to unreel it. The tether was never extended beyond 256 meters (840 feet). Investigators later determined that a protruding bolt had interfered with the unwinding mechanism.

In theory, tethers in space could serve a number of purposes. They might be used to move cargo to and from a space station or to lower scientific instruments into regions of the atmosphere that are too close to Earth for satellites. A tether linking one spacecraft to another at a lower, faster orbit would act like a slingshot, so tethers could be used to boost satellites into higher orbits. Tethers could even be used to swing two spacecraft in loops to create artificial gravity.

Tethers could also generate electricity. Dragging an electrically conductive cable through Earth's magnetic field would create an electric current, which could be used to power spacecraft. In the case of the TSS, scientists expected up to 5,000 volts of electricity to shoot down the cable to the shuttle. The partially unreeled tether generated less than 40 volts.

Although scientists were disappointed

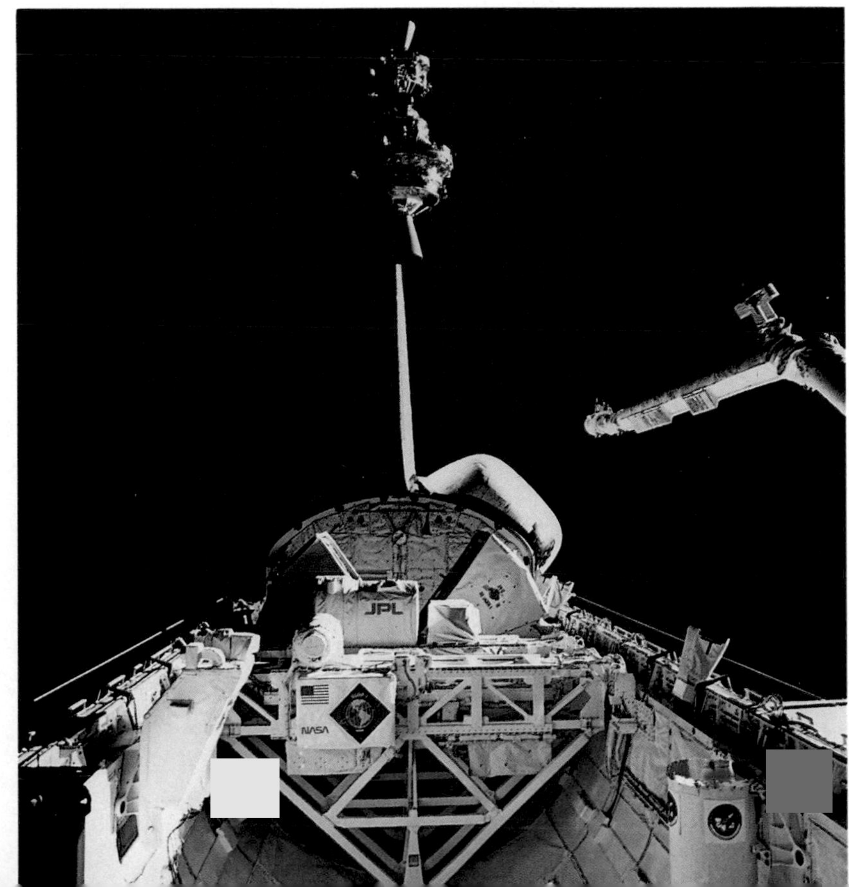

Lageos 2 (Laser Geodynamics Satellite 2) perches on a booster rocket during its release from the space shuttle Columbia in October 1992. The satellite—a 400-kilogram (900-pound) brass and aluminum ball only 60 centimeters (2 feet) in diameter—is covered with more than 400 reflective surfaces. By beaming light beams at the surfaces and measuring how long it takes for them to be reflected back, scientists can track the movement of crustal plates and other geological features on Earth.

by the TSS results, they were encouraged by some aspects of the exercise. They noted, for example, that the crew had been able to keep the partially extended tether from becoming tangled.

In April 1993, scientists successfully unreeled another 20-kilometer (12.4-mile) tether, this one made wholly of plastic fiber. The cord, connected to a small aluminum box, was unwound from an unmanned Delta II rocket. The tether was then cut, and the box and cord burned up as they entered the atmosphere.

New shuttle record. Columbia, the U.S. shuttle fleet's oldest member, was launched in June 1992 on NASA's longest shuttle mission to date. Extra tanks of hydrogen and oxygen, which provide fuel for generating electricity on the shuttle, enabled Columbia to stay in orbit almost 14 days. The previous shuttle record, just short of 11 days in January 1990, had also been set by Columbia.

During the mission, called the U.S. Microgravity Laboratory-1, astronauts performed experiments in the biological and physical sciences. The research was conducted chiefly in the Spacelab module, a laboratory carried in the shuttle's cargo bay. The crew grew crystals for use in electronics and medical research. They also ignited various materials to study the properties of fire and fire safety in space.

Other Columbia missions. In October 1992, Columbia carrried an Italian satellite, Lageos 2 (Laser Geodynamics Satellite 2), into orbit. The spherical satellite is covered with more than 400 reflective surfaces. By beaming lasers at the surfaces and measuring how long they take to be reflected back, geologists on Earth can precisely track the movement of Earth's *tectonic plates* (the rigid plates that make up the planet's crust).

Columbia flew again in April 1993, in the second shuttle mission chartered by Germany. (The first was in 1985.) These missions marked the only cases in which shuttle payload operations were controlled from outside the United States.

Endeavour, NASA's newest shuttle, flew two missions in late 1992 and early 1993. The shuttle was launched on its second flight in September 1992, in

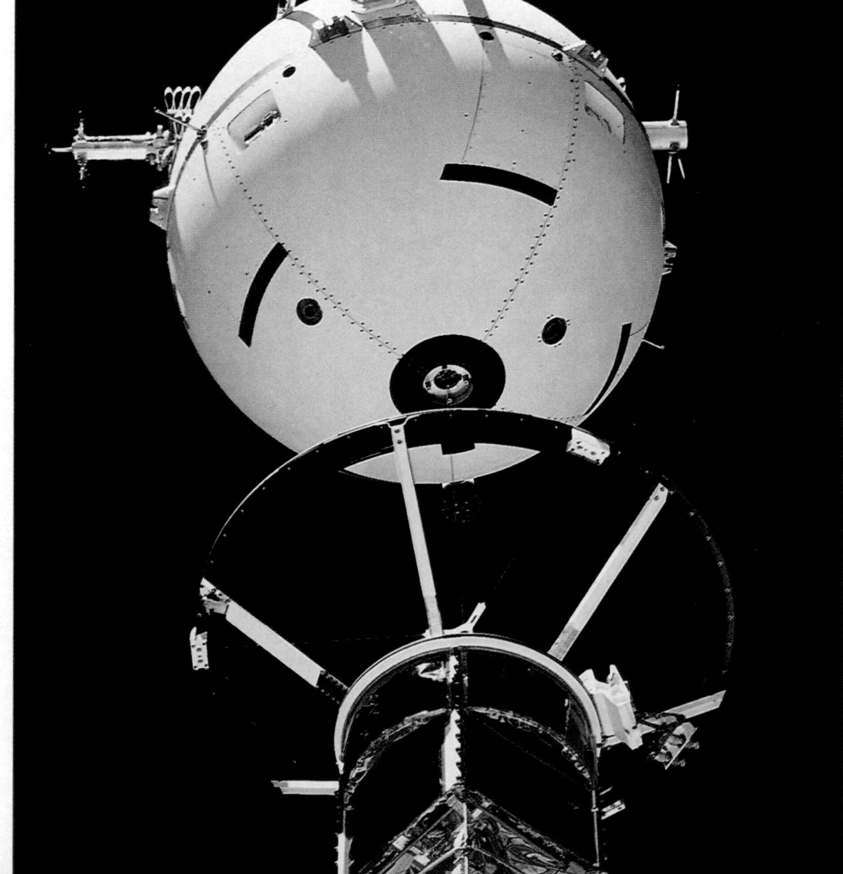

A spherical satellite is reeled out from the cargo bay of the U.S. space shuttle Atlantis by a cable thinner than a clothesline in a test of the Tethered Satellite System (TSS) in August 1992. The exercise was an eagerly awaited test of some possible uses of tethers in space, which could include moving cargo, generating electricity, and boosting satellites. The mission was less than successful, however, because the Atlantis crew was able to unwind the SST tether to only about 256 meters (840 feet), rather than the planned 20 kilometers (12.4 miles).

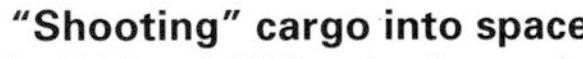

"Shooting" cargo into space

In 1992 and 1993, scientists at the Lawrence Livermore National Laboratory in Livermore, Calif., prepared to test a "supergun" designed to launch materials into space at low cost. For the test, the gun will be aimed at a target on the ground, *below.* Eventually, it will be pointed at the sky, *right.* In such a launch, a mixture of methane and air will be ignited in one end of a hydrogen-filled pipe called a pump tube. The explosion will drive a piston toward the launch tube, compressing the hydrogen. The compressed gas will then push the projectile through the launch tube at high speed.

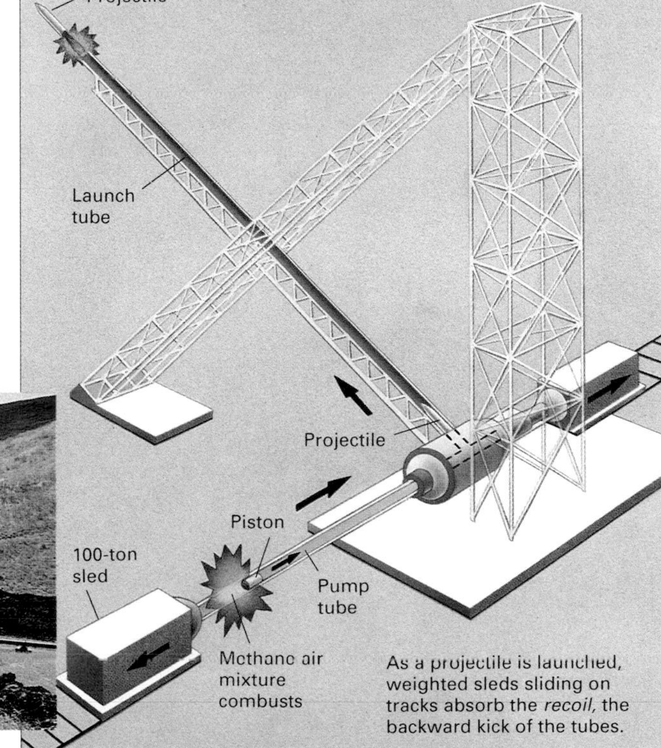

As a projectile is launched, weighted sleds sliding on tracks absorb the *recoil,* the backward kick of the tubes.

what was the first joint venture between NASA and the Japanese space agency. The seven-member crew included chemist Mamoru Mohri, the first Japanese astronaut to fly in space, and U.S. astronaut and physician Mae C. Jemison, the first black woman to do so. The astronauts performed materials science and life sciences research in the Spacelab.

In January 1993, Endeavour released a large data relay satellite, the sixth in NASA's series of Tracking and Data Relay Satellites. The satellites transmit data to Earth from the shuttles and from orbiting spacecraft. During the mission, two astronauts performed a four-hour spacewalk in the shuttle's cargo bay. The astronauts carried each other, lifted massive objects, and moved about the bay as they practiced techniques to be used in construction of the planned space station. Endeavour also carried an X-ray spectrometer, an instrument designed to measure the chemical composition of gases in space.

Discovery missions. Discovery went aloft in December 1992 on what was expected to be the shuttle fleet's last military mission. The U.S. Department of Defense had said it would launch future missions on unmanned rockets. The five-member crew deployed a classified defense satellite, then tested a laser receiver. The crew failed in an attempt to release a set of small metal balls into orbit, however. The balls were to have been used to help researchers improve the use of radar to track debris in space.

Discovery flew again in April 1993. On this mission, called Atlas 2, it carried instruments for studying the atmosphere and sun. Atlas 2 was the second of 11 missions designed to be flown, one per year, over a full 11-year solar cycle. In a solar cycle, the number of *sunspots* (dark spots on the sun's surface) rises and falls as the sun goes through alternate periods of turbulence and relative calm.

Mission to Mars. The Mars Observer, the first U.S. spacecraft sent to Mars since 1975, was launched from Cape Canaveral, Florida, by a Titan III rocket on Sept. 16, 1992. The spacecraft was to reach the red planet in August 1993 and enter an orbit over the Martian poles. It was to circle the planet for a full Martian year (687 Earth days), mapping its surface and collecting data on its atmosphere and weather.

Other space probes and telescopes yielded significant results in 1992 and 1993. The Hubble Space Telescope—an orbiting observatory launched in April 1990—continued to produce important data, despite a flaw in its main mirror. (In the Special Reports section, see FIXING HUBBLE'S TROUBLES.)

A second orbiting spacecraft, the Compton Gamma Ray Observatory, continued to monitor *gamma ray bursters,* brief bursts of gamma rays whose origins are one of the mysteries of astronomy. In April 1993, researchers studying data from the observatory announced that these bursts are more energetic than previously thought and appear to originate far beyond the Milky Way.

In May 1993, scientists announced that data from two American spacecraft, Voyager 1 and Voyager 2, had provided the first physical evidence of the boundary of the solar system. The spacecraft, which were launched in 1977, began detecting low-frequency radio signals in August 1992. In their May 1993 report, the scientists said that the signals were created when bursts of particles from the sun, known as solar wind, collided with cold gases in the *heliopause,* the region at the solar system's edge.

Russian space mirror. Despite Russia's huge economic and political problems, that nation kept its space program going in 1992 and 1993, including operations on the space station Mir. In February 1993, Russian scientists successfully deployed a "space mirror"—a thin, aluminum-coated plastic film some 20 meters (66 feet) in diameter. The mirror, which was unfurled from a cargo spacecraft attached to Mir, briefly sent a narrow beam of reflected sunlight across the Atlantic Ocean and Russia. In theory, such mirrors could light up cities at night and extend planting and harvesting seasons. More important, the exercise tested the concept of a "solar sail," in which *photons* (particles of light) striking the unfurled film could be used to propel spacecraft. [James R. Asker]

In WORLD BOOK, see SPACE EXPLORATION; SPACE TRAVEL.

Zoology

Ecologists have solved a puzzle about how a lizard called the house gecko has been able to take over the home territory of a closely related lizard, the mourning gecko. The house gecko is larger than the mourning gecko, and it simply scares the smaller native lizard away from food, according to a January 1993 report by scientists at the University of California, San Diego.

In the 1940's, house geckos began arriving on South Pacific islands, accidental stowaways on planes and boats from the Philippines and Indonesia. Scientists thought the house geckos would not threaten mourning geckos, because the newcomers had a disadvantage. To reproduce, they needed to find mates. The native species, on the other hand, are all female, and they reproduce without mating. But surprisingly, the scientists found that the new geckos quickly outnumbered the native lizards.

Both species easily enter buildings, attracted by insects swarming around electric lights at night. The California scientists simulated this situation by putting one or both species of gecko in 18 abandoned aircraft hangars at a naval air station on Hawaii. In half the hangars, a light was turned on, which attracted insects that geckos eat.

In the lighted buildings containing both species of geckos, the house geckos grew, while the mourning geckos wasted away. The researchers found that the house geckos tended to crowd around the light, eating the insects, and the mourning geckos stayed away. In the hangars without lights, both species thrived. Thus, the researchers said, well-lit buildings have accidentally given the invading lizards an advantage in overcoming the native species.

Foul-tasting fowl. Scientists have long known that some reptiles, frogs, fish, and insects produce or store terrible-tasting chemicals and poisons in their bodies, which prevents predators from eating them. In October 1992, an ecologist at the University of Chicago and his colleagues reported discovering the first birds known to be poisonous, three species of birds of the genus *Pitohui.* All three species—the hooded pitohui, the rusty pitohui, and the variable

The hooded pitohui's brightly colored feathers are poisonous, as are the bird's skin and some internal organs, according to a University of Chicago scientist who discovered the first known poisonous species of birds in 1992. The species include the hooded pitohui, rusty pitohui, and variable pitohui, all of which are native to New Guinea. The researcher made his discovery after he felt numbness and pain due to handling the birds.

pitohui—have a bitter substance called an alkaloid in their feathers, skin, and some internal organs. And all three species live in New Guinea.

Ecologist John P. Dumbacher made the discovery after he handled one of these birds for a research study and later felt a numbness in his lips and pain in his hands. He sent bird tissue to biologist John W. Daly and his colleagues at the Laboratory of Bioorganic Chemistry at the National Institutes of Health in Bethesda, Md. They performed chemical analyses and found an alkaloid that resembled a substance found in three poisonous species of the *Phyllobates* genus of frogs that live in South America. The hooded pitohui, a bright orange and black bird, was the most poisonous. Dumbacher has yet to discover if the bird makes the poison or extracts it from something in its diet.

Deadly ocean phantom. Biologists have found out why thousands, and sometimes millions, of fish die in shallow bays for no apparent reason. The culprit, they reported, is a newly discovered type of *dinoflagellate* (a tiny marine organism) that releases poison into the water. Fish pathologist JoAnn M. Burkholder and her colleagues at North Carolina State University in Raleigh reported this finding in July 1992. (See OCEANOGRAPHY.)

A good ear for crickets. Most flies have feathery antennae that detect very slight vibrations in the air so they can hear the buzz of other flies. But female brown flies have evolved special ears to hear cricket chirps, neurobiologists at Cornell University in Ithaca, N.Y., reported in November 1992.

Crickets, primarily the males, sing or chirp to attract mates. But the female brown fly uses the cricket's song to home in on the cricket and deposit maggots on the male cricket's back. The maggots then burrow into the cricket, feeding on it and killing the cricket in the process.

According to the researchers, the ears of the brown fly are at the front of the head. The ears have membranes covering an air-filled chamber containing a pair of organs that sense sound as the membranes vibrate. The male brown fly

Should the Wolf Return to Yellowstone?

Not since the 1920's has the once familiar and eerie howl of the gray wolf been heard anywhere in the vast reaches of Yellowstone National Park. Yellowstone, in northwestern Wyoming, was established in 1872 as America's first national park. As such, it was to be the protected home of all the native plants and animals that were living in the area. Among the native mammals still roaming Yellowstone's 898,000 hectares (2.2 million acres) are coyotes, beavers, cougars, grizzly and black bears, bison, big horn sheep, moose, elk, antelope, mule deer, ground squirrels, chipmunks, and tiny shrews—but not the gray wolf. The wolf was deliberately hunted to extinction in the area.

In the late 1980's and early 1990's, however, conservationists have been trying to bring the wolf back to the park. Their efforts have touched off a heated and deeply felt controversy with ranchers, who fear that the wolves will kill off their livestock.

Many Americans have long regarded the wolf as the nation's most dangerous predator. Even Theodore Roosevelt, who served as President of the United States from 1901 to 1909 and was known for his strong support of wildlife conservation, called wolves "the beast of waste and desolation." Others have called them killing machines.

The average wolf weighs close to 45 kilograms (100 pounds), and its powerful jaws and sharp teeth are capable of ripping open an elk's stomach in seconds or snapping a cow's tail off at the base with a single bite. Wolves hunt in packs of up to 10 animals. Together, they can bring down a large bison or moose with the precision of a trained assault squad.

With these characteristics, the wolf was once the target of livestock owners who wanted to eradicate these predators from the West. Beginning in the late 1800's, ranchers, hunters, and U.S. government agents in the American West shot, trapped, and poisoned wolves by the tens of thousands each year. In 1915, the federal government passed legislation calling for the removal of wolves from all federal lands, including national parks, where animals were usually protected. The law dealt the wolves in Yellowstone a final, fatal blow.

In 1963, federal action toward wolves began to swing in a new direction. A report ordered by Stewart L. Udall, then secretary of interior, recommended that the national parks be restored as much as possible to the ecological state they were in when explorers and settlers first arrived in North America. For Yellowstone in 1963, the only species lacking was the gray wolf. Conservationists began discussing how to return them to the park.

Another step toward recovery for the wolf occurred in 1973, when the government included the gray wolf in the newly enacted Endangered Species Act. The act designated wolves as endangered in every state except Alaska and Minnesota. Further, the act directed the National Park Service to use whatever procedures and methods necessary to restore the gray wolf to habitats such as Yellowstone. Anyone convicted of shooting a wolf could be fined up to $50,000 and face a 5- to 10-year prison sentence, a sharp contrast to the earlier policy of paying hunters for dead wolves. In 1980, the National Park Service approved the first plan to bring the gray wolf back to Yellowstone, but another seven years passed before a plan was formulated that called for the introduction of three mating pairs.

From the outset, wolf restoration plans faced stiff opposition from ranchers near the park, who feared the return of the wolf would mean certain death to their livestock. Hunters, too, opposed the plan, because they believed wolves would kill off favored game, such as elk, deer, and antelope. Other Wyoming residents considered the wolf a threat to people and pet dogs and cats living near the park. After all, they said, wolves would not be able to recognize the park's boundaries. The opponents rallied support from several key members of Congress from Western states and effectively prevented the plan's quickly taking effect.

Conservationists, however, argued that bringing back the wolf would restore an important component of the food chain in the park. Without natural predators such as the gray wolf, populations of elk, deer, and antelope grow unchecked. If they increase beyond the land's ability to support them, hundreds of animals will starve. Indeed, the populations of elk—30,000—and mule deer—3,000—in Yellowstone in 1993 were at record highs. Conservationists also say that, because the wolf preys on weak animals, its presence in the park would actually strengthen the park's herds.

Answering the charge that wolves attack livestock, conservationists pointed out that wolves kill only to eat. Attacks occurred during the late 1800's and early 1900's because settlers had killed off virtually all the region's bison, antelope, elk, and deer. There was little else for the wolf to eat.

Statistical evidence also showed that areas now inhabited by wolves have experienced only minor losses of livestock due to wolf predation. For example, in Minnesota, where 1,500 wolves share the land with more than 230,000 cattle and 20,000 sheep, only about 100 animals are lost to wolves each year. And a conservation group established a fund to compensate ranchers for such losses.

The average gray wolf weighs nearly 45 kilograms (100 pounds), and its powerful jaws and sharp teeth make it an efficient hunter. This hunting ability makes some people fearful of reintroducing the animal into Yellowstone National Park. Conservationists argue that wolves kill only to eat and will not harm livestock in the region.

The government responded to residents' fears by conducting environmental impact studies to find out how reintroducing the wolf would affect livestock and wildlife in and around the park. The first studies, completed in 1990, revealed that introducing wolves in Yellowstone would cause minimal losses to livestock and that the area's plentiful elk and mule deer would be the primary wild prey. Results of the final study were scheduled for January 1994. Depending on the study's outcome, wolves could be introduced within a few months thereafter. The goal of conservationists is to have introduced into Yellowstone 10 breeding pairs, which biologists predict would produce a total of 100 animals in three years.

Ironically, wolves may have returned to Yellowstone on their own. In August 1992, a wildlife film crew in the park photographed what appeared to be a gray wolf standing near two grizzly bears eating a freshly killed bison. Although biologists cannot be sure it was a wild wolf—it could have been a wolf-dog hybrid—the animal behaved like one. For example, it did not fear the grizzlies.

Then, in September, a hunter shot what he thought was a coyote. But upon close examination, it appeared to be a wolf. The U.S. Fish and Wildlife Service (FWS) tested the carcass and confirmed that the animal was indeed a wolf genetically related to wolves living in northern Montana.

Lone wolves have been sighted in or near Yellowstone since the 1960's. A lone wolf can easily travel 113 kilometers (70 miles) in a single day and thus be found long distances from its home territory, according to Ed Bang, wolf project leader at the FWS. For example, a wolf once tagged in Glacier National Park in Montana was later shot 837 kilometers (520 miles) north in Canada. But, even though a lone wolf can reach Yellowstone, Bang said, the chances of a single wolf surviving, finding another single wolf of the opposite sex, and establishing a pack are slim.

In the unlikely event that wolves are returning naturally, an interesting conservation dilemma arises. Any wolf returning on its own is protected by the Endangered Species Act, which means ranchers could not shoot a wolf that wandered outside the park even if it attacked their livestock. Wolves introduced to Yellowstone under the current restoration plan are designated as an "experimental population," which gives ranchers permission to shoot a wolf outside the park if it threatened livestock.

From the wolf's perspective, government protection might not be such a good thing after all. Returning to Yellowstone on its own could be the safest, though longest, route. [Robert C. Holl] has ears that are much smaller and less sensitive to sound than the female's—a rare occurrence in the animal kingdom, according to the researchers. They said the difference was not unexpected, however, because it is the female who must find the singing cricket in order to successfully reproduce. The female fly's ears were about 100 times more sensitive to a male cricket's song than were the female cricket's hearing organs.

Bullied bunnies. Many animals that live in close association with each other set up *pecking orders*—a hierarchy or order of dominance. A physiologist from the University of Bayreuth in Germany reported in July 1992 that rabbits have such a pecking order. He defined a rabbit at the top of the hierarchy as one that chases the others and always gets the right of way. The study found that the dominant rabbits had healthier hearts than the lower ranking rabbits.

The researcher discovered the pecking order of 20 wild European rabbits living in Australia by putting portable heart monitors on them. The monitors detected each animal's heart rate, which indicated how much or how little stress the rabbit was experiencing.

As the bullies chased the less dominant rabbits, the monitors relayed heartbeats to a computer via radio signals. The monitors showed that during chases or other aggressive behaviors, the heart rates of higher and lower ranking rabbits increased by about the same amount. But during calm periods, the heart rates of lower ranking rabbits were on average about 10 percent faster than the heart rates of the bullies.

Then, the scientist removed the bullies from the group, and monitored two rabbits that moved up the pecking order to the dominant positions. The heart rates of the rabbits gradually slowed down by about 8 percent. The study suggested that, in the long run, stress may make lower ranking rabbits less healthy than the bullies.

Playing in the rain. The more it rains, the more young baboons play with infant baboons. This was the finding reported in July 1992 by zoologist Louise Barrett and her colleagues at University College in London.

An organism identified in 1992 as the largest known bacterium dwarfs one-celled animals called paramecia, *below.* At 0.4 millimeters (0.015 inches) long, the bacterium is big enough to be seen by the unaided eye and offers scientists an opportunity to study bacterial cell structure with greater ease.

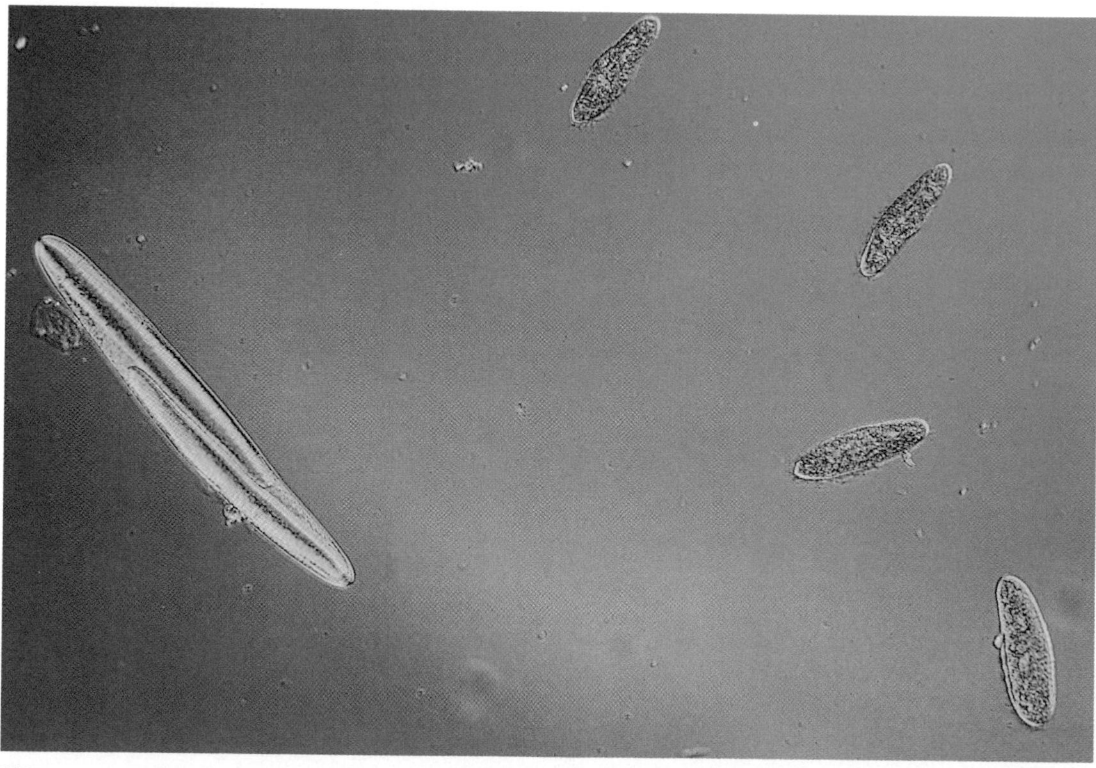

A tiny marmoset, *right*, captured by a Swiss biologist in the central Amazon region of South America in 1985 was identified as a new species of primate in 1992. The adult animal weighs about 370 grams (13 ounces) and is 20 centimeters (8 inches) long.

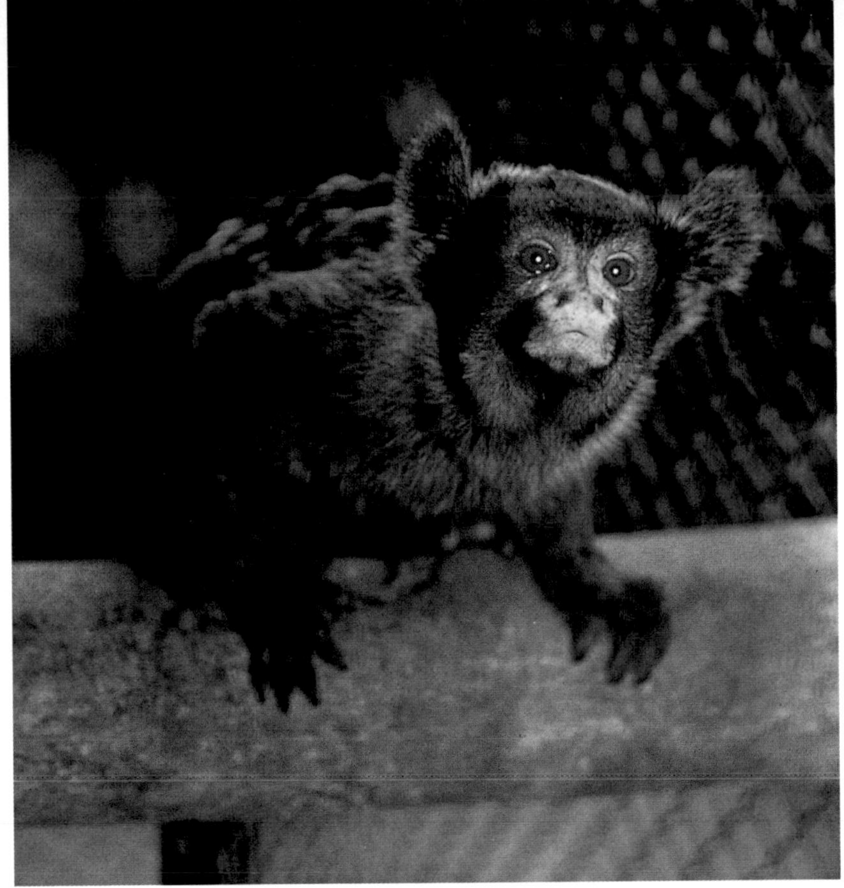

The researchers watched baboons in Ethiopia for many months, keeping track of how often the young baboons played with baby baboons. The scientists also measured the amount of grass available and recorded how much it rained. During wetter months, when grass was more plentiful and the animals thus had more to eat, the baboons tended to wrestle and chase each other in a rough-and-tumble way.

But as grass became scarce, the play changed. The young baboons did not run around as much and just tended to pull each others' tails and pretend they were biting each other.

The research suggests that weather can influence the social experiences of baboons, with consequences for the animals' development. For example, a baboon born during wet weather would probably experience greater interaction with young baboons than a baboon born during a dry season. Their different social experiences could affect what lessons they learn from their environment, and even influence their adult behavior, according to the scientists.

The brighter bird wins. Males of warbler species that live in dense woods in Kashmir, India, have evolved brighter feathers than their relatives that live in more open habitat. The forest dwellers also have developed patches of bright feathers that they display in unique ways to expand their territory in preparation for the mating season, according to a study reported in March 1993. The more prominent the patterns of bright feathers, the more successful the male is at gaining additional territory.

Evolutionary biologist Karen Marchetti of the University of California, Davis, studied the males of some similar species of warblers that breed in the forests and open habitat of Kashmir. All the species are green, but Marchetti found that some species in the dense forests also had bright patches of feathers in specific patterns. For example, one species had a yellow bar on the wings, another species had two yellow wing bars, another had two yellow wing bars plus a stripe on the head, and still another had two wing bars, a head stripe, and a patch on the rump.

A rare debut
A baby komodo dragon, *right,* peers from its leathery egg, one of 26 to hatch at George Mason University in Fairfax, Va., on Sept. 13, 1992. It was the first time komodo dragons, an endangered species of lizard, were bred successfully in captivity outside the animal's native Indonesia.

At 1 month old, the komodo dragon, *left,* was 40 centimeters (16 inches) long.

Marchetti found that in territorial displays the birds used behavior geared to emphasize their particular bright patch. The species with wing bars flashed its wings, the species with a head stripe tilted its head from side to side, and the species with the rump patch turned to flash its rump at a competitor.

To gauge the importance of the prominence of bright patches, Marchetti reduced the size of the wing bars in one species by covering them with green paint—the bird's body color—and enlarged the bars of a second group with yellow paint the same shade as the bar. The birds with the bigger bars always took a slice of the territory of the less prominently colored bird. The results were the same when Marchetti added a second stripe to the heads of some birds: They won more territory.

Defying the nesting odds. Wildlife ecologists now know why a bird called the lapwing builds its nests in the middle of farm fields. Even though the nests get destroyed when farmers plow the soil, the lapwings still fare better in the fields than when they build nests on the ground near trees, where crows and other birds are likely to snatch and eat the eggs. This was the finding that Ake Berg and his colleagues at the Swedish University of Agricultural Sciences in Uppsala reported in August 1992.

Berg and his team found that lapwings lost about 85 percent of their nests because of spring plowing and other farming operations. Despite such losses, the birds persist in building nests in fields, often in groups of up to 28 pairs. The ecologists reported that the birds that lost their nests had time to build a second nest and lay another clutch of eggs. More than half of the second clutches hatched.

Berg concluded that when the birds build nests near trees, the crows living there are a constant threat. On the other hand, once plowing is completed in the spring, the fields are the safest place to raise young. [Elizabeth J. Pennisi and Thomas R. Tobin]

In the Special Reports section, see BEWARE THE EXOTIC INVADERS; THE CASE OF THE MISSING SONGBIRDS. In WORLD BOOK, see ZOOLOGY.

Science You Can Use

In areas selected for their current interest, *Science Year* presents information that the reader as a consumer can use in understanding everyday technology or in making decisions—from buying products to caring for personal health and well-being.

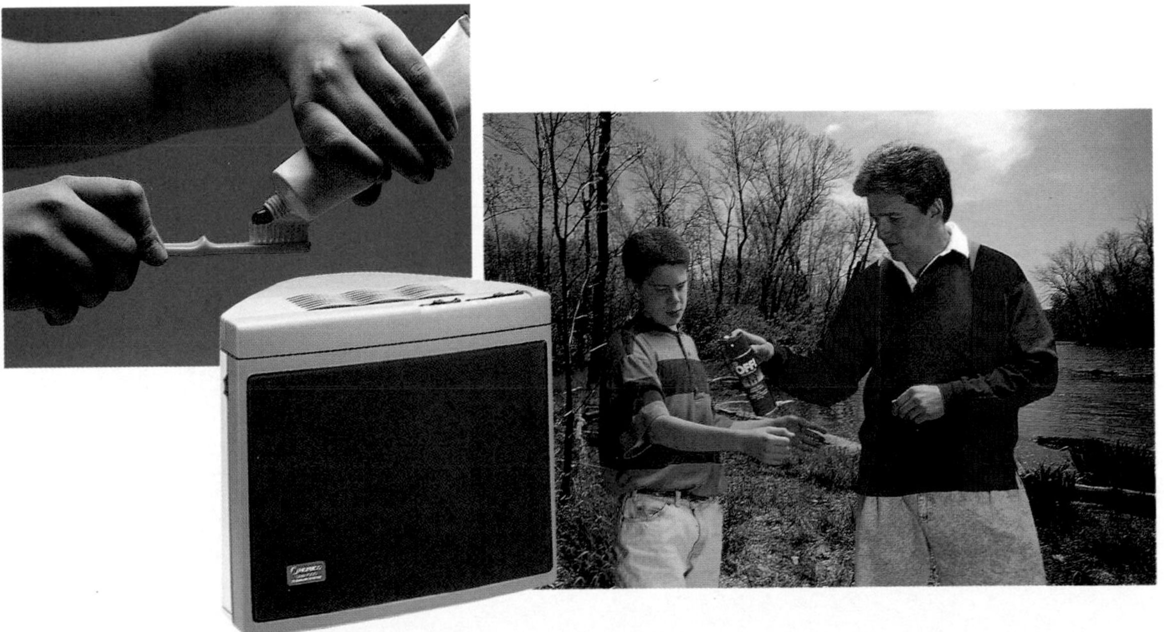

Choosing the Right Light

The electric light bulb, one of the most essential electric devices in the average home, may also be the most ignored. Most people think about them only when one burns out. But, in fact, carefully choosing the proper *lamp*—a less common but more accurate term than *light bulb*—for your home can help reduce your energy consumption as well as lower your electric bill and the amount of money you spend on replacement lamps. Using energy-efficient lamps, which need not be replaced often, can also reduce the amount of trash piling up in landfills. And, by using less electricity, consumers can help reduce the amount of pollution entering the atmosphere from electric power plants.

In 1993, consumers seeking energy-efficient lamps had a new choice—the electronic lamp, or E-lamp. The E-lamp uses only 25 percent as much electricity as a standard incandescent lamp and will last for 14 years if lit for four hours a day, according to its developers.

Most lamps are classified as either *incandescent* or *gaseous-discharge* lamps. Standard incandescent lamps, pear-shaped lamps typically found in table lamps, are the most common type of incandescent lamp. A less common incandescent used in the home is the tungsten-halogen lamp. Such lamps include the tiny, intense lamps used in some newer indoor fixtures. Higher-wattage tungsten-halogen lamps are used in patio lighting and outdoor spotlights.

Fluorescent lamps are the most commonly used gaseous-discharge lamp. Linear fluorescents are long tubes often found in kitchen fixtures and widely used in schools and offices. A recent addition to the fluorescent family is the compact fluorescent lamp, which can be used as an energy-efficient replacement for standard incandescents.

Incandescent and gaseous-discharge lamps differ in the way electric current is passed through the lamp. Incandescent lamps light up when electricity passes through a thin wire called a *filament.* As the filament heats up, it begins to *incandesce* (glow). Along with a filament, an incandescent lamp consists of a glass enclosure called a bulb, supports that hold the filament, and a base that makes contact with an electric power source. Over time, the filament, which is usually made of tungsten, evaporates. Eventually, the weakened filament breaks, and the lamp burns out. To lengthen the life of the filament, the bulb is filled with a mixture of nonreactive gases, usually argon and neon, that slow the filament's evaporation.

Standard incandescents are popular in part because they give off light that brings out red tones. As a result, walls and other objects—including pale skin—appear warmer. Incandescents' greatest appeal, however, may be their low purchase price, usually less than $1.

But standard incandescents are a fairly inefficient way to light your home because they last for a relatively short time—about 750 hours—and because only about 10 percent of the electrical power they use is transformed to visible light. The rest is wasted as heat. And standard incandescent lamps darken over time, as tungsten evaporating from the filament leaves a dark deposit inside the bulb. This deposit gradually reduces the amount of light given off.

Loss of light is less of a problem with tungsten-halogen lamps, which contain a small amount of iodine vapor in addition to argon and neon gas. The iodine vapor combines with the burnt-off tungsten that has evaporated from the filament. When the combined iodine and tungsten touch the hot filament, the tungsten sticks to the filament. The iodine then recirculates to combine with more evaporated tungsten. These substances reduce the dark deposit on the inside of the lamp. As a result, tungsten-halogen lamps produce more light over their lifetime than do standard incandescents. In addition, they may last almost twice as long under ordinary conditions. But their purchase price is up to 15 times higher.

Tungsten-halogen lamps are also

much more fragile than are standard incandescents. In fact, they may explode if not handled with care. An explosion may occur, in part, because the gas in the bulb is pressurized. Pressurizing raises the temperature within the bulb so that when it is lit, the iodine and tungsten can combine. In addition, tungsten-halogen lamps are made of quartz rather than glass to withstand the higher temperatures. If you touch the quartz bulb, the oil from your fingertips will interact chemically with the quartz when the lamp heats up, weakening its ability to contain the gases within. For this reason, packages containing these lamps usually contain a plastic glove to be used when changing them. Because of their potential for failure, tungsten-halogen lamps are best used in enclosed fixtures such as wall sconces.

Gaseous-discharge lamps rely on gas rather than a filament to transmit electric current through the lamp. The two most common types of fluorescent lamps—linear lamps and compact fluorescent lamps—contain mercury vapor and argon gas. Both of these lamps consist of a glass tube; metal enclosures for the ends of the tube; a connection to the power source; an electrical device called a *ballast*, which provides voltage to start the lamp and helps regulate the flow of current; and devices called *cathodes* at each end of the tube. The cathodes are coils of tungsten covered with chemicals that easily give off electrons, negatively charged subatomic particles.

As an electric current passes from one cathode to the other, the electrons shoot through the mercury vapor. The collision of the electrons with the mercury vapor causes the mercury atoms to release invisible ultraviolet rays. These rays are absorbed by phosphors, chemicals that coat the inner surface of the tube, which then glow or *fluoresce*, producing visible light.

Compact fluorescent lamps are smaller than linear fluorescents, and some are available with a screw-in base that contains a ballast. The screw-in base is the size of an incandescent base so that the lamp and ballast unit can be used in incandescent lamp sockets, such as the socket of a table lamp.

The new E-lamp, like a fluorescent lamp, produces visible light by the action of ultraviolet rays on an inner coat-

Kinds of lamps

Most electric lamps are classified as either *incandescent* or *gaseous-discharge* lamps, depending on whether they pass electric current through a wire or a gas to produce light. The electronic lamp, or E-lamp, introduced in 1993, is a close relative of the fluorescent lamp, the most common type of gaseous-discharge lamp. The E-lamp produces light by passing radio waves through a gas.

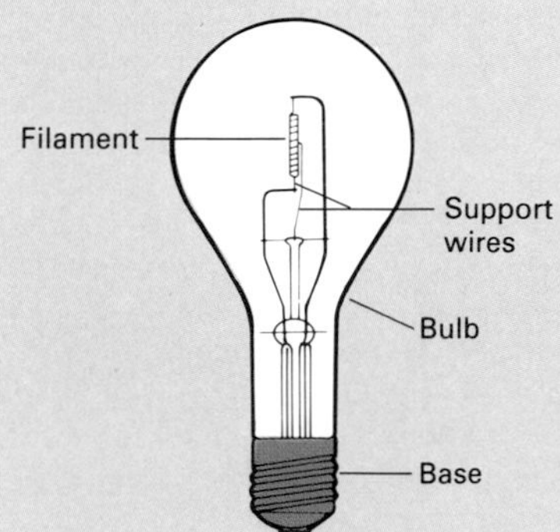

Incandescent lamps light up when electricity passes through a thin wire called a filament. As the filament heats up, it glows. The filament and its support wires are enclosed in a glass bulb. The bulb contains a mixture of gases that lengthen the life of the filament. The base makes contact with the electrical source.

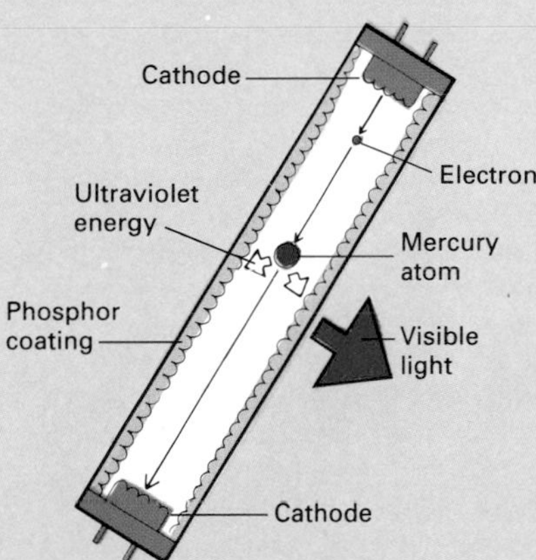

Fluorescent lamps are filled with mercury vapor and argon gas. Electric current passes through a cathode, which gives off subatomic particles called electrons. As these particles flow through the gas to a second cathode, they collide with mercury atoms, producing invisible ultraviolet light. This ultraviolet light is absorbed by a coating of phosphors, which glow, producing visible light.

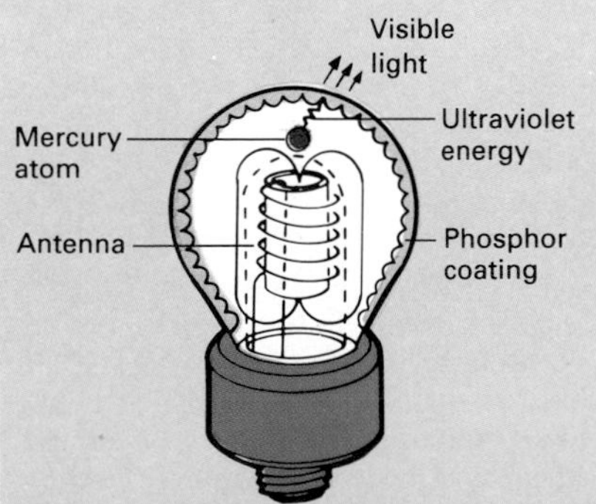

Electronic lamps contain a tiny antenna that emits high-frequency radio waves. The waves, like electrons shooting through a fluorescent lamp, collide with mercury atoms, causing them to release ultraviolet light. The ultraviolet light causes the phosphors to glow and produce visible light.

ing of phosphor. In this case, however, ultraviolet energy is created by high-frequency radio waves emitted by a tiny antenna inside the bulb. The radio waves cause the mercury vapor inside the bulb to release ultraviolet energy that acts on the phosphor to produce visible light.

The label on a lamp's package will provide you with the information you need to calculate how energy-efficient it is compared with other lamps. By federal law, the package labels for incandescent lamps must list the amount of light the lamp gives off, expressed in units called *lumens;* the amount of electricity the lamp uses, expressed as *watts;* and the average life span of the lamp, expressed as *rated hours.* Similar information may appear on fluorescent lamp packages, though not mandated by law.

One measure of a lamp's energy efficiency is its *lumens per watt* (LPW). The LPW compares the amount of light a lamp produces to the amount of electricity it uses. The higher the LPW, the less energy is used to produce a given amount of light. In other words, lamps with a high LPW rating give you more light for your energy dollars.

You can calculate a lamp's LPW rating by dividing its lumen rating by its rated wattage. A 60-watt incandescent lamp, which typically gives off 870 lumens, has an LPW of only 14.5. In contrast, a 40-watt fluorescent lamp, which converts more of its electric energy to visible light, is rated at 3,150 lumens and, therefore, has a LPW of 78.8—at least five times higher than a 60-watt incandescent.

A second measure of a lamp's cost-effectiveness is how often it must be replaced. A typical 60-watt incandescent has an average life of 750 hours. In comparison, a fluorescent lamp is rated to last 20,000 hours.

Incandescents marketed as "long-life" or "energy-efficient" lamps do, in fact, last longer and consume less electricity than standard incandescents do. But studies have shown that they create this saving by giving off significantly fewer lumens. Such lamps produce less light than standard incandescents do for the same amount of electricity consumed. "Soft white," pink, or other coated incandescents also produce less light than do clear incandescents because the coating absorbs some of the light.

Among fluorescents, energy efficiency depends on the colors produced by a lamp. "Deluxe cool white" and "deluxe warm white" fluorescent lamps may make red surfaces look more natural than they would under the light of ordinary "cool white" fluorescent lamps. But you pay a price for the more natural color appearance. Cool white lamps produce about 1.4 times more lumens than do deluxe cool white or deluxe warm white lamps.

Because they are so energy-efficient, compact fluorescent lamps may be a good replacement for standard incandescents. A 7-watt compact fluorescent lamp produces 400 lumens, for example, about half the lumens as one 60-watt incandescent lamp. But two 7-watt compact fluorescents will use only about 36 percent of the energy consumed by a single 60-watt incandescent. And although the purchase price of compact fluorescents is about 15 times higher than that of standard incandescents, the fluorescents last approximately 10 times longer. Because they do not have to be replaced frequently, they cost significantly less to use over their lifetime.

The E-lamp has the benefits of its fluorescent cousins. A 25-watt E-lamp has an LPW of approximately 70, about 5 times higher than that of a standard 60-watt incandescent, according to the lamp's developer. Such a drop in electricity consumption would produce substantial savings.

Experts report that a higher purchase price sometimes discourages consumers from buying more energy-efficient lamps. But purchase price is only one aspect of a lamp's price to consider. To calculate the cost of buying and using a particular lamp—for example, a 60-watt incandescent—first determine your basic electric rate by dividing the total amount of your monthly electric bill by the kilowatt-hours (kwh) used. (The average rate per kwh in the United States is 8 cents.) Then multiply the rate per kwh by the rated wattage of the lamp, in this case $0.08 × 60 = $4.80. This result is the cost of operating a 60-watt lamp for 1,000 hours.

Next, find the cost of the lamp itself over the same time by dividing the rated hours by 1,000 (750 ÷ 1,000 = 0.75) and then dividing the cost of the lamp

The right light

Electric lamps differ widely in their purchase price, the amount of electricity they use, and their life span. Carefully choosing the right lamp for your home not only can help to lower your electric bills but also can help to reduce the amount you spend to replace burned-out lamps.

Type of lamp	Advantages	Disadvantages
Incandescent	• Low purchase price—often less than $1 • Brings out red tones in colors, making objects, including pale skin, appear warmer	• About 90 percent of energy produced is in the form of wasted heat • Easily broken or damaged • Lasts only about 1,000 hours
Tungsten-halogen	• Provides more light over its life span than incandescent lamps • May last twice as long as standard incandescent lamps • Small, compact lamp allows light to be focused more easily	• Extremely fragile if mishandled • Costs about 15 times more than standard incandescent lamps
Linear fluorescent	• Uses about 20 percent as much electricity, produces 20 percent as much heat, and lasts up to 20 times longer than standard incandescent lamps	• Initial costs are higher than those of standard incandescent lamps • Uses more energy for dimming than incandescent lamps • Cannot be started and operated on voltage supplied by most electric outlets and so requires a bulky electrical component called a ballast
Compact fluorescent	• Produces light similar to that of standard incandescent lamps • Uses about 85 percent less electricity and lasts up to 10 times longer than standard incandescent lamps • Can be screwed into an incandescent lamp socket • Lamps that produce light of different tints available to meet design needs	• Initial costs are higher than those of standard incandescent lamps • Some lamps cannot be dimmed • Uses more energy for dimming than incandescent lamps
Electronic lamp	• Produces more light with less power than standard incandescents and fluorescents and lasts up to 20 times longer than standard incandescents	• Initial costs are higher than those of standard incandescent or fluorescent lamps

by the result ($0.89 ÷ 0.75 = $1.19). Finally, add this number to the cost of operating the lamp ($1.19 + $4.80 = $5.99). The sum is the true cost of using an 89-cent incandescent lamp for 1,000 hours.

In contrast, the cost of buying and operating an 18-watt fluorescent and its ballast, which provide an equivalent amount of light, is about $1.63. The cost for a 7-watt compact fluorescent lamp is about $1.90. Experts say that such significant savings—coupled with the sharp reduction in the amount of energy used—should provide consumers with a strong incentive to consider lamps other than the familiar incandescent. [Ronald N. Helms]

Chemistry in a Tube of Toothpaste

Throughout history, people have used many methods to clean their teeth. Some ancient people chewed on frayed twigs. Others rubbed their teeth with cloth or swabbed them in vinegar. The Egyptians may have been the first people to use toothpaste when they mixed powdered *pumice* (a porous rock formed in volcanic eruptions) with vinegar and applied it to their teeth in the 100's B.C. Some toothpastes still contain pumice, but science has since supplied other toothpaste ingredients that the ancient Egyptians never dreamed existed.

Toothpaste has two functions—cosmetic and hygienic. It improves one's appearance by removing stains from teeth. The hygienic function of toothpaste helps prevent tooth decay by removing food particles and plaque from teeth and delivering other decay-preventing substances such as fluoride. Plaque is a sticky substance that forms on teeth. It is made up primarily of common oral bacteria and products of those bacteria. Plaque also contains some saliva and components that dissolve from food onto the teeth. If plaque is not removed, the bacteria digest the sugar in foods and beverages to form a family of acids that slowly erode a tooth's *enamel* (hard outer coating).

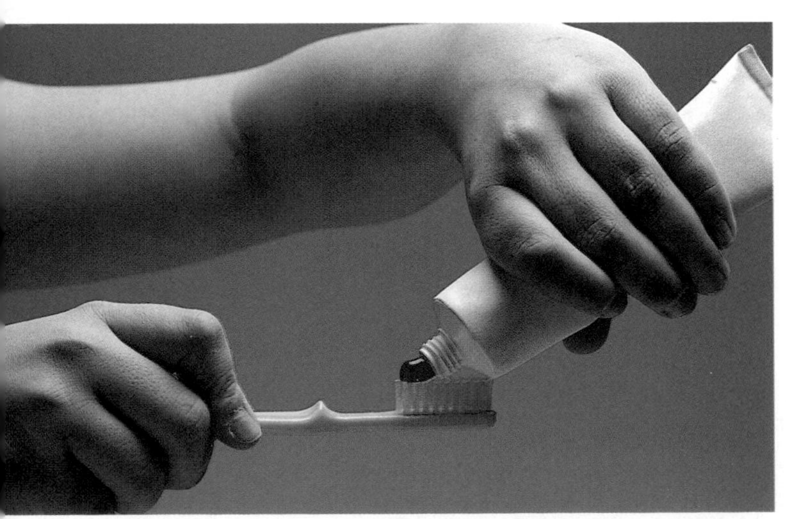

A major purpose of brushing is to prevent plaque from turning into tartar. Tartar, also called calculus, is a hard mineral substance that is chemically similar to tooth enamel. It forms along the gum line when calcium salts in saliva collect on the teeth along with dead bacteria in plaque. Tartar above the gum line does not hurt the teeth, but because it quickly absorbs stains and turns brown, it is a cosmetic problem. Tartar below the gum line can cause *gingivitis* (inflamed gums), which may lead to *periodontitis* (loss of bone that supports teeth) and eventually loss of teeth.

All toothpastes are mixtures of an abrasive, a detergent, a thickener, a *humectant* (moisturizer), water, and other ingredients such as flavoring agents, coloring, and sweeteners. Most toothpastes also contain a fluoride compound that helps prevent tooth decay. Some toothpastes contain compounds that fight tartar formation.

Other toothpastes contain desensitizing agents to reduce the sensitivity of tooth surfaces below the gum line. This sensitivity can develop if the gums recede or are worn away by brushing too hard with an extra firm toothbrush. Some toothpastes also contain a bleaching agent that may make teeth look whiter.

Abrasives provide the cleaning power in toothpastes. They give toothpaste a slightly gritty texture designed to polish teeth and remove plaque, food remnants, and stains.

The most commonly used abrasives are hydrated silica, calcium carbonate, aluminum oxides, and various phosphates of calcium or aluminum. Calcium carbonate is the compound of which chalk is made, but the form used in toothpaste is not simply ground-up chalk. The compound is chemically reformed into finer particles than those found in chalk. Likewise, sand contains pure silica, but the form used in toothpaste is gentler than that in sand. Harsher abrasives include anhydrous dicalcium phosphate, zirconium silicate, crystalline silica, and pumice. Those abrasives are commonly used by the dentist to remove tartar and stains. Some European toothpastes even contain abrasives made of tiny granules of hard plastic.

Most abrasives used in toothpaste pose no harm to teeth because tooth enamel is made almost entirely of *hydroxyapatite,* a form of calcium phosphate that is about as hard as the mineral topaz or the steel in most knives.

But toothpastes designed to remove tough stains, such as those from tobacco and coffee, may contain harsher abrasives than do regular brands. Some experts say that the frequent use of highly abrasive toothpastes over many years has the potential to damage tooth enamel.

Harsher abrasives may also contribute to the loss of cementum at the base of the teeth. Cementum is a thin layer of calcium, magnesium, and phosphorus that covers the *dentin* (the main material of which teeth are composed) below the gum line. If the cementum is worn away—from overly vigorous brushing or the aging process—the sensitive dentin is exposed. Repeated brushing with harsh abrasives may also wear away exposed dentin, increasing sensitivity.

Detergents, like abrasives, also provide cleaning power. They loosen food remnants and plaque so that the toothbrush can scrub them away. Detergents contain cleaning agents called *surfactants.* Surfactants are complex *molecules* (groups of linked atoms) that attach to stains at one end and to water at the other. The water pulls the surfactant and the stain away from the surface to which they are attached. In toothpaste, surfactants help pull food particles and stains away from teeth. Surfactants alone are not as effective as abrasives in removing plaque and stains, however.

Chemicals in toothpaste detergents can also help prevent tooth decay. The most frequently used toothpaste detergent, sodium lauryl sulfate, can slow the growth of some plaque-forming bacteria. Another detergent, sodium lauryl sarcosinate, may inhibit the chemicals that plaque-forming bacteria use to digest sugar.

Baking soda, one of the oldest forms of tooth cleaners, acts both as a gentle abrasive and as a detergent. In some toothpastes, it is also used to fight decay-causing microbes. Baking soda has the disadvantage of dissolving quickly in saliva and water, however. Once dissolved, baking soda is no longer abrasive, though it continues to function as a detergent.

How toothpaste works

Toothpaste ingredients may differ from one brand to another, but most have the same basic ingredients and work in similar ways. Common ingredients in toothpaste include abrasives, detergents, fluoride, and bleaching agents. Some toothpastes also contain tartar-control agents and desensitizing compounds.

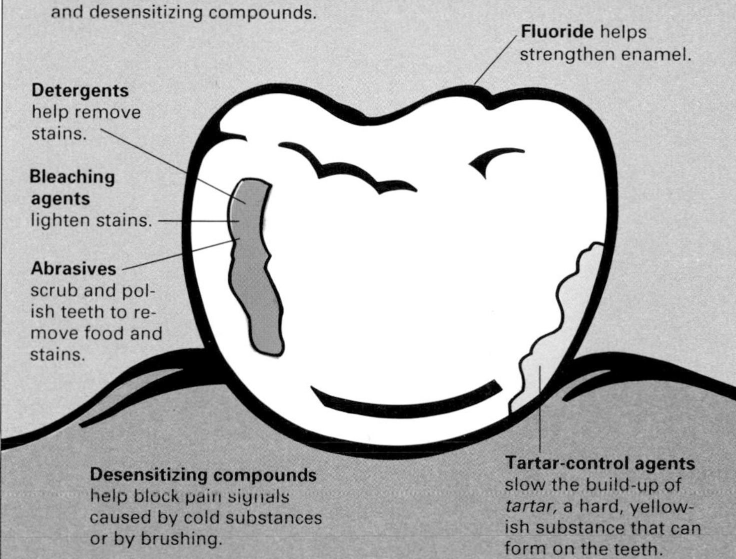
Fluoride helps strengthen enamel.
Detergents help remove stains.
Bleaching agents lighten stains.
Abrasives scrub and polish teeth to remove food and stains.
Desensitizing compounds help block pain signals caused by cold substances or by brushing.
Tartar-control agents slow the build-up of *tartar,* a hard, yellowish substance that can form on the teeth.

Fluoride—most frequently in the form of sodium fluoride, sodium monofluorophosphate, or stannous fluoride—combats tooth decay by strengthening enamel and, according to some researchers, by inhibiting bacteria's formation of acids that attack enamel. Although enamel is the hardest substance in the body, constant attack by the acids in saliva and plaque can dissolve it.

Fluoride's action is not completely understood, but chemists believe that the ingredient works through a series of chemical reactions called *remineralization,* in which fluoride is transformed into fluoride-rich compounds called *apatites* that chemically bond with enamel to strengthen it and help protect it from dissolving in acid.

Calcium can help strengthen enamel, but excess calcium can contribute to tartar build-up. Calcium that is naturally present in saliva can be a source of excess calcium. Tartar-control agents in toothpaste work by slightly changing the chemistry of saliva and teeth to inhibit the build-up of calcium on teeth. These changes inhibit calcium from crystallizing into tartar once the calcium binds to plaque. In this way, less tartar forms, and the tartar that does form is often easier to remove. Tartar-control agents in toothpaste do not work below the gum line and thus have no clear effect on gingivitis.

Toothpastes with ingredients designed to desensitize teeth are intended for people whose teeth are extremely sensitive to heat and cold and to brushing. As many as 1 in 7 adults develop sensitive teeth, usually after age 45. This typically happens when the gums recede, exposing the sensitive cementum and dentin below the gum line.

Desensitizing toothpastes work differently depending on the desensitizing compounds they contain. Potassium nitrate, strontium chloride, and sodium fluoride are common desensitizing compounds. Fluorides help rebuild a thin layer of cementum on the tooth's exposed root, a process similar to the way in which fluoride remineralizes enamel. Strontium chloride blocks tiny *tubules* (channels) that lead from root surfaces to the nerves. In doing so, the compound disrupts pathways that help produce pain signals. Although chemists are not completely certain how potassium nitrate works, they believe it also alters chemical signals in the tubules. It and the other desensitizing compounds must be used regularly to be effective.

Bleaching agents for whitening teeth date to at least the early 1900's. Today, the most common bleaching agent in toothpaste is a compound called carbamide peroxide, which in the mouth forms hydrogen peroxide. The hydrogen peroxide in turn breaks down into water and oxygen. The free oxygen is a powerful bleaching agent because it combines with materials in the stains and lightens their color. A few brands of toothpaste contain a different type of bleaching agent, citric acid, which is found in lemons and other citrus fruits and acts as a mild bleach.

Toothpastes do not contain enough bleaching agents to lighten heavy stains beyond the normal effect of abrasives, however. Products called home tooth-bleaching kits claim to offer stronger whitening power. But according to the American Dental Association (ADA), these products may be dangerous due to their high peroxide content. In 1989, the ADA warned that the regular use of home-bleaching kits may temporarily damage the soft tissues of the mouth, delay healing of already damaged tissue, or damage tooth pulp by traveling down tubules in the enamel to the pulp. None of these kits had received approval from the ADA as of mid-1993.

The remaining ingredients in toothpaste perform functions unrelated to cleaning the teeth. Toothpastes need something to keep the solid and the liquid ingredients together, and a binding agent performs this function. Most such agents are some form of cellulose, a gel, or an extract of seaweed. Without the binder, the toothpaste would separate into a liquid portion and a sort of mush and would have to be stirred, like paint, before each use.

In addition, moisturizers keep the toothpaste from drying out. The most commonly used moisturizers are glycerin and sorbitol.

Whenever the cap is left off the toothpaste tube, molds or bacteria can get inside. Most manufacturers add a preservative such as methylparaben or propylparaben to the toothpaste to prevent the microbes' growth.

To make toothpaste visually appealing, manufacturers add coloring agents, which usually are food colorings. Some coloring agents are natural, and others are artificial. White, pink, green, and blue are the favored colors of toothpaste. White toothpastes often contain titanium dioxide, a brilliant white pigment used in some paints, to make the toothpaste appear whiter and brighter.

Puzzled over which type of toothpaste to buy? The ADA considers most major brands of toothpaste to be about equally effective for people with healthy teeth and gums. One indication of effectiveness is the ADA seal. To receive the seal, toothpaste manufacturers must submit results of long-term scientific studies that prove their product's safety and effectiveness. Despite the ordinary appearance of your favorite toothpaste, a lot of science went into that dab of goo. [Peter Limburg]

Clearing the Air About Electronic Air Cleaners

Many families realize that their homes may harbor high levels of air pollution. Researchers reported in the 1980's and early 1990's that air pollutants are generally present in higher concentrations in indoor air than in outdoor air. Moreover, the news media regularly feature stories of "sick buildings," where employees complain of mysterious headaches, nausea, and dizziness. Indoor air pollution is named as a prime suspect in many of those cases.

Because people spend most of their time indoors, air pollution in the home has a greater potential impact on a person's health than does outdoor air pollution. Fortunately, there are a number of ways to improve the quality of indoor air—by controlling pollutants at their sources, by increasing ventilation, or by using air-cleaning devices, either housewide systems or portable tabletop models.

Air pollutants take two basic forms: gases and particles. Gaseous indoor air pollutants include combustion fumes produced by fuel burning in stoves and furnaces or by lighted cigarettes; chemical fumes from consumer products such as paints, glues, and cosmetics; and radon, a colorless, odorless gas that arises from the decay of radioactive elements in the soil and seeps into the house through the foundation. These pollutants can irritate a person's nose, eyes, throat, and lungs. Some of these gases, particularly radon, can even lead to cancer.

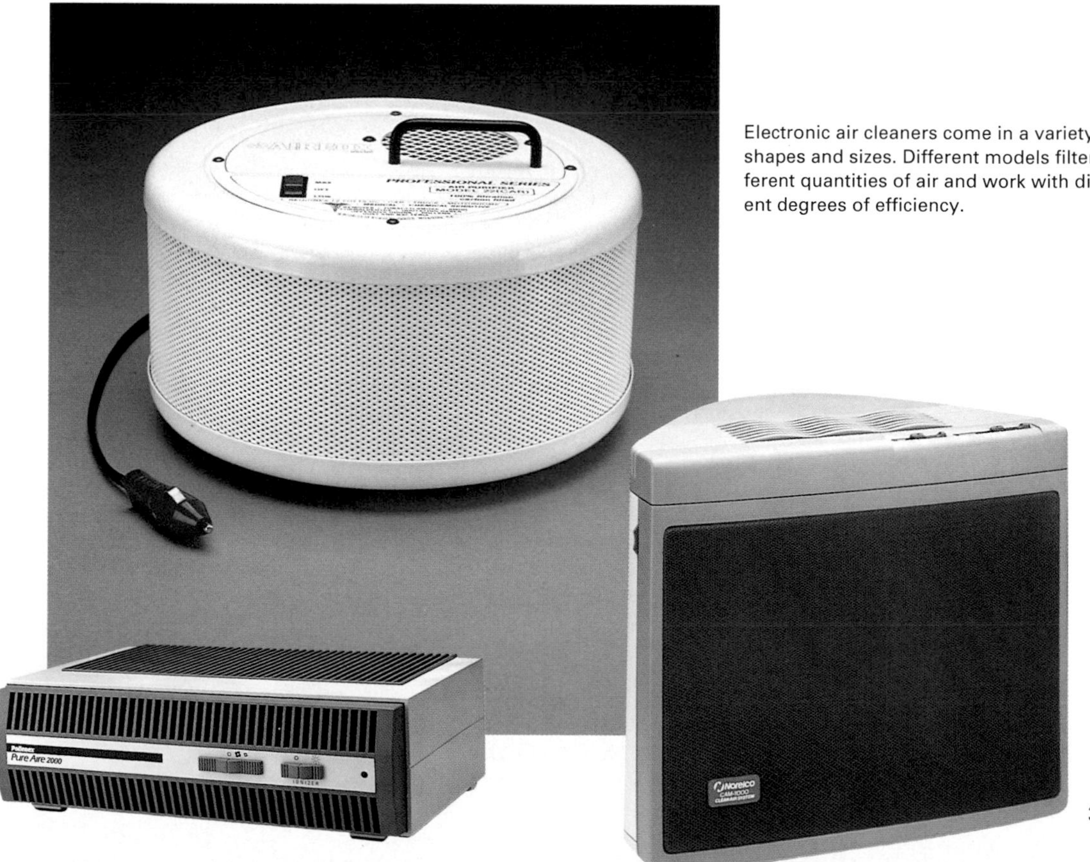

Electronic air cleaners come in a variety of shapes and sizes. Different models filter different quantities of air and work with different degrees of efficiency.

Gaseous air pollutants are a problem in many homes. However, air cleaners are unable to filter gases from the air. Gases can be eliminated only by preventing the pollutants from entering the house in the first place or by increasing the ventilation of the house to continually replenish indoor air with air from the outside. To help fight gaseous pollutants, many new houses are equipped with whole-house mechanical ventilation systems. These systems use vents to exchange indoor and outdoor air and prevent gaseous air pollutants from building up indoors. (The energy cost of whole-house ventilation systems can be reduced by using a heat recovery ventilator [HRV]. An HRV helps make incoming air the same temperature as outgoing air, thereby saving on heating and air-conditioning bills.)

Air cleaners can, however, filter particles from the air. Particulate air pollutants include soot and ash from burning fuel and lighted cigarettes as well as biological contaminants and asbestos. Biological contaminants are living organisms or the products of living organisms. They include pollen, mold, mildew, bacteria, and viruses. Biological contaminants can lead not only to respiratory irritation but also to allergic reactions and disease.

Asbestos is a group of soft, threadlike mineral fibers. One common form, formerly used in home construction, is now known to be a powerful cancer-causing substance. People should not tackle an asbestos pollution problem with air cleaners but should seek expert advice on removing or sealing asbestos.

There are four major types of air cleaners: mechanical filters, electrostatic precipitators, charged-media filters, and negative ionizers. Portable air cleaners usually measure about 0.3 to 0.6 meter (1 to 2 feet) on each side, and the smaller ones can easily rest on a table or desk in a room where air pollutants are concentrated. Running nonstop, they can typically clean a mildly polluted air volume of about 34 cubic meters (1,200 cubic feet).

Mechanical filters are the simplest air cleaner. They use an electric fan to pull air through a fiber filter that traps particles and removes them from the air. There are three grades of mechanical filters: panel filters, extended-surface filters, and high-efficiency particulate-arresting (HEPA) filters.

Panel filters are flat sheets of loosely packed fibers. They provide a basic level of air cleaning, generally removing no more than 50 percent of the particles they encounter.

Extended-surface filters are packed more densely than are panel filters, and they contain pleats or folds. Pleats give the filters a greater surface area and allow them to trap more particles. Extended-surface filters can remove more than 70 percent of the particles they encounter.

HEPA filters are even more densely packed and more tightly pleated than are extended-surface filters. They are capable of removing more than 99 percent of the particles they encounter.

Mechanical filters vary in price, depending on their effectiveness. At the high end, HEPA models cost $300 to $500 to purchase, plus $80 to $150 per year in energy costs and replacement filters.

Unlike mechanical filters, electronic air cleaners rely not on sheets of fibers to remove particles from the air but on the attraction between positive and negative electric charges. Particles in the air tend to be *neutral* (possessing no net electric charge). To make the particles electrically charged so that they can be more easily manipulated, electronic air cleaners add or take away negatively charged subatomic particles called *electrons* from the atoms that make up the particles. An air cleaner makes particles negatively charged by adding electrons to them, or positively charged by removing electrons from them. The process of removing electrons from particles is called *ionization.*

After the airborne particles have become charged, the cleaner can remove them from the air by passing them near oppositely charged plates or filters. Opposite charges are attracted to each other, so the plates or filters will attract the particles, pulling them from the air.

Electrostatic precipitators, for example, use a fan to draw air past special ionizing wires. The wires remove some of the electrons from particles that pass nearby, leaving the particles with a positive electric charge. The fan blows these particles past columns of negatively charged metal plates, which attract the

How air cleaners work

All air cleaners use fans to pull particles through various filters. The devices differ in the makeup of their main filter, but they generally share three other filters: a prefilter, to weed out larger particles before the main filter; a post filter, to remove particles the main filter has missed; and a carbon filter, to help neutralize odors.

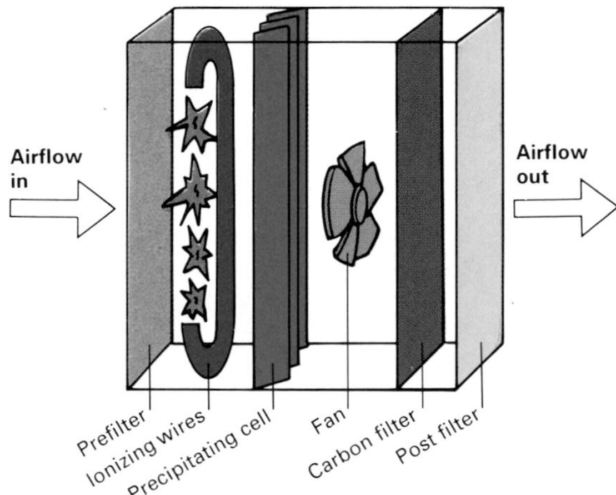

Electrostatic precipitator

An electrostatic precipitator filters particles using the principle of *electrostatic attraction,* the attraction that opposite electrical charges have for one another. A fan pulls air through a set of ionizing wires, which give passing particles a positive electric charge. The particles then flow between negatively charged metal plates called precipitating cells. These plates pull the positively charged particles out of the air.

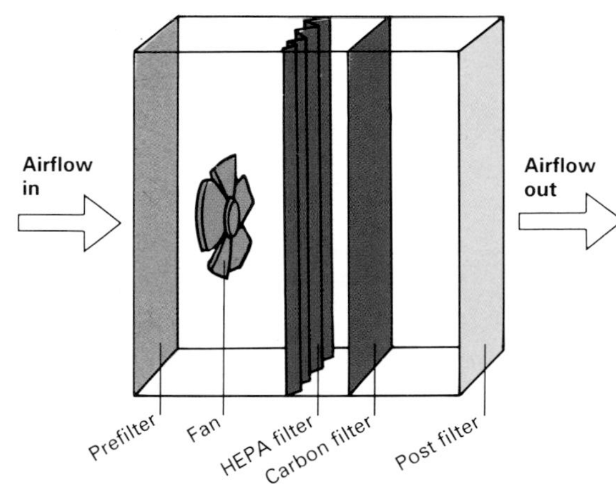

High-efficiency particulate-arresting (HEPA) filter

In a HEPA filter, the main filter is made up of many sheets of tightly packed glass fibers. The sheets are pleated to increase their surface area and so trap more particles than a flat filter would.

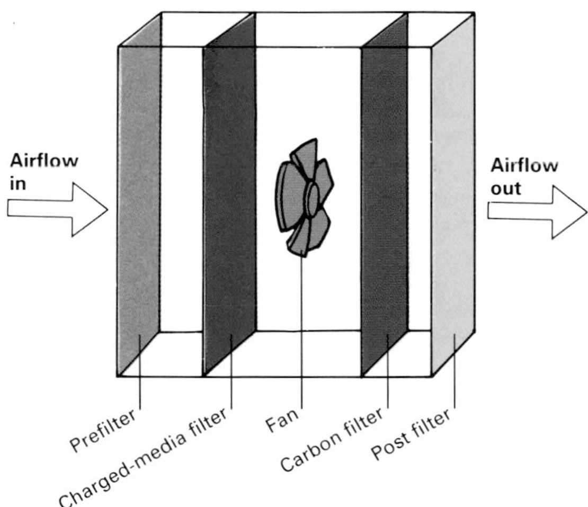

Charged-media filter

A charged-media filter uses a combination of mechanical and electrical forces to capture air pollutants. The main filter is made up of polyester and cellulose fibers that physically trap particles. The filter also has an electrostatic charge, which helps attract particles to it.

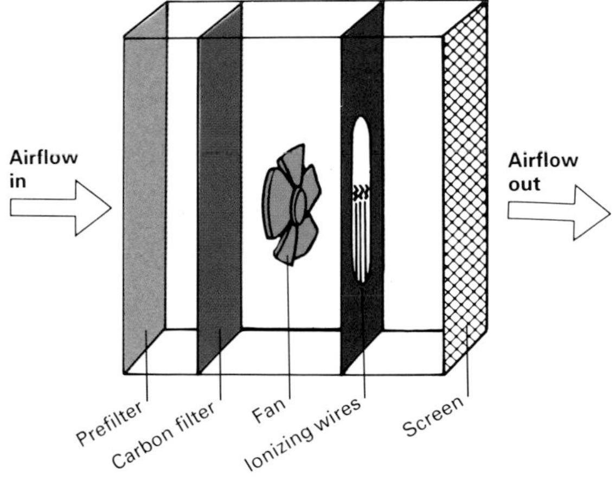

Negative ionizer

A negative ionizer uses electrostatic attraction to cause airborne particles to drift out of the air and attach to objects. Inside the ionizer, needlelike wires give a negative electric charge to the particles in the air flowing through the device. After the charged particles leave the ionizer, they are attracted to walls and other surfaces in the room.

positively charged particles and thus remove them from the air.

The force of electrical attraction is strong, so electrostatic precipitators are generally the most effective of all air cleaners. But they are also the most expensive, costing $300 to $600, plus $60 to $100 per year in electricity and replacement parts. And the metal plates must be cleaned regularly.

Another portable electronic air cleaner, the *charged-media filter,* is essentially a mechanical filter, but the filter has an electrostatic charge to help it attract particles. Electrostatic charges produce weak electric fields that attract both charged and uncharged particles. Electrostatic charges are responsible, for example, for the "dust bunnies" that sometimes form under furniture.

Charged-media filters, which cost $60 to $180, are more effective than simple panel filters. But they are less effective than electrostatic precipitators, because the charged filter does not pull particles out of the air as effectively as do the charged metal plates.

One other portable electronic air cleaner is the *negative ionizer.* This device has metal ionizing needles that add electrons to particles pulled past them by a fan. But the negatively charged particles are not then removed from the air by the device itself. Instead, the device blows the particles out into the room, where they leave the air by attaching themselves to positively charged spots on the room's walls, carpeting, and furniture. Thus, negative ionizers do not strictly clean the air but only promote the attachment of air pollutants to other objects. The approach saves on filters and maintenance costs for the device. But these savings come at the expense of the cleanliness of the room that contains the negative ionizer. Negative ionizers cost $50 to $150.

Regardless of which air cleaner you may buy, its effectiveness depends on the amount of air handled by the device and the efficiency of the device at removing particles from the air. As an aid to consumers, the Association of Home Appliance Manufacturers (AHAM) rates air cleaners by a *clean air delivery rate* (CADR). The CADR is calculated by multiplying the unit's airflow rate by its cleaning efficiency. For example, a unit that handles an air flow of 100 cubic feet per minute (100 cfm) and removes 50 percent of the particles from that air would have a CADR of 50 cfm (100 cfm × 50% = 50 cfm). A unit that handles twice as much air but removes only half as many particles would have the same CADR (200 cfm × 25% = 50 cfm).

If you decide to shop for a portable air cleaner, look for an AHAM seal giving the device's CADR rating. The higher the CADR, the more effective the air cleaner. CADR's vary widely, from 20 for panel filters to 300 for electrostatic precipitators and HEPA filters.

However, cost and efficiency should not be your only criteria for an electronic air cleaner. Portable air cleaners, for example, can be noisy, since they use powerful fans to push air through their filters. Experts also point out that some electronic air cleaners generate *ozone,* a chemical that is itself a respiratory irritant and an indoor air pollutant. For this reason, some electronic air cleaners contain a carbon filter, which absorbs some of the ozone produced by the device. The carbon filter also helps remove odors from the air.

Portable air cleaners can be effective at attacking indoor air pollution that is concentrated in a specific room or area. However, if indoor air pollution is a problem in several rooms, homeowners may want to consider installing an air-cleaning system in ducts throughout the house. Ducted systems filter much more air than do portable models and therefore remove many more particles. However, they cost $500 to $800 or more to purchase and install, plus annual costs of $120 to $200 for electricity and replacement filters.

Ducted air cleaners also may restrict air flow. If air cleaners are used in heating or air-conditioning ducts, for example, homeowners may need to install booster fans in the ducts to help circulate air. Otherwise, the restricted air flow may lower the efficiency of the home's heating or cooling system.

The most effective ways for homeowners to eliminate indoor air pollution are to keep the pollutants out of the house in the first place and to maintain adequate ventilation. But when those actions fall short, portable air cleaners may help homeowners breathe a little more easily.

[John H. Morrill]

Repelling Bugs That Bite

Just as surely as summer brings a host of outdoor pleasures, it also brings swarms of mosquitoes, flies, ticks, and other biting, disease-spreading insects. For many years, people have used insect repellents on their skin and clothes to ward off the unwelcome advances of these pests.

Such products, when used properly and in the right amounts, can not only make outdoor life much more pleasurable but also reduce the risk of acquiring certain insect-borne diseases, such as the tick-transmitted Lyme disease. But the misuse or overuse of these chemicals can be ineffective and possibly even dangerous, according to the United States Environmental Protection Agency (EPA), which determines the safety and effectiveness of chemical insect repellents.

Repellents work by discouraging insects from alighting on skin or clothes treated with these substances. Scientists do not know exactly how repellents accomplish this task, but through experimentation they have learned that hungry, blood-feeding insects, such as mosquitoes, ticks, and chiggers, are attracted to warm, moist areas and to carbon dioxide, a gas that human beings and animals exhale. Researchers have also noted that insects seem to zero in on their warm-blooded hosts by smelling lactic acid, a compound produced by muscles during exercise and also released through the skin.

Some scientists believe that repellents simply cover up or mask the scent of these chemicals to prevent insects from detecting our presence. Other researchers think repellents act on insects' odor receptors to desensitize them to the fragrance of body chemicals. In 1988, for example, entomologist Edward E. Davis, working at SRI International in Menlo Park, Calif., reported that his studies of mosquitoes

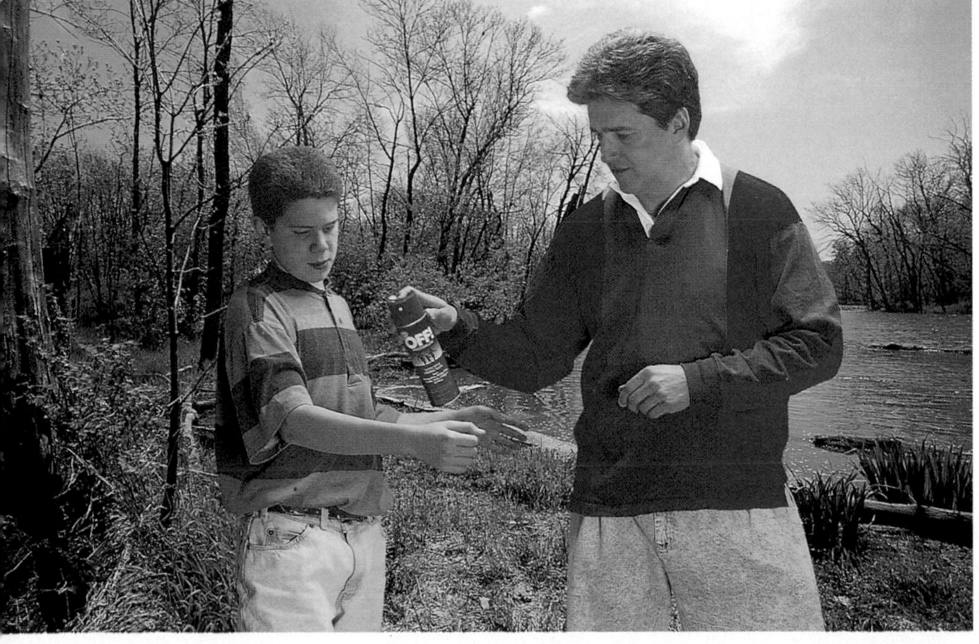

A light coating of insect repellent on exposed skin helps deter mosquitoes, ticks, and other biting pests. When used properly, insect repellents can make outdoor life more enjoyable while they reduce the risk of acquiring insect-borne diseases.

Compounds that repel insects

Ingredient	Trade names	Target insects	Characteristics
N,N-diethyl-meta-toluamide (deet)	Ben's, Cutter, Muskol, Off!, Repel, Skedaddle, Skeeter Stop, Ultrathon	Mosquitoes, black flies, gnats, chiggers, fleas, ticks	• Regarded as most effective all-around repellent • Greasy and sticky, especially at higher concentrations • May dissolve synthetic materials • May cause rashes and allergic reactions • Has been linked to a brain disorder in children when swallowed or repeatedly used in high concentrations
Permethrin	Permanone Tick Repellent	Ticks, mosquitoes, and numerous insects	• Ineffective on skin; can be used only on clothing • Not available in all states • Poisons rather than repels insects
Citronella	Natrapel	Mosquitoes, black and biting flies, midges	• Must be applied often

supported the theory that repellents interfere with an insect's odor receptors. Davis found that certain nerve cells on mosquito antennae were more active than others in the presence of lactic acid, which indicated that particular insect odor receptors are sensitive to this chemical. The rate at which the nerves fired decreased sharply in the presence of a repellent. But exactly how repellents achieve this effect remains unclear.

Unlike insecticides, insect repellents do not kill pests, nor are they effective when sprayed in the open air. Repellents also will not deter bees, wasps, or other stinging insects that do not feed on human body fluids. Because these pests are attracted to flowers and flowery scents, people can reduce their risk of being stung by not wearing perfumes or clothing with colorful floral patterns when spending time outdoors.

Since the time of the ancient Egyptians, some 5,000 years ago, people have searched for the perfect insect-repelling potion. Over the centuries, such substances as smoke, animal grease, plant extracts, and even a diet rich in garlic, have been used to deter different insects, with various degrees of success.

Modern insect repellents, which come in the form of lotion, creams, sprays, sticks, and towelettes, tend to be far more effective than folk preparations. Repellents were originally discovered—and have continued to evolve—through trial-and-error studies in which human volunteers apply a chemical to their arms and bravely stick them into cages full of voracious mosquitoes. Scientists also conduct field tests by applying repellents to themselves and entering areas heavily infested with mosquitoes or other blood-sucking insects.

In the 1950's, scientists discovered that one chemical was vastly superior to the many thousands that had been previously tested. This clear, oily, sticky substance is officially called N,N-diethyl-meta-toluamide but is commonly referred to as *deet.* This compound is still regarded as the most effective mosquito repellent on the market. Deet also protects against other blood-feeding insects, including black flies, gnats, chiggers, fleas, and ticks.

Deet is found in many popular brands of repellents, such as Off!, Repel, REI Jungle, Cutter, and Muskol. The deet in these products ranges in concentration from 5 percent to 100 percent. Although deet is considered to be the best overall repellent, commercial products that use deet also often contain small amounts of additional repellents, which may be more effective against particular pests. Repellent formulas also contain inactive ingredients, such as spray propellants or cream bases.

Up to a point, the more deet in a repellent formula, the more effective and long-lasting the product will be. For summer backpackers or backyard barbecue enthusiasts, a repellent containing 30 percent deet, applied thinly and evenly to all exposed skin, should provide protection for about three hours. However, those who swim or sweat heavily will need to reapply repellents more frequently.

The 100 percent deet formulations do not seem any more effective than low-concentration formulas. However, because sweat and water do not easily wash away the 100 percent solutions, these products can protect against insects for as long as 10 hours. Such formulas are most appropriate for people who are camping out in jungles, marshes, or other heavily infested areas.

As effective as they are, deet repellents have their drawbacks. They can have a greasy, sticky feel, particularly in high concentrations. And while spraying clothing with repellents provides added protection, deet can dissolve certain synthetic fabrics, such as rayon, spandex, and acetate. It can also erode plastic watch crystals.

Deet may cause mild rashes and allergic reactions. In addition, the EPA reported in 1989 that it had received several reports of possible toxic effects of deet in children. These effects ranged from nausea to coma and followed repeated and excessive applications of deet. But because the number of such reports was small (fewer than 10) and because no connection to deet had been proven, the agency stated that deet's benefits "may far outweigh any risk" from exposure to it.

Nevertheless, some researchers and pediatricians recommend that repellents containing more than 30 percent deet should not be used on children. The EPA advises consumers to follow carefully the directions on deet product labels. It also warns against using deet products near the eyes or mouth, and against applying them to children's hands (which usually find their way to eyes and mouths). The EPA also says to use only enough of a repellent to cover *exposed* skin and clothing, and to wash off the repellent upon returning indoors.

Before deet became the most popular insect repellent, another chemical, known as Rutgers 612, was widely used. The compound was named after Rutgers The State University of New Jersey in New Brunswick, where its repellent properties were discovered, and after the number of the bottle that contained the chemical during its initial testing. Rutgers 612 is a clear oil effective at driving away mosquitoes and biting flies. It does not, however, last as long as deet, nor does it repel as many types of insects.

Rutgers 612 had been the major ingredient in several brands of insect repellents since World War II (1939-1945). But after scientific studies had linked the chemical to birth defects in laboratory animals, the EPA required its manufacturer to reregister the product. To do this, the maker would have had to perform new tests on the product proving its safety. The manufacturer chose not to reregister Rutgers 612 and stopped selling it in 1990.

Another type of clear oil, known as dimethyl phthalate (DMP), has also been used as an insect repellent. Like the Rutgers 612 manufacturer, the makers of DMP stopped its production after questions arose about its safety and the EPA required its reregistration.

The compound acts powerfully against mosquitoes, chiggers, and ticks. In the tropics, DMP has been used successfully over the years to repel the mosquito species that spreads yellow fever. When applied to clothing, the chemical is stable and long-lasting, remaining active for 11 to 22 days unless the clothing is washed. But on the skin, it can break down in the presence of sweat. Large doses of DMP can irritate the eyes and *mucous membranes* (the linings of body passages and cavities).

A repellent called Permanone is designed for use on clothing and contains

Selecting and using an insect repellent

Insect repellents are available as sprays, sticks, lotions, creams, and towelettes, with the concentration of active ingredients ranging from 5 percent to 100 percent. Experts say that adults, but especially children, should use the lowest concentration that is effective. The product should be applied sparingly, reapplied only when necessary, and never used near the eyes or mouth or on broken or sunburned skin.

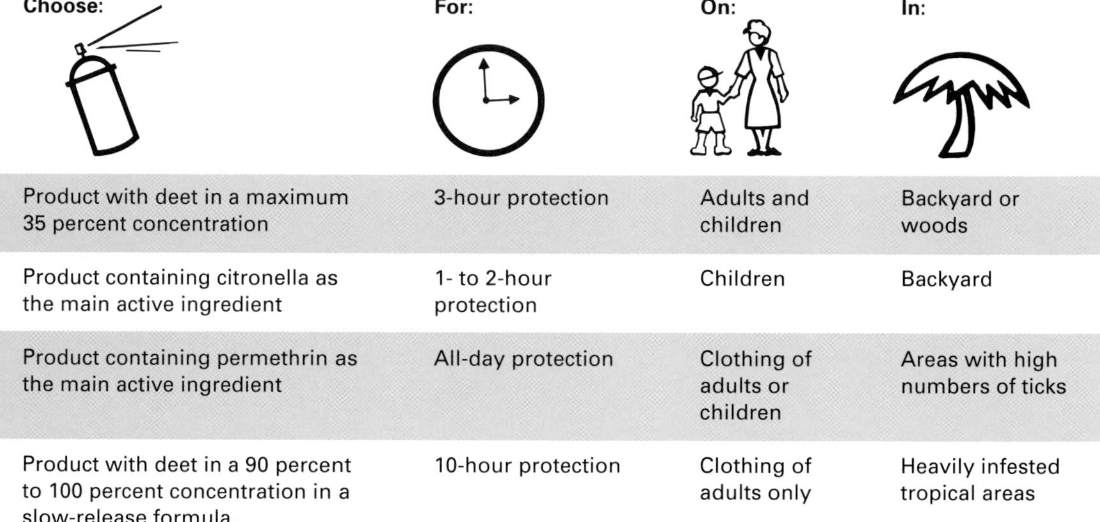

Choose:	For:	On:	In:
Product with deet in a maximum 35 percent concentration	3-hour protection	Adults and children	Backyard or woods
Product containing citronella as the main active ingredient	1- to 2-hour protection	Children	Backyard
Product containing permethrin as the main active ingredient	All-day protection	Clothing of adults or children	Areas with high numbers of ticks
Product with deet in a 90 percent to 100 percent concentration in a slow-release formula.	10-hour protection	Clothing of adults only	Heavily infested tropical areas

the chemical permethrin. This product primarily targets ticks and is registered in 21 states where ticks tend to be most prevalent.

A number of naturally occurring substances have also gained popularity as insect repellents. One of these, *citronella,* is an oil extract of the lemon-scented citronella grass. Citronella candles have long been used as a mosquito repellent on patios and porches. Because citronella is fairly mild, it is recommended for children. It must be reapplied to the skin more often than deet. While citronella is effective against mosquitoes, black flies, biting flies, and midges, it is not effective against ticks. Natrapel is a brand-name citronella product that can be applied to the skin and contains 10 percent citronella.

Some other common substances have garnered a strong following among many people. These "unofficial" repellents include eucalyptus oil, baby oil, and Avon Products Incorporated's Skin-So-Soft bath oil. No scientific studies have yet shown how these substances work, nor to what extent.

Hoping to find even more effective ways to avoid the bite of the bug, researchers are hot on the trail of new repellents. For example, investigators at International Flavors & Fragrances, Inc., located in Union Beach, N.J., have discovered that two chemicals—a synthetic compound with a rose-apple aroma and a pleasant-smelling chemical that occurs naturally in tea and several flower oils—drive away mosquitoes and houseflies. The finding is unusual because most floral scents attract insects. The researchers hope to use such fragrances in air fresheners or odor strips that would double as insect repellents. Such airborne repellents would work best in enclosed rooms.

Of course, the best way to avoid being bitten by bugs is to avoid the bugs themselves. Mosquitoes are almost everywhere in summer, but they especially like swamps, marshes, and any damp, low-lying areas. And remember, early evening is prime feeding time for mosquitoes and many other pests. So even if you applied repellent earlier in the day, you shouldn't let your guard down as night falls and the critters start looking for dinner. [Gordon Graff]

How Electronic Locks Work

Have you ever stood outside your front door on a cold night, hunting through your key chain with numb fingers for the right key? Or have you left a key under the doormat for the plumber, knowing that someone might find the key, duplicate it, and come back later to rob your home? Experts predict that in the 1990's, a growing number of people will eliminate these problems by replacing ordinary locks and keys with electronic entry systems, also known as electronic locks.

Instead of opening by means of keys, most electronic locks are programmed to open when activated by a code. In an electronic lock for residential use, the code typically consists of a series of digits that the user enters on a numbered keypad or by turning a dial. In the electronic locks used by most businesses, the code is typically carried on a small plastic card.

Electronic locks came into use in the 1970's as a way to increase security for hotels and their guests. With ordinary locks, a thief might check into a hotel, keep the room key, and return a few days later to steal the belongings of a guest who was occupying the same room. Changing a door's mechanical lock each time the key disappeared was expensive, so some hotel operators replaced their keys with encoded cards that could be read by an electronic mechanism mounted in the door. After a guest checked out of the hotel, the room card was destroyed and the door mechanism was recoded for another card.

Today, electronic locks are growing in popularity as advanced microchips, like those that power today's personal computers, make these locks cheaper and more secure. Business owners especially like electronic locks because they allow them to control access to areas where security is important, such as rooms where expensive equipment or confidential records are kept. For example, a company can make particular codes valid only for certain doors or for

Electronic locks, unlike ordinary mechanical locks, open when activated by a card/key or other code-carrying device. These locks are widely used in factories, offices, hotels, and other businesses.

a specified period of time. Many electronic locks can be programmed to accept 500 or more codes. In addition, some electronic lock systems can keep a computer record of each time a door is opened and the code that was used to open it.

Most electronic locks work in basically the same way. Typically, they have three main components: a device for receiving a code; a computer or microprocessor for processing the code; and a latch or other locking mechanism. The lock is set up by programming a code or codes into the memory of the microprocessor. The microprocessor stores these codes as a series of 0's and 1's, the binary digits that make up computer language. A person who wants to open the door enters a code into the receiving device, which translates the code into the 0's and 1's readable by the computer. If the entered code matches the code in the memory of the microprocessor, an electrical signal tells the locking mechanism to release the lock.

Electronic locks differ chiefly in the types of codes they use. In the most basic electronic locks—and the type that is most frequently found in homes—the entry code typically consists of a three- or four-digit combination. To open the door, you enter the combination on a keypad or by turning a dial. The dial works like that on a conventional combination lock: You turn it to the right to enter the first digit, to the left for the second digit, and so on. If the entered code matches a code stored in the microprocessor's memory, a signal unlocks the door.

In more sophisticated types of electronic locks, plastic cards called card/keys carry the code. The cards can be encoded in several ways.

The simplest of these systems is that developed for hotels in the 1970's, and still in use today. In these locks, each card is punched with tiny holes that form a unique pattern. A hotel guest inserts the card into a slot in the room door, where a scanner shines a light across the card. The scanner converts the pattern of light passing through the holes into electrical pulses that are sent to a central computer or, sometimes, to a microprocessor in the lock. If the code is valid for the room, the computer or microprocessor sends a signal that opens the door.

In the card/keys most commonly used in offices, the code is stored on a magnetic strip, like those that carry data on automated teller machine (ATM) cards and credit cards. The strip consists of a plastic base coated with metallic particles that act as tiny magnets. Special equipment, called an en-

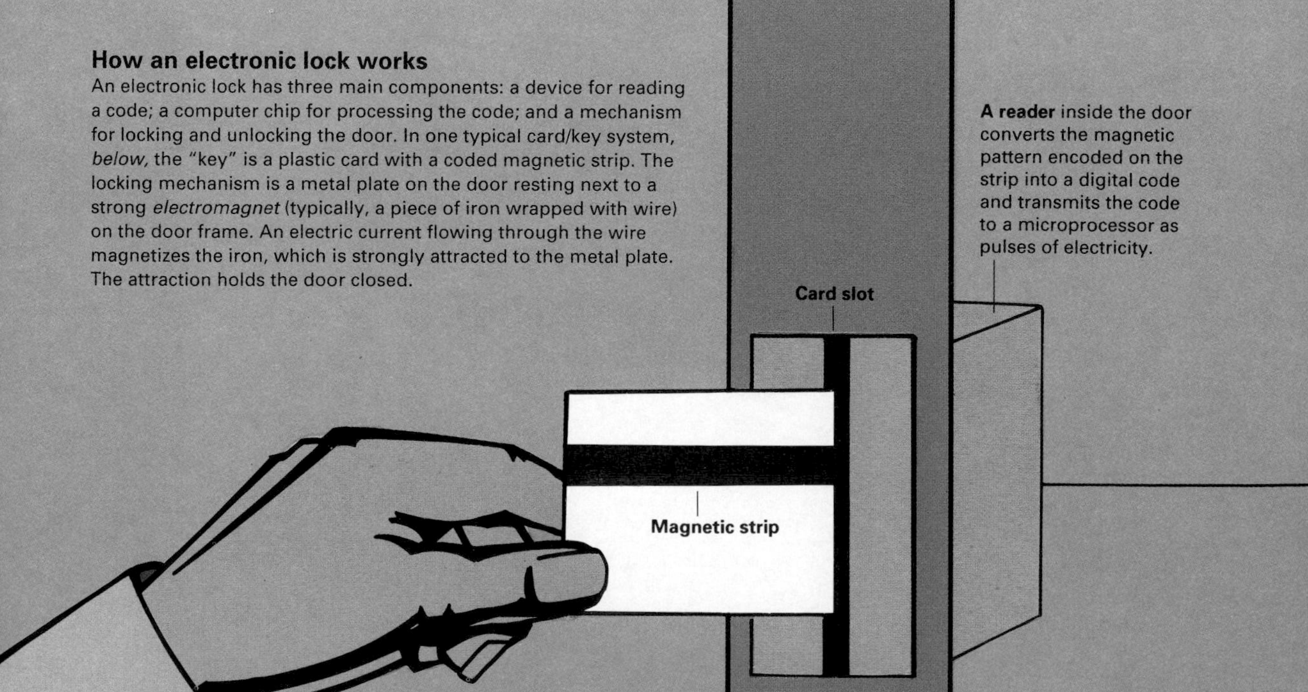
A reader inside the door converts the magnetic pattern encoded on the strip into a digital code and transmits the code to a microprocessor as pulses of electricity.
Card slot
Magnetic strip

How an electronic lock works

An electronic lock has three main components: a device for reading a code; a computer chip for processing the code; and a mechanism for locking and unlocking the door. In one typical card/key system, *below,* the "key" is a plastic card with a coded magnetic strip. The locking mechanism is a metal plate on the door resting next to a strong *electromagnet* (typically, a piece of iron wrapped with wire) on the door frame. An electric current flowing through the wire magnetizes the iron, which is strongly attracted to the metal plate. The attraction holds the door closed.

coder, reverses the direction of magnetism in some of the particles, creating a distinctive magnetic pattern that represents the code.

When the user inserts the card into a slot on the door, a special card reader translates the magnetic pattern into electrical signals corresponding to 0's and 1's. The lock's microprocessor then compares the code to that stored in its memory. If the codes match, the microprocessor sends a signal unlocking the door.

Some card/keys store the entry code in the form of a bar code, like those used to label products sold in supermarkets. A bar code consists of a printed row of lines and bars that have varying widths. The user inserts the card into a slot, where an optical scanner shines a laser beam across the code. The bars and intervening spaces reflect different amounts of light, which the scanner then translates into the 0's and 1's readable by the microprocessor.

Another type of card/key uses tiny radio transmitters like those found in garage-door openers. The transmitter sends out a series of radio waves of varying frequencies (wavelengths per second) in a pattern representing the code. A radio receiver mounted in or near the door picks up the radio signals when the user passes the card in front of it.

Some electronic entry systems work with specially coded metal keys instead of cards. In one such system, each key has a unique pattern of tiny holes much like those in the coded cards used by many hotels. The holes are covered by a plastic coating so that they cannot be copied. A mechanism inside the door reads the pattern of holes by passing infrared light, which is invisible to the human eye, through them.

Even more advanced electronic entry systems have a microchip inside a card or key, as well as on the door. These "smart cards" and "smart keys" can be programmed to keep a record of their use or to give the holder access to an area for a specific time, eliminating the need to program individual doors.

In the most sophisticated electronic locks—suitable for some high-security uses in maximum-security prisons and defense installations—the code-entering device is the user's own body. These locks are known as biometric entry systems, and their entry "code" consists of a unique characteristic, such as the person's voice, fingerprint, signature, or even the patterns of blood vessels in the eyes.

In most biometric systems, a computer stores an image of some characteristic in a central database. If the characteristic is a fingerprint, the user places a

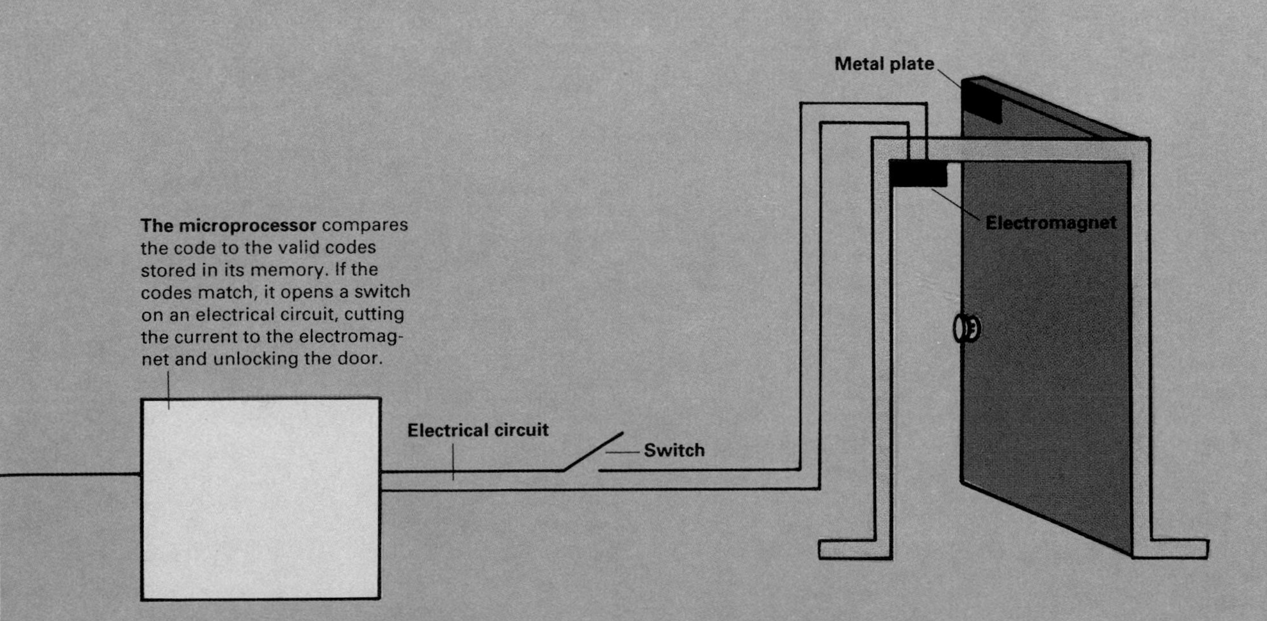

The microprocessor compares the code to the valid codes stored in its memory. If the codes match, it opens a switch on an electrical circuit, cutting the current to the electromagnet and unlocking the door.

finger against a plastic plate, and a scanner connected to the computer "reads" the fingerprint. If the print matches the computer image, the computer releases the door.

Smart cards and keys can take biometric locks a step further and eliminate the need for a central computer database of images in these systems. The microprocessor inside each card or key can store an image for the individual carrying it. To gain entry, a user inserts the card into a reader, for example, then places a finger against a plate as in ordinary biometric locks. A microprocessor in the lock compares the two images. If the image stored in the card matches the person's fingerprint, the door opens.

After a microprocessor or computer accepts a code, it signals the locking mechanism to unlock the door in two main ways. In some systems, it does this by closing a switch that completes an electrical circuit, allowing current to flow through electrical wires. The current then operates an electrical device that releases the lock.

Other systems use an electromagnetic locking device. In many of these locks, the door is held closed by a metal plate resting next to a strong electromagnet mounted at the top of the door. The electromagnet consists of a piece of iron wrapped with wire. When an electric current flows through the wire, the iron is temporarily magnetized and strongly attracted to the metal plate. The attraction holds the door tightly closed.

In such systems, entering the correct code causes the microprocessor to open an electrical switch, turning off the current. The piece of iron is then no longer magnetized and the user can pull the door open.

As yet, experts warn, most electronic locks for residential use have not proved to offer better protection against break-ins than do deadbolts and other high-security mechanical locks. And even these less-secure electronic locks may be more expensive than ordinary mechanical locks. A simple electronic lock designed for homes sells for up to $100, while one offering as much security as the best mechanical locks may cost $300 or more. In constrast, a deadbolt can be installed for well under $100.

In considering a lock for home or office use, it's important to know how the lock is programmed. Many electronic locks for home use can be programmed easily through the keypad or dial. Others must be disassembled each time you want to change the code. Some card/key systems require a computer to reprogram a lock to accept a new code. In other systems, the lock can be reprogrammed simply by inserting a recoded card into the door slot.

It's also important to know how the lock is powered. Some electronic locks get their power from the household or building current. With such locks, an electric power failure could lock you out of your home or office. A lock with an electromagnetic locking mechanism would have the opposite problem in a power outage. The magnet would shut off, releasing the lock and leaving the door open.

Many electronic locks for home use have their own power source, usually about four AAA batteries. This feature assures that the system will remain active in case of an electric power failure. Battery-powered locks are also cheaper and easier to install because they don't require special wiring. However, you must remember to check periodically to make sure the batteries in the lock are fresh.

Battery-powered locks are not suitable for most businesses, whose doors might be opened hundreds of times a day. So much use would run down the batteries. But many electronic locks designed for business use come with battery backups.

Building managers often connect electronic locks for stairwell and corridor doors into the building's fire alarm system. In case of a fire, the alarm automatically deactivates the locks, allowing people to move in and out freely. These locks are considered safer than conventional mechanical locks, which can jam and trap people inside a smoke- or fire-filled area.

Many electronic locks also come with a conventional key override, allowing the lock to be opened with a key as well as a code. This assures access in case of battery failure or an electrical problem. It also keeps you from being locked out if you forget the entry code—as long as you have remembered to carry the key.

[John Rhea]

World Book Supplement

Three new or revised articles reprinted from the 1993 edition of *The World Book Encyclopedia.*

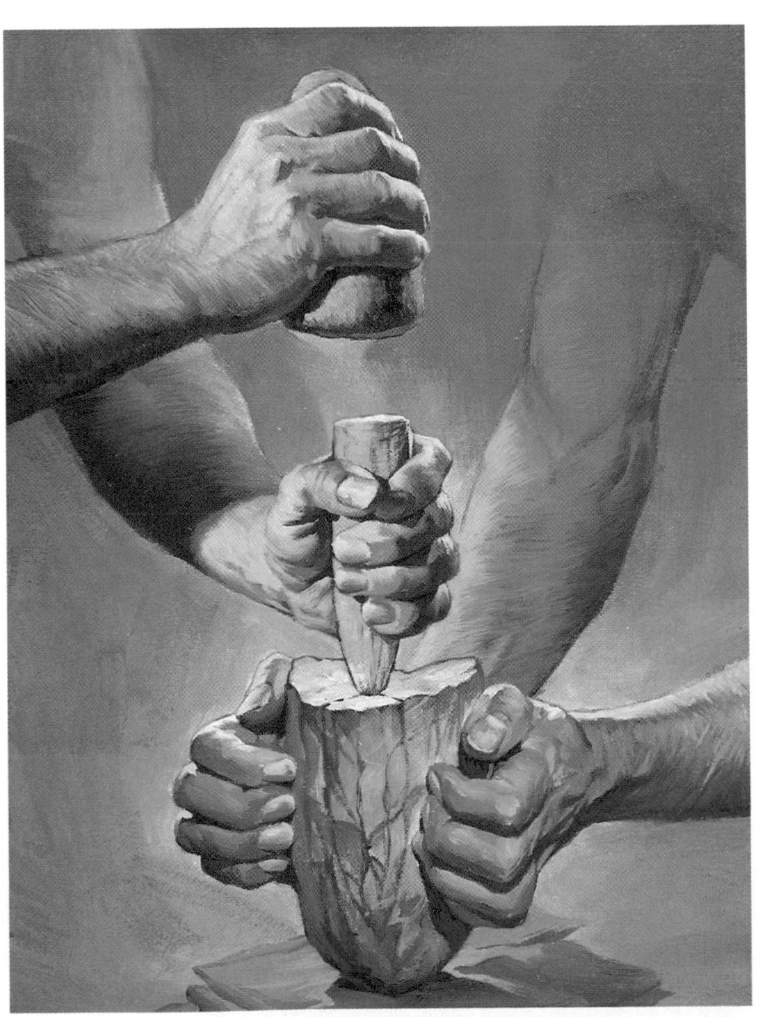

© Roger Ressmeyer, Starlight

© John Bova, Photo Researchers

NASA

Telescopes enable us to view distant objects and study the universe far beyond earth. Telescopes vary greatly in size from those used by amateur astronomers, *above,* to huge instruments in observatories, *top right.* Some telescopes orbit the earth in satellites, *bottom right.*

Telescope

Telescope is an instrument that magnifies distant objects. Astronomers use telescopes to study the planets, stars, and other heavenly bodies. Without telescopes, we would know little about the universe beyond our own planet.

Telescopes vary in shape and size from huge bowl-shaped reflectors that measure up to 1,000 feet (305 meters) across to small binoculars and gunsights. Binoculars are actually two telescopes joined side by side. In most telescopes, a lens or mirror is used to form an image of an object. The image may be viewed through an eyepiece or recorded on photographic film or by electronic devices.

The most familiar telescopes are *optical telescopes.* These instruments, like our eyes, see visible light. But objects in space give off many other kinds of radiation that people cannot see, such as radio waves and X rays. Astronomers use other kinds of telescopes to observe this radiation.

J. Roger P. Angel, the contributor of this article, is an astronomer and Director of the Mirror Laboratory at the University of Arizona's Steward Observatory. He has developed techniques for making huge, lightweight mirrors for some of the world's most powerful optical telescopes.

The Dutch optician Hans Lippershey probably made the first telescope in 1608, when he mounted two glass lenses in a narrow tube. Within a year, the Italian astronomer Galileo built a similar device and became the first person to use a telescope to study the sky. Galileo soon made discoveries that revolutionized astronomy. For example, he discovered that several moons revolve around Jupiter. In 1668, the English astronomer Isaac Newton built a telescope that used a mirror. Today, most large research telescopes use mirrors instead of lenses.

What telescopes do

Telescopes produce a clear image of objects too far away to be seen by the unaided human eye. A good pair of binoculars, for example, makes objects appear about 10 times larger or nearer than they actually are. Telescopes used by amateur astronomers enable them to see about 100 times more detail than the unaided eye can see.

Even large, powerful telescopes in observatories cannot reveal much more detail because the earth's atmosphere blurs the images of stars and other heavenly bodies. To escape this blurring, scientists launched the Hubble Space Telescope to operate above the atmosphere. A telescope's ability to see objects in fine detail is called its *resolution.*

Telescopes can also detect extremely faint objects. In optical telescopes, this ability depends on the amount of light the telescope can collect. The larger the telescope's light-gathering lens or mirror, the more light the telescope can collect. Large research telescopes can gather about 1 million times more light than the human eye can and so detect objects about 1 million times fainter.

Visible light is only one of the many kinds of *electromagnetic radiation* that reach the earth from space. This radiation moves through space in patterns called *waves.* Different types of radiation differ in their *wavelength.* Wavelength is the distance between the crest of one wave and the crest of the next. The chief types of electromagnetic radiation—in order of increasing wavelength—are gamma rays, X rays, ultraviolet rays, visible light, infrared rays, and radio waves. When electromagnetic radiation interacts with matter, it takes on the character of particles called *photons* in addition to its wave-like character. Each photon carries a specific amount of energy. The photons of radiation with the shortest wavelengths have the highest energy. Astronomers use special telescopes with electronic detectors to make images of invisible forms of electromagnetic radiation.

Some types of electromagnetic radiation, including visible light and certain radio waves, pass through the atmosphere and can be studied from the earth. But the atmosphere blocks other types of radiation—particularly ultraviolet rays, X rays, and gamma rays. Astronomers use telescopes aboard satellites to observe these forms of radiation.

Telescopes with devices called *spectrometers* enable astronomers to study individual wavelengths of electromagnetic radiation. These devices spread and separate the wavelengths of radiation to form a pattern called a *spectrum.*

Astronomers use spectrometers to determine the temperature and chemical composition of stars, planets, and gas clouds. Scientists also use spectrometers to calculate how fast an object is approaching or moving away from the earth.

Optical telescopes

Optical telescopes vary greatly in size. Binoculars may have lenses about 1 inch (2.5 centimeters) in diameter. A huge observatory telescope may have a mirror 236 inches (6 meters) in diameter. But both telescopes operate according to the same optical principles.

How optical telescopes work. Optical telescopes use a lens or mirror to collect and focus light waves. Each wave from a faint star is so weak that it can only be detected if its energy is concentrated by a lens or mirror. A lens or mirror makes the crest of a wave come together at a point called the *focus.* Waves from stars in different locations in the sky meet at different focuses, but all of the focuses lie at an equal distance from the lens or mirror in an area called the *focal plane.* The distance from a lens or mirror to the focus is called the *focal length.*

In the simplest telescopes, astronomers place photographic film at the focal plane to record images of objects in space. For direct observation, images can be magnified by an eyepiece. Most eyepieces consist of two small lenses. A viewer focuses the telescope by adjusting the eyepiece to change the distance between the eyepiece and the light-gathering lens or mirror. The eyepiece also has a focal length. The magnifying power of a telescope can be found by dividing the focal length of the lens or mirror by the focal length of the eyepiece.

Types of optical telescopes. There are three main types of optical telescopes: (1) refracting telescopes, (2) reflecting telescopes, and (3) refracting-reflecting telescopes.

Refracting telescopes, also called *refractors,* have a large lens called an *objective lens*—or simply an *objective*—at one end of a long, narrow tube. The objective is *convex* (curved outward) on both sides so that the middle of the lens is thicker than the edges. The glass slows the light waves as they pass through the lens. A wave is slowed most in the middle of the lens, where the glass is thickest. The lens thus causes the entire crest of the wave to arrive at the focus at the same time.

Some important optical telescopes

Telescope	Location	Operated by	Number of mirrors	Diameter of mirrors (in feet)	(in meters)
Bolshoi Teleskop Azimutal'ny (BTA)	Zelenchukskaya, Russia	Special Astrophysical Observatory	1	19 3/4	6
Cerro Tololo Inter-American Observatory	Cerro Tololo, Chile	National Optical Astronomy Observatories	1	13 1/2	4.1
Hale Telescope	Palomar Mountain, California	California Institute of Technology	1	16 3/4	5
Keck Telescope	Mauna Kea, Hawaii	California Association for Research in Astronomy	1*	33	10
Multiple Mirror Telescope (MMT)	Mount Hopkins, Arizona	University of Arizona	6†	6	1.8

*The telescope's reflecting surface consists of 36 six-sided mirrors measuring 6 feet (1.8 meters) across.
†The six mirrors have a light-gathering power equal to a single 14 3/4-foot (4.5-meter) diameter mirror.

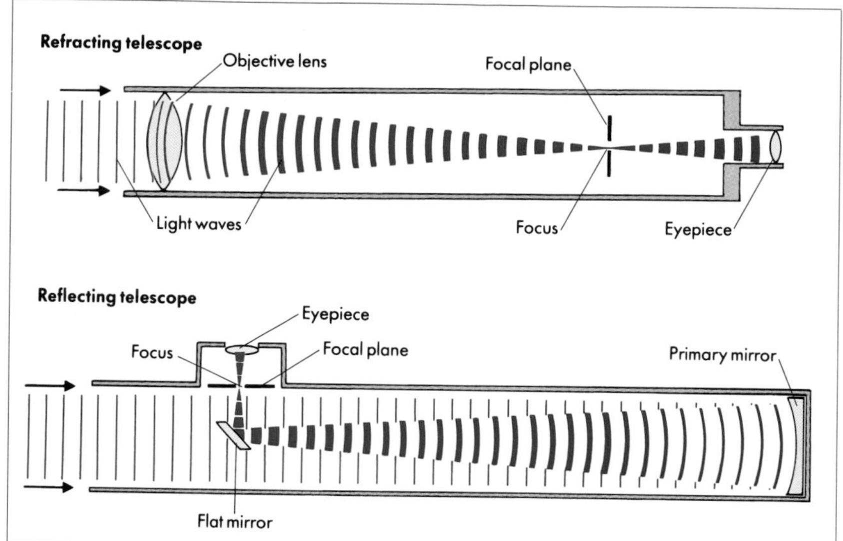

WORLD BOOK illustrations by Oxford Illustrators Limited

How optical telescopes work

Optical telescopes form an image of a star or other heavenly body in two ways. In a refracting telescope, light waves enter a glass lens, which concentrates each wave crest at a point called the focus. There, an image is formed that can be viewed with an eyepiece. In a reflecting telescope, a bowl-shaped mirror reflects the light waves to a focus. This design, called a Newtonian telescope, uses a small, flat mirror to reflect the light to an eyepiece at the side.

Refractors with a magnifying eyepiece invert the image so that it appears upside down. Astronomical observations do not require an upright image. But telescopes used to observe objects on the earth, such as binoculars, gunsights, and surveying equipment, use additional lenses or prisms to turn the image right side up again.

Galileo made all of his discoveries using refracting telescopes. Galileo's instruments and other early refractors, however, produced images with rainbow coloring around the edges called *chromatic aberration.* This coloring appeared because a lens slows blue light more than red, giving blue light a shorter focal length. When white light, which consists of all colors, passes through a lens, only one color focuses correctly.

Astronomers found that gently curved lenses made chromatic aberration less noticeable. But these lenses had long focal lengths and required extremely long tubes. To reduce chromatic aberration, some early telescopes stretched more than 200 feet (60 meters) long. In the mid-1700's, however, astronomers discovered they could make a compound lens of two different types of glass that had a short focal length and almost no chromatic aberration.

Reflecting telescopes, also called *reflectors,* use bowl-shaped mirrors instead of lenses. The mirror, called the *primary mirror,* has a surface shaped so that any line across the center of the mirror is a *parabola,* a curve like the path of a ball batted high in the air. A mirror with that shape, called a *parabolic mirror,* reflects light rays to a sharp focus in front of itself. There, a second mirror reflects the rays to an eyepiece.

Astronomers generally prefer reflecting telescopes to refracting telescopes. The weight of a large lens can cause it to bend and become distorted. But a large, heavy mirror can be supported from behind. As a result, mirrors can be made much larger than lenses and, thus, can gather more light. In addition, parabolic mirrors are useful because they can collect infrared and some ultraviolet rays as well as visible light.

The English astronomer Isaac Newton designed one of the first reflectors in 1668 to avoid chromatic aberration caused by lenses. He used a small, flat mirror to reflect light from the primary mirror to an eyepiece at the side of the telescope tube. In 1672, a French telescope maker known only as Cassegrain designed a telescope using a small convex mirror in front of the primary mirror. The small mirror reflected the light through a hole in the primary mirror to an eyepiece behind it. This design, called a Cassegrain telescope, is used most frequently by astronomers today for optical and infrared telescopes.

The mirrors of early reflecting telescopes were shaped like a section of a sphere. A spherical mirror is much easier to polish than a parabolic mirror, but it does not focus light properly. Astronomers developed techniques to make parabolic mirrors in the early 1700's. Early mirrors were made of *speculum metal,* a heavy mixture of copper and tin that tarnished easily and required repeated polishing. In the mid-1800's, the German chemist Justus von Liebig learned how to deposit a thin coating of silver on glass to produce a brilliantly reflecting surface. When the surface tarnished or dulled, the mirror could be recoated without having to polish it.

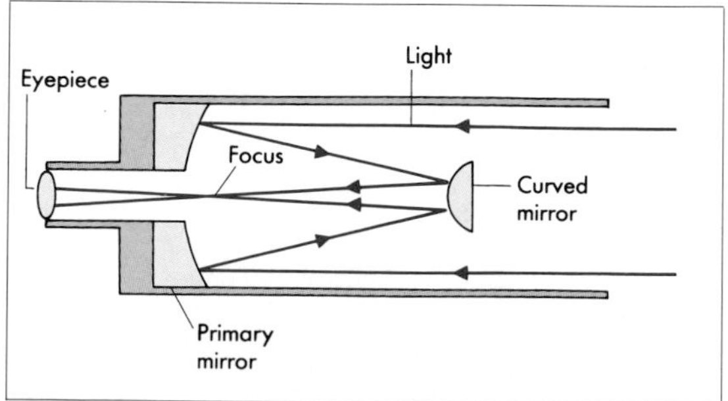

WORLD BOOK illustration by Oxford Illustrators Limited

A Cassegrain reflecting telescope uses a primary mirror with a hole in the center to reflect light to a smaller, curved mirror. The smaller mirror reflects the light through the hole, where it may be viewed with an eyepiece or recorded.

Today, all reflecting telescopes have glass mirrors, most of which are coated with aluminum.

The first telescope to use a large glass mirror was built in 1908 on Mount Wilson, near Pasadena, Calif. It has a Cassegrain design, and its mirror measures 60 inches (1.5 meters) in diameter. Most large telescopes today are modeled after this telescope. One of the largest is the Hale Telescope, an instrument with a 200-inch (5-meter) mirror on Palomar Mountain in southwestern California. Another telescope of the same type at Zelenchukskaya, near Mineralnyye Vody in southwestern Russia, has a mirror 6 meters (20 feet) in diameter.

The Multiple Mirror Telescope on Mount Hopkins, near Tucson, Ariz., has a much different design. Completed in 1979, the telescope has six 72-inch (1.8-meter) mirrors that work together. For a single mirror to collect as much light, it would have to be 176 inches (4.5 meters) in diameter.

Refracting-reflecting telescopes, also called *catadioptric* telescopes, have a large lens at the front end of the tube and a large mirror at the rear. These telescopes use spherical mirrors instead of parabolic mirrors. The lens, however, refracts light rays slightly to correct the reflective errors caused by a spherical mirror.

Bernhard Schmidt, a German optician, invented the catadioptric telescope in 1930. The telescope forms images of a larger region of the sky than is possible with any other telescope design. Astronomers have used large Schmidt telescopes to photograph the entire sky.

Recording images produced by optical telescopes. Astronomers often use photographic plates or film to record the images formed by an optical telescope. If film is exposed to a dim star or other object for a long time, a bright picture results. For this reason, photographs of the sky taken through a telescope reveal many details that cannot be seen with the eye. In the mid-1970's, electronic detectors called *charge-coupled devices* (CCD's) began to replace film. CCD's convert light into an electric charge, which is used to form images on a computer screen. CCD's produce better pictures than photographic plates or film because they are far more sensitive to light. Using CCD's, astronomers can see extremely faint galaxies in almost any part of the sky. Some of these galaxies are so far away that their light began its journey toward the earth long before the earth formed about $4\frac{1}{2}$ billion years ago.

Limitations of optical telescopes. Astronomers can see galaxies across the universe with big optical telescopes. The images are blurred, however, by the earth's atmosphere. Wind and daily heating and cooling of the atmosphere create pockets and swirls of warm and cool air. These differences in air temperature affect the direction and speed of light through the air. As a result, the waves of starlight arrive at the focus at slightly different times, spoiling the image. When atmospheric conditions cause little blurring, astronomers say "seeing" is good.

Since the late 1970's, astronomers have discovered that they can improve seeing by insulating and cooling observatory domes. But only a telescope operating above the atmosphere can escape all blurring. One such telescope, an orbiting observatory called the Hubble Space Telescope, was launched into orbit in 1990. Although a flawed mirror has prevented the telescope from working as well as scientists hoped, the Hubble has produced sharper, more finely detailed images than any telescope on the earth.

Ground-based telescopes may someday overcome atmospheric blurring with instruments called *adaptive optics.* These devices will measure and correct the timing error of light arriving across a mirror to restore a sharp focus. Adaptive optics are extremely difficult to make because air currents move and change so quickly. Computers must calculate and correct the timing error several hundred times a second. Astronomers developed the concept of adaptive optics in the 1950's. But they did not begin building such systems until the early 1990's, after scientists developed the advanced technology the systems required.

Radio telescopes

Radio telescopes collect and measure faint radio waves given off by objects in space. An American engineer, Karl G. Jansky, discovered radio waves from space in 1931. In the late 1930's, Grote Reber, another American engineer, built the first bowl-shaped radio telescope and operated it in his backyard. Early radio telescopes found that the sun and the center of our galaxy were strong sources of radio waves. They also detected strong radio waves coming from dark areas of space. These sources were discovered to be the remains of exploded stars and a rare type of distant galaxy. Since then, astronomers using radio telescopes have discovered objects in space that had been missed by optical telescopes. These discoveries include giant clouds of gas molecules; *pulsars,* collapsed stars that send out regular pulses of radio waves; and *quasars,* extremely distant starlike objects that produce enormous amounts of radiation.

How radio telescopes work. Most radio telescopes use a large parabolic reflector, often called a *dish antenna* or simply a *dish,* to collect radio waves from space. The dish has the same shape as the parabolic mirror of a reflecting telescope. Radio waves, however, are much longer than light waves. As a result, a radio telescope's dish need not be polished or shaped as accurately as the mirror of a reflecting telescope. But it must be much larger in diameter to focus the long radio waves. The reflector focuses the waves onto an antenna that changes them into electric signals. A radio receiver amplifies these signals and records their strength at different frequencies and from different directions as data on a tape. The data are analyzed by a computer, which combines the signals from the receiver. The computer then uses the signals to draw a picture of the source of the radio waves or to analyze the radio spectrum and chemical composition of the source.

Large radio telescopes are also used as giant radar systems to map the surfaces of the moon and the planets. Astronomers send powerful radio waves to the moon or planet and then record the radio echoes that bounce back. Astronomers call this *radar mapping.*

Types of radio telescopes. In most radio telescopes, motors turn the reflector toward any source of radio waves in the sky. The largest moving dish measures 330 feet (100 meters) across. Astronomers can use a single large fixed dish to study radio signals from a faint object. The world's largest radio telescope is a fixed dish built into a bowl-shaped valley near Arecibo, Puerto

E-Systems

The Very Large Array radio interferometer, located near Socorro, N. Mex., is one of the world's most powerful radio telescopes. It consists of 27 radio telescope reflectors, each measuring 82 feet (25 meters) in diameter.

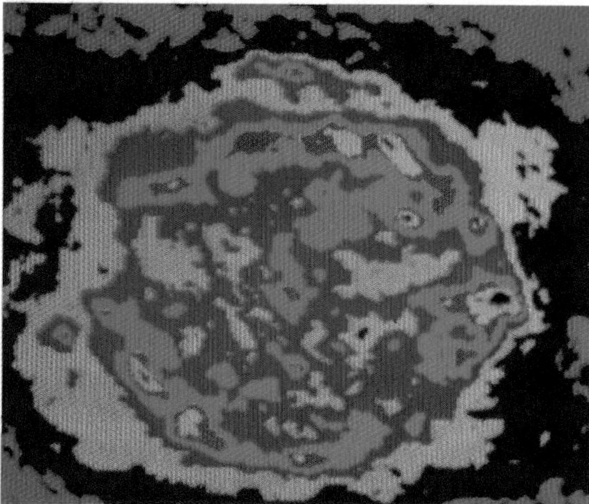

© Dan McCoy, Rainbow

A radio image of the remains of an exploded star was created by a computer from radio waves collected by the Very Large Array radio interferometer. This image shows a star that exploded about 300 years ago in the constellation Cassiopeia.

Rico. The dish is 1,000 feet (305 meters) in diameter. It is often used for locating and measuring pulsars.

Astronomers produce extremely sharp radio images by combining signals from many radio dishes spread over large distances. At a central station, computers electronically combine the radio signals from various locations, introducing time delays between the signals from the different dishes. These delays cause the signals from a radio wave to come together at the same time and reinforce each other, just as a light wave is concentrated at the focus by a mirror or lens. Radio telescopes connected in this way are called a *radio interferometer.* The longer the *baseline* (distance) between the telescopes, the better the resolution of the interferometer. Astronomers use interferometers to make radio maps of the sky.

The most powerful radio interferometer is called the Very Large Array (VLA). It stands on a high plain near Socorro, N. Mex. It has 27 dishes, each measuring 82 feet (25 meters) in diameter. Another important interferometer, the Very Long Baseline Array (VLBA), was scheduled for completion in late 1992. The system consists of 10 reflectors located across the United States from Hawaii to the Virgin Islands. Scientists expect the VLBA to provide the sharpest radio images ever produced.

Infrared telescopes

An infrared telescope collects *infrared* (heat) rays from objects in space. Most infrared telescopes are reflecting optical telescopes with an infrared detector instead of an eyepiece. Any object at room temperature gives off huge amounts of infrared rays because of the heat it holds. As a result, astronomers must design infrared telescopes so that heat from the telescope itself does not interfere with the radiation from space. They also must cool parts of the telescope to extremely low temperatures to detect infrared rays from the coldest sources, which are very faint. Some infrared waves from space pass through the atmosphere. But water vapor and carbon dioxide in the air block many others. For this reason, astronomers install infrared telescopes on mountaintops where the air is thin and dry. They also send infrared telescopes above the earth's atmosphere aboard high-flying planes or satellites.

In 1961, American physicist Frank J. Low built the first infrared detector sensitive enough for use in astronomy. The device, called a *bolometer,* was an extremely cold electronic thermometer in a vacuum. When infrared rays hit the bolometer, it warmed up and gave off an electric signal. Today, infrared telescopes use electronic devices called *array detectors,* similar to CCD's, to form infrared images on a computer screen.

An infrared telescope operated in orbit aboard the Infrared Astronomical Satellite (IRAS) from January to November 1983. Liquid helium cooled the entire telescope —its mirrors, detectors, and tube—to a temperature only a few degrees above absolute zero (−273.15 °C). IRAS detected rings of dust around the star Vega and other nearby stars that might be solar systems in the process of formation. Astronomers believe a similar ring of dust around the sun developed into the planets.

Other telescopes

Electromagnetic radiation with short wavelengths has the highest photon energies among all the kinds of electromagnetic radiation. These forms of radiation include ultraviolet rays, X rays, and gamma rays. Because of their high energy, these rays cannot be easily reflected by a mirror, as light can. As a result, the telescopes used to observe these forms of radiation often look quite different from other telescopes. Another difference is that—except for the least energetic ultraviolet rays—the atmosphere absorbs these rays before they reach the earth. To study these high-energy forms of radiation, astronomers must send telescopes above the atmosphere on rockets or satellites.

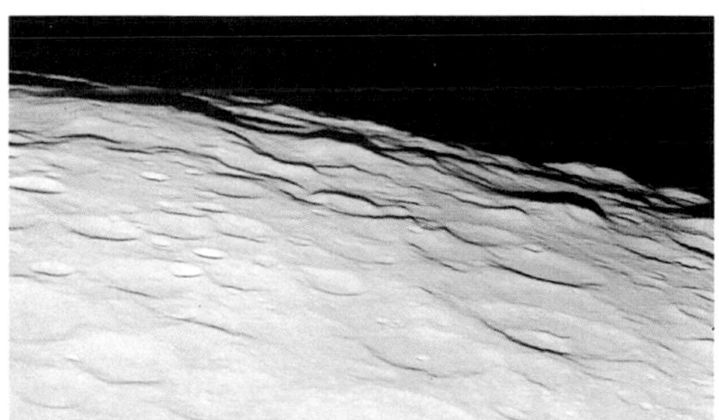

Steward Observatory, University of Arizona

An infrared image of the moon, *above,* shows the warmest parts of the moon's surface as bright areas. The dark areas are partially or completely hidden from sunlight and are cooler.

Ultraviolet telescopes. Astronomers use reflecting telescopes in space with electronic detectors to study most wavelengths of ultraviolet rays, which can be reflected just like visible light. But the shortest wavelengths, called *extreme ultraviolet rays,* are harder to reflect. Extreme ultraviolet rays can only be reflected off a mirror at a small angle called a *grazing incidence.* This characteristic of the rays resembles how stones can be skipped along the surface of a pond.

Ultraviolet telescopes enable astronomers to study extremely hot objects in space, including quasars and stars called *white dwarfs.* Astronomers also use ultraviolet telescopes to study how stars form and the composition of gas between stars and galaxies.

X-ray telescopes. X rays have shorter wavelengths and higher energy than ultraviolet rays. X rays that are not absorbed or scattered by matter pass straight through many materials. But longer wavelength X rays, like extreme ultraviolet rays, can be reflected at a grazing incidence. Astronomers have found that some objects in the universe give off much of their energy in the form of X rays. These X-ray sources include the centers of galaxies and clouds of hot gas between galaxies.

The simplest X-ray telescopes use an arrangement of iron or lead slats instead of mirrors. The slats block all X rays except those from one line across the sky. The X-ray photons then enter a detector filled with an X-ray absorbing gas, where they are counted. By scanning the sky, these telescopes can locate X-ray sources.

During the 1970's, slatted X-ray telescopes discovered many sources of X rays in space. Astronomers now know that many of the brightest X-ray sources are certain *double stars*—that is, a pair of stars orbiting each other. In these pairs, one of the stars has collapsed and become a small, dense star called a *neutron star* or a *black hole*—an invisible object with such powerful gravitational force that not even light can escape its surface.

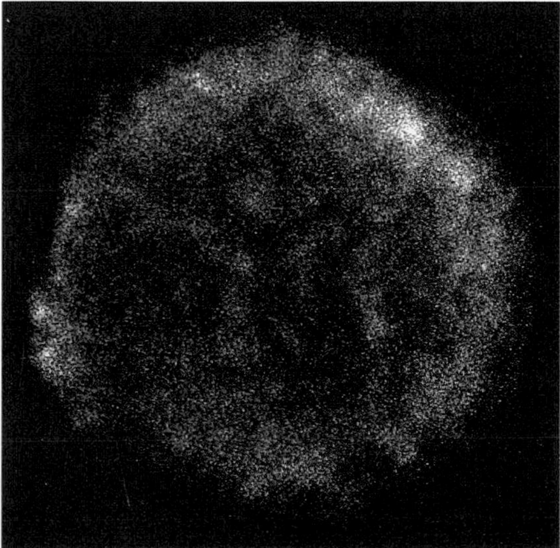

NASA

An X-ray image taken by the Rosat satellite in 1992 shows the remains of a star that exploded in 1572. The explosion was so violent that it was clearly visible to the unaided eye.

An X-ray telescope aboard Rosat, a satellite launched in 1990, reflects X rays off curved mirrors at a slight angle called a *grazing incidence.* This method of focusing X rays resembles how stones can be skipped along the surface of a pond.

WORLD BOOK illustrations by Oxford Illustrators Limited

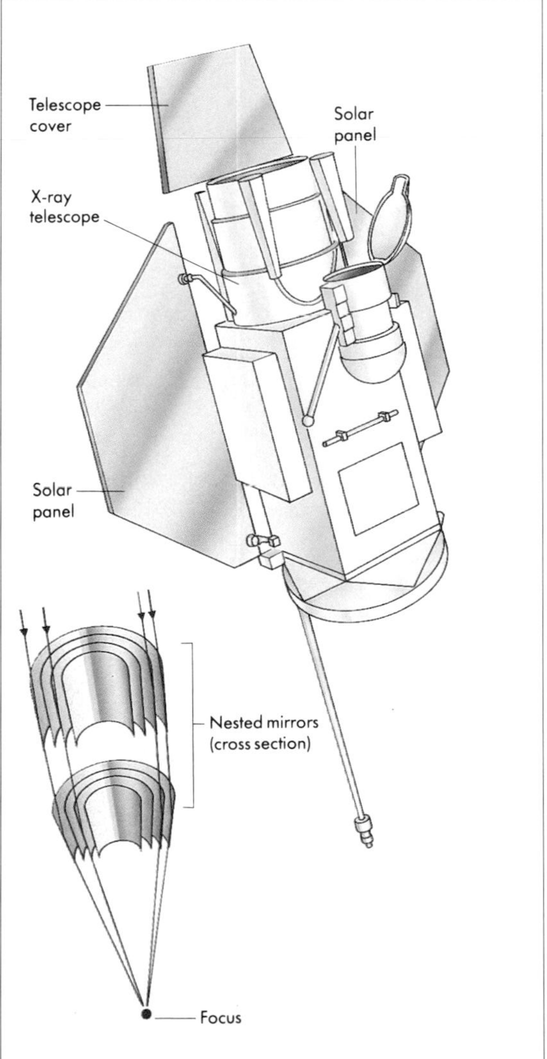

X rays occur when gas from a star falls into the neutron star or black hole.

Gamma ray telescopes. Gamma rays have the shortest wavelength and highest energy of any electromagnetic radiation. When a gamma ray photon collides with an atom while passing through matter, it may knock electrons loose from the atom or even break up the nucleus of the atom. These collisions can produce a shower of subatomic particles and low-energy radiation. The shower travels in the same direction as the original gamma ray and is detected with devices called *scintillators.* When radiation or particles from a shower hit a scintillator, the instrument produces a flash of light that can be recorded. By measuring the shower, scientists can calculate the energy level of the gamma ray and the direction of its source. Gamma ray telescopes on the Compton Gamma Ray Observatory, a satellite launched in 1991, have enabled scientists to learn more about some of the least understood objects in the universe, including pulsars and quasars. Many of these high-energy objects are strong sources of gamma rays.

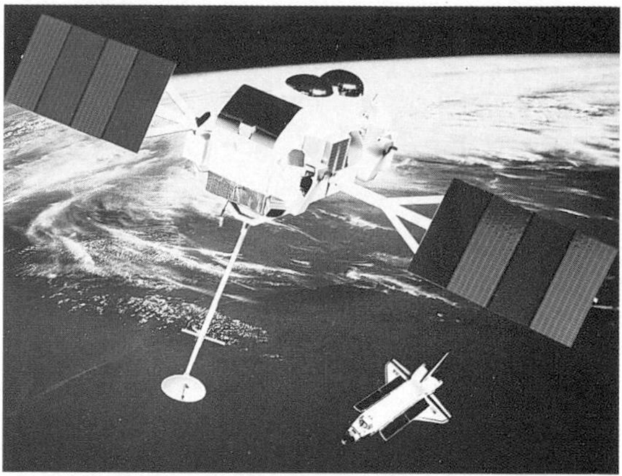

NASA

The Compton Gamma Ray Observatory, a satellite launched in 1991, carried instruments called *scintillators* to detect gamma rays, measure their intensity, and locate their sources.

Recent developments in telescopes

In the 1980's, astronomers began work on a new generation of ground-based optical telescopes larger than any built before. Astronomers hope to use the telescopes to unlock many secrets of the universe, including how planets, stars, and galaxies form. These telescopes will use sensitive optical and infrared detectors and adaptive optics to produce extremely sharp images. In addition, the telescopes will have larger, more accurate mirrors than any instruments built before. Several breakthroughs in mirror design have enabled astronomers to make large mirrors that do not bend or become distorted under their own weight.

One new design is the segmented mirror, used in the Keck Telescope on the island of Hawaii, completed in 1992. The Keck's light-gathering mirror consists of 36 *hexagonal* (six-sided) mirrors mounted close together. The mirrors form a reflecting surface 33 feet (10 meters) in diameter. A second Keck Telescope of the same design is due to begin operating at the site in 1996.

Some projects involve linking two or more telescopes to collect more light. A project called the Very Large Telescope (VLT) will consist of four telescopes with mirrors 27 feet (8.2 meters) in diameter. Used together, the four telescopes will have the light-gathering power of a single mirror with a diameter of 52 feet (16 meters). The VLT's mirrors will consist of thin disks of glass supported by hundreds of computer-controlled devices. The devices, called *actuators,* will make continuous adjustments to maintain the mirrors' proper shape. The European Southern Observatory, led by astronomers from several European nations, is building the VLT near Antofagasta, Chile. The project is scheduled to start operations in the late 1990's.

Astronomers at the University of Arizona have made huge glass honeycomb mirrors using a mold filled with hundreds of hexagonal blocks. Melted glass covers the blocks and fills spaces between them. The blocks are removed after the glass cools, leaving a stiff glass structure that is light enough to float on water.

The Columbus Telescope, scheduled for completion in 1997, will also use honeycomb mirrors. The instrument will consist of two telescopes, each with a mirror $27\frac{1}{2}$ feet (8.4 meters) in diameter, mounted side by side like a giant pair of binoculars. The Columbus Telescope is a joint project of Italian and American astronomers. It will stand on Mount Graham in southeastern Arizona. Several other telescopes with honeycomb mirrors were under construction in the mid-1990's.

Both honeycomb mirrors and the VLT's thin disk mirrors are made by a new technique called *spin-casting,* developed in the mid-1980's. Spin-casting replaces the costly, laborious process of grinding a mirror to the proper parabolic shape. Instead, a huge rotating oven spins molten glass at a carefully controlled rate. The liquid glass flows into a shape that is nearly perfect for a telescope mirror.

J. Roger P. Angel

© Roger Ressmeyer, Starlight

The Keck Telescope, on Mauna Kea on the island of Hawaii, has a huge segmented mirror measuring 33 feet (10 meters) across. The mirror consists of 36 smaller mirrors.

© Jay H. Matternes

Prehistoric people are the ancestors of modern human beings. This illustration is an artist's idea of how one type of prehistoric people, *Homo erectus,* may have learned to use fire. Cultural advances, such as the use of fire and clothing, helped *Homo erectus* spread to much of the world.

Prehistoric people

Prehistoric people are human beings who lived before writing was invented about 5,500 years ago. Writing enabled people to record information they wished to save, including descriptions of events in their lives. In this way, the invention of writing marked the beginning of history. The period before human beings learned to write is called *prehistory,* and people who lived during this period are known as *prehistoric people.*

Most scientists believe the first human beings lived about 2 million years ago. But early humans probably arose from ancestors who first lived more than 4 million years ago. These prehuman ancestors were small, humanlike creatures who walked erect. This article will discuss both prehistoric people and their near ancestors.

Scientists first discovered evidence of prehistoric people during the mid-1800's. Most of this evidence consisted of ancient, sharp-edged tools that prehistoric people had made of stone. The first fossilized bones of prehistoric people were also found during this time.

As scientists collected more fossils of prehistoric people, they began to form a clearer picture of what these early people looked like. For example, scientists learned from fossil evidence that early human beings had smaller brains than most modern men and women have. This evidence indicated to many scientists that humans had *evolved*—that is, modified their physical structure over time. Scientists developed a set of ideas about human origins called the theory of human evolution. This theory states that as the environment of the prehistoric world changed, the prehuman ancestors of prehistoric people went through a series of changes that resulted in the first human beings. They, in turn, evolved into modern human beings.

Today, many kinds of scientists work together to learn about prehistoric people. Archaeologists search for and examine such physical evidence as pottery and tools to help explain how prehistoric people lived. Botanists study the remains of prehistoric plants, and zoologists analyze fossils of prehistoric animals that lived during the time of prehistoric people. Geologists study the layers of rock in which fossils are found. All these scientists are called *anthropologists* if their chief concern is the study of human physical and cultural development.

Evidence of prehistoric people—such as fossils, tools,

Alan E. Mann, the contributor of this article, is Professor of Anthropology at the University of Pennsylvania and coauthor of Human Biology and Behavior.

and other remains—is rare and often fragmented. Evidence of the earliest types of prehistoric people is the most difficult to find. Anthropologists must base their theories about prehistoric people's way of life on this extremely limited evidence. As a result, scientists cannot yet present a detailed picture of early human life. In addition, new discoveries sometimes disprove theories that scientists already hold. Scientists then must rethink their ideas about the lives of our prehistoric ancestors.

Prehuman ancestors

Most scientists believe that human beings and apes—such as chimpanzees and gorillas—share a common ancestor. To support this theory, scientists point out that the fossilized remains of ancient humanlike beings and apes reveal many similarities, including similar brain sizes. In addition, scientists have conducted studies comparing the physical structure, blood, and genetic material of modern humans with those of apes. These studies show that people are more similar to apes than to any other type of living animal.

The ancestors of human beings probably began evolving separately from the ancestors of apes between about 10 million and 5 million years ago. This evolutionary split marks the beginning of the development of *hominids.* Hominids are members of the scientific family made up of human beings and early humanlike ancestors. Most anthropologists believe that the first hominids were humanlike creatures called *australopithecines* (pronounced *aw STRAY loh PIHTH uh seenz*).

Where and when they lived. The australopithecines first appeared between $5\frac{1}{2}$ million and 4 million years ago in Africa. Fossil evidence suggests that these creatures became extinct between 2 million and 1 million years ago, about when the first humans appeared.

Scientists have discovered australopithecine fossils at sites in eastern and southern Africa. Because these are

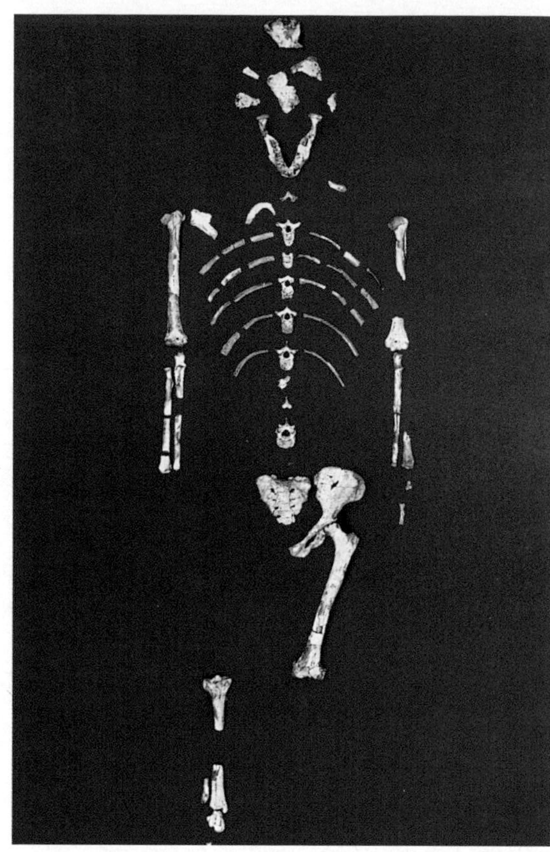

Institute of Human Origins

The skeleton of "Lucy," a prehuman ancestor, is the most complete australopithecine fossil that scientists have found. This creature lived about 3 million years ago.

Where remains of prehistoric people have been found

The earliest fossils and other remains of prehistoric people have been found in Africa, Asia, and Europe. Most scientists believe that our closest prehuman ancestors originated in Africa, and prehistoric people later spread to other parts of the world.

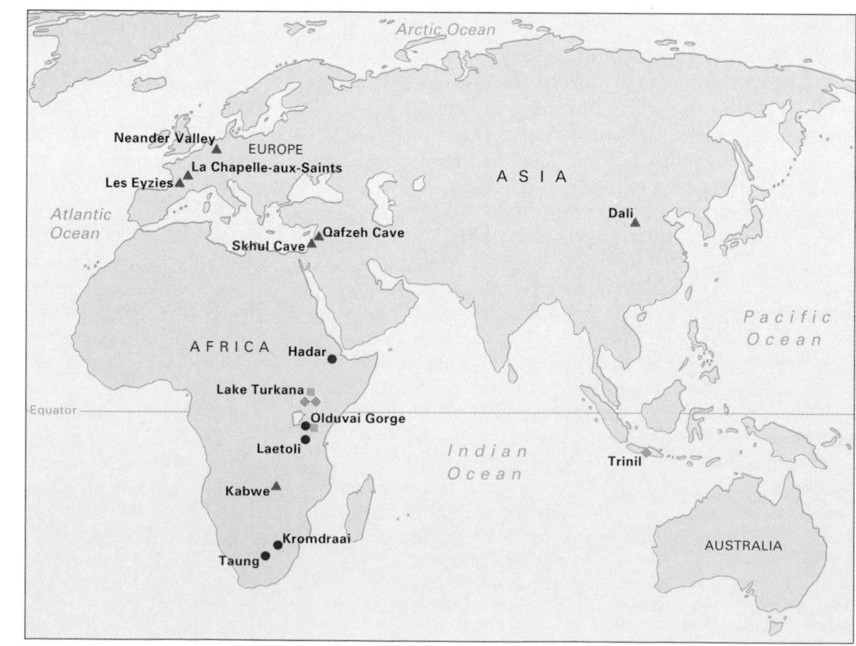

WORLD BOOK map

the oldest examples of hominid fossils, most scientists generally believe that the hominid family originated in Africa and prehistoric people later spread out into other parts of the world.

What they looked like. The australopithecines looked very different from modern human beings. In some ways, such as in their facial features, they may have resembled chimpanzees. However, they could stand upright and walk on two legs, and their canine teeth were much smaller and less pointed than those of apes. These features identify australopithecines as members of the hominid family and separate them from apes.

The australopithecines had large faces that jutted out. Their brains were about one-third the size of modern human brains. Their molars were large, flat, and suitable for grinding food. Anthropologists believe from the shape of these creatures' teeth that they ate such foods as fruits, vegetables, nuts, seeds, and insects.

Types of australopithecines. The australopithecines were members of the genus *Australopithecus* (southern ape). According to differences in the shape of the creatures' jaws and teeth and the size of their brains, scientists have divided the genus *Australopithecus* into four species: (1) *A. afarensis,* (2) *A. africanus,* (3) *A. robustus,* and (4) *A. boisei.*

The earliest species of *Australopithecus* was *A. afarensis,* which appeared in eastern Africa by 4 million years ago. The most complete australopithecine fossil scientists have found is a partial skeleton of a female *A. afarensis.* The fossil was found at Hadar, Ethiopia. Scientists estimate that this creature, nicknamed "Lucy," was slightly more than $3\frac{1}{2}$ feet (107 centimeters) tall and that it weighed approximately 60 pounds (27 kilograms). *A. afarensis* had about the same size brain that a chimpanzee has.

By about $2\frac{1}{2}$ million years ago, *A. africanus* replaced *A. afarensis.* Scientists have found fossils of *A. africanus* at several sites in South Africa. These creatures had rounder skulls and slightly larger brains than those of *A. afarensis,* but they were otherwise not much different.

Many scientists believe that an evolutionary split occurred among the australopithecines during the time of *A. africanus.* This split resulted in the appearance of an additional evolutionary line, separate from *A. africanus,* that led to *A. robustus* and *A. boisei.* Scientists refer to these two species as the *robust* australopithecines. They had larger molars and more powerful jaws than the other two species of *Australopithecus.* But their brain size was about the same as that of *A. africanus.* The earlier two species are called *gracile* (slender) australopithecines. The robust australopithecines probably became extinct about 1 million years ago.

© John Reader, SPL from Photo Researchers

Footprints of a prehuman ancestor were found at Laetoli, in Tanzania. These footprints, which were fossilized in volcanic ash, provide evidence that early hominids walked upright.

The first human beings

Most anthropologists believe that the first human beings evolved from a gracile australopithecine about 2 million years ago. The oldest tools that scientists have found date from about $2\frac{1}{2}$ million years ago. But because no hominid fossils were found with these tools, scientists do not know whether an australopithecine or an early human made them.

Most prehistoric tools that have been found and studied are made of stone. As a result, this period of time is called the Stone Age. Early toolmakers may also have used wood and other materials, but none of those tools have survived. The Stone Age lasted from the first use of stone tools until bronze replaced stone as the chief toolmaking material. In some areas, this occurred about 3000 B.C. The first part of the Stone Age is called the Paleolithic Period. This period lasted until about 8000 B.C., after people had started farming. Even after some people learned to raise food by farming, many others continued to live by gathering wild plants and by hunting. These Stone Age hunters and gatherers who lived after 8000 B.C. are called Mesolithic people. Farmers from this period are called Neolithic people.

Homo habilis is considered by anthropologists to be the oldest human species. These prehistoric people lived in Africa about 2 million years ago. The Latin word *homo* means *human being. Habilis* means *handy* or *skillful.* Anthropologists have found important fossils of *Homo habilis* at sites east of Lake Turkana in northern Kenya and in Olduvai Gorge in Tanzania.

Homo habilis' brain was much larger than that of an australopithecine, but only about half the size of a modern human brain. *Homo habilis* also had smaller molars and a less protruding face than the australopithecines had. Some fossil evidence indicates that *Homo habilis* males were much larger than *Homo habilis* females. This difference, known as *sexual dimorphism,* appears among many modern primates. Scientists have also detected such a difference among the australopithecines. Among modern human beings, however, sexual dimorphism is less extreme.

WORLD BOOK illustration by Nathan Greene

© Lee Boltin

© Lee Boltin

Prehistoric tools

Prehistoric tools were made chiefly of stone. Neanderthals made a variety of tools, including the hand axe, *top left,* and the scraper, *bottom left.* To form these tools, they used a hard object—such as a rock or bone—to chip pieces away from a carefully selected stone. Later prehistoric people used more complicated toolmaking techniques, such as the one shown on the far left. This process, employed by Upper Paleolithic people, required the cooperation of two individuals. The toolmakers used two hard objects as a hammer and chisel to split bladelike slivers from a large stone. In this way, they could create many useful tools from a single stone.

Many anthropologists believe *Homo habilis* made the first tools. Some of the earliest known tools have been found with *Homo habilis* fossils. These devices were sharp-edged stones used for cutting, scraping, and chopping. Prehistoric people made them by striking one piece of stone with another, chipping pieces away to produce a cutting edge. These first tools were extremely crude, but over time early human beings began to craft tools of a finer quality. Later toolmakers started using mallets of wood or bone to tap away small chips of stone, producing a straight, sharp cutting edge.

Scientists believe *Homo habilis* ate meat in addition to fruits, insects, and plants. Archaeologists have found animal bones buried with stone tools from the time of *Homo habilis.* Many of the bones show scratch marks that were probably made by the cutting action of stone tools. These marks indicate that *Homo habilis* used tools to butcher game and to scrape meat off bones. But scientists do not know whether these early humans killed large animals themselves or merely ate the meat after the animals had been killed by predators.

Homo erectus. Fossil evidence indicates that by about $1\frac{1}{2}$ million years ago, *Homo habilis* had evolved into a more advanced human species. Scientists call this species *Homo erectus.* The term *Homo erectus* refers to the upright posture of these creatures. One of the best examples of *Homo erectus* that scientists have found is a nearly complete fossil skeleton of a boy who was probably about 12 years old. The skeleton, which is more than $1\frac{1}{2}$ million years old, was found west of Lake Turkana in northern Kenya.

Homo erectus probably stood slightly more than 5 feet (150 centimeters) tall. These creatures had thick skulls, sloping foreheads, and large, chinless jaws. Their skulls had a *browridge,* a raised strip of bone across the lower forehead. *Homo erectus* also had smaller molars, a smaller face, and a less protruding face than *Homo habilis* had. The brain size of early *Homo erectus* was only slightly larger than that of *Homo habilis.* During the course of *Homo erectus* evolution, however, brain size increased considerably. It eventually reached a size just slightly smaller than that of a modern human brain. Fossil evidence indicates that *Homo erectus* males were larger than *Homo erectus* females.

The earliest *Homo erectus* fossils have been found in Africa, where these prehistoric people probably remained until about 1 million years ago. Many scientists believe that prehistoric people had begun to migrate out of Africa by that time. Anthropologists have found fossil bones of *Homo erectus* that date from about 1 million years ago on the island of Java, in Indonesia. *Homo erectus* tools from the same time have been discovered in southern Europe and Asia. By about 500,000 years ago, *Homo erectus* had spread into northern Asia.

Homo erectus was probably the first human being to master the use of fire. These people may also have been the first to wear clothing. Scientists believe that as *Homo erectus* moved into northern areas and faced cold winters, fire and clothing became necessary. Archaeologists have not found any traces of early clothing, but it was probably made from animal hides. The oldest evidence of the use of fire was found in a cave that *Homo erectus* occupied about 500,000 years ago near what is now Beijing, in northern China. Stone tools and the remains of more than 40 *Homo erectus* individuals were found in the cave, along with burnt animal bones surrounded by thin layers of ash.

Homo erectus was a more skillful toolmaker than

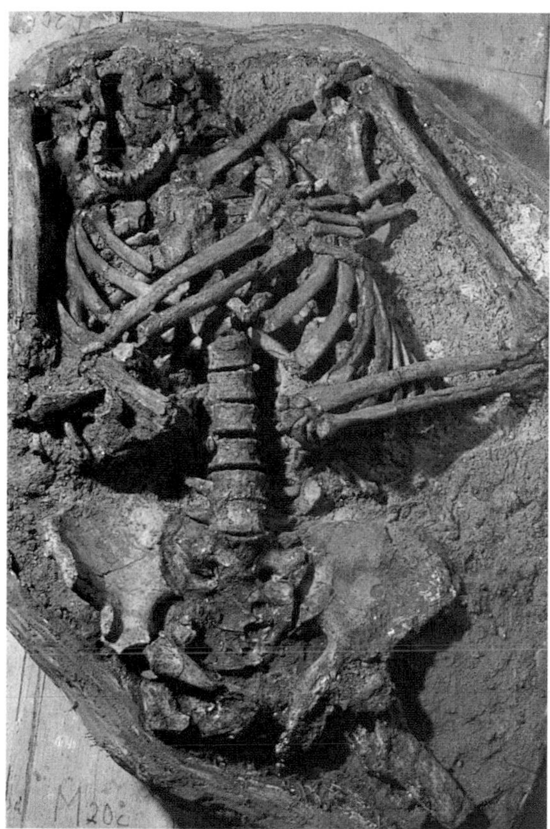

Ofer Bar-Yosef, Harvard University

A Neanderthal grave from Kebara Cave in Israel contains a human skeleton that is about 60,000 years old. Such graves suggest that the Neanderthals were the first to bury their dead.

Homo habilis. For example, *Homo erectus* created double-edged cutting tools called *hand axes* out of stone. These early human beings probably used hand axes for many tasks, such as shaping wood or bone and cutting up meat. The bones of large animals, including mammoths, have been found at *Homo erectus* sites. But scientists do not know if these people actually hunted big game. They may have collected the remains of animals that had been killed by predators. The main foods in the *Homo erectus* diet were probably fruits, vegetables, nuts, seeds, insects, and small animals.

Early *Homo sapiens*

Between about 400,000 and 300,000 years ago, *Homo erectus* evolved into a new human species called *Homo sapiens.* Because evolution took place gradually during this time, anthropologists have found it difficult to say precisely when *Homo sapiens* first appeared. Anthropologists disagree on whether certain fossil specimens from this period are *Homo sapiens* or *Homo erectus.*

The term *Homo sapiens* means *wise human being.* All people living today belong to this species. But early *Homo sapiens* differed greatly from modern people.

The first *Homo sapiens* strongly resembled *Homo erectus.* The main difference between the two was that *Homo sapiens* had a higher and more rounded skull. However, like *Homo erectus,* the first *Homo sapiens* individuals had large faces that protruded around the mouth and nose. They also had big browridges and low, sloping foreheads. These people lacked a chin, a feature found only in the modern type of human beings.

The brain size of early *Homo sapiens* varied over a wide range. Some of these people had brains that were similar in size to those of late *Homo erectus.* Others had brains nearly as large as modern human brains.

Early *Homo sapiens* were about as tall as modern human beings. They were solidly built with powerful muscles and were probably much stronger than modern people. The differences in size between males and females that is so well marked in earlier hominids appears to be reduced in *Homo sapiens.*

Homo sapiens were the first prehistoric people to inhabit large areas of Europe. Anthropologists have found important *Homo sapiens* fossils in England, France, Germany, Greece, and Italy. *Homo sapiens* fossils have also been discovered in many parts of Asia and Africa.

Some of the most important evidence of *Homo sapiens'* way of life comes from a site called Terra Amata, which lies near Nice, France, along the coast of the Mediterranean Sea. Terra Amata was a settlement occupied by what some anthropologists believe was a group of *Homo sapiens* about 250,000 years ago or earlier.

At Terra Amata, scientists found evidence of tentlike structures that a group of *Homo sapiens* probably built for shelter. Further evidence indicates that this group stayed for periods of time at Terra Amata to hunt and gather food during a yearly round of various campsites. Prehistoric people did not form permanent settlements until farming began about 11,000 years ago. But the studies at Terra Amata suggest that earlier people created temporary settlements at specific locations based on their knowledge of food sources.

Neanderthals were a type of early *Homo sapiens* who lived in parts of Europe and the Middle East from about 130,000 to 35,000 years ago. Different types of early *Homo sapiens* occupied other parts of Africa, Europe, and Asia during this period. Neanderthals have become the most widely known of the early *Homo sapiens* mainly because they were the first prehistoric people to be discovered. The term *Neanderthal,* also spelled *Neandertal,* comes from the Neander Valley near Düsseldorf, Germany. The first Neanderthal fossils that scientists identified as prehistoric people were found there in 1856.

The Neanderthals were large and muscular. Like other early *Homo sapiens,* they had protruding faces, large browridges, and low foreheads. Most of them also lacked a chin. However, the Neanderthals had large brains. Their average brain size was larger than that of modern human beings.

Some Neanderthals lived in Europe during the Ice Age, when sheets of ice covered many northern parts of the world. These Neanderthals developed qualities that enabled them to cope with harsh winter conditions. Archaeologists have found most evidence of Neanderthals in caves, where many of these people lived to escape the extreme cold. But archaeologists have also discovered sites where Neanderthals camped in the open. These sites provide evidence that the Neanderthals pitched large circular tents around a central hearth area.

The development of prehistoric human beings

The skulls of prehistoric people changed dramatically over time. By studying the fossilized skulls of our prehuman and early human ancestors, scientists have gained valuable information about these creatures.

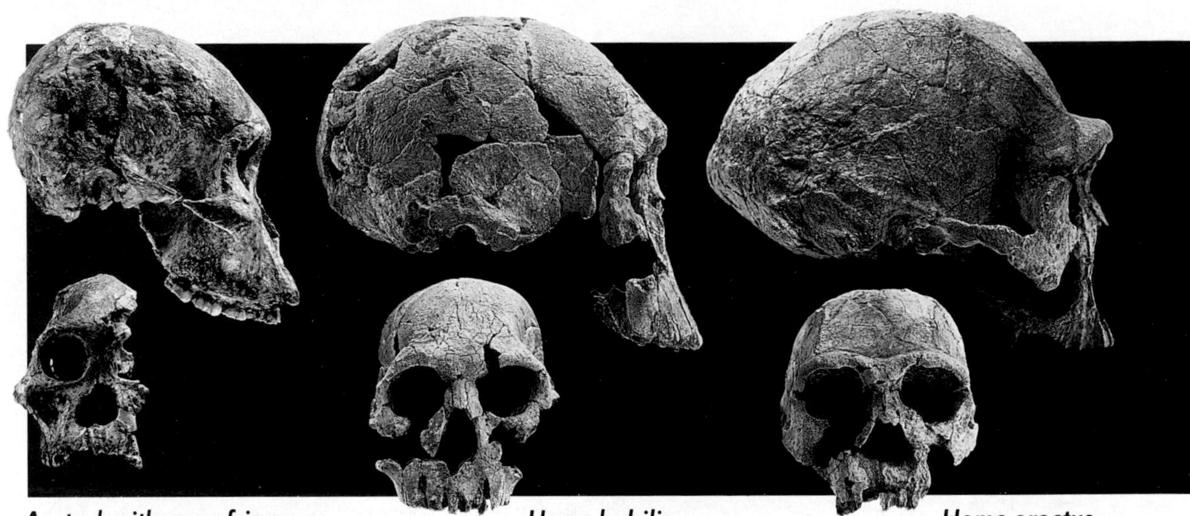

Australopithecus africanus
Transvaal Museum, Pretoria (© David L. Brill)

Homo habilis
National Museums of Kenya, Nairobi (© David L. Brill)

Homo erectus
National Museums of Kenya, Nairobi (© David L. Brill)

The evolution of human beings took place gradually over millions of years. The illustrations below are an artist's impression of how some of the major species of prehuman and early human ancestors may have looked.

© Jay H. Matternes

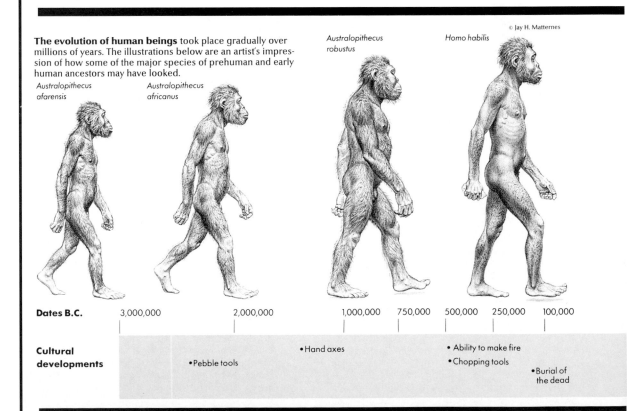

Australopithecus afarensis

Australopithecus africanus

Australopithecus robustus

Homo habilis

Dates B.C. 3,000,000 | 2,000,000 | 1,000,000 | 750,000 | 500,000 | 250,000 | 100,000

Cultural developments

- Pebble tools
- Hand axes
- Ability to make fire
- Chopping tools
- Burial of the dead

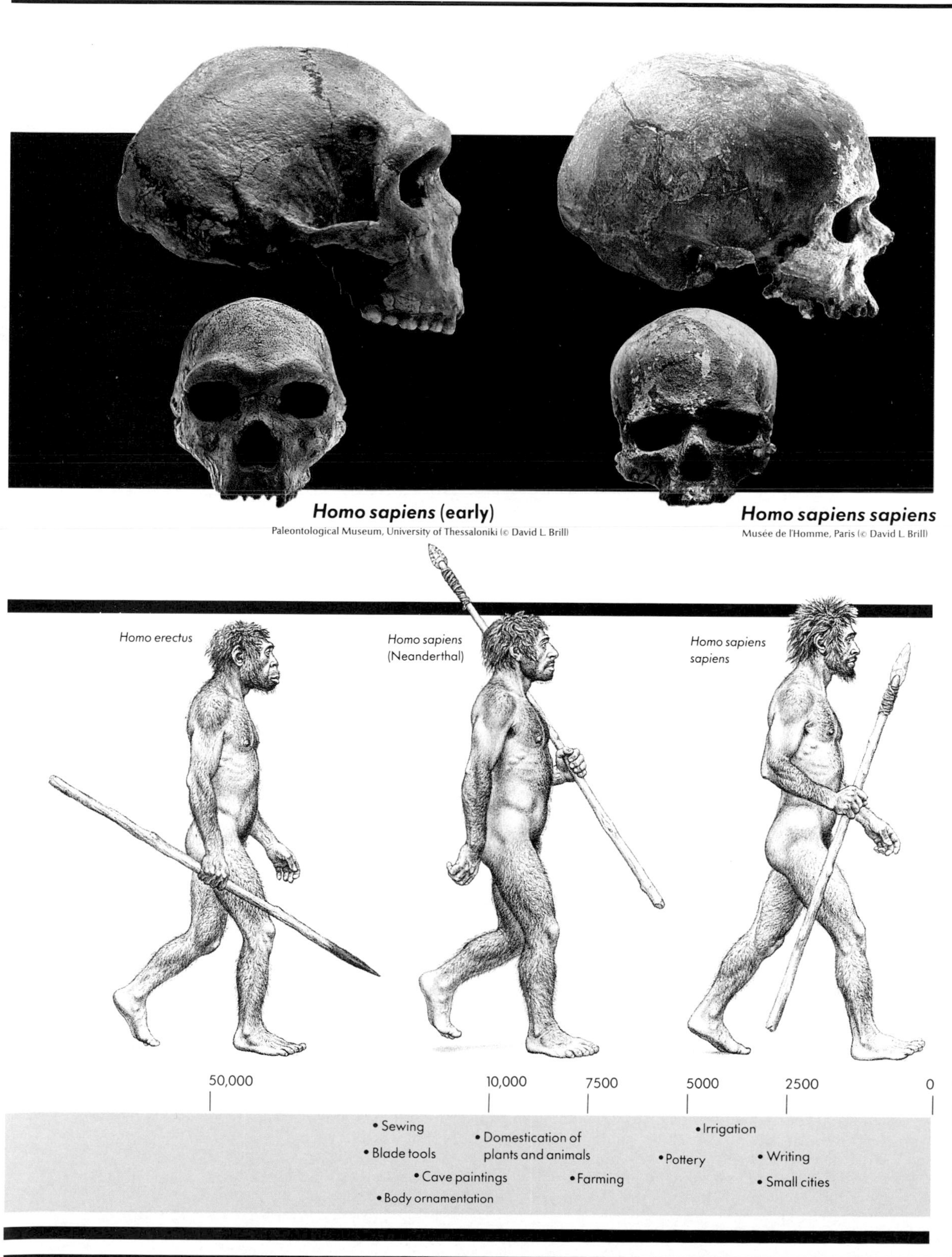

Homo sapiens (early)
Paleontological Museum, University of Thessaloniki (© David L. Brill)

Homo sapiens sapiens
Musée de l'Homme, Paris (© David L. Brill)

Ronald Sheridan Photo Library

J. Bergson, Art Diapo

Cave paintings were one of the earliest forms of art. Prehistoric people had begun producing them by about 20,000 years ago. Many cave paintings of deer, *left,* were found at Lascaux, a cave in southwestern France. The handprints of prehistoric people, *above,* decorate the walls of Gargas Cave in southern France. These prints show that some of the prehistoric artists had lost parts of their fingers.

The tent covering probably consisted of hides, leaves, or bark supported by wooden posts and secured to the ground by stakes made from animal bones.

The Neanderthals were more skilled hunters and toolmakers than earlier prehistoric people. The bones of many animals have been found at Neanderthal sites. Some of the bones indicate these people sometimes hunted such large animals as horses, reindeer, and mammoths. But they were more successful in capturing hares and other small animals. The Neanderthals made a variety of stone tools. They used these tools to butcher animals, prepare vegetable foods, scrape animal hides, and carve wood. They also made sharp, pointed tools that may have been spearheads.

The Neanderthals were the first human beings known to have buried their dead. In Neanderthal sites throughout Europe and the Middle East, archaeologists have uncovered the carefully buried skeletons of women, men, and children. Anthropologists do not understand why the Neanderthals adopted this custom.

The rise of modern human beings

The first prehistoric people with modern human features appeared about 100,000 years ago in either the Middle East or Africa. These people had a chin, a high forehead, and a smaller, less-protruding face than earlier *Homo sapiens* had. The first physically modern human beings also lacked the large browridge of earlier people and had a higher and more rounded skull. Scientists classify modern human beings as *Homo sapiens sapiens,* a subspecies of *Homo sapiens.*

Anthropologists are fairly certain that the first modern human beings evolved from earlier types of *Homo sapiens.* But scientists have had difficulty understanding the precise evolutionary relationship between modern humans and early *Homo sapiens.* For example, fossil evidence shows that Neanderthals lived in Europe and the Middle East after the modern type of human beings appeared. This evidence makes it difficult for scientists to determine whether Neanderthals were the ancestors of modern Europeans or were a related type of early human being that became extinct.

The question of human races is related to the origin of modern human beings. Most anthropologists today reject the idea that the human population can be divided into biologically defined races. Physical features of modern human beings change gradually from one region to another, making it difficult to draw a dividing line between them. However, anthropologists have observed that groups of people who have lived in certain parts of the world for many thousands of years tend to differ in appearance from groups in other parts of the world. These differences are probably adaptations to local environments. For example, people whose ancestors have lived for generations in sunny climates tend to have dark skin. Dark pigment helps protect the skin from sunburn and reduces the risk of skin cancer.

Anthropologists have developed two main theories to explain the origin of modern human beings and the development of what are sometimes called "races"—that is, the physical differences among populations in different regions. These theories may be referred to as (1) the multiple origins theory and (2) the single origin theory.

The multiple origins theory. Some anthropologists believe that the origin of separate human populations began with the migration of *Homo erectus* out of Africa about 1 million years ago. According to this theory, *Homo erectus* split into separate populations in Africa, Asia, and Europe. These groups evolved according to their different environments and developed different physical characteristics. Eventually, *Homo erectus* in each geographic area evolved into a form of *Homo sapiens* unique to the area. These multiple types of *Homo sapiens,* in turn, became the ancestors of the so-called modern human races.

The best evidence supporting this theory comes from a series of skulls found in Indonesia and Australia. In age, these skulls span a period beginning about 1 million years ago and lasting until the appearance of physically modern human beings. All the skulls show similar features that are characteristic of that part of the world. These fossils appear to represent a population that continuously evolved over time and resulted in modern Southeast Asian people.

The single origin theory. Other anthropologists disagree with the multiple origins theory and claim that separate modern human populations had a common ancestor much more recently. According to this single origin theory, modern human beings—*Homo sapiens sapiens*—first appeared in either Africa or the Middle East between 200,000 and 100,000 years ago. This modern type of human being then spread to other parts of Africa, Asia, and Europe, replacing the older populations of *Homo sapiens* who were living there. All other populations of early human beings, such as Neanderthals, became extinct. According to this theory, the development of different physical characteristics in today's so-called racial groups began with the spread of *Homo sapiens sapiens* from Africa or the Middle East.

Some of the best fossil evidence that supports this theory comes from cave sites in Israel. At two of these sites, called Qafzeh and Skhul, archaeologists excavated fossil skeletons of modern-looking human beings that date from about 100,000 years ago. But at a nearby site called Kebara, a Neanderthal skeleton that dates from about 60,000 years ago was found. Supporters of the single origin theory point out that it is difficult to place the Neanderthals as ancestors of modern human beings if they were known to have lived after modern human beings first appeared. Therefore, another group—the first modern humans from either Africa or the Middle East—must have replaced the Neanderthals.

Some scientists also support this theory through use of genetic evidence from living people. Molecular biologists have gained a greater understanding of human evolution by studying the rate of change of human genetic material. By calculating this rate of change, some scientists have concluded that all living human beings must have evolved from one physically modern human ancestor who lived about 200,000 years ago. In one version of this theory, the common ancestor—who has become known as African Eve—was a Stone Age woman in Africa. Although most scientists accept the conclusion of genetic studies that modern human beings originated in Africa, many believe this common ancestor appeared earlier than supporters of the single origin theory claim.

Cultural development of modern human beings

Fossil evidence indicates that the cultural activities of the first physically modern humans were similar to those of other *Homo sapiens* who lived during that time. For example, the modern-looking human beings from the 100,000-year-old sites of Qafzeh and Skhul were found with the same kinds of stone tools that Neanderthals used at sites nearby. Thus, the appearance of modern human beings did not represent a sudden change in life style or culture from the earlier populations.

Throughout the early stages of human evolution, the rate of cultural change among prehistoric people was extremely slow. Stone tools and other products of human skill remained unchanged for many thousands of years. However, about 35,000 years ago, the rate of cultural change began to accelerate rapidly. This later period is generally referred to as the Upper Paleolithic.

During the Upper Paleolithic, prehistoric people made an extraordinary number of advances in their way of life. The best-known type of human beings from this period are the Cro-Magnons. The Cro-Magnons lived in Europe, the Middle East, and North Africa from about 40,000 to 10,000 years ago. Scientists believe they resembled modern Europeans.

The improvement of tools was one of the major accomplishments of the Cro-Magnons and other Upper Paleolithic people. After 35,000 years ago, new tool types and methods of manufacture appeared at a rapid pace. Stone tools made during this time were much more refined and complex in design. Toolmakers invented many new devices to serve specialized carving, cutting, and drilling functions. Tools made from bone, ivory, and animal horns also became widely used. Archaeologists have found harpoons, fish spears, and needles made from bone that date from this period. These tools suggest the introduction of many new activities, such as sewing close-fitting clothes and fishing with improved equipment.

Upper Paleolithic fossil sites also indicate that these people had become skillful hunters. Some sites hold the remains of thousands of animals. In addition, the bones of mammoths, horses, and reindeer are common, suggesting these people hunted large animals successfully.

The appearance of art was one of the most spectacular developments of the Upper Paleolithic. The oldest works of art that archaeologists have found date from this period. Furthermore, the practice of creating art seems to have spread rapidly—especially in Europe.

Some of the oldest artworks from the Upper Paleolithic were ornaments, such as beads made from polished shells. After about 20,000 years ago, prehistoric people began to produce a variety of artwork. They excelled at carving—creating beautiful sculptures of animals and people, usually from ivory or bone. They also made engravings of people, fish, birds, and other animals on bone, ivory, and stone. The Upper Paleolithic people also sculpted clay, ivory, and stone figurines of women, which may have represented fertility.

A number of caves in Europe are covered with paintings, drawings, and engravings from the Upper Paleolithic. Most distinctive of these are the paintings, which appear on the cave walls and ceilings. Most of the paintings are of the animals early people probably hunted, including bison, mammoths, and horses. Some of the paintings show animals that have been speared.

Many of the paintings are of a high artistic quality. Paleolithic artists used three basic colors: black, red, and

Important fossils of prehistoric people

Fossil	Location	Date found	Discovered or identified by	Importance
		Australopithecus		
Taung child	Taung, South Africa, near Vryburg	1924	Raymond A. Dart (South African)	First australopithecine discovered
Kromdraai hominid (formerly *Paranthropus robustus*)	Kromdraai, South Africa, near Johannesburg	1938	Robert Broom (South African)	First *A. robustus* discovered
OH 5, nicknamed "Zinj" (formerly *Zinjanthropus boisei*)	Olduvai Gorge, Tanzania	1959	Mary D. Leakey (British)	First East African australopithecine found
"Lucy"	Hadar, Awash River Valley, Ethiopia	1974	Donald C. Johanson (U.S.)	Most complete australopithecine skeleton
Laetoli fossil footprints	Laetoli, Tanzania, near Lake Eyasi	1978	Mary D. Leakey (British)	Evidence that australopithecines walked erect
		Homo habilis		
Olduvai Gorge *Homo habilis*	Olduvai Gorge, Tanzania	1960	Jonathan Leakey (Kenyan)	First *H. habilis* found
ER-1470	Lake Turkana (formerly Lake Rudolf), Kenya	1972	Richard E. F. Leakey (Kenyan)	Oldest known *H. habilis* skull
		Homo erectus		
Java fossils (formerly *Pithecanthropus erectus*)	Trinil, Indonesia, on the Solo River, island of Java	1891	Eugène F. T. Dubois (Dutch)	First *H. erectus* found
East Turkana *Homo erectus*	Lake Turkana, Kenya	1975	Richard E. F. Leakey (Kenyan)	Earliest *H. erectus* found, dating from $1\frac{1}{2}$ million years ago
Nariokotome boy	Lake Turkana, Kenya	1984	Kamoya Kimeu (Kenyan)	Nearly complete skeleton of a boy, the most complete *H. erectus* found
		Homo sapiens		
Kabwe fossil (formerly Rhodesian man)	Kabwe (also known as Broken Hill), Zambia	1921	Arthur Smith Woodward (British)	Early African *H. sapiens*
Dali skull	Dali, Shanxi Province, China	1978	Wu Xinzhi (Chinese)	Early Asian *H. sapiens*
Neanderthal	Neander Valley, Germany, near Düsseldorf	1856	Johann K. Fuhlrott (German)	First fossil recognized as remains of prehistoric people
"Old Man" of La Chapelle-aux-Saints	La Chapelle-aux-Saints, France, near Brive	1908	Amédée Bouyssonie and Jean Bouyssonie (French)	Most complete Neanderthal skeleton
Skhul skeletons	Skhul Cave, Mount Carmel, Israel	1931-1932	Theodore D. McCown and Hallam L. Movius (U.S.)	Early modern humans (*H. sapiens sapiens*), dating from 100,000 years ago
Qafzeh skeletons	Qafzeh Cave, near the Sea of Galilee, Israel	1933-1975	René Neuville and Bernard Vandermeersch (French)	Early *H. sapiens sapiens*, dating from 100,000 years ago
Cro-Magnons	Les Eyzies, France, near Brive	1868	Louis Lartet (French)	First Cro-Magnon skeletons discovered

yellow. They obtained these pigments from natural sources including charcoal, clay, and such minerals as iron. Often, the artists painted animals on a part of the cave wall where there was a natural swelling, which created a three-dimensional effect.

The development of speech. No one knows when or how spoken language developed. However, many anthropologists think that human beings may have first begun to speak sometime during the Upper Paleolithic. These scientists believe that the many cultural developments which occurred at this time—especially the appearance of art—may be related to the development of speech. The beginnings of speech, the creation of artwork, and the making of complex tools all required advancements in human intelligence and cooperation.

The spread of settlement. Prehistoric people spread into new areas during the Upper Paleolithic. Cultural and technological advances enabled them to migrate to such places as Australia, the Pacific Islands, and North and South America.

Perhaps as early as 50,000 years ago, people used boats to reach Australia. About 20,000 years ago, people from Australia and Asia began to colonize the Pacific Islands. These people employed sophisticated navigational systems that involved knowledge of the stars, water currents, and wind direction. They also used simple navigational instruments.

By 30,000 years ago, human beings had spread to the cold, harsh tundra of northeast Asia. At that time, the Bering Strait was a land bridge that connected Asia and North America. Most scientists believe that prehistoric people crossed this land bridge and were living in North America by 15,000 years ago. Eventually, early modern human beings populated all of North and South America.

The last Ice Age ended about 10,000 years ago. As the vast sheets of ice receded, the environment of many prehistoric people changed and greatly affected their way of life. In some areas, such as Europe, forests began to spread across the land. The people of these areas learned to hunt new species of animals and gather new varieties of plants from these forests. In other parts of the world, people began to experiment with methods of controlling their supply of food. This led to the beginning of farming.

The rise of agriculture, according to most scientists, began in the Middle East about 11,000 years ago, or 9000 B.C. The first farmers lived in a region called the Fertile Crescent, which covers what is now Lebanon and parts of Iran, Iraq, Israel, Jordan, Syria, and Turkey. At first, these people probably did not depend entirely on the crops they raised. But as they improved their methods, farming became their most important source of food. The earliest plants grown in the Middle East were probably barley and wheat. Early farmers in the Middle East eventually raised cattle, goats, hogs, and sheep.

The first farmers originated in areas where there were enough wild plants and animals to provide food for large populations. As a result, people often settled in permanent villages for years at a time. At the end of the Ice Age, the climate became warmer and affected the food supply. New plants, such as grains, replaced older plants. Scientists believe that Upper Paleolithic people were able to remain in permanent settlements because they discovered how to control these new plants and increase the amount of food in their area. They learned that they could plant seeds from the plants that they ate. They also learned that they could domesticate animals, perhaps by capturing young ones from the wild and raising them. In time, people began to depend on these planted crops and domestic animals for a steady supply of food.

By about 7000 B.C., agriculture had developed independently in Asia and southern North America. In what are now Thailand and southern China, farmers grew breadfruit, bananas, and rice, while people in what became Mexico raised beans, corn, and squash.

People were herding cattle and growing grain in northern Africa by 6000 B.C. By that time, people had also begun to farm in the Indus River Valley of what is now Pakistan. By 4000 B.C., farming had begun in the Huang He Valley of China. Farming spread throughout most of Europe by 3000 B.C. Farming in most parts of North and South America began after prehistoric times. Food was probably more plentiful in these areas, so farming did not become necessary until later.

Changes in life style. Prehistoric farmers, called *Neolithic* people, had a way of life that differed greatly from that of Upper Paleolithic people. In some ways, farming made life easier. It provided a steady supply of food and enabled people to stay in one place for a long time. However, farmers also had to work longer and harder than did hunters and gatherers.

Prehistoric farmers set up villages near their fields and lived there as long as their crops grew well. Most fields produced good crops for only a few years. The land then became unproductive because continuous planting used up nutrients in the soil. The early farmers did not know about fertilizers that could replace these nutrients. They shifted their crops to new fields until none of the land near their village was fertile. Then they moved to a new area and built another village. In this way, farmers settled many new areas.

Prehistoric farmers built larger, longer-lasting settlements than the camps that Paleolithic people had built. In the Middle East, for example, early farmers constructed their houses of solid, sun-dried mud. Dried mud was much more resistant to weather than the materials earlier people used, such as skins and bark. The early farmers also learned to build fences to confine and protect their livestock.

The end of prehistoric times. Neolithic people made inventions and discoveries at an even faster rate than did the people of the Upper Paleolithic. Early farmers developed a number of useful tools. These implements included sickles to cut grain, millstones to grind flour, and polished stone axeheads.

By about 11,000 B.C., people had discovered how to make pottery. Before that time, they used animal skins or bark containers to hold water. To boil water, early cooks had to drop hot stones into the water, because they could not hang animal skins or bark over a fire. Pottery containers enabled people to hold and boil water easily. After the rise of agriculture, people used pottery to store grain and other food.

No one knows when people made the first objects out of metal. But metals became important only after metalworkers learned to make bronze, a substance hard and durable enough to make lasting tools. People of the Middle East made bronze as early as 3500 B.C. The Bronze Age began when bronze replaced stone as the chief toolmaking material. In some areas, such as the Near East, the Bronze Age began about 3000 B.C.

The development of farming was an important step toward the rise of civilization. As farming methods improved and food became more plentiful, many people were freed from the jobs of food production. These people developed new skills and trades. In addition, the abundant food supply enabled more people to live in each community. In time, some farming villages became cities. The first cities appeared by about 3500 B.C. These cities were the birthplaces of modern civilization.

Archaeologists believe writing was invented about 3500 B.C. in cities in the Tigris-Euphrates Valley in what is now Iraq. People then learned to record their history, and prehistoric times came to an end.

Studying prehistoric people

Since the mid-1800's, scientists have accumulated much information about prehistoric people and their ways of life. This information has enabled scientists to piece together a general picture of how hominids have evolved. Scientists have used various methods to obtain and study information about prehistoric people.

Studying fossils of prehistoric people has provided anthropologists with much of their most valuable information. Human fossils give direct evidence of what prehistoric people looked like, what they ate, how long they lived, and how their lives were different from ours.

Unfortunately, fossil bones and teeth are extremely rare. This is because certain unusual conditions must be present for a fossil to form. An animal or person must be buried soon after death, and minerals from the soil must gradually replace the bony material to create a fossil. Also, the bones and teeth that do become fossils have often been broken, damaged, or distorted by the weight of the deposits in which they were buried. Soft tissues—such as skin, hair, and internal organs of prehistoric people—decay without leaving any fossil remains. As a result, scientists cannot determine certain characteristics of early humans, such as the color of their skin or the texture of their hair.

After a fossil is discovered, scientists first determine whether it came from an adult or a child, and whether the individual was a male or female. Anthropologists then compare the fossil with similar structures from extinct hominids, living human beings, and apes. These examinations enable scientists to more fully understand the specimen's place in human evolution. Later, anthropologists study the fossil further to determine a relationship between the individual's physical structure and its way of life.

Examining prehistoric sites is another method anthropologists use to gain information about prehistoric people and their way of life. These sites may be places where prehistoric people camped, made tools, butchered animals, or buried their dead.

The excavation of prehistoric sites is a complicated process. Archaeologists carefully scrape away soil, sand, or rock to reveal tools, bones, pottery, and other evidence of early human life. They make detailed notes and maps, often using computers, to record the exact position of all important items. They later use these records to precisely reconstruct the layout of the site. By studying the objects they find and the layout of the site, scientists learn how prehistoric people used the site and how these activities fit into their way of life.

Scientists have found much evidence of prehistoric people's lives at sites up to 100,000 years old. These sites have produced ancient tools, pottery, artwork, bits of clothing, traces of dwellings, and evidence of food, such as animal bones and plant material. These abundant clues enable anthropologists to form a fairly detailed picture of early people's lives. Unfortunately, scientists have found little of this type of evidence at sites older than 100,000 years. As a result, we know much less about hominids who lived before that time.

Placing prehistoric people in time is one of the most important elements of learning about human ancestors. To understand the significance of a newly discovered hominid fossil, scientists must determine how that hominid relates to others that have already been studied. This relationship can be determined by dating the newly found fossil—that is, by determining when it was alive.

Scientists have traditionally dated fossils by studying the deposit in which the fossil was found. Based on knowledge of geological history, scientists can determine the age of the deposit. They then interpret this information to provide an approximate age for the fossil.

Alan E. Mann, University of Pennsylvania

Anthropologists dig for objects left behind by prehistoric people. The scientists above are digging at St.-Césaire, France, a site occupied by Neanderthals about 35,000 years ago.

Newer dating methods are much more accurate. These methods are based on the fact that certain *radioactive isotopes* (unstable forms of chemical elements) decay at a known rate and form different isotopes. By measuring the amount of each isotope in a fossil, scientists can determine how long the decay has been going on and therefore how old the fossil is. The most commonly used dating methods of this type are *radiocarbon dating* and *potassium-argon dating.*

Molecular analysis of living people is a more recent method of study that scientists have used to better understand prehistoric people. This method largely involves the study of genetic material called DNA.

Molecular biologists in this field extract DNA from living people and compare it with DNA from other people and from other living primates. Scientists also study the rate of change that DNA appears to go through during evolution. Scientists often use computers to help analyze data obtained through this research. From these studies, scientists have gained valuable knowledge about the relationship between living people and their ancestors. In many cases, however, anthropologists have disagreed on how to best interpret the results of these studies.

Alan E. Mann

Related articles in ***World Book*** include:

Anthropology
Archaeology
Australopithecus
Bronze Age
Cave dwellers
Cro-Magnons
Evolution
Fossil
Heidelberg jaw
Homo erectus
Homo habilis
Java fossils
Johanson, Donald C.
Lake dwelling
Leakey family
Megalithic monuments
Neanderthals
Peking fossils
Piltdown hoax
Pleistocene Epoch
Races, Human
Ramapithecus
Stone Age
Swanscombe fossil
Zinjanthropus

Questions

When did the first human beings live?
What are *hominids*?
From what type of australopithecine did the first humans evolve?
Who were the first prehistoric people to master the use of fire?
When did scientists first find evidence of prehistoric people?
How do scientists use computers to study prehistoric sites?
Who were the first prehistoric people to inhabit large areas of Europe?
How did early human beings make tools out of stone?
Who were the Neanderthals?
Why did some prehistoric people become farmers?

Additional resources

Early Humans. Ed. by Phil Wilkinson. Knopf, 1989. For younger readers.

Encyclopedia of Human Evolution and Prehistory. Ed. by Ian Tattersall and others. Garland, 1988.

Gowlett, John. *Ascent to Civilization: The Archaeology of Early Man.* Knopf, 1984.

The Human Revolution: Behavioural and Biological Perspectives on the Origins of Modern Humans. Ed. by Paul Mellars and Chris Stringer. Princeton, 1989.

Lambert, David, and the Diagram Group. *The Field Guide to Early Man.* Facts on File, 1987. Also suitable for younger readers.

Leakey, Richard E. *Human Origins.* Lodestar, 1982. Also suitable for younger readers.

Lewin, Roger. *In the Age of Mankind: A Smithsonian Book of Human Evolution.* Smithsonian Bks., 1988.

Sattler, Helen R. *Hominids: A Look Back at Our Ancestors.* Lothrop, 1988. For younger readers.

Wenke, Robert J. *Patterns in Prehistory: Humankind's First Three Million Years.* 3rd ed. Oxford, 1990.

NASA

An artificial satellite is designed to carry out a specific mission. One major type of satellite, called a communications satellite, performs the task of relaying information between different points in space and on the earth. The Tracking and Data Relay Satellite (TDRS), *left,* carries out this mission, as evidenced by its many antennas. The umbrella-shaped and dishlike structures are antennas, as are the spike-shaped objects extending from the box in the middle of the satellite.

Satellite, Artificial

Satellite, Artificial, is a manufactured object that continuously orbits the earth or some other body in space. Most artificial satellites orbit the earth. People use them to study the universe, help forecast the weather, transfer telephone calls over the oceans, assist in the navigation of ships and aircraft, monitor crops and other resources, and observe movements of military equipment on the ground.

Artificial satellites also have orbited the moon, the sun, Venus, and Mars. Such satellites mainly gather information about the bodies they orbit.

Strictly speaking, manned spacecraft in orbit—space capsules, space shuttle orbiters, and space stations—are artificial satellites. So, too, are orbiting pieces of "space junk," such as burned-out rocket boosters and empty fuel tanks that have not fallen to the earth. This article does not deal with these kinds of satellites. For information on manned spacecraft, see **Space exploration.**

Artificial satellites differ from *natural satellites,* natural objects that orbit a planet. The earth's moon is a natural satellite. See **Satellite.**

The Soviet Union launched the first artificial satellite, Sputnik 1, in 1957. Since then, the United States and many other countries have developed, launched, and operated satellites. Today, more than 2,000 satellites are orbiting the earth.

Satellite orbits

Satellite orbits have a variety of shapes. Some are circular, while others are highly *elliptical* (egg-shaped). Orbits also vary in altitude. Some circular orbits, for example, are just above the atmosphere at an altitude of about 155 miles (250 kilometers), while others are more than 20,000 miles (32,200 kilometers) above the earth. The greater the altitude, the longer the *orbital period*—the time it takes a satellite to complete one orbit.

A satellite remains in orbit because of a balance between two factors: (1) the satellite's *velocity* (speed at which it would travel in a straight line), and (2) the gravitational force between the satellite and the earth. Were it not for the pull of gravity, a satellite's velocity would send it flying away from the earth in a straight line. But were it not for velocity, gravity would pull a satellite back to the earth.

To help understand the balance between gravity and velocity, consider what happens when a small weight is attached to a string and swung in a circle. If the string were to break, the weight would fly off in a straight line. However, the string acts like gravity, keeping the weight in its orbit. The weight and string can also show the relationship between a satellite's altitude and its orbital period. A long string is like a high altitude. The weight takes a relatively long time to complete one circle. A short string is like a low altitude. The weight has a relatively short orbital period.

Many types of orbits exist, but most artificial satellites travel in one of three types: (1) *high altitude, geosynchronous;* (2) *sun-synchronous, polar;* and (3) *low altitude.* Most orbits of these three types are circular.

A high altitude, geosynchronous orbit lies above the equator at an altitude of about 22,300 miles (35,900 kilometers). A satellite in this orbit travels around the earth's axis in exactly the same time, and in the same direction, as the earth rotates about its axis. Thus, as seen from the earth, the satellite always appears at the same place in the sky overhead. To boost a satellite into a high altitude, geosynchronous orbit requires a large, powerful launch vehicle.

Important satellites

Date orbited	Name	Accomplishments
Scientific research satellites		
1957 Oct. 4	**Sputnik 1**	First artificial satellite.
1957 Nov. 3	**Sputnik 2**	First satellite with an animal (a dog) aboard.
1958 Jan. 31	**Explorer 1**	First U.S. satellite; discovered earth's radiation belts.
1958 Mar. 17	**Vanguard 1**	Discovered earth is slightly pear-shaped; first solar-powered satellite.
1959 Jan. 2	**Luna 1**	First satellite in orbit around the sun.
1959 Aug. 7	**Explorer 6**	Returned first images of earth taken from a satellite.
1960 Aug. 19	**Sputnik 5**	First satellite to return animals (two dogs) from orbit.
1962 Mar. 7	**OSO**	Orbiting Solar Observatory, used to study the sun.
1962 Apr. 26	**Ariel 1**	First international satellite (American and British).
1968 Dec. 7	**OAO A2**	Carried telescopes to study distant stars.
1977 Aug. 12	**HEAO-1**	Used to survey and search for X-ray sources.
1979 Feb. 18	**Sage**	Used to study how atmospheric dust and aerosols affect sunlight.
1979 Oct. 30	**MAGSAT**	Mapped the magnetic field near the earth.
1980 Feb. 14	**SMM**	Observed the sun during a period of maximum solar activity.
1983 Jan. 25	**IRAS**	Performed all-sky survey for sources of infrared radiation in outer space.
1989 Nov. 18	**COBE**	Mapped remnants of energy released during the early moments of the universe.
1990 Apr. 5	**Pegsat**	First satellite launched from an airplane in flight.
1990 Apr. 25	**HST**	Very high-resolution optical telescope; largest in orbit.
1991 Apr. 7	**GRO**	Measures gamma rays originating in outer space and maps their sources.
Weather satellites		
1960 Apr. 1	**Tiros 1**	First successful weather satellite.
1964 Aug. 28	**Nimbus 1**	First satellite to monitor cloud cover and observe conditions over most of the earth's surface.
1970 Dec. 11	**NOAA-1**	Provided high-resolution images of weather conditions.
Communications satellites		
1958 Dec. 18	**Score**	First voice transmission from a satellite.
1960 Aug. 12	**Echo 1**	First communications satellite to relay messages by reflection.
1960 Oct. 4	**Courier 1B**	First communications satellite to relay messages electronically.
1962 July 10	**Telstar 1**	Relayed first TV pictures across the Atlantic Ocean.
1963 July 26	**Syncom 2**	First functional satellite in a geosynchronous orbit.
1965 Apr. 6	**Early Bird**	First commercial communications satellite.
1982 Sept. 28	**Intelsat 5 F-5**	Carries 12,000 voice channels and two television channels simultaneously.
1983 Apr. 5	**TDRS**	Relays communications between other satellites and earth.
1983 Apr. 11	**RCA Satcom 6**	First fully transistorized nonmilitary satellite.
Navigation satellites		
1960 Apr. 13	**Transit 1B**	First navigation satellite.
1961 June 29	**Transit 4A**	First satellite with a nuclear power supply.
1989 Feb. 14	**GPS 2-1**	First satellite in a network called the Global Positioning System.
Earth observation satellites		
1966 Feb. 3	**ESSA 1**	First operational environmental survey satellite.
1972 July 23	**ERTS-1 (Landsat 1)**	First satellite to map earth's resources continuously and comprehensively.
1976 May 4	**LAGEOS-1**	Measures movements in the earth's crust.
1983 Apr. 17	**Rohini 3**	Mapped resources for India.
1990 Jan. 22	**SPOT 2**	Maps earth's resources at high resolution.
Military satellites		
1959 Feb. 28	**Discoverer 1**	First satellite in a polar orbit.
1960 May 24	**Midas 2**	Designed to detect enemy missile launches.
1960 Aug. 10	**Discoverer 13**	First object recovered from orbit.
1976 Apr. 30	**NOSS-1**	Monitors locations of ships.
1987 June 20	**DMSP**	Provides weather data for military operations.

A sun-synchronous, polar orbit passes almost directly over the North and South poles. A slow drift of the orbit's position is coordinated with the earth's movement around the sun in such a way that the satellite always crosses the equator at the same local time on the earth. Because the satellite flies over all latitudes, its instruments can gather information on almost the entire surface of the earth. One example of this type of orbit is that of the NOAA-H satellite, which monitors the weather. The altitude of the orbit is 540 miles (870 kilometers), and the orbital period is 102 minutes. When the satellite crosses the equator, the local time is always either 1:40 a.m. or 1:40 p.m.

A low altitude orbit is within the earth's atmosphere, but the highest layer, where there is almost no air to cause drag on the spacecraft and slow it down. Because the orbit is so low, less energy is required to launch a satellite into it than would be needed to place the same satellite into either of the other two main types of orbit. Satellites that point toward deep space and provide scientific information generally operate in this type of orbit. The Hubble Space Telescope, for example, operates at an altitude of about 380 miles (610 kilometers), with an orbital period of 97 minutes.

Types of artificial satellites

Artificial satellites are classified according to their mission. There are six main types of artificial satellites: (1) scientific research, (2) weather, (3) communications, (4) navigation, (5) earth observation, and (6) military.

Scientific research satellites gather data for scientific analysis. These satellites are usually designed to perform one of three kinds of missions. (1) Some gather information about the composition and effects of the

Satellite orbits

Most artificial satellites travel in one of the three types of orbits shown at the right. A *high altitude, geosynchronous orbit* is above the equator at an altitude of about 22,300 miles (35,900 kilometers). A *sun-synchronous, polar orbit* passes almost directly over the North and South poles several hundred miles above the earth. A *low altitude orbit* is in an almost airless level of the atmosphere that begins about 300 miles (480 kilometers) above the earth.

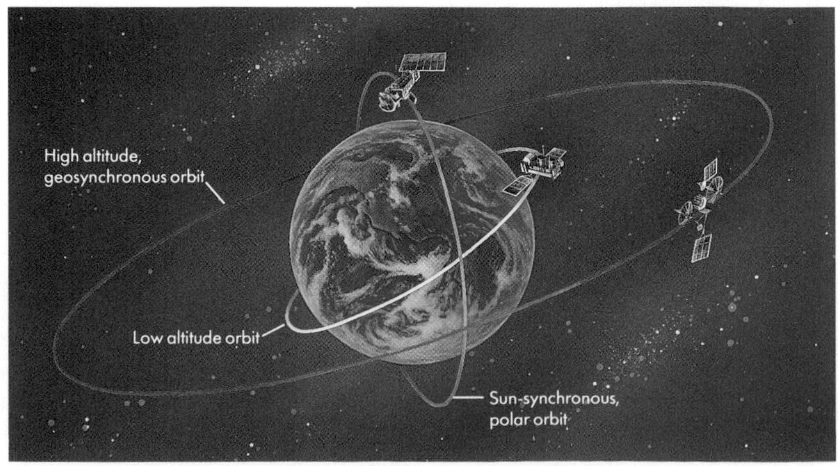

WORLD BOOK illustration by Ken Tiessen, Koralik Associates

Walter A. Bohan Co.

A weather satellite called the Geostationary Operational Environmental Satellite observes atmospheric conditions over a large area to help scientists study and forecast the weather.

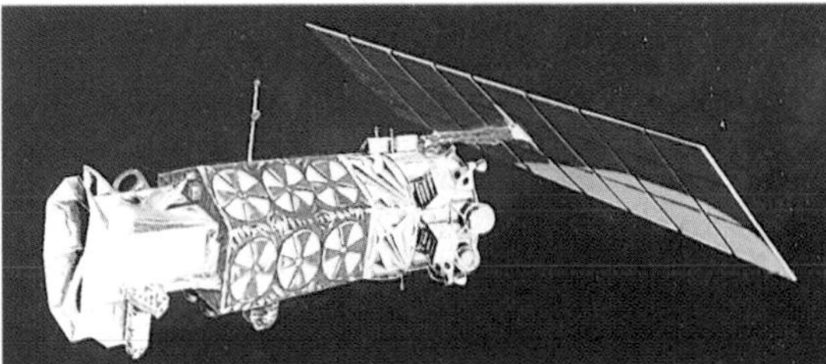

U.S. Air Force

A military satellite in the United States Defense Meteorological Satellite Program provides weather information that could be used to advantage in wartime.

NASA

A scientific satellite, the Compton Gamma Ray Observatory, detects *gamma rays,* high-energy waves given off by such objects as supernovas and quasars, and by matter near black holes.

space near the earth. These satellites are placed in a variety of orbits. (2) Other satellites record changes in the earth and its atmosphere. Many of these satellites travel in sun-synchronous, polar orbits. (3) Still others observe planets, stars, and other distant objects. Most of these satellites operate in low altitude orbits. Scientific research satellites also orbit other planets, the moon, and the sun.

Weather satellites help scientists study weather patterns and forecast the weather. Weather satellites constantly observe the atmospheric conditions over large areas.

Some weather satellites travel in a sun-synchronous, polar orbit, from which they make close, detailed observations of weather over the entire earth. Their instruments measure cloud cover, temperature, air pressure, precipitation, and the chemical composition of the atmosphere. Because these satellites always observe the earth at the same local time of day, scientists can easily compare weather data collected under constant sunlight conditions.

The network of weather satellites in these orbits also function as a search and rescue system. They are equipped to detect distress signals from all commercial, and many private, planes and ships.

Other weather satellites are placed in high altitude, geosynchronous orbits. From these orbits, they can always observe weather activity over nearly half the surface of the earth at the same time. These satellites photograph changing cloud formations. They also produce *infrared images,* which show the amount of heat coming from the earth and the clouds. Infrared pictures reveal weather patterns even at night.

Communications satellites serve as *relay stations,* receiving radio signal messages from one location and transmitting them to another. A communications satellite can relay several television programs or many thousands of telephone calls at once. Communications satellites are usually put in a high altitude, geosynchronous orbit over a *ground station.* A ground station has a large dish antenna for transmitting and receiving radio signals. Countries and commercial organizations such as television broadcasters and telephone companies use these satellites continuously.

Navigation satellites enable operators of aircraft, ships, and land vehicles anywhere on earth to determine their locations within 100 feet (30 meters). These satellites send out radio signals that are picked up by a computerized receiver carried on a vehicle.

Navigation satellites operate in networks, and signals from a network can reach vehicles anywhere on the earth's surface. The receiver on board a vehicle calculates its distance from at least three satellites whose signals it has received. It then uses this information to determine the vehicle's location.

Earth observation satellites are used to map and monitor our planet's resources. They follow sun-synchronous, polar orbits. Under constant illumination from the sun, they take pictures in different colors of visible light and in infrared radiation. Computers on the earth combine and analyze the pictures. Scientists use earth observation satellites to locate mineral deposits, to determine the location and size of freshwater supplies, to identify sources of pollution and study its effects, and to detect the spread of disease in crops and forests.

Military satellites include weather, communications, navigation, and earth observation satellites used for military purposes. Some military satellites—often called "spy satellites"—can detect the launch of missiles, the course of ships at sea, and the movement of military equipment on the ground.

The life and death of a satellite

Building a satellite. Every satellite carries special instruments that enable it to perform its mission. For example, a satellite that studies the universe has a telescope. A satellite that helps forecast the weather carries cameras to film the movement of clouds.

In addition to such mission-specific instruments, all satellites have basic *subsystems,* groups of devices that help the instruments work together and keep the satellite operating. For example, a *power subsystem* generates, stores, and distributes a satellite's electric power. This subsystem may include panels of solar cells that gather energy from the sun. *Command and data handling subsystems* consist of computers that gather and process data from the instruments and execute commands from the earth.

A satellite's instruments and subsystems are designed, built, and tested individually. Workers install them on the satellite one at a time until the satellite is complete. Then the satellite is tested under conditions like those that the satellite will encounter during launch and while in space. If the satellite passes all tests, it is ready to be launched.

Launching the satellite. Space shuttles carry some satellites into space, but most satellites are launched by rockets that fall into the ocean after their fuel is spent. Many satellites require minor adjustments of their orbit before they begin to perform their function. Built-in rockets called *thrusters*—some as small as a mechanical pencil—make these adjustments. Once a satellite is placed into a stable orbit, it can remain there for a long time without further adjustment.

Performing the mission. Most satellites operate under the direction of a *control center* that is located on the earth. Computers and human operators at the control center monitor the satellite's position, send instructions to its computers, and retrieve information that the satellite has gathered. The control center communicates with the satellite by radio. Ground stations transmit and receive the radio messages. These stations are located beneath the satellite's orbit or elsewhere within the satellite's range.

A satellite does not usually receive constant direction from its control center. It is like an orbiting robot. It controls its solar panels to keep them pointed toward the sun and keeps its antennas ready to receive commands. Its instruments automatically collect information.

Satellites in a high altitude, geosynchronous orbit are always in contact with the earth. Ground stations can contact satellites in low orbits as often as 12 times a day. During each contact, the satellite transmits information and receives instructions. Each contact must be completed during the time the satellite passes overhead—about 10 minutes.

If some part of a satellite breaks down, but the satellite remains capable of doing useful work, the satellite owner usually will continue to operate it. In some cases, ground controllers can repair or reprogram the satellite. In rare instances, space shuttle crews have retrieved and repaired satellites in space. If the satellite can no longer perform usefully and cannot be repaired or reprogrammed, operators from the control center will send a signal to shut it off.

Falling from orbit. A satellite remains in orbit until its velocity decreases and the gravitational force pulls it down into a relatively dense part of the atmosphere. A satellite slows down due to the friction of air particles in the upper atmosphere and the gentle pressure of the sun's energy. When the gravitational force pulls the satellite down far enough into the atmosphere, the satellite rapidly compresses the air in front of it. This air becomes so hot that most or all of the satellite burns up.

Sovfoto

Sputnik 1, the first artificial satellite, was launched by the Soviet Union in 1957. It transmitted radio signals that were received on earth.

History

In 1955, the United States and the Soviet Union announced plans to launch artificial satellites. On Oct. 4, 1957, the Soviet Union launched Sputnik 1, the first artificial satellite. It circled the earth once every 96 minutes and transmitted radio signals that could be received on the earth. On Nov. 3, 1957, the Soviets launched a second satellite, Sputnik 2. It carried a dog named Laika, the first animal to soar in space. The United States launched its first satellite, Explorer 1, on Jan. 31, 1958, and its second, Vanguard 1, on March 17, 1958.

In August 1960, the United States launched the first communications satellite, Echo I. This satellite reflected radio signals back to the earth. In April 1960, the first weather satellite, Tiros I, sent pictures of clouds to the earth. The U.S. Navy developed the first navigation satellites. The Transit 1B navigation satellite first orbited in April 1960. By 1965, more than 100 satellites were being placed in orbit each year.

Since the 1970's, scientists have created new and more effective satellite instruments and have made use of computers and miniature electronic technology in satellite design and construction. In addition, more nations and some private businesses have begun to purchase and operate satellites. By the early 1990's, more than 20 countries owned satellites. About 2,000 satellites were operating in orbit. John E. Oberright

Additional resources

Bendick, Jeanne. *Artificial Satellites: Helpers in Space.* Millbrook, 1991. For younger readers.

Chetty, P. R. K. *Satellite Technology and Its Applications.* 2nd ed. TAB, 1991.

Herda, D. J. *Research Satellites.* Watts, 1987. *Operation Rescue: Satellite Maintenance and Repair.* 1990. Both for younger readers.

Hudson, Heather E. *Communication Satellites: Their Development and Impact.* Free Pr., 1990.

Index

How to use the index
This index covers the contents of the 1992, 1993, and 1994 editions of *Science Year,* The World Book Science Annual.

Each index entry gives the last two digits of the edition year, followed by a colon and the page number or numbers. For example, this entry means that information on the E-lamp may be found on pages 310 through 313 of the 1994 *Science Year.*

When there are many references to a topic, they are grouped alphabetically by clue words under the main topic. For example, the clue words under **Earthquakes** group the references to that topic under five subtopics.

An entry in all capital letters indicates that there is a Science News Update article with that name in at least one of the three volumes covered in this index. References to the topic in other articles may also be listed in the entry.

An entry that only begins with a capital letter indicates that there are no Science News Update articles with that title but that information on this topic may be found in the editions and on the pages listed.

The indication (il.) after a page number means that the reference is to an illustration only.

The "see" and "see also" cross references indicate that references to the topic are listed under another entry in the index.

Index

A

B

C

D

E

F

G

H

I

J

K

L

M

P

T

Acknowledgments

The publishers of *Science Year* gratefully acknowledge the courtesy of the following artists, photographers, publishers, institutions, agencies, and corporations for the illustrations in this volume. Credits should read from top to bottom, left to right on their respective pages. All entries marked with an asterisk (*) denote illustrations created exclusively for *Science Year*. All maps, charts, and diagrams were prepared by the *Science Year* staff unless otherwise noted.

2 Superconducting Super Collider Laboratory; NASA
3 Walter Jaffe, Leiden Observatory and Holland Ford, Johns Hopkins University/STScI, and NASA; Paul Sereno, University of Chicago; H. Groskinsky; I. K. Curtis
4 Cornell University; Telegraph Colour Library from FPG
5 St. Johns Expeditions (Franklin Viola); Jessie Cohen, National Zoological Society
10 NASA; Bill Dyer, Cornell Laboratory of Ornithology
11 Ray Webb, Oxford Illustrators Limited*; Gerry Ellis, The Wildlife Collection
12 Space Telescope Science Institute/NASA; Walter Jaffe, Leiden Observatory and Holland Ford, Johns Hopkins University/STScI, and NASA
15 Roberta Polfus*
16 Roberta Polfus*; Bob Hanisch, STScI/NASA; Bob Hanisch, STScI/NASA; Marshall Space Flight Center
17 Roberta Polfus*; Bob Hanisch, STScI/NASA; Bob Hanisch, STScI/NASA
18 STScI/NASA
22 STScI/NASA; STScI/NASA; C. R. O'Dell, NASA
26 Alan Dressler, Carnegie Institution, Augustus Oemler, Yale University, James E. Gunn, Princeton University, Harvey Butcher, Netherlands Foundation for Research in Astronomy, and NASA; STScI/NASA
28 W. Greene, Visual Resources for Ornithology (VIREO); Bill Dyer, Cornell Laboratory of Ornithology; B. Schorre, VIREO; B. Schorre, VIREO; J. R. Woodward, VIREO
31 B. Schorre, VIREO
33 Clyde Morris, U.S. Forest Service
34 Wayne Ferran, Department of Biological Sciences, University of California at Santa Barbara
36 Cornell Laboratory of Ornithology; Doug Nilson, Kern River Research Center; Deanna K. Dawson, U.S. Fish and Wildlife Service
37 L. Walkinshaw, VIREO; Norvia Behling
40 Sidney A. Gauthreaux, Clemson University
41 D. C. Twichell; J. Stasz, VIREO
42 Marc Maritsch/Nancy H. Sandburg
44 Ray Webb, Oxford Illustrators Limited*
48 Bill Ballenberg, © National Geographic Society; Christopher Donnan
50 Christopher Donnan
51 Donna McClelland
52 Christopher Donnan
53 Donna McClelland; Christopher Donnan (3)
54 Christopher Donnan
56 Drawing: F. Caycho—photo: G. Reparez; Christopher Donnan; Christopher Donnan
58 "ALIENS" © 1986 Twentieth Century Fox Film Corporation. All rights reserved.
61 "ALIENS" © 1986 Twentieth Century Fox Film Corporation. All rights reserved; Nancy Lee Walter*
62 © 1980 Lucasfilm, Ltd. All Rights Reserved; Nancy Lee Walter*
63 Nancy Lee Walter*
64 Sygma; Nancy Lee Walter*
67 © 1980 Lucasfilm, Ltd. All Rights Reserved; Nancy Lee Walter*
68 STAR TREK Copyright © 1992 Paramount Pictures. All Rights Reserved; Nancy Lee Walter*
70 © Walt Disney from Shooting Star; Nancy Lee Walter*
72 Bradley Arden, Panos Pictures; © Philippe Mazellier
76-77 Carol Brozman*
80 *Los Angeles Daily News* from Gamma/Liaison; Peter Glynn
81 Jeremy Hartley, Panos Pictures; Robert Wink, Pierre's Photo Lab
83 James B. Richardson; Rob Dunbar, Rice University
84 T. Ruzala, National Oceanic and Atmospheric Administration; X. Yan, V. Klemas, D. Chen, EOS, "Transactions of the American Geophysical Union," Vol. 73, No. 4, p. 41, 1992, © AGU
85 Maureen Small, Duke University
88 Jack Jennings; Tim Parker*
90 Tim Parker*
92 Carol Brozman*
93 James P. Blair, © National Geographic Society
94 Tim Parker*
96 Mike Andrews, Earth Scenes; Don Enger, Earth Scenes
97 Uniphoto Inc.; Harold Hungerford; E. R. Degginger
99-100 Tim Parker*
102 Cornell University
106 David Malin, Anglo-Australian Telescope Board
107 Tom Herzberg*
108 Tom Herzberg*; Tom Herzberg*; Dana E. Backman, Franklin and Marshall College
109 Tom Herzberg*
110 Roger Ressmeyer, Starlight
111 Seth Shostak, SETI Institute
112 Chris Hildreth, Cornell University
113 Cornell University
116 Nick Davies, Oxford Illustrators Limited
119-121 John Milner Associates
124 Ted A. Rathbun, University of South Carolina
127 Ted A. Rathbun, University of South Carolina; Jerome Rose, Arkansas Archeological Survey; Michael L. Blakely, Howard University (Victor Krantz, Smithsonian Institution)
128-129 Jerome S. Handler, Southern Illinois University
132 Stouffer Enterprises, Inc. from Earth Scenes
136 Bruce Davidson, Animals Animals; Patti Murray, Animals Animals
137 Gerry Ellis, The Wildlife Collection
140 Gerry Ellis, The Wildlife Collection; Klaus Uhlenbut, Animals Animals
141 Commercial Diving and Marine Services; J. R. Williams, Earth Scenes
142 U.S. Department of Agriculture
146 Telegraph Colour Library from FPG
149 Carol Brozman*
151 Larry Mulvehill, Photo Researchers; EEG Systems Laboratory
152 Hank Morgan, SS from Photo Researchers; M. E. Raichle, Washington University School of Medicine (4)
155 Hank Morgan, Rainbow; S. J. Williamson, New York University
156 Michael Haglund, George Ojemann, and Daryl Hochman, Department of Neurological Surgery, University of Washington
158 Denis Le Bihan, National Institutes of Health; Peter A. Bandettini, Biophysics Research Institute, Medical College of Wisconsin; Denis Le Bihan, National Institutes of Health
160-164 Joe Van Severen*
165 Raymond Measures, University of Toronto; Joe Van Severen*; Joe Van Severen*
167 Joe Van Severen*
169 Joe Van Severen*; Joe Van Severen*; Center for Intelligent Materials and Structures, Virginia Polytechnic Institute and State University
170 Brian S. Thompson and Mukesh V. Gandhi, Intelligent Materials and Structures Laboratory, Michigan State University
171 Intelligent Materials and Structures Laboratory, Michigan State University; Intelligent Materials and Structures Laboratory, Michigan State University; Joe Van Severen*
174 © Royal Observatory, Edinburgh (David Malin, Anglo-Australian Observatory)
176 Bibliothèque Nationale; Granger Collection
178 J. R. Eyerman, *Life* Magazine © 1950 Time, Inc.
181 Comstock; NASA; NASA/ESA
183 NASA
184 M. Seldner, B. Siebers, E. J. Groth and P. J. E. Peebles, 1977 "Astronomical Journal," Vol. 82, p. 249; Smithsonian Astrophysical Observatory

186	Rob Wood, Stansbury, Ronsaville, Wood, Inc.*
188	Patrice Loiez, CERN/SPL from Photo Researchers; Fermilab National Accelerator Laboratory
190	Rob Wood, Stansbury, Ronsaville, Wood, Inc.*
194-195	Photo: NASA—art: Roberta Polfus*
196	© Roger Ressmeyer, Starlight
197	Superconducting Super Collider Laboratory
198	Space Telescope Science Institute
200	John Underwood, Purdue University; NASA
201	© Steve Kirk, *Discover* Magazine; Superconducting Super Collider Laboratory
202-203	U.S. Department of Agriculture
204	John Underwood, Purdue University; Itzhak Wolf, Volcani Institute, Israel
206	East News from Sipa Press
207	© Li Tianyuan, *Discover* Magazine
209	Chip Clark, Smithsonian Institution
210	Fred Wendorf, Southern Methodist University
211	St. Johns Expeditions (Franklin Viola)
212	University of Pennsylvania Excavations at Caesarea
213	Fanny Broadcast, Gamma/Liaison
214	AP/Wide World
215	Giovanni Lattanzi
217	J. J. Hester, Arizona State University/NASA
218	National Radio Astronomy Observatory/Associated Universities, Inc.
220	D. Jewitt and J. Luu, University of Hawaii
221	NASA
223	H. Ford, Johns Hopkins University/STScI, the Faint Object Spectrograph IDT/NASA
225	Richard Ellis, Durham University/NASA; Roberta Polfus*
231	Johann N. Bruhn, University of Missouri at Columbia; Jeffry Mitton, University of Colorado; Johann N. Bruhn, University of Missouri at Columbia
232	U.S. Department of Agriculture
235	EO Inc.; Apple Computer, Inc.
236	Apple Computer, Inc.
238	Davidson & Associates
239	Knowledge Adventure, Inc.
240	AP/Wide World; AP/Wide World; Carnegie Institution of Washington
241	Carnegie Mellon University; AP/Wide World; UPI/Bettmann
245	David Cameron Duffy
247	Philips Electronics; Sony Corporation
249	Advanced Photovoltaic Systems, Inc.
251	Epipress/Sygma; AP/Wide World
254	© Steve Kirk, *Discover* Magazine
255	Paul Sereno, University of Chicago; University of Chicago Hospitals
257	Beverly Koller, University of North Carolina at Chapel Hill; J. Snouwaert, University of North Carolina at Chapel Hill
260	H. Groskinsky
263	I. K. Curtis
266	Ziolo, Xerox
267	Michael J. Sailor, University of California at San Diego
269	Thomas A. Lippert, Gamma/Liaison
271	Jamie A. Grifo, Cornell Medical Center; Barbara Cousins*
274	National Center for Atmospheric Research/University Corporation for Atmospheric Research/National Science Foundation
275	Allan Tannenbaum, Sygma; National Oceanographic and Atmospheric Administration; Nick Galante, Gamma/Liaison; NASA
278	Judith Salerno
280	CERN
281	T. Crosby, Gamma/Liaison
282	Barbara Cousins*
285	Jet Propulsion Laboratory
289	Uniax Corporation
290	Superconducting Super Collider Laboratory
292	David Sanders, *Arizona Daily Star*
294	AP/Wide World
297	Westinghouse Electric Corporation
299-300	NASA
301	Lawrence Livermore National Laboratory; Roberta Polfus
303	J. P. Dumbacher, University of Chicago
305	Tim Davis, Allstock
306	Esther Angert and Norman Pace, Indiana University
307	Russell A. Mittermeier
308	Jessie Cohen, National Zoological Park
309	Ralph J. Brunke*; Norelco Consumer Products Company; S. C. Johnson & Son, Inc.
311	Julie Pace*
314	Ralph J. Brunke*
315	Julie Pace*
317	Aireox Research; Pollenex Corporation; Norelco Consumer Products Company
319	Julie Pace*
321	S. C. Johnson & Son, Inc.
325	Ralph J. Brunke*
326	Julie Pace*
329	Nathan Greene*

World Book Encyclopedia, Inc., provides high-quality educational and reference products for the family and school. They include THE WORLD BOOK~RUSH-PRESBYTERIAN-ST. LUKE'S MEDICAL CENTER~MEDICAL ENCYCLOPEDIA, a 1,072-page fully illustrated family health reference; THE WORLD BOOK OF MATH POWER, a two-volume set that helps students and adults build math skills; THE WORLD BOOK OF WORD POWER, a two-volume set that is designed to help your entire family write and speak more successfully; and the HOW TO STUDY video, a presentation of key study skills with information students need to succeed in school. For further information, write World Book Encyclopedia, Inc.; 2515 E. 43rd St.; P.O. Box 182265; Chattanooga TN 37422-7265.